SPATIAL DECISION SUPPORT SYSTEMS

PRINCIPLES AND PRACTICES

SPATIAL DECISION SUPPORT SYSTEMS

PRINCIPLES AND PRACTICES

RAMANATHAN SUGUMARAN
JOHN DEGROOTE

CRC Press is an imprint of the
Taylor & Francis Group, an **informa** business

CRC Press
Taylor & Francis Group
6000 Broken Sound Parkway NW, Suite 300
Boca Raton, FL 33487-2742

First issued in paperback 2019

ISBN-13: 978-1-4200-6209-0 (hbk)
ISBN-13: 978-0-367-86439-2 (pbk)

Library of Congress Cataloging-in-Publication Data

Sugumaran, Ramanathan.
Spatial decision support systems / authors, Ramanathan Sugumaran, John Degroote.
p. cm.
"A CRC title."
Includes bibliographical references and index.
ISBN 978-1-4200-6209-0 (hardcover : alk. paper)
1. Decision support systems. 2. Geographic information systems. I. Degroote, John. II. Title.

T58.62.S84 2011
658.4'03--dc22 2010028317

Visit the Taylor & Francis Web site at
http://www.taylorandfrancis.com

and the CRC Press Web site at
http://www.crcpress.com

Dedicated to Grandpa …

Contents

Foreword

Geographic information systems (GIS) have been under continuous development for several decades. By now, they are both well known and widely used, and have become integral elements of information technology applications in a wide variety of domains. In its simplest form, GIS software enables users to address a variety of questions that have two root forms: what are the attributes associated with a place and which places have one or more specified attribute(s)? Such systems are particularly helpful when they are used to obtain results for simple queries or to address structured problems that have a well-defined solution process that can be specified and followed as a sequence of steps.

But many problems, particularly those that have a contested public policy component, are neither well structured nor clearly defined. In such cases, different interest groups may not only fail to agree on a solution process for a problem, they may fail to agree on fundamental aspects of its formulation. Consequently, there is no prescriptive process that can be followed to yield a solution. Spatial decision support systems (SDSS) are designed and implemented to address this class of semistructured problems with advanced analytical tools that help people explore a problem, learn about it, and use the information gained to arrive at improved decisions.

This timely book begins with coverage of basic geospatial data handling concepts, methods, and materials. It places the development of SDSS concepts within a historical framework of development and treats important system components with a level of detail that is appropriate for students who may have different backgrounds or be at different stages of intellectual development. Coverage then moves on to demonstrate how these components can be assembled into flexible collections that are used to address particular types of applications. It is here, with the illustration of different component assemblages, that the book coheres by demonstrating how an SDSS can be implemented in the form of a traditional desktop system or using distributed, web-based services. This is done in a way that should prove instructive to both students and their teachers.

I sincerely hope that you enjoy reading and learning from this book and that it will lead you to contribute new insights. I came away from it wishing that the book had been available to me many years ago when I was beginning to struggle with the SDSS concepts that now seem rather straightforward after having read these chapters.

Marc P. Armstrong
Professor and Chair, Department of Geography, The University of Iowa

Preface

Spatial decision support systems (SDSS) are designed to help decision makers solve complex spatially related problems and provide a framework for integrating (a) analytical and spatial modeling capabilities, (b) spatial and nonspatial data management, (c) domain knowledge, (d) spatial display capabilities, and (e) reporting capabilities. The use of SDSS in academic and business communities is increasing. For example, businesses are using sophisticated SDSS to analyze customer information for marketing, customer relationship management, and generating business intelligence to gain competitive advantage. Organizations are also using SDSS for traditional problems such as determining plant locations, where typically only ZIP code information is used. There is also growing interest from planners and managers of resource assessment, environmental analysis, geological exploration, remote sensing, business analyses, soil science, public health, and hazard analysis in developing spatial models and SDSS to support managerial decision making. As the use of SDSS proliferates, there is a great demand for SDSS-related publications, especially books that could be used for training students as well as professionals.

It is evident from the previous examples that there is tremendous interest in the design and deployment of SDSS in various domains. Research on SDSS is also on the rise, which is evidenced by the number of conferences discussing this topic as well as special issues of journals. In addition, there are an increasing number of professional training courses that aim to discuss the fundamentals of SDSS and their applications. With this increased interest and development of SDSS, there is a great need for a comprehensive book that covers the fundamentals of SDSS as well as advanced design concepts for building SDSS. However, currently no such book is available for students, planners and managers, and the research community. Most of the existing materials on SDSS are book chapters, conference proceedings, and journal articles. Many of these are domain or application specific and do not provide a comprehensive treatment of SDSS. In addition to research by the academic community, there have been a number of important developments from vendors and the practitioner community. Hence, there is tremendous opportunity and need for a comprehensive book on SDSS. The primary goal of the authors is to provide a thorough overview concerning the current state of the art in SDSS technology and their application from an interdisciplinary perspective.

The collection in this book consists of four major parts, each addressing different topic areas in SDSS. Part 1, consisting of Chapters 1 and 2, primarily presents an introduction to SDSS and the evolution of SDSS.

Chapter 1 provides an introduction to the importance of spatial decision making and discusses how SDSS supports the spatial decision-making process. The purpose of Chapter 2 is to detail the evolution of SDSS from decision science and geographical information science perspectives.

Part 2 covers the different components of SDSS. Chapter 3 focuses on the spatial database management and spatial analysis capabilities of geographical information science (GIS) software. Chapter 4 focuses on the other components of SDSS, including the model base, user interface, stakeholders, and knowledge components. The focus of Part 3 is the design and implementation of SDSS. Chapter 5 provides an overview of the range of existing SDSS software configurations and covers software that can be used to construct new SDSS. Chapter 6 investigates techniques and technologies for building new SDSS while Chapters 7 and 8 provide examples of desktop and Web-based SDSS development and implementation. In the final part, Chapter 9 provides an overview of SDSS applications from various domains or disciplines with numerous detailed case studies provided. Chapter 10 addresses both technical and organizational challenges that affect the success or failure of SDSS uptake. The chapter concludes by documenting some of the likely future directions of SDSS.

The intended audiences for this book are students as well as professionals working in all decision and geosciences application domains including, but not limited to, resource assessment, environmental analysis and assessment, geological exploration, remote sensing, business analyses, soil science, public health, and hazard analysis. This book will also be of interest to researchers, planners, and managers involved in urban and regional planning. This book will be suitable for teaching at different levels. It will be easy for instructors to adopt because of the organization of its content, which starts with a basic introduction and progresses to advanced step-by-step implementation of SDSS. It also includes creative projects and exercises that instructors can use in introductory or graduate-level courses. This book can also be used by professional trainers that offer short training courses on various aspects of SDSS and their application.

Ramanathan Sugumaran
University of Northern Iowa

John DeGroote
University of Northern Iowa

Acknowledgments

Many people have contributed directly or indirectly to the completion of this book and need to be acknowledged and thanked. First, Taisuke Soda, who was a former Acquisitions Editor of CRC Press, needs to be thanked for encouraging our book proposal and getting approval from the publisher. The authors would also like to thank Professor Vijayan Sugumaran, Oakland University, Michigan, who initially put forth the idea of writing this book. This book is the result of his initiative and encouragement. Though initially he was a co-author of this book, due to unforeseen circumstances and prior commitments, he was unable to continue in that capacity.

Secondly, we acknowledge our debt of gratitude to the University of Northern Iowa GeoInformatics Training, Research, Education, and Extension Center (GeoTREE) staff and students. Particularly we would like to thank Scott Larson, Matt Voss, Alexander Savelyev, and Associate Director for the GeoTREE Center Dr. Andrey Petrov for their critical editing of and valuable content suggestions for the book. This book has benefitted greatly from the efforts of these individuals who contributed advice, gave feedback on materials, or helped in testing different software. We have learnt much from discussion and debates with these contributors. The SDSS examples in Chapters 7 and 8 were developed by a number of current and former GeoTREE Center staff. The ArcGIS-based desktop SDSS from Chapter 7 and the ArcGIS Server-based example from Chapter 8 were developed by Dr. Yanli Zhang, currently an Assistant Professor of Water Resources/Spatial Science in the Arthur Temple College of Forestry and Agriculture at Stephen F. Austin State University. The Microsoft Excel Spreadsheet-based AHP SDSS example and the SpreadsheetSDSS Plug-in discussed in Chapter 7 were developed by Dossay Oryspayv. Alexander Savelyev is the primary developer of the OpenSDSS software, which was described in Chapter 7, with contributions from Dossay Oryspayv. Jonathan Voss developed the web-based SDSS using open-source technology described in Chapter 8. Dmitry Ershov developed the web-interface for the SDSS web-portal described in Chapter 9. Matt Clover carried out literature searches and collated many of the articles that were recorded in the SDSS database. Many individuals helped us in administrative matters and in editing, proofreading, and preparation. Our special thanks go to Scott Larson, Jane Gillen (former GeoTREE Administrative Assistant), and Holly Bokelmen for proofreading the manuscript, organizing references, and formatting figures and tables. Thanks also to University of Northern

Iowa for providing time for Dr. Sugumaran to partially write this book through Professional Development Assignment.

Third, the publication of this book could not have been possible but for the efforts by a large number of individuals working at CRC Press. We thank Irma Shagla, Editor for Environmental Sciences & Engineering of CRC Press for her encouragement, copy editing, and for not giving up on us. We also thank the production team, particularly Stephanie Morkert, who transformed the manuscript into a book.

Finally, Dr. Sugumaran would like to thank his family for their support during the process including his wife Vanitha, and his sons Sriram (elder son) and Srivishnu (younger son) for their unfailing support and love. John DeGroote would especially like to thank his wife Joan for her patience and support, and kids Emma and Kieran for providing joy at home.

Authors

Dr. Ramanathan Sugumaran is Professor of Geography and Director of GeoTREE Center at the University of Northern Iowa. He has over nineteen years of research experience in remote sensing, geographic information systems (GIS), Global Positioning Systems (GPS), and spatial decision support systems (SDSS) with applications for natural resources and environmental planning and management. Dr. Sugumaran has served as PI or Co-PI on over $5 million worth of research grants funded by the National Aeronautics and Space Administration (NASA), Raytheon Corp., the National Oceanic and Atmospheric Administration (NOAA), the U.S. Department of Defense (DOD), the U.S. Department of Agriculture (USDA), Missouri Department of Natural Resources (MDNR), the U.S. Department of Transportation (DOT), and the U.S. Fish and Wildlife Service. He has also published numerous journal articles and presented more than one hundred papers at national and international conferences. Dr. Sugumaran has two PhDs—a PhD in geography from the University of Edinburgh in the United Kingdom and one from the University of Baroda, India. For the past ten years, he has developed and taught several courses and advised more than twenty students on their masters theses. Dr. Sugumaran has also been a recipient of several academic awards that include the outstanding graduate faculty teaching award, Outstanding Scholar award, and Veridian Community Engagement Award.

John DeGroote is a GeoInformatics Scientist at the GeoTREE Center at the University of Northern Iowa. He has been actively applying geospatial technologies for environmental and natural resource applications for nine years. He has experience working on a wide range of issues with a diverse set of investigators including hydrologists, soil scientists, ecologists, and economists. He has extensive experience in developing custom GIS and SDSS applications, using programming and database development, for use by researchers and environmental managers. John has authored or co-authored numerous peer-reviewed articles concerning the use of geospatial technologies for a variety of application domains. He has also presented research at numerous national and international conferences.

Authors

[illegible] and [illegible] Ph.D. [illegible] [illegible] [illegible] [illegible] [illegible] [illegible] [illegible] [illegible] applications [illegible] [illegible] environmental [illegible] [illegible] [illegible] funded by the National Aeronautics and Space Administration (NASA), Raytheon Corp., the National Oceanic and Atmospheric Administration (NOAA), the U.S. [illegible] (DOD), the U.S. Department of [illegible] (USDA), Minnesota Department of Natural Resources (MDNR), the U.S. [illegible] [illegible] (DOT), and the U.S. Fish and Wildlife Service. [illegible] [illegible] [illegible] [illegible] the United Kingdom [illegible] University of [illegible] [illegible] developed and taught several courses and [illegible] [illegible] [illegible] [illegible] [illegible] [illegible] [illegible] [illegible] Award [illegible]

John [illegible] [illegible] the University of [illegible] [illegible] [illegible] [illegible] [illegible] [illegible] economists [illegible] experience in developing [illegible] GIS and [illegible] [illegible] and database development [illegible] [illegible] environmental [illegible] [illegible] [illegible] numerous peer-reviewed articles [illegible] [illegible] [illegible] application domains [illegible] [illegible] [illegible] [illegible]

Abbreviations

AGNPS: Agricultural Non-Point Source Pollution Model
AHP: Analytic Hierarchy Process
AI: artificial intelligence
AML: Arc Macro Language
ANN: artificial neural networks
API: application programming interfaces
AVHRR: Advanced Very High Resolution Radiometer
AvIMS: ArcView Internet Map Server
CA: cellular automata
CAD: computer-aided design
CLIPS: C Language Interface Production System
COM: Component Object Model
CORBA: Common Object Request Broker Architecture
DBMC: database management component
DBMS: database management system
DCOM: Distributed Component Object Model
DDE: Dynamic Data Exchange
DEM: digital elevation model
DLL: dynamic-link libraries
DNR: Department of Natural Resources
DSS: decision support systems
EDSS: environmental decision support systems
EMDS: Ecosystem Management Decision Support
ES: expert systems
ESRI: Environmental Systems Research Institute
FEMA: Federal Emergency Management Agency
GA: genetic algorithms
GADS: geo-data analysis and display system
GDAL: Geospatial Data Abstraction Library
GIS: Geographic Information Systems
GML: Geography Markup Language
GPS: Global Positioning Systems
GRASS: Geographic Resource Analysis Support Systems
GUI: graphical user interface
HSPF: Hydrological Simulation Program-Fortran
ILWIS: Integrated Land and Water Information System
KMC: knowledge management component
KML: Keyhole Markup Language
LiDAR: Light Detection and Ranging

MCA: multi-criteria analysis
MCDA: multi-criteria decision analysis
MCDM: multi-criteria decision making
MCE: multi-criteria evaluation
MMC: model management component
NDVI: Normalized Difference Vegetation Index
NOAA: National Oceanic and Atmospheric Administration
NTF: National Transfer Format
OGC: Open Geospatial Consortium
OLE: Object Linking and Embedding
OWA: ordered weighted averaging
PSS: planning support systems
QGIS: Quantum GIS
RFID: radio frequency identification
RIKS: Research Institute for Knowledge Systems
RMI: remote method invocation
RS: remote sensing
SAGA: System for Automated Geoscientific Analyses
SC: stakeholder component
SDLC: systems development life cycle
SDSS: spatial decision support systems
SML: Spatial Modeler Language
SOAP: Simple Object Access Protocol
SWAT: Soil and Water Assessment Tool
TIGER: Topologically Integrated Geographic Encoding and Reference System
uDig: User-friendly Desktop Internet GIS
UNI: University of Northern Iowa
VPN: virtual private network
WCS: Web Coverage Service
WFS: (OGC) Web Feature Service
WLC: weighted linear combination
WMS: Web Map Service
WSDL: Web Services Description Language
XML: Extensible Markup Language

1

Introduction

Learning Objectives

- Be introduced to spatial decision-making processes.
- Learn how decision support tools can aid these processes.
- Be introduced to spatial decision support systems and learn some basic definitions.

1.1 Introduction

Location, location, and location.

—Harold Samuel

This long reiterated maxim, credited to Lord Harold Samuel in 1944, was spoken to stress the importance of location in relation to property and real estate in London. However, it can be applied to many aspects of life and society. Maps have always formed an integral part in decision-making processes as witnessed by elementary maps drawn on the walls of caves thousands of years ago. In 1854, John Snow, a doctor in London, created a map that provided evidence that the Soho, London, cholera epidemic originated from a single water well. This work was credited with forming the beginning of the science of epidemiology and demonstrated the value of spatial information in addressing a real-world problem. Over the last few decades, the use of locational or geographical information has exploded in commercial, governmental, academic, and individual enterprises. With the evolution of ever more powerful computing hardware and software, the ability to capture, manage, and spatially analyze geographical information has grown tremendously. Further, the ability to incorporate locational information into decision-making processes has been democratized through the use of inexpensive Global Positioning Systems (GPS) and navigation systems as well as Web-based neo-geographical services provided by sites such as

Google Maps, Google Earth, Yahoo! Maps, and MapQuest. These and many other Web sites use underlying databases of spatial information and processing algorithms to provide information to individuals and businesses about directions, locations of property for sale, locations of businesses such as hotels or restaurants, or weather forecasts for a particular place.

In conjunction with the evolution of these geographically democratizing technologies, there has been an exponential growth in the use of spatially related information to support commercial, governmental, and academic decision-making processes for situations more complex than just deciding where to have dinner. Issues such as environmental management, land use planning, transportation design, commercial or public welfare service provision, and emergency/hazard management cut across administrative, institutional, and stakeholder settings, which are couched within a complicated spatial matrix. For example, imagine a situation in which there is a desire to improve the water quality in a popular recreational lake that has excessive nutrient levels. The lake is fed by a river, which in turn receives water from numerous tributaries and sub-watersheds. There are various land uses, varying topography, and a wide range of people with different interests and priorities (i.e., economic, environmental, recreational) across the lake's watershed. These kinds of issues present a level of complexity that requires tools to aid in the decision-making process. This book investigates spatial decision support systems (SDSS), which are a class of tools used to address complicated spatial decision problems. The book provides an overview of the evolution of SDSS, the technological underpinnings of SDSS development, examples of SDSS applications, and the challenges facing the successful development and application of SDSS.

The purpose of this chapter is to provide an introduction to the importance of spatial decision making and how SDSS supports the spatial decision-making process. This chapter is organized into five sections. The first section explains spatial decision making and associated complexities. The second section describes spatial decision-making processes, while the third section demonstrates the need for a computer-based support system to assist the spatial decision-making process. The fourth section provides basic definitions of SDSS and related systems, and the final section outlines the remaining content of this book.

1.2 Spatial Decision Making

1.2.1 What Are Spatial Decisions?

A decision can be defined as a choice that is made between two or more alternatives. Individuals have to make many decisions every day. The potential choices in a decision are formed after defining certain minimum objectives, and alternatively, more demanding objectives. There are many examples of tools to help people make decisions, such as cost-of-living calculators for planning a move to a new city or retirement calculators that aid people in deciding how to invest for retirement. Institutional or organizational decision making is often much more complex, but individuals are still charged with making those decisions. There are greater resources in these decision-making situations but also a greater range of constraints, alternatives, and possible decisions. These people have to identify management goals and in turn determine a range of choices or alternatives that can meet those goals. From a range of disciplines, there have been a variety of decision aids and tools designed to provide a more systematic approach to making organizational decisions. In many cases, spatial characteristics and attributes are crucial to the decision-making situation. Imagine that you are planning a day of running errands and you have to plan your itinerary. Most people will want to plan their trip along an efficient route, but might also make sure they get to eat lunch at their favorite restaurant. Thus, you need to process various pieces of information, including locational information, and make a choice about your route that meets your goals as fully as possible. Today, with online tools such as Google Maps (Figure 1.1) or MapQuest, decision aids can be used to help support the routing decision.

Now imagine a delivery driver who has to make ninety-four deliveries the following day across a medium-sized city taking into account traffic patterns. The sheer number of deliveries, as well as traffic considerations, makes it difficult for that individual to decide on the exact route that is most efficient. In these cases, computer routing applications are often used in order to help plan efficient routes. The location of customers and businesses is clearly important in the business world. Many consumers have noticed how businesses ask for your address or ZIP code when purchasing an item. They are collecting locational information about customers, for inclusion in databases that can be used to support decisions about how to deploy their resources. Examples of spatially contingent decision making could include siting problems, such as where to put a retail store, a landfill, a community center, or other facility; allocation decisions, such as how many police officers to deploy to a certain neighborhood; or resource status decisions, such as when and where to control for a pest species such as mosquitoes.

Geographic information is crucial to decision making by all manner of organizations. It is estimated that 80% of data used by managers and decision makers is geographically related (Worrall 1991). The amount of spatial information collected, managed, and analyzed has grown greatly over the last few decades. There has also been tremendous growth, from the 1990s to the present, in the sale and utilization of both hardware and

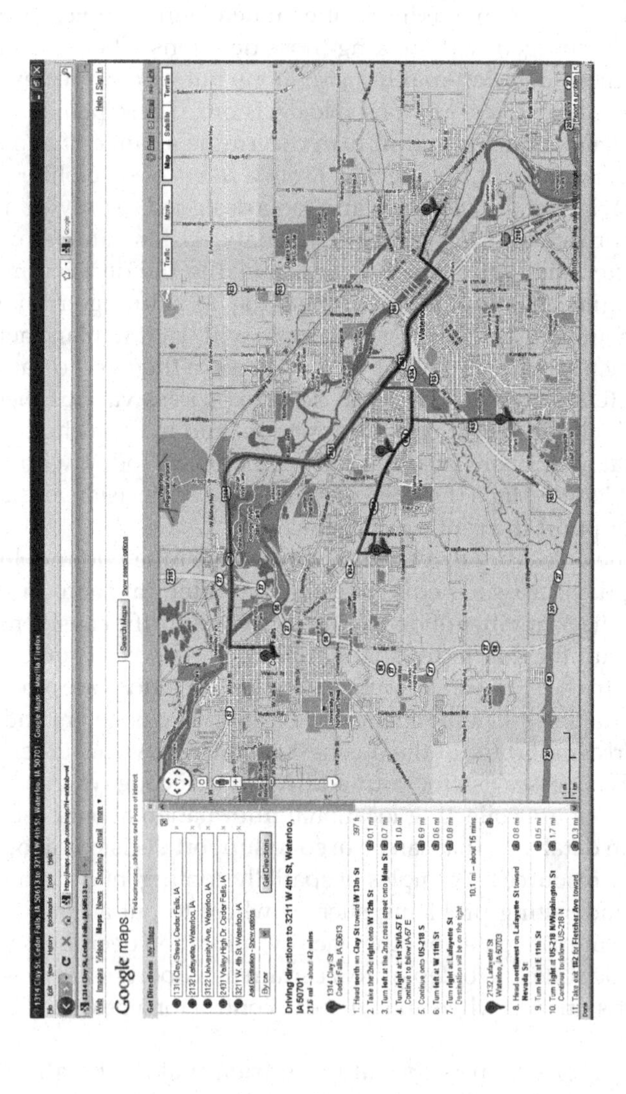

FIGURE 1.1
(See color insert following page 74.) An example of routing in Google Maps.

software related to spatial information. This growth is presently continuing with increased use of spatial information at all levels of government, business, and academics.

1.2.2 Types of Spatial Decisions

There is a tremendous range of spatially dependent decisions for various organizations throughout the world. However, these can be organized into general categories depending on perspective. For example, the Committee on the Geographic Foundation for Agenda 21 (Jensen et al. 2002) listed three spatial decision categories in relation to sustainable development: (1) resource allocation decisions, (2) resource status decisions, and (3) policy decisions. Resource allocation decisions require value judgments as well as logistical and technical considerations. An example of a resource allocation decision could be where to place limited air quality monitoring equipment in order to efficiently collect data to understand the risk of exposure to residents of a city. Resource status decisions often require timely spatial information. Examples of resource status decisions could include those made in relation to crop condition, timber harvesting, or potential disease vector populations. The use of dynamic data, such as that collected from real or near real-time remote, sensing systems, or GPS-collected field data, is important in relation to resource status decisions. The implications, including the spatial implications of policy decisions, represent their final decision category. For example, if a state government provides significant tax and business incentives for the development of wind energy, where are the most likely places for these developments to occur? In another study, Kemp (2008) organized types of spatial decisions into four categories: (1) site selection, (2) location allocation, (3) land use selection, and (4) land use allocation. Site selection is a very common activity for business, government, and individuals and requires the consideration of many spatial factors. A city locating a new park might want the parcel(s) of land to be accessible to children and adults, be of a certain minimum size, and have acceptable geology or soil conditions. Location allocation decisions are ones in which a main goal of the decision is to site something in order to have optimal allocation. For example, it would be ideal to locate a medical clinic in order to minimize travel times for the largest possible number of patients. Land use selection is the opposite of site selection, in that for a given a parcel of land, what would be an ideal use for it? This could be dependent on the zoning designation of the parcel, potential surrounding customer base for a business, or physical parameters and limitations of the land for certain kinds of development. Kemp's final category, land use allocation, is for when there are a variety of parcels of land that are best used for certain purposes. A good example of this is planning and zoning activities where land might need to be allocated for a variety of purposes, including business or residential development, open space, or transportation corridors.

1.2.3 Spatial Decision-Making Problems

Spatial decision making is often complex and requires information produced from many sources and interpreted by a variety of decision makers in relation to different goals and objectives. Simon (1960) characterized decisions as being structured (programmable), semistructured, or unstructured (nonprogrammable), with the latter representing those that suffer from a lack of knowledge, large search space, need for data that cannot be quantified, and so on (Carlson 1978). Spatial decisions are often described as semistructured, meaning that they fall between structured and unstructured. These semistructured problems are often multidimensional, have goals and objectives that are not completely defined, and have a large number of alternative solutions (Gao et al. 2004). Spatial decision problems are also often characterized by uncertainty and conflicts between the various stakeholders interested in the process (Wang and Cheng 2006). Important aspects of the decision alternatives and the potential outcomes also can vary spatially, adding to the complexity. The great complexity involved in spatial decision making suggests the use of automated or computer-based techniques. However, there is usually not a single solution that meets all objectives for all stakeholders (Xiao 2007).

The difficulties of spatial decision making can be understood better with some examples. Let us return to the lake water quality issue that was mentioned earlier. An example lake in the state of Iowa is shown in Figure 1.2. This lake has had consistently high nutrient and bacteria levels. The pollution levels are driven by non-point source pollution that comes from the agricultural land in the watershed (a watershed is the land area that drains into the lake) as well as urban areas that surround the lake. The watershed contains a variety of land uses and conditions, including intensive agriculture. A major goal of some stakeholders is the improvement in these water quality parameters as they adversely affect the ecological system of the lake, which in turn hurts recreational opportunities in the lake and also economic potentials. The farmers, on the other hand, attempt to maximize profit, which often leads to the use of fertilizers, manure, and agricultural chemicals. In addition, there are houses that ring the lake that use chemicals to keep their lawns looking nice. All of these practices lead to runoff of pollutants into the lake and water quality problems. There are numerous government agencies at different levels that have an interest in the lake and the land in its watershed. These include the city government that receives complaints regarding water quality; the state natural resource management agency, which manages a state park on the lake, manages the lake's fishery,

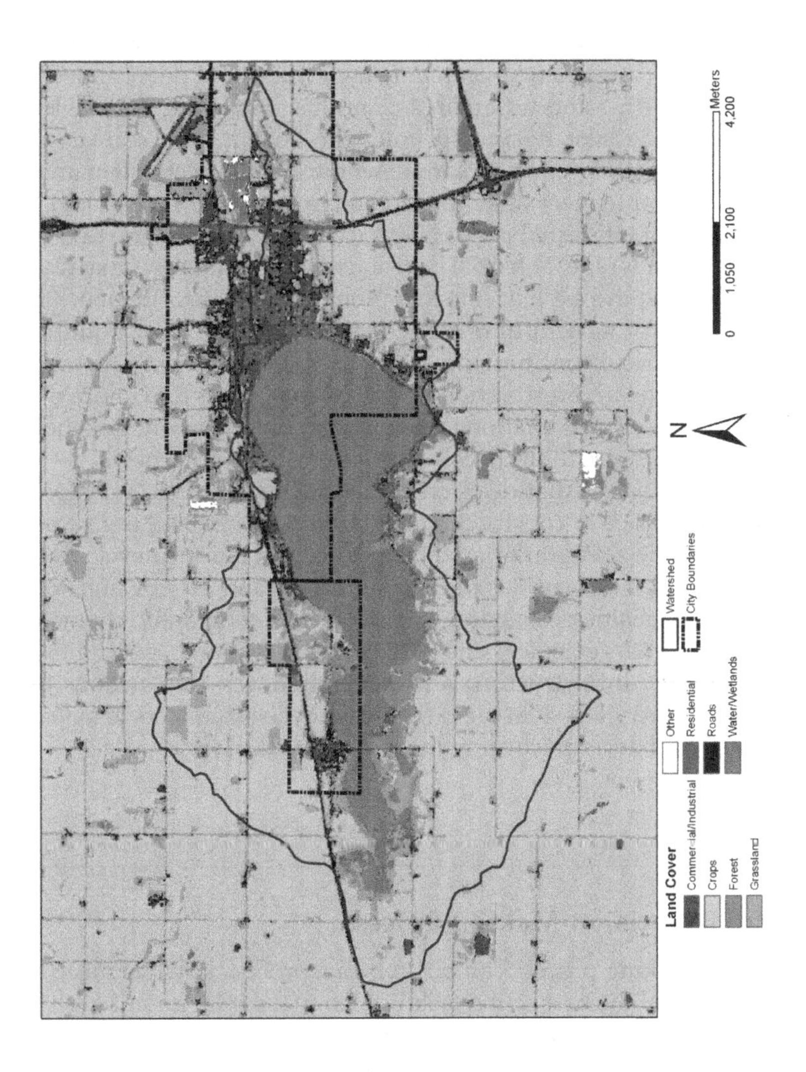

FIGURE 1.2
(See color insert following page 74..) Lanc use patterns and watershed and city boundaries for a watershed with lake water quality problems.

and also monitors the state's surface waters; as well as a federal land management agency that assists farmers with land management in the area. There are also organized public interest research groups, such as environmentalists, who are interested in restoring wetlands in the lake's watershed and in improving the lake water quality. There are numerous types of individuals who have a stake in the policies affecting the lake, including the farmers who don't want their profits cut by overregulation of their practices, anglers who want a fishery that is not deteriorated by pollution from farms and urban areas, families who want to be able to safely swim in the lake, businesspeople who benefit from recreational activities, and finally agency personnel who want to effectively manage the land and water to meet stakeholder needs and wants. Thus, you have a large number of interested parties with varying goals that are tied to a varied spatial landscape. Potential decisions made to improve lake water quality would require explicit consideration of spatial information, such as the locations of agricultural areas that lead to the most pollution, where might agricultural best practices be most beneficial, or where might possible wetland restoration be placed to achieve maximum benefit. These types of considerations are complex, involve many stakeholders, and would require multidisciplinary scientific approaches.

Let's take another spatial decision example. Imagine that a company is looking for a place to construct a new building to expand (e.g., a new supermarket). The selection of a suitable piece of land for development involves several factors, such as the availability of land, cost of land, location of existing businesses, zoning regulations, demographic characteristics, physical and geological characteristics of the land, proximity to utilities and other infrastructure, and city ordinances. In situations like this, a computer-based system can be useful to decision makers in making quick and critical decisions during the site selection process.

1.3 Spatial Decision-Making Process

The decision-making process can be seen as a process in which decision makers try to find the best action (solution) to move from an initial situation to a desired goal situation. A general overview of the spatial decision-making process is given in Figure 1.3. Simon (1960) suggested that the decision-making process can be seen as being structured in three phases: intelligence, design, and choice. The intelligence phase includes formulation of the problem and the search for information relevant to finding solutions to the problem. The design phase involves the compilation and analysis of data and information to work toward a solution. The final

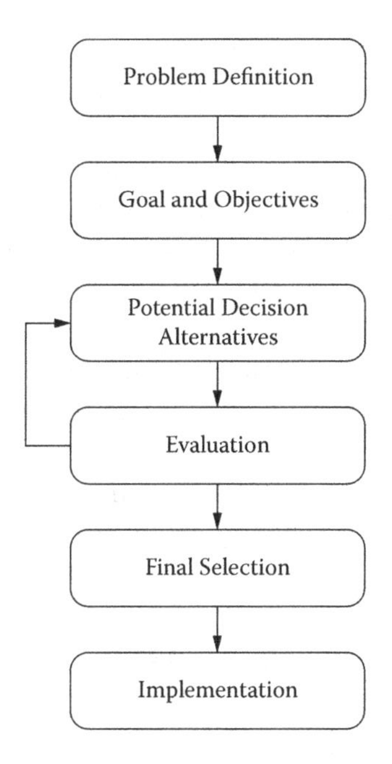

FIGURE 1.3
General spatial decision-making process.

phase is the choice phase in which selection from alternatives is made (Feeney and Williamson 2002). These phases do not necessarily progress linearly as there is usually a return to previous phases after gathering new knowledge or after the generation of new ideas. This overall decision-making process may also be modeled as an adaptive process that consists of subprocesses (or phases, stages), such as problem identification and goal specification, generation of alternative actions, identification of consequences of actions, and selection of one alternative over the others (Huber 1989).

There are intricacies and attributes particular to complex spatial decision-making processes. Problems or issues requiring spatial considerations are complex, multidimensional, characterized by aspects of uncertainty, and usually involve numerous concerned stakeholders. These characteristics make it unlikely for the decision-making process to proceed linearly. Rather, it is likely to follow an iterative process with various interactions between groups. These groups require adequate information in order to participate in the decision-making process. Rarely is there enough or exactly the right kind of information. Malczewski (1999) categorized information used in spatial decision-making processes into "hard" and "soft" categories. *Hard information* is that which is derived from reported facts, quantitative estimates, or systematic opinion surveys, while *soft information* is that which is based on opinions, priorities, or preferences of decision makers, or based on ad hoc surveys or comments. Both sets of information are likely always considered in spatial decision-making situations.

Keller (1997) listed five steps governing the spatial decision-making process: (1) identifying the issue, (2) collecting the necessary data, (3) defining the problem, including objectives, assumptions, and constraints, (4) finding appropriate solution procedures, and (5) solving the problem by finding an optimal solution. In our watershed example, the problem would be formally recognized by the Iowa Department of Natural Resources (DNR) when they declare the lake an impaired water body as required by the United States Environmental Protection Agency given certain water quality attributes. Meetings between stakeholders, with the Iowa DNR likely serving as the lead agency, could lead to the development of a comprehensive database containing spatial and nonspatial data regarding the watershed. Again, through meetings between stakeholders, an overall objective as well as constraints could be identified. The overall objective might be defined as reducing the environmental (and negative economic) consequence of pollution while minimizing the effect on the economic capabilities of agricultural and other interests within the watershed. There are almost always multiple objectives arising from different stakeholder perspectives, and defining the relationships between the objectives and quantifying them in common terms, such as monetary amounts, is necessary. Ideally, Keller says the objectives could be minimized into a single overall objective. Finding the appropriate solution procedure, from step 4 above, is likely the most difficult step in the spatial decision-making process. As the human and natural processes that occur in the real world are not always easily defined, it is usually necessary to define multiple scenarios or scenario structures. For example, within the watershed, the amount of soil erosion depends on the land use practices, which are controlled by economic factors such as the price of crops or the demand in the housing market. Such complicated situations require some model or combination of models to help evaluate these different scenarios effectively. In conjunction with all stakeholders, at this stage a set of software tools in the form of a formalized SDSS, which would work with the database from step 2, should be developed. With various stakeholders involved, the SDSS would then be used to seek an optimal solution. Although this is presented as a linear path, in reality, in order to be successful, iteration between the steps in the process is required.

1.4 Need for Decision Support Systems

As has been demonstrated, spatial decision situations are often complex, multidisciplinary, and usually involve many stakeholders. Due to the wide variety of interested parties, it is also important to build support or justification for the decision that is made (Ingram 1973). To meet this type of requirement, relevant information concerning the issue must be acquired and organized to support problem analysis. In complex decision situations, the decision-making process is often iterative, interactive, and participative (Goel 1999). The process is iterative as alternative actions are analyzed and information gained is used to guide further analyses. It is interactive and participative because a variety of information must be incorporated and a variety of stakeholders must participate in the process. As spatial decision-making situations are often complex and ill-structured, individuals cannot process all of the necessary information. There are human cognitive deficiencies in memory and analysis abilities, and in order to address complex spatial problems or issues, support systems are often necessary and useful. These systems can help describe the evolution of the issue or system, provide knowledge-based formulation of possible actions, simulate consequences or actions of decision possibilities, and assist in the formulation of implementation strategies (Chen and Gold 1992).

The complicated nature of spatial decisions and the requirement for the accumulation, management, and analysis of a variety of data sets make it necessary to utilize computer-based tools. There are several tools, technologies, or systems such as Geographic Information Systems (GIS), decision support systems (DSS), expert systems (ES), remote sensing (RS), and spatial decision support systems (SDSS) available to support spatial decisions. Geographic information systems have been defined many times in the literature. Example definitions include "a computer system for capturing, storing, querying, analyzing, and displaying geospatial data" (Chang 2009, p. 1), and "a group of procedures that provide data input, storage and retrieval, mapping and spatial analysis for both spatial and attribute data to support the decision-making activities of the organization" (Grimshaw 2000, p. 33). Thus, GIS can be considered as a set of software tools that are used to create, manage, display, and analyze spatial data for the purpose of supporting modeling, investigation, and understanding of the real world. GIS software is used in a wide variety of disciplines, including all levels of government, across many business types, and in various academic disciplines for a variety of purposes.

Geographic information systems are a very useful technology because they provide utility for creating, managing, and analyzing a variety of spatial and nonspatial datasets. In our lake watershed example, spatial data on land use, land ownership, planning and zoning, soils, topography, hydrologic features, recreational lands, location of any drainage or discharge outlets to the lake, and other information can be organized in a GIS database. Ancillary data on crop prices, land use practices, and other information could also be included in the database. Malczewski (1999) discussed how, in the intelligence phase of the decision-making process, a search of the decision environment must be carried out and data acquired, stored, retrieved, and managed. The construction of a spatially enabled database allows for exploratory analysis of the problem environment to be carried out using GIS. In the watershed example, it would be possible to query the GIS database to demonstrate areas where high-intensity agriculture has taken place on steep slopes. This could provide an indication of areas of potential erosion problems. However, this type of information might be subject to uncertainty, such as whether the land use/cover map is out of date, spatially precise, or whether it captures the exact farming practices that have taken place on given pieces of land. The GIS database does not necessarily capture the complexity of the situation as the data itself might not be detailed enough. While GIS software tools are very useful for spatial decision making, they lack the ability to adapt the decision makers' knowledge to the analysis, and the software out of the box often is not flexible enough to allow for the analysis logic to be articulated (Goel 1999). A detailed description and examination of GIS are given in Chapter 3.

Decision support systems (DSS) have been developed over the course of the last several decades across many disciplines to support decision making. They incorporate modeling or analysis along with database management systems and user interfaces for aiding the user. They also sometimes incorporate knowledge or expert systems. Using computers for decision support became practical with the development and evolution of computers over the last few decades. The concept originated in the 1960s and growth accelerated in the 1970s. Much of the development in DSS has come from the business world, in which applications such as accounting and financial models and executive information systems were developed (Power 2008). The number and diversity of DSS have grown significantly with greater computing power and ever greater amounts of data. Decision support systems often do not account for or handle spatial aspects of decision making, and thus extension of the concept of DSS to SDSS has been necessary.

Expert or knowledge-based systems are often built into DSS and SDSS. These systems are meant to incorporate knowledge within a DSS in order to provide humanlike reasoning within the system. This provides the advantage of utilizing computing power, which can carry out many calculations or process tremendous amounts of data, while at the same time building in humanlike reasoning ability to develop useful scenarios. As a result, complex problems can be analyzed and what-if analyses can be carried out with the aid of computing power and organizational and domain knowledge (Courtney et al. 1987; Özbayrak and Bell 2003).

Remote sensing has been defined many times; two definitions are provided here: "the science, technology, and art of obtaining information about objects from a distance" (Aronoff 2005, p.2) and "the practice of deriving information about the Earth's land and water surfaces using images acquired from an overhead perspective, using electromagnetic radiation in one or more regions of the electromagnetic spectrum, reflected or emitted from the Earth's surface" (Campbell 2008, p. 6). Remote sensing is a very common and extremely valuable way of developing usable geospatial data for GIS and SDSS applications. The main platforms for collection of remote sensing data are satellites and airplanes. There are many software systems that are designed for processing, analyzing, and discerning information from imagery that is collected from remote sensing platforms. The primary benefits from remotely sensed imagery include the possibility of mapping features of the Earth's surface using manual interpretation and automated processing, repeated temporal recording of the Earth's surface for time-series analysis of changes, recording of meteorological conditions across large areas and over short time periods, and recording of wavelengths invisible to the human eye. As with GIS, the number of remote sensing instruments and use of imagery have grown significantly over the last few decades. Commercial satellite imagery has grown greatly as an industry over the last decade to supplement the many satellites operated by national governments.

Modeling is a very general term that will be used in different capacities throughout this book. The two main objectives of modeling in GIS, and often SDSS, are to describe and predict (Aronoff 2005). A model can be defined as a representation of one or more processes in the real world, and is often are designed as a computer program (Maguire et al. 2005). Examples of modeling with a spatial component could include mapping ideal travel paths on a road network based on distance and average speeds; mapped estimation of erosion potential in a watershed based on rainfall amounts, soil properties, and slope; or derivation of maps of land cover change over time based on a series of remotely sensed imagery.

All of these technologies can play a crucial role in the development of SDSS. The GIS software often plays a fundamental and central role in SDSS. However, in order to truly support the spatial decision-making process, GIS functionality must be extended or joined with other technology, such as DSS and ES, in order to form true spatial decision support systems (SDSS).

1.5 Definition of SDSS

The use of SDSS has grown dramatically over the last few decades but there is still no universally accepted definition. There is some uncertainty even about the definition of DSS, which have a longer history than SDSS. Some authors in the past have characterized GIS as SDSS, using the simplest perspective that they are computer tools that can be used to support decisions (Keenan 2003). However, using a more stringent idea of SDSS, there are more recent conclusions that GIS alone do not qualify as SDSS (Keenan 2006; Keenan 2003; Sugumaran et al. 2007; Sugumaran and Bakker 2007). Malczewski (1999, p. 281) defined SDSS as "an interactive computer based system designed to support a user or group of users in achieving a higher effectiveness of decision making while solving a semi-structured spatial decision problem." In an earlier definition, Densham (1991, p. 405) stated that SDSS are "explicitly designed to provide the user with a decision-making environment that enables the analysis of geographical information to be carried out in a flexible manner." Leipnik et al. (1993, p.1) defined SDSS as "integrated environments, which utilize the databases that are both spatial and non-spatial models, decision support tools like expert systems, statistical packages, optimization packages, and enhanced graphics to offer the decision makers a new paradigm for analysis and problem solving." In a nutshell, SDSS are integrated computer systems that support decision makers in addressing semistructured or unstructured spatial problems in an interactive and iterative way with functionality for handling spatial and nonspatial databases, analytical modeling capabilities, decision support utilities such as scenario analysis, and effective data and information presentation utilities. As SDSS are multifaceted technologies that are manifested in many varieties, it is useful to look at some of the common traits that characterize them.

1.6 SDSS Characteristics

Although these definitions convey an overall idea about the nature of SDSS, it is necessary to characterize what qualifies as an SDSS more thoroughly with some characteristics that they must have (see Figure 1.4). Kemp (2008) states that SDSS are systems that combine analytical tools with functions available in GIS as well as models for evaluating various options. She also mentions the presence of multi-criteria evaluation

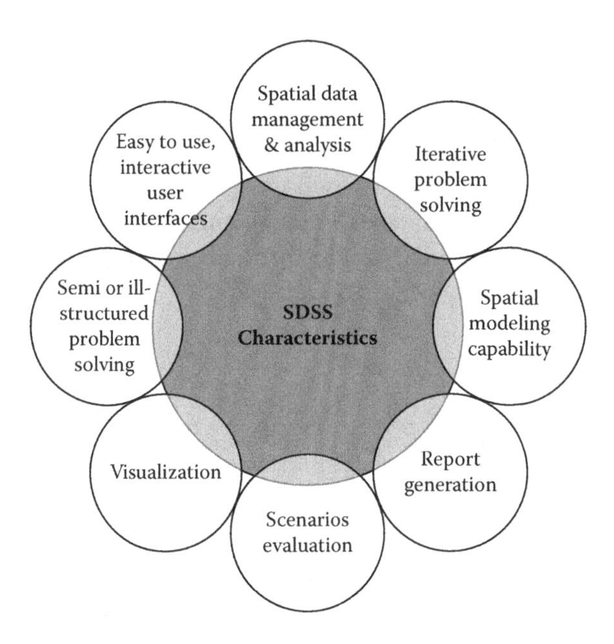

FIGURE 1.4
Characteristics of SDSS.

techniques for analyzing decision options and sensitivity analyses for testing the robustness of decision recommendations. Goel (1999) discussed numerous traits that characterize SDSS: they are designed to solve ill-structured problems, they have user interfaces, they have the ability to flexibly combine models and data, they contain tools to help users explore solution space to aid in the generation of feasible solutions/alternatives, and they can provide an interactive and recursive problem-solving environment. While GIS provides modeling capabilities, they are not usually sufficient or directly applicable to unstructured spatial decision problems. GIS, while able to allow spatial exploration of the solution space, do not usually have sufficient flexibility for interactive and recursive problem solving. In addition, GIS software is developed for spatial problems, while complex decision problems often involve both spatial and nonspatial aspects. As Keenan (2003) points out, an SDSS must cater to the overall problem representation, which will allow the user to not only incorporate the geographic data but also include structures and functionality for addressing the logical view of the problem. So, for example, in the watershed problem, the spatial data and GIS functionality are useful and necessary, but other aspects such as the cost of instituting best management practices, the possibility of new and stricter environmental regulations, and the projected future development must be taken into account. In summary, an SDSS must be built to be flexible to accommodate various stakeholder preferences and restrictions and allow for effective user interaction in an iterative problem-solving environment. To meet these requirements, custom software is often developed with easy-to-use graphical user interfaces and functionality for spatial database management and analysis, scenario evaluation, modeling, visualization through maps, graphs, tables, and report generation.

1.7 Types or Flavors of SDSS

A wide variety of SDSS have been under development for the last three decades, and there is a continuing evolution in the technologies being developed. The evolution of SDSS will be described in detail in later chapters but, in general, has followed or lagged slightly behind that of computing and also the development of GIS software. Since the 1980s, SDSS has been influenced by computing technology with a greater number of applications being developed based on greater processing power being available, development environments capable of providing user-friendly applications, object-oriented programming languages, and with the proliferation of Web-based capabilities. In the 1980s and for much of the 1990s, workstation and desktop GIS systems were most often used in the design of SDSS with ArcInfo software being the most commonly utilized. As SDSS technology evolved, the inclusion of intelligent components, such as expert-based or knowledge-based systems, was seen. As more user-friendly GIS software and flexible development environments were developed, a greater number of SDSS applications were built for a wider variety of applications. This period was in the latter half of the 1990s and into the 2000s. With the growing ubiquity of the Internet and the development of mapping and spatial analysis functionality in Web environments there have been, in recent years, SDSS fully or partially developed with Web technologies. The need for technologically feasible collaborative spatial decision-making tools was stressed by Karacapilidis et al. (1995). With technological improvements in networking and with Web geospatial services development, collaborative and participative SDSS are becoming more technologically feasible and common. The history and evolutionary drivers of SDSS development will be discussed in detail in Chapter 2.

1.8 Content of This Book

The use of SDSS in the academic and business community is increasing. For example, businesses are using sophisticated SDSS to analyze customer information for marketing, customer relationship management, location analysis, and generating business intelligence to gain competitive advantage. Besides the business community, there is growing interest from scientists, planners, and managers involved in resource assessment, environmental analysis, geological exploration, remote sensing, business analyses, soil science, medical geography, and hazard analysis in developing spatial models and SDSS to support managerial decision making. As the use of SDSS proliferates, there is going to be a greater demand for SDSS-related publications, especially books that could be used for training students as well as professionals.

It is clear that there is tremendous interest in the design and deployment of SDSS in various domains. Research on SDSS is also on the rise, which is evidenced by the number of conferences discussing this topic as well as special SDSS-related issues of journals (e.g., Li et al. 2003; Malczewski 2004; Andrienko et al. 2006; Balram et al. 2009). In addition, there are an increasing number of academic and professional training courses that aim to discuss the fundamentals of SDSS and their applications. With this increased interest in and development of SDSS, there is a great need for a comprehensive book that provides the fundamentals of SDSS as well as advanced design concepts and tools for building SDSS. However, currently no such book is available for students, planners and managers, and the research community. There are only a few books that include chapters on SDSS (e.g., Leung 1997 and Malczewski 1999). These books were published more than a decade ago and did not provide a broad treatment of the subject. Since research on SDSS is carried out by a number of disciplines, such as decision sciences, geosciences, and environmental sciences, there need to be interdisciplinary approaches for designing and deploying SDSS. This book will provide a broad-based coverage of SDSS that will be useful to a wide range of disciplines and will also provide thorough coverage of the important technologies that are useful in SDSS development. There is tremendous opportunity and need for a comprehensive book on SDSS that provides a complete overview and the current state of the art in SDSS technology and its application from an interdisciplinary perspective.

The material of this book is organized in four parts with nine chapters (Figure 1.5). We refer the reader to the list of acronyms on page xix. These acronyms will be listed freely throughout the text of the book. The chapters that follow and their contents are listed here:

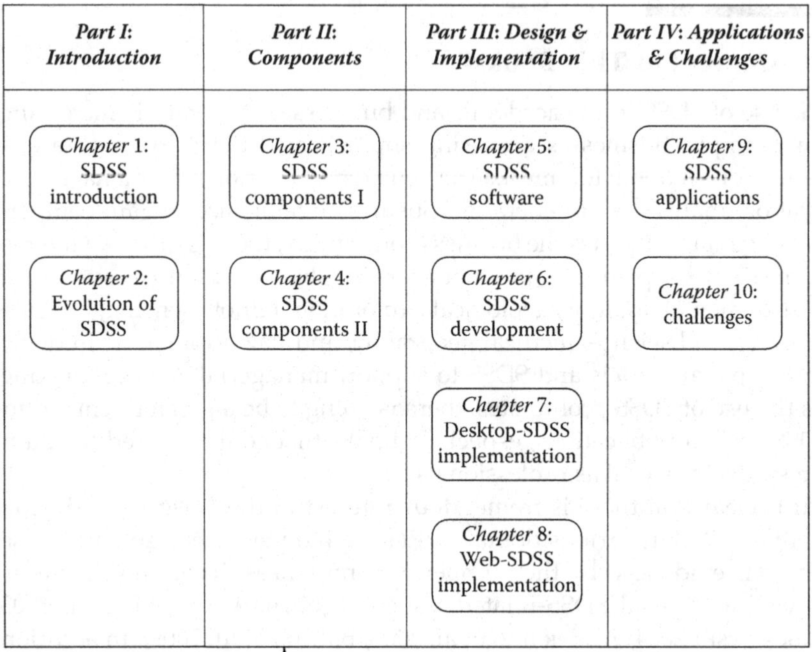

Part I: *Introduction*	*Part II:* *Components*	*Part III: Design & Implementation*	*Part IV: Applications & Challenges*
Chapter 1: SDSS introduction	*Chapter* 3: SDSS components I	*Chapter* 5: SDSS software	*Chapter* 9: SDSS applications
Chapter 2: Evolution of SDSS	*Chapter* 4: SDSS components II	*Chapter* 6: SDSS development	*Chapter* 10: challenges
		Chapter 7: Desktop-SDSS implementation	
		Chapter 8: Web-SDSS implementation	

FIGURE 1.5
Structure and chapter overview of the book.

Part 1: Introduction

This part covers the introduction and progression or evolution of SDSS.

- **Chapter 1, "Introduction":** As mentioned earlier, the purpose of this chapter has been to provide an introduction to the importance of spatial decision making and how SDSS supports the spatial decision-making process.
- **Chapter 2, "Evolution and Trends in SDSS":** The purpose of Chapter 2 is to discuss how SDSS evolved or progressed from decision science and geographical information science. This chapter investigates major technological and anthropogenic drivers of SDSS development and recounts the evolution of SDSS over the last several decades. Specific focus is given to the evolution of SDSS from decision support systems and GIS technology. The evolution of SDSS is traced through a comprehensive examination of the literature over the last several decades. Historical as well as present trends in SDSS development and use are traced.

Part 2: SDSS Components

- **Chapter 3, "Components of SDSS I: Geographic Information Systems":** Chapter 3 introduces the main components of SDSS, including the database, model base, user interface, stakeholder, and knowledge components. However, as GIS plays a central role in many SDSS, Chapter 3 mainly focuses on this technology. The chapter introduces the history of GIS development, definitions of GIS, spatial data models, attribute data, data exploration and visualization, and spatial processing and analysis.
- **Chapter 4, "Components of SDSS II":** Chapter 4 focuses on the remaining components that make up spatial decision support systems. The model management component is discussed in detail with both generic modeling systems such as weighted linear combination, artificial neural networks, cellular automata, and genetic algorithms, and application-specific models such as hydrological models being reviewed. The evolution of user interfaces in SDSS is discussed with state-of-the-art examples provided from the present day. The importance of stakeholders is also stressed in this chapter. Finally, the knowledge management or expert systems component is discussed with some real-world examples provided.

Part 3: Design and Implementation of SDSS

- **Chapter 5, "SDSS Software":** This chapter covers a broad range of types of SDSS software that are available for use. The chapter begins by looking at some of the broad functional classes of SDSS. Then the chapter moves on to discuss specific SDSS software including problem-specific SDSS for given application areas (e.g. overweight vehicle permitting SDSS), general domain-level SDSS, such as for land use planning or agricultural support, and generic SDSS modules that can be utilized for a variety of spatial decision-making situations.
- **Chapter 6, "Building SDSS Software":** Chapter 6 investigates techniques and technologies for building new SDSS. The most common techniques for developing SDSS are investigated. The various techniques of coupling multiple programs to form a single system are examined. In addition, the technique of fully embedding all components into a single piece of software, usually the GIS software is explored. The various technologies that support SDSS development, such as programming languages and development environments are detailed.

- **Chapter 7, "Building Desktop SDSS":** In this chapter, potential SDSS development processes are examined. Then several specific SDSS development examples are provided including a Microsoft Excel based SDSS and an ArcGIS extension SDSS. All of the necessary steps including programming code are provided to guide the user in the development of example SDSS software.
- **Chapter 8, "Building Web-Based SDSS":** In this chapter we will provide some background and cover some of the important issues that need to be considered for Web-based SDSS. We then provide two examples of building Web-based SDSS tools. The first example provides step-by-step instructions for developing an ArcGIS server application. The second example details the development of a Web-based SDSS using entirely open source software.

Part 4: Applications and Challenges

- **Chapter 9, "SDSS Applications":** Chapter 9 reviews a variety of SDSS application examples from a range of disciplines using a variety of techniques. The chapter provides an overview of some of the uses of SDSS as documented in the scientific literature over the last several decades. A summary of application areas in which SDSS have been used is provided. Detailed descriptions of specific SDSS application examples from a range of disciplines are also provided.
- **Chapter 10, "SDSS Challenges and Future Directions":** Chapter 9 provides a discussion of some of the present challenges in effective SDSS development and use. This chapter addresses both technical and organizational challenges that affect the success or failure of SDSS uptake. The chapter concludes by documenting some of the future directions of SDSS.

References

Andrienko, G., N. Andrienko, P, Jankowski, Pand A. MacEachren. 2006. Workshop on visualization, analytics and spatial decision support. *GIScience Conference*, Münster.

Aronoff, S. 2005. *Remote sensing for GIS managers*. Redlands, CA: ESRI Press.

Balram, S., S. Dragicevic, and R. Feick. 2009. Collaborative GIS for spatial decision support and visualization. *Journal of Environmental Management* 90(6):1963–1965.
Campbell, James B. 2008. *Introduction to remote sensing*. New York: Guilford Press.
Carlson, E. D. 1978. An approach for designing decision support systems. *ACM SIGMIS Database* 10(3):3–15.
Chang, Kang-tsung. 2009. *Introduction to geographic information systems with data files CD-ROM.* New York: McGraw-Hill.
Chen, J., and C. M. Gold. 1992. Research directions for spatial decision support. Paper presented at The International Colloquium on Photogrammetry, Remote Sensing and Geographic Information Systems, Wuhan, China.
Courtney, J. F., Jr., D. B. Paradice, and N. H. A. Mohammed. 1987. A knowledge-based DSS for managerial problem diagnosis. *Decision Sciences* 18(3):373–399.
Densham, P. J. 1991. Spatial decision support systems. In *Geographical information systems: Principles and applications*, ed. J. Maguire, M. S. Goodchild, and D. W. Rhind, 403–412. London: Longman Publishing Group.
Feeney, M. E., and I. Williamson. 2002. The role of institutional mechanisms in spatial data infrastructure development that supports decision making. *Cartography Journal* 312:21–37.
Gao, S., D. Sundaram, and J. Paynter. 2004. Flexible support for spatial decision making. Paper presented at the 37th Annual Hawaii International Conference on System Sciences, Honolulu, Hawaii.
Goel, R. K. 1999. Suggested framework (along with prototype) for realizing spatial decision support systems (SDSS). Paper presented at Map India 1999 Natural Resources Information System Conference, New Delhi, India.
Grimshaw, D. J. 2000. *Bringing geographical information systems into business,* 2nd edition. New York: John Wiley & Sons, Inc.
Huber, O. 1989. Information-processing operators in decision making. In *Process and structure in human decision-making,* ed. H. Montgomery and O. Svenson, 3–21. New York: John Wiley & Sons, Inc.
Ingram, H. M. 1973. Information channels and environmental decision-making. *Natural Resource Journal* 1:150–169.
Jensen, J. R., K. Botchway, E. Brennan-Galvin, C. Johannsen, C. Juma, and A. Mabogunje. 2002. *Down to Earth: Geographic information for sustainable development in Africa.* National Research Council, Washington, DC.
Karacapilidis, N., D. Papadias, and M. Egenhofer. 1995. Collaborative spatial decision-making with qualitative constraints. Paper presented at 3rd ACM International Workshop on Advances in Geographic Information Systems, Baltimore, Maryland.
Keenan, P. B. 2003. Spatial decision support systems. In *Decision making support systems: Achievements, trends and challenges for the new decade*, ed. M. Mora, G. Forgionne, and J. N. D. Gupta, 28–39. Hershey, PA: Idea Group Publishing.
Keenan, P. B. 2006. Spatial decision support systems: a coming of age. *Control and Cybernetics* 35(1):9–27.
Keller, C. P. 1997. Unit 57—Decision-making using multiple criteria. http://www.geog.ubc.ca/courses/klink/gis.notes/ncgia/u57.html (accessed November 10, 2009).
Kemp, Karen. K. 2008. *Encyclopedia of geographic information science*. Thousand Oaks, CA: SAGE Publications, Inc.

Leipnik, M. R., K. K. Kemp, and H. A. Loaiciga. 1993. Implementation of GIS for water resources planning and management. *Journal of Water Resources Planning and Management* 119(2):184–205.

Leung, Y. 1997. *Intelligent spatial decision support systems*. Berlin: Springer.

Li, Zhilin, Qiming Zhou, and Wolfgang Kainz. 2003. Advances in spatial analysis and decision making. Lisse, The Netherlands: Swetz & Zeitlinger Publishers.

Maguire, D. J., M. Batty, and M. F. Goodchild. 2005. *GIS, spatial analysis, and modeling*. Redlands, CA: ESRI Press.

Malczewski, J. 1999. *GIS and multicriteria decision analysis*. New York: John Wiley & Sons, Inc.

Malczewski, J. 2004. GIS-based land-use suitability analysis: a critical overview. *Progress in Planning* 62(1):3–65.

Marble, Duane F., Hugh W. Calkins, and Donna J. Peuquet. 1984. *Basic readings in geographic information systems*. Williamsville: SPAD Systems, Ltd.

Özbayrak, M., and R. Bell. 2003. A knowledge-based decision support system for the management of parts and tools in FMS. *Decision Support Systems* 35(4):487–515.

Power, D. J. 2008. Decision support systems: A historical overview. In *Handbook on decision support systems 1*, ed. F. Burnstein and C. W. Holsapple, 121–140. Leipzig: Springer-Verlag Berlin Heidelberg.

Simon, H. A. 1960. *The new science of management decision*. Englewood Cliffs, NJ: Prentice-Hall.

Sugumaran, R., and B. Bakker. 2007. GIS-based site suitability decision support system for planning confined animal feeding operations in Iowa. In *Emerging spatial information systems and applications*, ed. B. N. Hilton, 219–239. Hershey, PA: Idea Group.

Sugumaran, R., S. Ilavajhala, and V. Sugumaran. 2007. Development of a web-based intelligent spatial decision support system WEBSDSS: A case study with snow removal operations. In *Emerging spatial information systems and applications*, ed. B. N. Hilton, 184–202. Hershey, PA: Idea Group.

Wang, L., and Q. Cheng. 2006. Web-based collaborative decision support services: Concept, challenges and application. Paper presented at ISPRS Technical Commission II Symposium, Vienna.

Worrall, L. 1991. *Spatial analysis and spatial policy using geographic information systems*. New York: Belhaven Press.

Xiao, N. 2007. Considering diversity in spatial decision support systems. Paper presented at Geocomputation 2007, Ireland.

2

Evolution and Trends in SDSS

Learning Objectives

- Understand the drivers of SDSS development.
- Explore the influence of decision science and geographic information science on SDSS evolution.
- Learn about the historical trends in the use and application of SDSS.

2.1 Introduction

In Chapter 1, we introduced the importance of spatial decision-making processes and explained various terms related to spatial decision support systems (SDSS). The purpose of this chapter is to discuss how SDSS have evolved or progressed from the decision science and geographic information science disciplines. This chapter is organized in three sections. The first section explains some of the major drivers in the evolution of decision support systems (DSS), Geographic Information Systems (GIS), and SDSS. The second section provides background and perspectives on the evolution of SDSS from DSS and GIS. The final section explains the progression of SDSS from simple, stand-alone desktop SDSS to advanced service-based SDSS.

2.2 Origins of SDSS

As mentioned in the previous chapter, SDSS are "explicitly designed to provide the user with a decision-making environment that enables the analysis of geographical information to be carried out in a flexible manner" (Densham 1991, p. 405). In the past three decades, SDSS have experienced tremendous growth and evolved from stand-alone desktop applications to Web-based and service-based SDSS. A plethora of names have been utilized in relation to systems that could potentially be considered SDSS, including multi-criteria SDSS, group SDSS, environmental DSS, Web-based SDSS, planning support systems, policy support systems, and collaborative SDSS. The focus of this chapter is to detail how SDSS have evolved while noting some important milestones.

Research on SDSS has mainly originated from two different disciplines—DSS and GIS (Keenan 2006; Peterson 1998; Sugumaran and Sugumaran 2007). Figure 2.1 demonstrates the overall progression of SDSS from GIS and DSS. Each discipline provided a unique contribution to the growth of SDSS, and they are discussed in Sections 2.3 and 2.4. Before investigating the contributions of GIS and DSS, it is important to understand the drivers or factors that governed the evolution of these disciplines. The following section provides an overview of some of the major drivers that assisted in the development of different decision support technologies such as DSS, GIS, and SDSS.

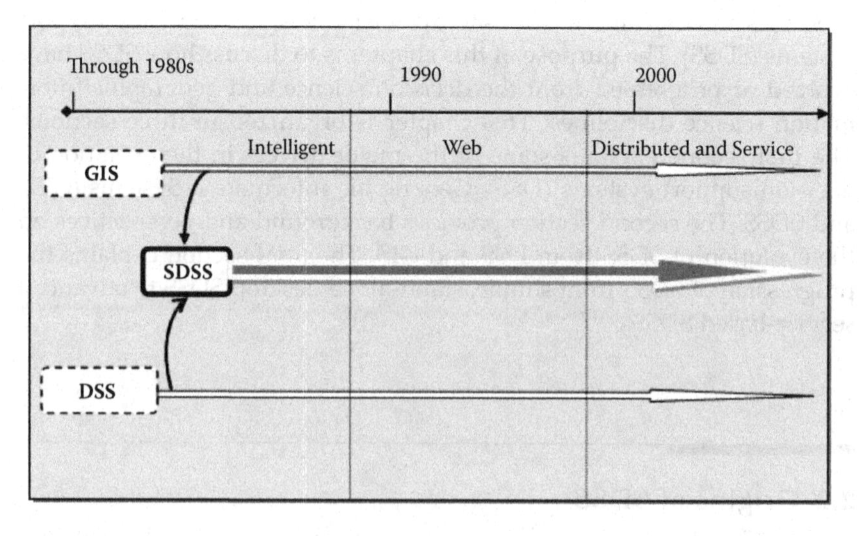

FIGURE 2.1
Evolution of decision technologies (GIS, DSS, and SDSS).

2.3 Core Drivers for the Development of Spatial Decision Support Technology

During the past three decades, significant progress has been made in the field of information systems, including the area of spatial decision support technologies. There have been many drivers spurring this progress, including advancements in information technology (IT) and communication technology (CT), the variety of users, application domains and experts in those domains, computer cost, developers/analysts, advancements in spatial sciences, commercial incentives in spatial industries, data affordability, data types, and data availability (Keenan 2003; Malczewski 2006). A schematic representation of the major driving forces involved in the progression of SDSS is shown in Figure 2.2, and a brief description of each driving factor is explained in the following section.

2.3.1 Information and Communication Technology

Although there have been several drivers enabling the progression of DSS, GIS, and SDSS, information and communication technology (ICT) has possibly had the largest impact. The continuous cost decline of computing power in concert with the rapid expansion of computer power have been the dominant technological drivers for the successful implementation of GIS, DSS, and SDSS (Peterson 1998). As suggested by Moore's Law (the idea that the number of transistors that can be placed on a circuit would

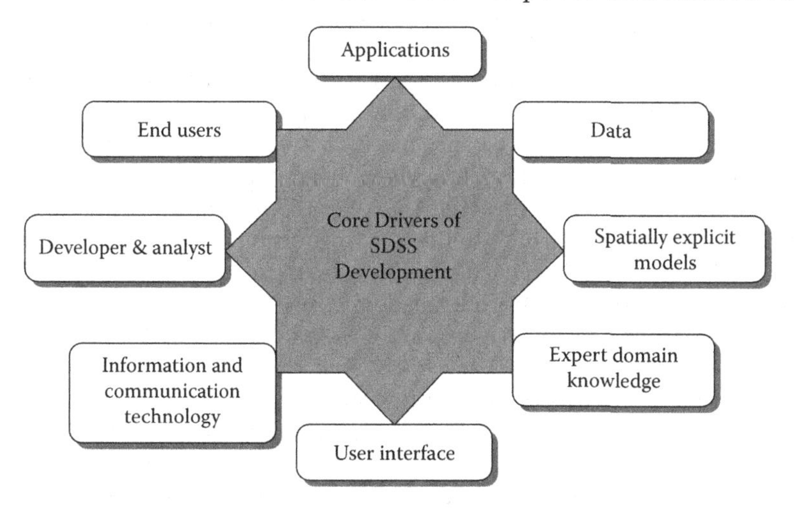

FIGURE 2.2
Core drivers for the evolution of spatial decision support systems.

double approximately every two years—Figure 2.3) from 1965, the power of computing per unit cost has grown exponentially. This exponential growth has allowed more and more computing power to be concentrated in smaller and smaller packages, leading to the move from mainframe computers to workstations to personal computers. Computing advances have also facilitated advancement in other drivers, such as the types and number of users, user interfaces, and application domains. With the ever-expanding computing power and increased affordability came a greater variety of users from a tremendous range of commercial, government, and academic disciplines. To meet the needs of these users, an expanded range of software applications and functionalities within these different disciplines were developed. In addition, over time these applications were developed with more user-friendly and intuitive user interfaces. This transition was seen in GIS technology with a move from command-line-driven systems to graphical user interface (GUI)-driven systems by the 1990s. Further enabling IT technologies such as the Internet, intelligent agents, multimedia, Web services, markup languages, and ontologies have also played a major role in the progression of SDSS over the last decade or so. The increased presence of the World Wide Web (WWW) along with advanced network technologies such as the Java language, ActiveX controls, Common Object Request Broker Architecture (CORBA), and the Distributed Component Object Model (DCOM) have also helped in the evolution of SDSS. In more recent years, communication technologies such as wireless or mobile devices and radio frequency identification (RFID) have also played a role in the progression of SDSS.

2.3.2 Spatial Data Availability

The growth of spatial decision support technology has also been driven by the availability and accessibility of spatially related data. Over the last few decades, there has been a surge in data availability. In the 1980s and 1990s, there was a huge effort to convert hard-copy maps to digital data through scanning and digitizing. Also over the last decades, the amount of spatial data derived from satellite and airborne remote sensing as well as Global Positioning Systems (GPS) has grown greatly. Many institutions, including commercial, nonprofit, academic institutions, as well as all levels of government, have invested in GPS equipment for the purpose of building spatial databases. They have used this GPS equipment to collect information on diverse sets of features such as the location of infrastructure (e.g., road conditions, fire hydrants), biological conditions (e.g., endangered plants, animal movements), and surveying purposes for land development. There has been an explosion of airborne and satellite remote sensing imagery over the last few decades. National governments from developed countries, led by the United States, have traditionally been

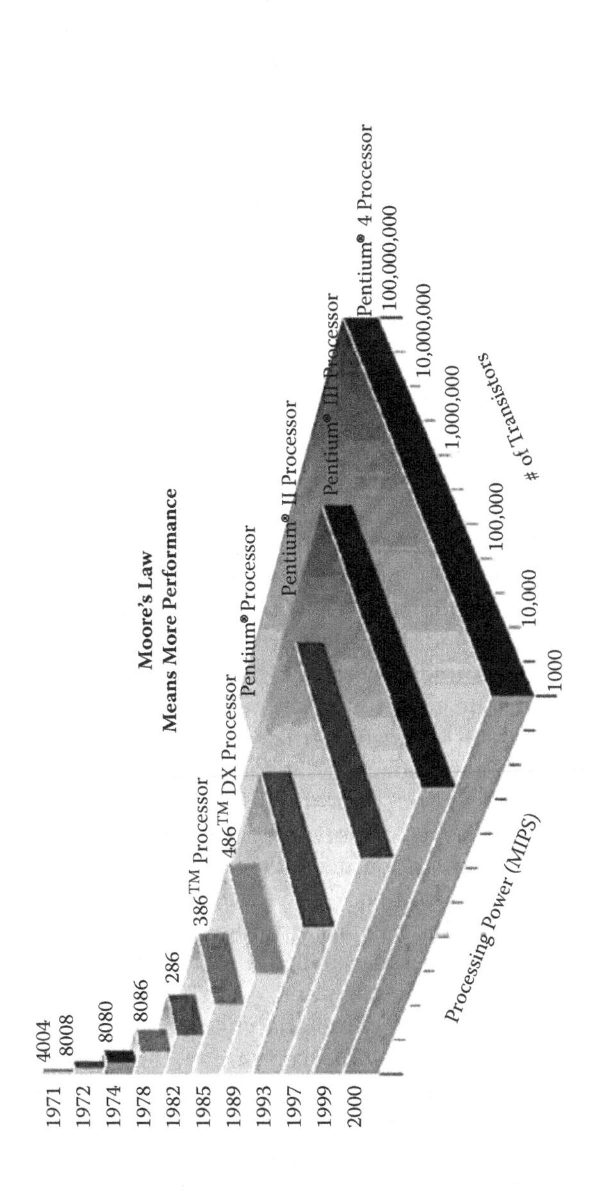

FIGURE 2.3
A graphical representation of the history of transistors as predicted by Moore's Law. (Adapted from http://www.developers.net/intelisnshowcase/view/127)

the only ones with enough resources to develop satellite remote sensing systems. However, in recent years numerous commercial companies (e.g., GeoEye) and more national governments (e.g., Thailand) have developed satellite remote sensing systems. One of the major reasons for this surge is the need for spatial data expressed by planners and managers from government agencies and business organizations (Keenan 2003). One estimate shows that up to 80% of data needed for the activities of business and government is spatially related (Worrall 1991). However, with the development of new data from new technology, new ideas and data needs are envisioned, leading to the demand for even more data. The availability and accessibility of spatial data have facilitated the growth of SDSS over the last few decades. For a more detailed description of spatial data collection, management, and analysis processes, please see Chapter 3.

2.3.3 Applications

Improved computer processing power and affordability led to the expansion of spatial applications into a variety of disciplines. Into the early 1980s, GIS and spatial processing software use was generally limited to those with high levels of computing and spatial sciences expertise. Primarily, this use was in academics and some government agencies. With the widespread development of desktop or personal computer-based systems with more user-friendly and standardized interfaces, the variety and number of GIS and SDSS consumers increased greatly. Many organizations, such as municipal and state governments, as well as a greater number of academic departments, started to recognize the potential of using spatial software such as GIS for managing data on spatial features, mapping purposes, and for addressing analytical questions. As spatial software became more greatly entrenched in organizations, their potential as decision support mechanisms was recognized. One of the first major reported SDSS publications was Densham and Armstrong's 1987 paper "A Spatial Decision Support System for Locational Planning: Design, Implementation and Operation." This article appeared in the Auto Carto Proceedings. Interestingly, in their paper, Densham and Armstrong pointed out that spatial information systems lacked the analytical modeling functionality necessary to fully support spatial decision-making processes. They argued that it was necessary to integrate the spatial information system with a separate modeling system to constitute an SDSS. As GIS technology has evolved to include greater analytical functionality, the debate as to whether GIS by themselves constitute an SDSS has continued (Murphy 1995; Keenan 2006). The early Densham and Armstrong study marked the beginning of a trend that is still continuing—the use of SDSS for an ever-expanding list of application areas. At the Geographic/Land Information Systems 1989 (GIS/LIS '89) conference, articles appeared detailing SDSS use for oil or hazardous chemical spill response (Gould 1989), real estate planning (Peterson 1989), and land use planning (van der Vlugt 1989). The range of SDSS application areas grew quickly to include urban, natural resource management, agriculture, defense, transportation, and many others. Thus advances in the technology helped drive applications in a more diverse set of disciplines, which in turn led to further technical evolution driven by the variety of functions needed by different disciplines. A more detailed investigation of the history of SDSS follows in this chapter, while in Chapter 9 a discussion of a range of example SDSS applications from a wide range of disciplines is provided.

2.3.4 Users, Developers, and User Interfaces

Spatial decision support systems can be utilized by individuals, groups, or entire enterprises. The evolution of hardware, software, and networking technology has led to a movement from individual expert-driven SDSS use to the inclusion of a much broader set of stakeholders. Earlier hardware and software configurations of SDSS generally were characterized by fairly complicated command-line-driven GIS and modeling software. These systems required someone with significant experience in the spatial sciences and also often in computer programming because the person who developed and programmed the SDSS was likely the person who would operate it. In the 1990s, the introduction of graphical user interface-based computing and the development of GIS and modeling software with more intuitive graphical user interfaces greatly increased the number of organizations using these types of software. In addition, the number of individuals able to use and understand at least the basic functionality of GIS rose greatly during this time period as more training and formal educational opportunities became available. The number of students enrolling in courses associated with GIS and related topics at universities and colleges saw an upsurge in this time period. In addition, GIS software in the 1990s began to come packaged with programming languages that allowed the development of customized user interfaces and analysis routines. Using these systems, an expert GIS user and programmer could develop customized applications that could be used by people with much less expertise. These functionalities also allowed the development of routines that could interact with other software applications such as modeling programs. This allowed for the building of SDSS applications with GIS software at their core.

The rapid growth in Internet and intranet applications, networking technology, and bandwidth improvements has led to the development of group or participatory SDSS, especially in the 2000s. Group SDSS evolved to address unstructured spatial problems, which involve many stakeholder groups. For example, Nyerges et al. (2006) developed a group-based collaborative SDSS called WaterGroup. The system was meant to enable stakeholder groups to participate in the solution of conjunctive water resource administration decision problems.

2.3.5 Spatially Explicit Modeling

As has been established previously, SDSS techniques have evolved over time for application in a wide range of disciplines. The complexity of spatial decisions and the uncertainty inherent in information used for spatial decision-making processes have led to the inclusion of a wide variety of techniques in SDSS. Some of the earliest spatial approaches to land suitability analysis used hand-drawn overlays, which evolved into digital GIS overlay operations (Keenan 2006) in the 1960s. Automated spatial overlay and other spatial analytical operations developed in GIS software were powerful but still fell short in the ability to capture or represent many physical or biological processes in the environment and also human preferences and criteria relevant to spatial decisions. These shortcomings led to various SDSS developments that usually incorporated functionality from GIS software coupled with other techniques that frequently had evolved separately from GIS. Techniques such as location allocation, multi-criteria decision analysis (MCDA), hydrological and environmental modeling, artificial intelligence, agent-based modeling, and other approaches have been incorporated in SDSS often in concert with GIS. In many cases, the modeling techniques were not developed originally for inclusion in SDSS but rather were coupled later with GIS to formulate the SDSS. Early applications of SDSS utilized already existing models such as location allocation (Densham and Armstrong 1987; Armstrong et al. 1991), hydrologic (Grayman et al. 1992), and environmental (Engel et al. 1993) models, and used computer programming to make them interact with spatial databases and spatial processing routines. There were also many applications in which the modeling routines were built into the GIS framework using some customization programming. Early examples included a GIS-based expert SDSS for examining locations for a water well (Crossland 1990), a wetland value assessment model (Ji 1993), and a land use model for analyzing phosphorus runoff in the Lake Okeechobee watershed (Negahban et al. 1995). As technology progressed, it became easier to develop customized functionalities within GIS software or to couple GIS with external modeling programs.

2.3.6 Expert Domain Knowledge

The complexity of issues in which SDSS are utilized often calls for domain or expert knowledge to be built into the SDSS to assist users. These systems attempt to capture expert knowledge, which can then be used in automated fashion within the system. By building domain or expert knowledge into an SDSS, more effective decisions by users regarding data selection, model selection, and scenario evaluation can be made. Sikder (2008) pointed out that GIS are inherently limited in their capacity for integrating knowledge-based systems. He also pointed out that these limitations have led to the development of spatial expert systems in the form of knowledge acquisition modules, domain-specific knowledge bases, and rule-based inference engines. There have been many efforts to build expert systems or domain knowledge into SDSS. In 1995, Jain et al. built a knowledge-based system in the form of a decision tree algorithm using the Lisp language for siting livestock production facilities in environmentally sound locations. Another example was presented by Bellamy and Lowes (1999), who incorporated rules used for controlling the choice of regression and other models in an SDSS used for sustainable grazing management. Girvetz and Shilling (2003) coupled ArcView 3.2 with the knowledge base development program NetWeaver to build a fuzzy logic knowledge base into an SDSS for analyzing road system impact in a national forest. Further discussion and investigation of the modeling and knowledge components in SDSS will be investigated in later chapters, especially in Chapter 4.

2.4 DSS-Based Evolution

The first part of this section summarizes the general evolution of DSS and the second part explains specifically how DSS evolved into SDSS. The concept of decision support systems (DSS) originated with the work of Gorry and Scott-Morton (1971) in the early 1970s. They defined a DSS as an interactive computer system that helps decision makers solve unstructured or semistructured decision problems using data and models. Then, in the 1980s, Alter (1980) expanded the DSS framework and provided the first concrete DSS examples. The common DSS framework in the 1970s and 1980s primarily consisted of the integration of three major components (a) data management, (b) model management, and (c) dialog/interface management. The data management component used relational or other database technology to handle data that could be utilized in the system. The model management component handled analytical modeling capabilities, which utilized data controlled by the data management component. The dialog/interface component provided the mechanism for interaction between user and the system and allowed use of models and viewing of all outputs. Most of the earlier work on DSS focused on supporting individual decision makers (Alter 1980; Shim et al. 2002). As group support software matured, the traditional DSS was augmented with communication capabilities to create group decision support systems, which enabled geographically dispersed group members to work on complex unstructured problems and evaluate different scenarios (Nunamaker1989; Dickson et al. 1993).

The next phase in the progression of DSS was influenced by advancements in artificial intelligence, particularly in the late 1980s and early 1990s. During this time, expert systems added a new dimension to DSS (Holsapple and Whinston 1996). These knowledge-based DSS enabled users to analyze relatively complex problems and perform what-if analyses with the aid of organizational and domain knowledge (Courtney et al. 1987; Dutta 1996; Özbayrak and Bell 2003). Developments in the knowledge-based DSS area were drawn upon in creating intelligent SDSS. For example, Li et al. (2005) described an SDSS that used C Language Interface Production System (CLIPS; an expert system software) for supporting fuzziness and uncertainty in evaluating risk and insurance pricing in typhoon-affected areas in China.

The Web revolutionized application development in the 1990s. The ubiquitous nature of the Web and ease in using Web browser interfaces has facilitated the deployment of applications over the Web. Early work on Web-based decision support included an electronic marketplace Web-based decision support system called DecisionNet (Bhargava et al. 1995; Bhargava and Tettelbach 1997), which facilitated services between consumers (users of DSS) and providers (providers of DSS services). Several research and development efforts followed and myriad Web-based DSS were developed over the next decade (Barlishen and Baetz 1996; Bertolotto et al. 2001; Zhu et al. 2001; Wild and Griggs 2004). Hence, the next stage in DSS progression was Web-based DSS, which allowed delivery of appropriate data and models to managers or decision makers using a Web browser (Power and Kaparthi 2002; Liou et al. 2007). Using Web-based DSS, organizations can provide DSS capability to managers over a proprietary intranet, to customers and suppliers over a virtual private network (VPN), or to any stakeholder over the Internet (Sikder and Gangopadhyay 2004; Delen et al. 2007). Sikder and Gangopadhyay (2002) discussed the design and implementation of a Web-based SDSS that parallels the collaborative decision-making process typically supported by group support software discussed in the DSS literature.

The next and ongoing phase in DSS progression has been service-based DSS, which are based on software components that are accessed through the Internet. Component-based software development and Web services-based application development are maturing, and researchers are exploring ways to incorporate them into DSS and SDSS architecture and design (Lepreux et al. 2003; Di 2005; Wang and Cheng 2006; Zhao et al. 2007; Wu et al. 2004; Ray 2007). Component-based DSS development is based on existing code rather than development from scratch. Lepreux et al. (2003) discussed a phased approach for developing a component-based DSS that integrates the development and use of business and DSS components in collaboration with the users of the DSS. This method has been applied to design a DSS for investments in the French railway infrastructure (Lepreux et al. 2003). Mobile tools, mobile e-services, and wireless Internet protocols have been instrumental in expanding the capability and accessibility of DSS (Shim et al. 2002; Earle and Keen 2000). Wang and Cheng (2006) discussed a standardized framework for Web-based collaborative decision support services that facilitates information exchange and sharing of knowledge and models between various entities within an organization or across organizations. This framework provides metadata services, geodata services, and geoprocessing services to help collaborative decision making. Ray (2007) explained the development of a spatial Web-services-based SDSS at the Delaware Department of Transportation. This system helps in the management of the movement of oversized vehicles and integrates several components and services for analyzing vehicle characteristics, managing locations and routes, and providing graphical representation in a spatial map server.

In recent years, several researchers have reviewed the history or evolution of DSS in detail (Powell 2001; Power 2008; Bhargava et al. 2007). Some researchers detail a general evolution or overview of DSS. For example, Power (2002) provided a brief history of DSS in his book *Decision Support Systems*. Other researchers documented domain-specific evolution. For example, Segrera et al. (2003) reviewed the evolution of DSS architectures particularly for natural resources applications. Bhargava et al. (2007) described progress in Web-based DSS. For more in-depth consideration, we recommend readings at the end of this chapter.

2.4.1 DSS to SDSS

As mentioned earlier, DSS traditionally support the decision-making process using three major components: a database, a model base, and a user interface. Spatial decision support systems are an extension of this DSS concept, with spatial data used for the analysis of decisions (Densham and Goodchild 1988; Keenan 2005; Jarupathirun and Zahedi 2005). Although DSS research and applications have a rich history, only recently have they commonly incorporated spatial data and analysis (Keenan 2003). This is despite the fact that it has been reported that 80% of the data needed for activities of business and government are spatially related (Worrall 1991). This lack of spatial data use was mainly due to a lack of knowledge and skills in relation to spatial data models, spatial analysis or interpretation techniques, and cartographic skills. The ability to effectively utilize spatial technologies through the mid-1990s required expertise and was generally restricted to organizations willing to invest in the software and hardware infrastructures as well as the human resources necessary. The teaching of GIS and related spatial technologies also traditionally fell in geography departments with strong interest from environmental, physical, and ecological sciences. Until recent years, there was limited interest from business and economic academic arenas. In the last decade, this has started to change significantly with organizations recognizing the great commercial potential in the use of spatial data. Keenan captured this increased use of spatial data in his 2006 paper about the SDSS evolution. As Keenan is a faculty member in a business school, this paper was from a business perspective and reflected the advancement of DSS software to include spatial capabilities. The greater uptake in business was supported by a review of SDSS literature conducted for this book, which showed several business-related SDSS papers, book chapters, conference sessions, and workshops appearing in the 2000s (Ray 2007; Sugumaran and Sugumaran 2007; Sugumaran and Mobley 2002; Jankowski et al. 2001). Ray (2007) detailed the development of a Web-services SDSS that was built within the overall IT infrastructure at the Delaware Department of Transportation. Sugumaran and Sugumaran (2003) discussed the role of intelligent agents and GIS Web services in spatial decision support systems, and finally, Sugumaran and Mobley (2002) demonstrated the importance of spatial regression integration into a healthcare DSS.

2.5 GIS-Based Evolution

Similar to the DSS-based evolution section, the first part of this section summarizes the general evolution of GIS and the second part demonstrates how GIS evolved into SDSS. The first systems that are now called geographic information systems date back to the 1960s when computers were becoming available to large academic and government institutions (Malczewski 2004). The first large-scale use of a GIS-type system was the Canadian Land Inventory Project in the 1960s, which attempted to perform analyses to determine areas in different land uses and the possible future uses of different land areas (Keenan 2006). Later, several systems were developed, including a system called SYMAP, which was developed by the Harvard Laboratory for Computer Graphics and Spatial Analysis in the mid-1960s. The SYMAP system evolved into a family of related systems (CALFORM, SYMVU, GRID, POLYVRT, and ODYSSEY) throughout the 1960s and into the 1970s (Malczewski 2004). According to Malczewski (2004), computer hardware technology and the theoretical advances in spatial sciences motivated GIS growth in the 1960s and 1970s. In the late 1960s and early 1970s, GIS development was limited by cost and technical constraints—particularly by portability of software and data, high maintenance costs, difficulty in updating systems, the lack of distributed access, and the complexity of the command-line interfaces (Malczewski 2004; Meaden and Kapetsky 1991). While the first generation of GIS provided some modeling capability, they were inadequate for supporting any type of business decision making (Ozernoy et al. 1981; Pittman 1990).

The overarching technical evolutionary factors leading to increased GIS development in the 1980s and 1990s were the huge leaps made in computer hardware and processing speed (Malczewski 2004). In 1982, Environmental Systems Research Institute (ESRI), the largest GIS software company in the world, launched its first commercial GIS software called ArcInfo. The personal computer-based ArcInfo was introduced in 1986. There was a switch from workstation-based GIS to systems based on personal computers in the 1990s. Also during the period of the late 1980s and 1990s, in order to deal with complex decision situations, intelligent systems such as expert systems (ES) or knowledge-based systems were integrated into GIS to constitute *intelligent GIS*. Several applications were developed utilizing intelligent GIS in this time period, including by Kirkby (1996) for identifying and managing dry land salinization and by Moller-Jensen (1997) for classifying urban land cover.

The next phase (second half of 1990s) in the GIS-related development was shaped by the tremendous growth of the Web. GIS and mapping technologies began being presented in Internet-based technologies in the second half of the 1990s and into the 2000s. In the second half of the 1990s, numerous commercial (e.g., ESRI's ArcIMS and GeoMedia's WebMap) as well as open source (e.g., University of Minnesota MapServer) Web mapping platforms were introduced. These applications were adopted by many business, government, and academic organizations in the late 1990s and into the 2000s. These technologies were mainly used for data sharing and visualization in the early years of their existence. However, by the early 2000s these technologies began to show up more regularly in spatial decision making–related literature (Dragicevic et al. 2000; Rinner and Jankowski 2002; Sugumaran et al. 2003). These technologies continue to be used, but continuous advancements in technology have meant that in recent years the capabilities are expanding beyond simply Web mapping and data visualization.

Increasingly, GIS is moving away from only a standard desktop GIS with some Web mapping capabilities. Although these platforms are still extensively used, GIS functionalities are now being developed as distributable components for delivery through the Web, on mobile devices, or through other distributed networked technologies. These developments will permit many-to-many communications and facilitate distributed spatial problem solving. Numerous technologies can be utilized in this type of computing, such as mobile devices, Global Positioning Systems (GPS), real- or near real-time remotely sensed imagery delivery systems, and wireless communications for Internet GIS access. One prime example is using mobile devices enabled with GPS and GIS for real-time mapping applications such as forest fire mapping (Bolstad 2005). These types of distributed GIS architectures are constructed by partitioning client and server sides of an application into self-contained units that can interoperate across networks, integrating languages, applications, tools, and operating systems (Tsou and Buttenfield 2002). There are several examples of component-based GIS services software being developed in commercial (e.g., ESRI's ArcGIS Server, Intergraph's GeoMedia WebMap) and noncommercial (e.g., MapServer, GeoTools) sectors. These technologies have also begun to recently appear in GIS textbooks (e.g., Bolstad 2005) and in spatial science-related literature (e.g., Yue et al. 2009), and will continue to grow in use. There are a tremendous number of textbooks, articles, and other resources dedicated to GIS. At the end of the chapter, we recommend materials for further reading regarding GIS and its evolution.

2.5.1 GIS to SDSS

Although GIS are characterized by many attributes that are crucial to SDSS, such as having spatial data management and analysis capabilities, they are, in general, not considered SDSS. The major deficiencies in GIS limiting their characterization as SDSS are the lack of analytical modeling capabilities (Segrera et al. 2003) and their inability to present effective scenario evaluation techniques. Malczewski (1999) pointed out that GIS, in general, do not provide the tools for presenting choices and priorities in regard to evaluating conflicting criteria and goals. These deficiencies limit the effectiveness of GIS in solving semi- or ill-structured spatial decision problems (Densham 1991). A good example to illustrate the modeling deficiency would be the prediction of wildfire movement. Many useful functions could be carried out using GIS software in relation to wildfire management, including spatial data management and analysis as well as mapping. However, if the land managers and emergency personnel wanted to investigate possible deployment of resources depending on different weather scenarios, they likely would need to rely on some analytical modeling capabilities. These analytical models would have mathematical algorithms that could process information on topography, vegetative fuel, weather forecasts, and other information to predict possible future courses of the fire. The GIS software out of the box would not be able to handle these types of tasks. However, with some software development, the fire modeling algorithms and scenario development tools using real-time weather data could either be incorporated into the

GIS or linked to the GIS to form an SDSS. Indeed, there have been several examples of these types of SDSS developed (e.g., Guarniéri and Wybo 1995; Bonazountas et al. 2007). In later chapters, many examples will be given in which GIS are coupled or joined to other modeling or multi-criteria evaluation software to form an SDSS. Malczewski (1999) thoroughly covered GIS and multi-criteria decision analysis integration in his 1999 book, *GIS and Multicriteria Decision Analysis.*

2.6 SDSS Progression

This section provides a historical overview of how SDSS evolved over the last four decades, with an emphasis on how different drivers, such as advances in computing and communication technologies and uptake of spatially related technologies by various domains, contributed. In order to review the progression or evolution of SDSS development, an extensive literature review was carried out. Multiple searches using Web-based search engines and electronic libraries and databases were used to locate a wide variety of articles dealing with SDSS. The literature search used the following search terms: *spatial decision support system, spatial and DSS, GIS-based decisions support system, GIS and DSS, multi-criteria spatial decision, spatial decision making,* and *spatial decision support.* In total, 451 articles published during this time period were reviewed. Among these are a number of review articles that provide detailed coverage on specific SDSS-related aspects.

The development of SDSS has, to a large extent, entailed integrating analytical/decision models with GIS to produce systems capable of solving spatial problems. In many ways, the history of SDSS has mirrored the history of GIS, although with a lag of some years. Although SDSS concepts have arisen from the decision science discipline, most SDSS applications have been GIS-driven. Based on the literature review conducted for this book, we have divided SDSS evolution into three phases (the 3 I's): (a) introduction, (b) integration, and (c) implementation.

These broad evolutionary phases are represented in Figures 2.4 and 2.5 and are discussed in detail below. Broadly, these phases are the introductory phase, in which the concept of SDSS was introduced and prototype SDSS were being developed; the integration phase, in which many new technologies were being integrated in SDSS and development was growing quickly; and finally the implementation phase, in which the growth and use of SDSS really became widespread. Although these phases are useful for discussion purposes, there were not sharp temporal boundaries between them. The major categories of systems within each phase are described in the following sections.

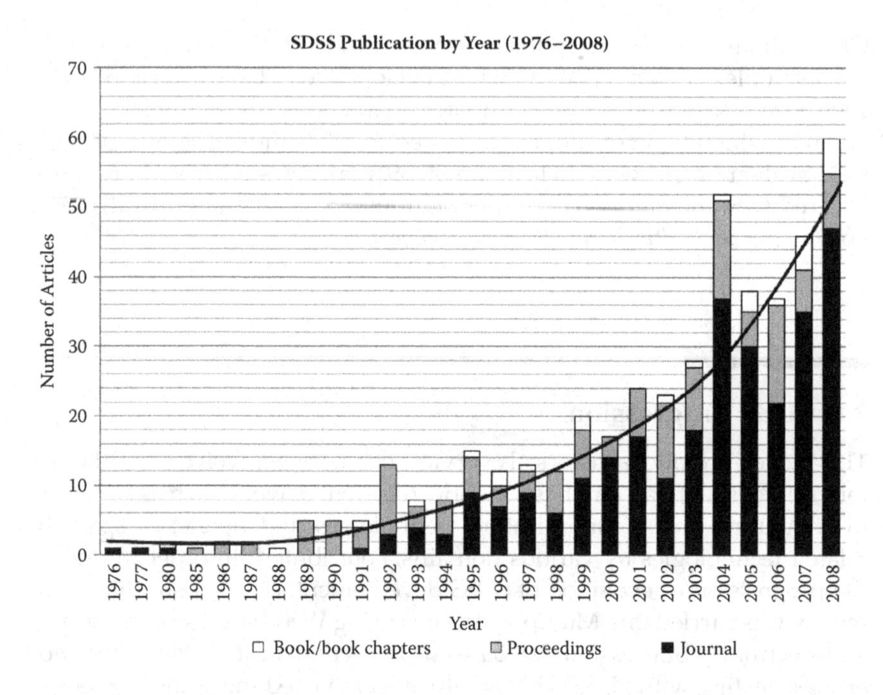

FIGURE 2.4
Number of SDSS publications per year as compiled from the literature by authors.

The development of SDSS began in earnest in the 1970s, slowly carried on in the 1980s, experienced rapidly accelerating growth in the 1990s, and continued this growth in the 2000s. Figure 2.4 shows the trend of SDSS publications over time based on the review of literature described above. The evolution of SDSS technology has followed that of computing, with a greater number of applications being deployed with the development of greater computer processing power, with more user-friendly development environments such as object-oriented programming, and with the proliferation of Web-based capabilities. In a review of literature on GIS-based multi-criteria decision analysis, Malczewski (2006) found that 70% of reviewed articles were published after 1999. This is similar to the literature review detailed in this book, which showed that approximately 72% of SDSS publications came in the 2000s. In the second half of the 1990s, customizable, more user-friendly GIS desktop applications were developed, allowing for much greater flexibility in specific user-designed applications. The flexibility and power in desktop development environments has continued to grow in the 2000s, and this has led to the greater number of SDSS applications as seen in the reviews by Malczewski (2006) and the literature review here. The great number of SDSS publications in the 2000s was also influenced by the increase in Web-based applications. An

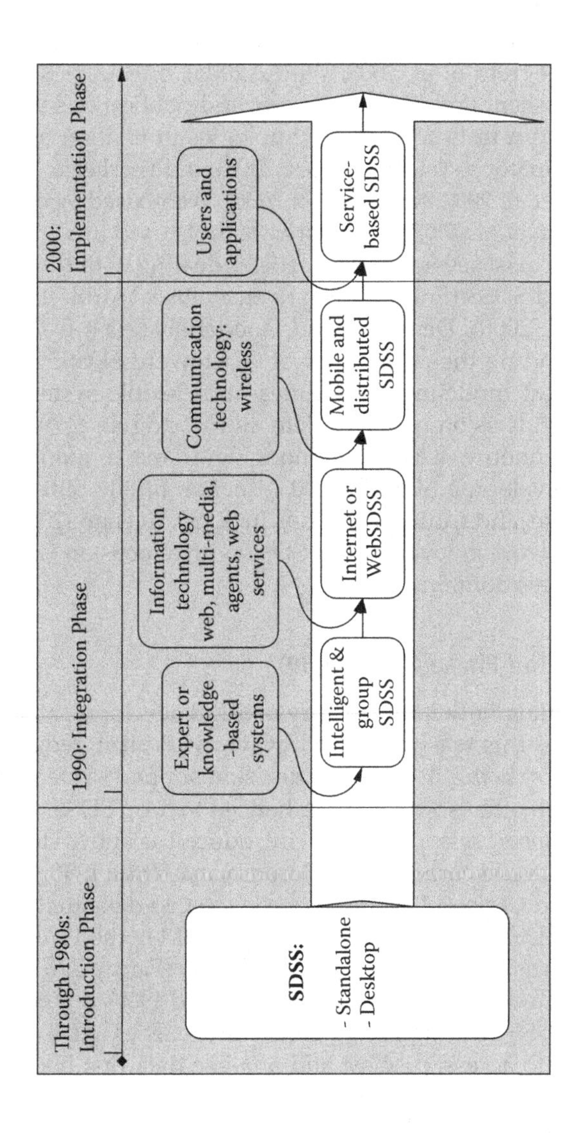

FIGURE 2.5
SDSS evolution. (Modified from Sugumaran and Sugumaran 2007.)

overview of Web-based SDSS was provided in Sugumaran and Sugumaran (2007).

There is a range of review articles available that cover specific aspects of SDSS, such as application domains or technologies used. Overviews of the use of SDSS for specific application areas have been detailed for aquaculture (Nath et al. 2000), agriculture (Colby and Johnson 2002), land use planning (Segrera et al. 2003; Witlox 2005), business (Keenan 2005), and forestry (Baskent and Keles 2005). Several publications were reviews focusing on certain techniques or technologies in relation to SDSS, such as public participatory systems (Sieber 2006), multi-criteria decision systems (Ascough et al. 2002; Malczewski 2006), Web-based systems (Rinner 2003; Sugumaran et al. 2007), and software tools used in SDSS (Yan et al. 1999). There also exist several review articles dealing with the relationship between GIS and SDSS (Enache 1994; Keenan 1996; Marble 1999; Murphy 1995; Nath et al. 2000). Densham and Goodchild (1989) were early proponents of extending the capabilities of GIS toward SDSS, which would include analytical modeling capabilities and flexible system design in order to support decision making. Neto and Rodrigues (1999) attempted to develop a taxonomy of SDSS methodologies and to identify effective strategies for developing SDSS. In 2003, Keenan briefly outlined the history, current state, and future directions in SDSS. Keenan (2006) provided in a subsequent work an overview of SDSS with a focus on GIS and applications to business domains.

2.6.1 Introduction Phase (1976–1989)

We label the initial phase in the history of SDSS development as the introduction phase as this was the period in which explicit geographic components were first being included in decision support systems, GIS were being joined with other software or enhanced to form SDSS, and the term *spatial decision support systems* was first introduced as known to the authors (Dobson 1983). Several early articles (Barbosa and Hirko 1980; Carlson et al. 1977; Mantey and Carlson 1980) focused on the geo-data analysis and display system (GADS), which was developed by IBM in the 1970s, and issues such as user interaction and algorithm inclusion (Figure 2.6). The work of Patrick Mantey and Eric Carlson in the 1970s and 1980s on the GADS was likely the first SDSS documented in the literature. The GADS enabled noncomputer users to access, display, and analyze data that had geographic content and meaning. The GADS was one of the earliest examples of what might now be called SDSS, although this term was not used to describe the system. It utilized spatial data in an interactive problem-solving environment and was used in numerous case studies, such as urban development, police officer allocation, and the design of school boundaries (Carlson et al. 1977). Through the 1980s there were fifteen articles identified as SDSS

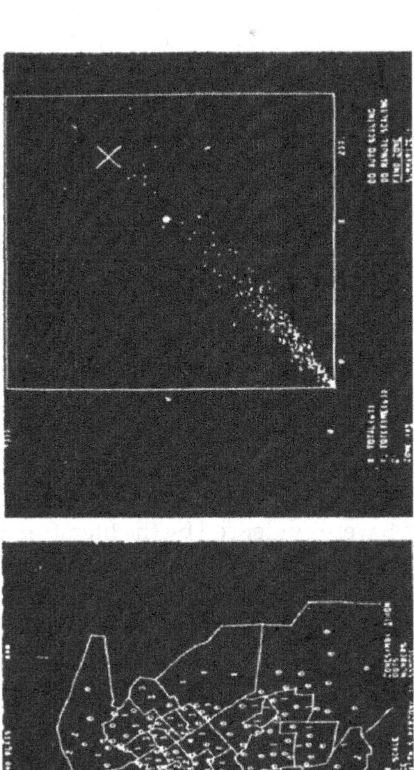

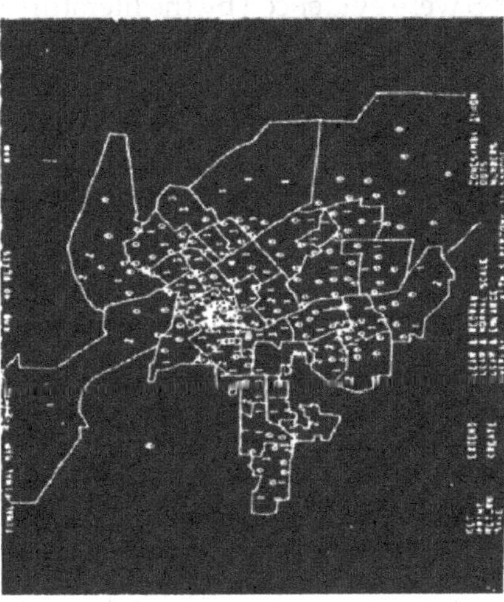

FIGURE 2.6
GADS interfaces with resultant map and graph. (From Carlson, E. D., et al. *MIS Quarterly* 1 (1), 1977. With permission.)

related. Although not termed as an SDSS, a decision support system (DSS) known as the Generalized Planning System (GPLAN) was designed for area-wide water quality planning and made use of spatial information on watersheds, point and nonpoint sources of pollutants, and land use plans (Holsapple and Whinston 1976). In 1985, Hopkins and Armstrong (1985) used the term *spatial decision support systems* in their description of a two-tiered framework of analytical and cartographic data and processing structures. In 1986, Armstrong et al. (1986) described an architecture for SDSS that went beyond existing spatial information systems' capabilities in supporting decision making. In 1989, five publications all appearing with the term *spatial decision support system* in the title appeared, demonstrating the adoption of the terminology. Three of these 1989 publications were from the Geographic Information Systems/Land Information Systems (GIS/LIS) 1989 proceedings. The majority of publications throughout the 1980s were in conference proceedings, with the exception of two early publications in *MIS Quarterly*, another in a planning journal, and two book chapters. There was an evolution in the literature from conceptual papers in the early years to more concrete implementation-driven articles later. The development of SDSS throughout the 1980s was often conceptual or commonly reviews of potential techniques (Carlson et al. 1977; Armstrong et al. 1986; Barbosa and Hirko 1980). Although the utility and initial acceptance of SDSS were evidenced by the literature though the end of the 1980s, there were a limited number of specific applications, mainly due to the fact that most spatial software systems needed large amounts of computing power, memory, and hard disk space. Furthermore, expensive hardware and software at this time required large budgets. In addition, the applications developed were characterized by GIS and other software that required a significant amount of expertise and were run only from command line interfaces.

In summary, the introductory phase of SDSS development was characterized by the definition of conceptual frameworks for SDSS, prototype SDSS development, desktop or workstation SDSS with single users, and command line-driven user interfaces.

2.6.2 Integration Phase (1990–2000)

The integration phase witnessed many new technologies being integrated in SDSS and the development of SDSS growing quickly although often only in prototype formats. During this phase, three major areas of development or evolution were occurring in relation to SDSS: (1) expansion from single-user SDSS to group SDSS and collaborative SDSS, (2) inclusion of intelligent components in SDSS, and (3) the beginning of Web-based SDSS.

The accelerated growth in SDSS publications from the early 1990s was directly related to advances in desktop computing power, reductions in the cost of desktop systems, development and proliferation of user-friendly GIS and other software, and the greater accessibility of software development environments. Advances in these technologies led to an increase in the development and application of SDSS and to the integration of new techniques and technologies such as user-friendly interfaces, spatial models, intelligent components, and eventually Web-based delivery platforms into SDSS architectures. The algorithms and spatial processing functions within GIS required significant computer processing speed and memory, which before the 1990s were generally not available in desktop computers. The use of computer-based GIS began in the 1960s (Keenan 2006), but these applications were only accessible to those with powerful computers. The applications in the 1980s generally relied on systems running on workstation computers, which were not user friendly and relied on experts to perform any spatial operations (Densham and Armstrong 1987; Mantey and Carlson 1980). In the 1990s, the most common GIS software used in SDSS applications was ESRI's ArcInfo. Other GIS software used included Geographic Resource Analysis Support Systems (GRASS), MapInfo, IDRISI, and TransCad. Software such as ArcInfo and GRASS were command line-driven and frequently run on UNIX workstations and consequently required a significant level of computer science and spatial science expertise. In the mid to late 1990s, software with more user-friendly interfaces and increasingly robust functionality were developed, allowing GIS usage to reach a much larger audience. In the latter half of the 1990s, numerous SDSS implementations utilized ESRI's ArcView software, which possessed user-friendly interfaces as well as a proprietary development language called Avenue. This software was originally developed to be an application for viewing spatial data, but evolved into a more robust GIS when ESRI realized the tremendous potential number of users for an easier-to-use GIS package compared to their workhorse GIS, ArcInfo. The interfaces a user would see when using ArcInfo Workstation and ArcView are shown in Figure 2.7. In ArcInfo, the user must be familiar with a wide range of commands and the related syntax to run operations in the command line interfaces. The ArcView GIS software introduced standardized graphical user interfaces and greatly opened up the number of potential GIS users and, subsequently, the number of SDSS developments.

A wide variety of development languages and environments were used in the development of SDSS in the 1990s. Although ArcView and similar GIS programs with GUIs and development languages were gaining traction, the most common GIS and application development languages for SDSS in the 1990s were ArcInfo and the Arc Macro Language (AML), which was used for the development of extended modeling routines. ArcInfo and AML were used in an SDSS incorporating air quality modeling within

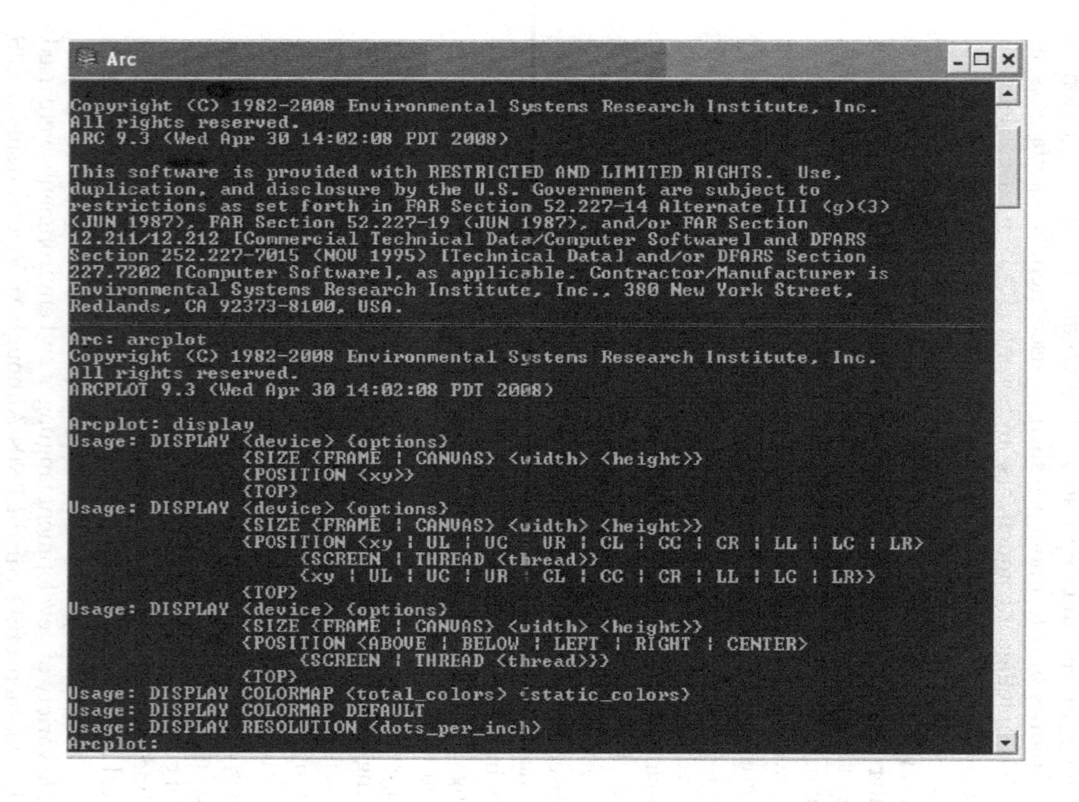

FIGURE 2.7a
ArcInfo workstation interface. (Carlson, E. D., Grace, B. F. and J. A. Sutton. 1977. Case studies of end user requirements for interactive problem-solving systems. *MIS Quarterly* 1(1):51–63.)

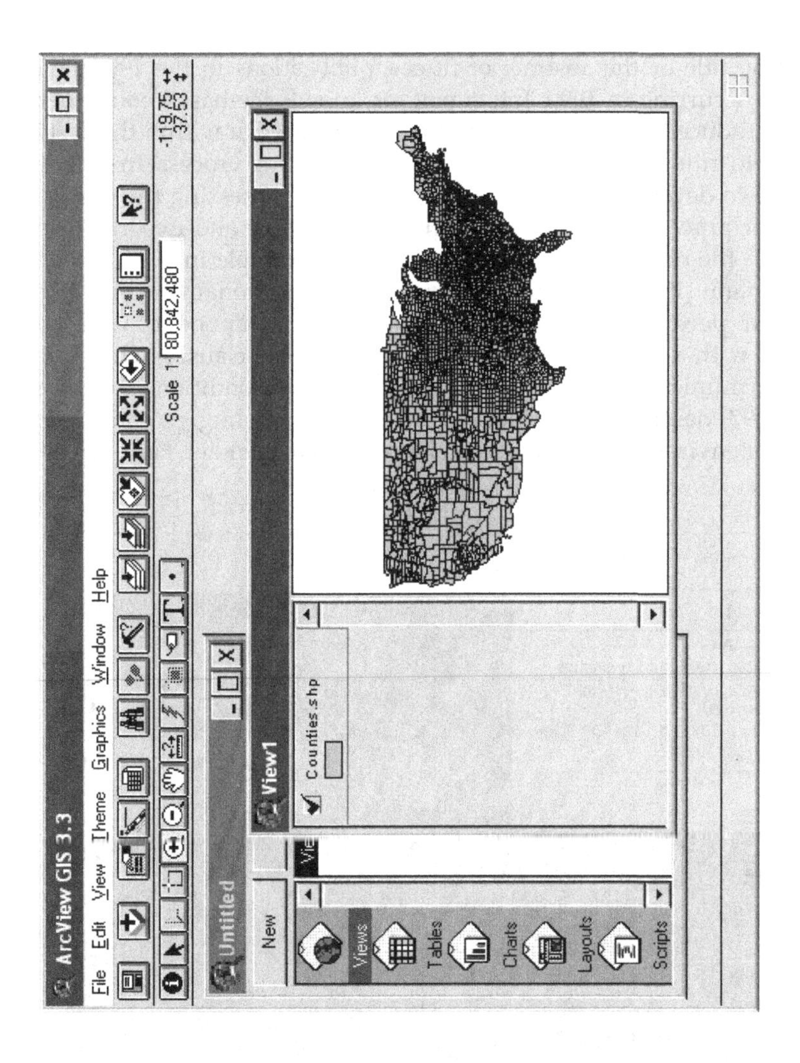

FIGURE 2.7b
ArcView 3.3 interface.

ArcInfo for Taiwan (Chang et al. 1997), and for coupling separate modeling capabilities, as did Jain et al. (1995) when they coupled ArcInfo with the Nitrate Leaching and Economic Analysis Package (NLEAP) model (Figure 2.8). Approximately sixty SDSS applications in the 1990s utilized ArcInfo software.

In the 1990s the use of intelligent components in SDSS became more common. The terms *expert* or some form of *intelligent* were included in either the title or the abstract of fifteen publications in the 1990s, with the first occurring in 1990. Intelligent- or knowledge-based components were introduced in SDSS to build domain knowledge within the system in order to guide users through the decision-making process. In order for an SDSS to develop legitimate scenarios, while harnessing the power of computer processing power, specialized knowledge and expertise were built into the decision processes in SDSS. For example in 1995, Gheorge and Vamanu (1995) combined expert systems functionality with GIS for a nuclear power emergency SDSS. The expert components were constructed with a large number of If … Then … Else clauses, which were meant to mimic an expert's judgment given certain conditions. Rodrigues et al. (1997) described a multiagent system for modeling geographic elements for environmental analysis in land use management. Ferrand (1996)

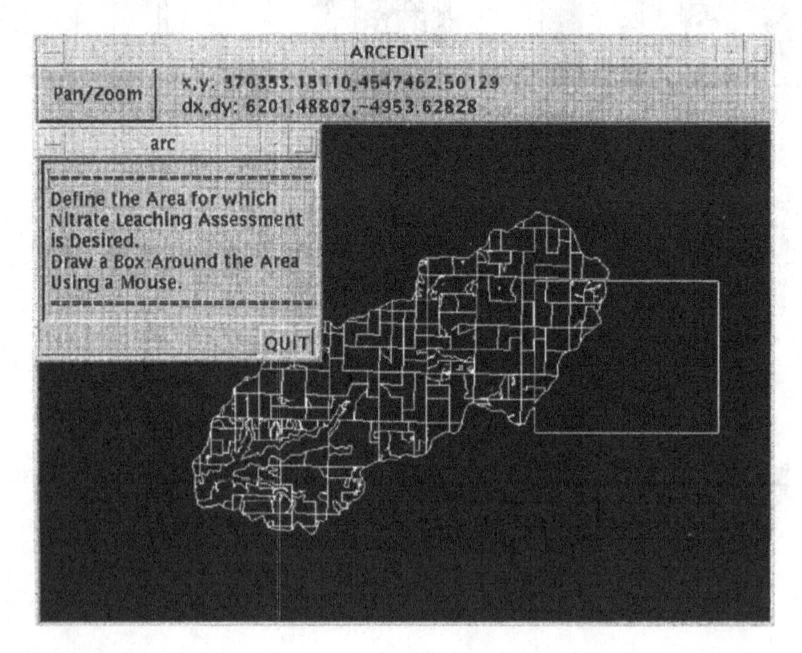

FIGURE 2.8
A desktop SDSS developed for livestock production planning. (Jain, D. K., Tim, U. S. and R. Jolly. 1995. A spatial decision support system for livestock production planning and environmental management. *Applied Engineering in Agriculture* 11(5):711–719.)

reported on a system used to solve complex spatial optimization problems in an attempt to minimize environmental impact of infrastructure development using agent-based modeling that incorporated knowledge bases. Bellamy and Lowes (1999) built expert knowledge about vegetation states, soil erosion risk, and management information into a sustainable grazing management SDSS. Matthews et al. (1999) used an existing knowledge-based system development environment (Gensym's G2) to include declarative rule-based systems and neural/Bayesian networks in a GIS-coupled SDSS. Expert or intelligent SDSS components will be discussed in more detail in Chapter 4.

Research and development concerning collaborative or public participatory SDSS was instigated in the 1990s. The very nature of spatial decision-making situations usually means that multiple stakeholder groups or individuals have a stake in any potential outcome. As pointed out by Jankowski et al. (1997), by the 1990s a movement from hierarchical decision making to a flatter structure in which workgroups would dominate organizational structures occurred. The development of technologies to support multigroup decision making became more widely available in the 1990s (Jankowski 1997). Group decision support technologies for spatial decision-making situations were being investigated actively by the mid-1990s (Armstrong 1993; Armstrong 1994; Armstrong and Densham 1995). Armstrong (1994) pointed out that there was a mismatch between group-based decision making and single-user GIS software, arguing for expansion to encompass the group decision-making processes. Jankowski et al. (1997) described the implementation of a collaborative SDSS known as Spatial Group Choice. This software was developed to utilize GIS functionality and to provide specialized features that would support alternative generation, modeling, evaluation, and cartographic display functions. The software was developed to be used in a face-to-face meeting environment. Spatial Group Choice used multi-criteria decision modeling techniques as well as consensus-building tools applied in an environmental restoration project. In their 2001 book, *Geographic Information Systems for Group Decision Making*, Jankowski and Nyerges (2001) detailed applications of group systems for healthcare management, transportation improvement, and habitat restoration.

In comparison to the 1980s, the 1990s produced a greater number (and proportion) of publications that described concrete implementations of SDSS as opposed to conceptual studies. The publications shifted from mainly occurring in conference proceedings to those in peer-reviewed journals. In the 1990s, the majority of articles demonstrated an implementation of an SDSS. In the first half of the 1990s, the majority of the literature was published in conference proceedings, while in the second half of the decade, journal publications constituted more than half of the publication total, demonstrating the furthering acceptance of SDSS in scientific literature. Books published on topics such as decision support systems, GIS for business, and environmental GIS applications appeared during the 1990s and discussed various aspects of SDSS.

2.6.3 Implementation Phase (2000s)

The growth of SDSS has continued in the 2000s and has also diversified based on advances in Web, networking, and component-based technologies. The number of SDSS applications continued to grow rapidly after the year 2000. We have termed this phase the implementation phase because the greatest number of publications exhibited concrete applications of SDSS technology. The majority of publications after 1999 demonstrated a concrete implementation of SDSS to a specific problem domain. The movement from command-line-driven UNIX-based GIS software to personal computer-based GIS software with user-friendly graphical user interfaces (GUIs) that began in the second half of the 1990s continued into the 2000s. In the last five years of the 1990s, ArcInfo was still the most common software used in SDSS implementations, while ArcView was becoming used more frequently toward the end of the 1990s. In the first five years of the 2000s, ArcView became the most commonly used GIS software with ArcInfo dropping to second place. The popularity of ArcView and its Avenue customization language appealed to many academics, businesses, and agencies that developed SDSS specifically for their own purposes. Advancement in desktop GIS software continued in the 2000s with the most prominent GIS software being used in SDSS changing to the next-generation ESRI package ArcGIS after 2004. A large number of other commercial GIS software (e.g., IDRISI, MapInfo, ILWIS), freely available GIS software (e.g., CommonGIS, GeoTools), or user-developed software with spatial processing/display functionality were also used in SDSS applications, but these other GIS software programs were applied in no more than 5% of the studies in the 2000s.

In the 2000s, there has also been tremendous advancement in the development of Web-based spatial technologies as well as component- or service-based spatial technologies. These advancements have increasingly been implemented into a variety of SDSS applications. In the latter half of the 1990s and early part of the 2000s, Web-based mapping systems were being adopted, with ESRI's ArcIMS, the University of Minnesota's MapServer, and GeoMedia WebMap being utilized for spatial data presentation via the Web. These systems were often used for information presentation with limited analytical functionality. One of the earliest SDSS utilizing Web-based spatial display and analysis was described by Wan et al. (1999). Their customized system used their own Java-based WebGIS for spatial display and analysis in conjunction with a multi-criteria analytical module used for conducting an economic analysis of the Beijing–Kowloon railway corridor. The increased use of Web-based mapping capability foreshadowed the increased use of Web-enabled SDSS in the coming years. Web-enabled SDSS benefits include not only accessibility, efficient distribution, efficient administration, and cross-platform flexibility, but also data storage advantages (Molenaar et al. 2001; Sugumaran et al., 2004).

In the past 10 years, numerous research articles on Web-based SDSS have appeared in the literature. For example, a Web-Based Spatial Decision Support System (WEBSDSS) prototype developed by Sugumaran et al. (2004) prioritized management of local watersheds on the basis of environmental sensitivity using a multiple-criteria evaluation model with a weighted linear combination method for the City of Columbia, Missouri, USA. Qiu et al. (2002) developed a Web-based watershed hydrologic SDSS for St. Charles County, Missouri, which used the Hydrologic Simulation Program FORTRAN (HSPF) model to simulate the predicted runoff at a user-defined outlet point along the stream network. In another study, Bhargava and Tettelbach (1997) presented a Web-based system that supported consumers in finding the best options to dispose of recyclable materials using a route-finding algorithm with time and cost/benefit constraints. Compas and Sugumaran (2004) developed a Web-based urban growth model and visualization tool for St. Charles County's planning and zoning department to use in urban growth planning and management (Figure 2.9). They successfully implemented a complex multi-criteria evaluation tool, the Analytical Hierarchy Process (AHP), to model end user's decision-making processes and generate model weights. They also discussed the issues associated with advanced model implementation on the Web. There have been numerous other examples of Web-based (ArcIMS, ArcGIS Server, GeoTools, SpringWeb GIS, Minnesota MapServer, etc.) and mobile applications (ArcPad) being used in SDSS either alone or in conjunction with other software (e.g., Chen and Daoliang 2008; Wang and Cheng 2006; Carver et al. 2001). The arrival of numerous other Web-based GIS software (e.g., GeoMedia WebMap, MapXtreme, and GeoMatica WebServer) leads to the expectation that the number of Web-based SDSS will grow greatly over the coming years. With the advent of Web-based SDSS, applications involving public participation have increased with the first appearing in 1999 and continuing in the 2000s (e.g., Jarupathirun and Zahedi 2007). Due to the increased use of Web-based spatial applications, there have been numerous publications that have investigated issues (e.g., data exchange, software, model sharing) of implementing SDSS on the Web (Wang and Cheng 2006, Sugumaran and Sugumaran 2007). Sugumaran and Sugumaran (2007) concluded that developments in intelligent agent technology, ontology-based information systems, knowledge-based systems, GIS Web Services, data warehousing and analytical processing, and Web technologies will have a huge impact on future SDSS development.

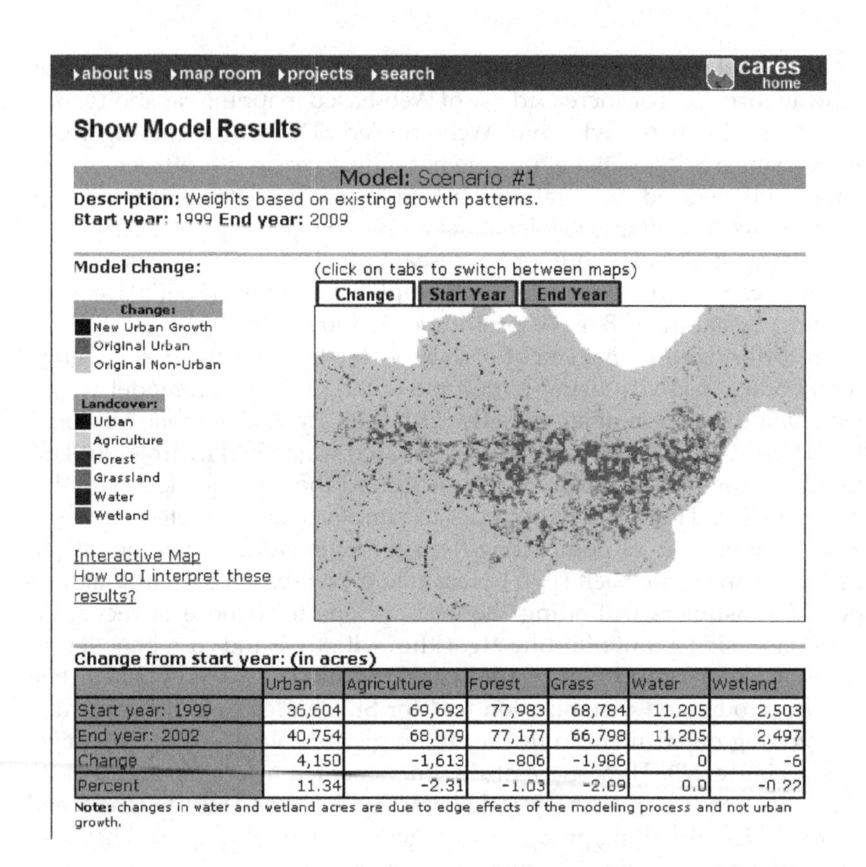

FIGURE 2.9
(See color insert following page 74.) Web-based SDSS developed for urban growth prediction. (Compas, E. and R. Sugumaran. 2004. Urban growth modeling on the web: A decision support tool for community planners. Paper presented at the 27th Annual Applied Geography Conference, St. Louis.)

In addition to the valuable potential of adding Web-based technologies to SDSS development, other advances in technologies and standards expand possibilities in SDSS development and application. Advances in wireless technology, as well as software development on GPS-enabled mobile devices, have given rise to the development of SDSS that utilize mobile technologies. These SDSS provide access to spatial data as well as decision support applications using handheld devices from remote locations. One such example is the development of integrated mobile geospatial information services to support and help optimize field-based management tasks for border security agents (NASA 2005). Another example is the development of spatial decision support software called First Response, which provides access to spatial and nonspatial data, 3D visualization capabilities, and historical information on search and rescue operations on mobile devices. In Australia, the tracking of locust plagues uses GPS-connected palmtop computers that communicate directly with a GIS server. Field information is fed, along with real-time climatic and other data, into a locust development model that is used by land managers to decide on control strategies (Deveson and Hunter 2008).

Spatial decision support systems functionalities can be modularized and implemented as components or services that could be subscribed to or embedded in other applications. These services can be executed at the provider's site to alleviate incompatibility problems. For example, Yeh and Qiao (2004) developed a component-based approach for implementing a knowledge-based planning support system. Ideally, a service-based SDSS provides ubiquitous access to *spatial computational services* from a variety of devices. Taking it one step further, these components can actually act as spatial Web services, and users can compose a set of these services to achieve a particular functionality. Web services technology is supported by several key protocols and standards, such as Extensible Markup Language (XML), Web Services Description Language (WSDL), Simple Object Access Protocol (SOAP), and Universal Description, Discovery and Integration (UDDI). Service-based SDSS can be effective in minimizing the cognitive load on end users because of its ability to deal with heterogeneity in hardware as well as software components that may be written using different languages. These types of SDSS provide interoperability by seamlessly taking care of the translations that need to be performed for different components or services to work together. For example, Jung and Sun (2006) built a GIS services Web site, *Location Analysis of Business Decision Support System in Taipei city (LABDSSiT)*, for evaluating potential locations of convenience stores in Taipei, Taiwan (Figure 2.10). Server GIS technologies, such as ESRI's ArcGIS Server, provide spatial analysis and processing services over the Internet, allowing greater possibilities for a wider range of nonexperts to use these functionalities and allowing the development of Web-based SDSS. It is expected that these types of applications will grow greatly in the coming years.

2.7 Related and Important Literature

There are many crucial publications dealing with SDSS as a whole or at least some related aspect of SDSS. However, there have been no comprehensive SDSS books published to date. Some of the important published materials are listed at the end of the chapter.

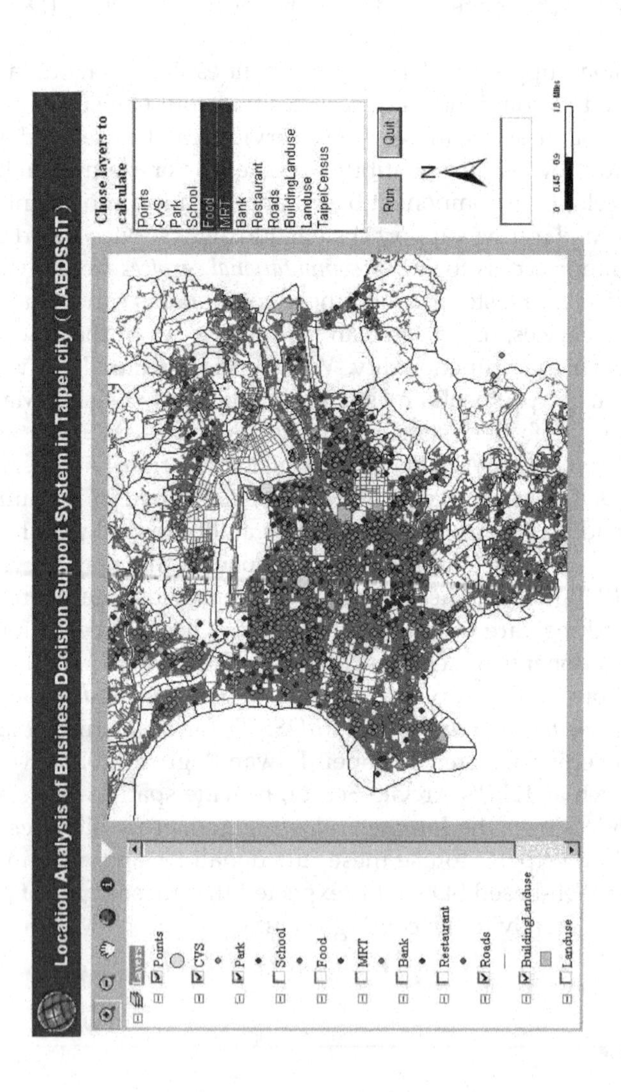

FIGURE 2.10
(See color insert following page 74.) Service-based SDSS example. (Jung, C. T. and C. H. Sun. 2006. Development of a web-based spatial decision support system for business location choice in Taipei City. Paper presented at the ESRI 2006 User Conference, San Diego.)

2.8 Important Contributors to SDSS Development

There have been a number of seminal figures who helped shape the evolution of SDSS and also its acceptance in a variety of scientific, business, and other disciplines. We cannot cover all of the contributors to SDSS development, but provide an overview of some important contributions. The work of Patrick Mantey and Eric Carlson in the 1970s and 1980s on the Geodata Analysis and Display System (GADS) was likely the first SDSS documented in the literature. The GADS enabled non-computer users to access, display, and analyze data that had geographic content and meaning. Early work from Dr. Marc Armstrong and his colleagues (Paul Densham, Gerard Rushton, and others) from the University of Iowa helped to define and codify the concept of spatial decision support systems as something beyond Geographic Information Systems. They also provided some of the earliest examples of SDSS development (Hopkins and Armstrong 1985; Armstrong et al. 1986; Densham and Armstrong 1987; Armstrong 1988; Armstrong and Lolonis 1989; Armstrong and Densham 1990, Armstrong et al. 1991). Dr. Piotr Jankowski and Dr. Timothy Nyerges were leaders in the development of collaborative or public-participatory SDSS. They helped develop concepts behind effective collaborative SDSS and also demonstrated real-world applications in the 1990s and 2000s (Jankowski et al. 1997, 2001, 2006; Nyerges et al. 2002, 2006) that helped guide the continued development of effective collaborative systems. Dr. Jacek Malczewski has been a leading researcher in the use of spatially enabled multi-criteria evaluation systems, which have constituted a significant proportion of SDSS applications. Malczewski has published several useful overviews of the techniques and technology (Malczewski 1999, 2004, 2006) as well as specific case studies (Malczewski et al. 1997, Malczewski et al. 2003). From a business perspective, several important figures influenced the development of SDSS. Peter Keenan produced important work on the use of GIS for business decision support and the use of SDSS for business for more than a decade (Keenan 1996, 1998, 2003, 2005, 2006, 2008). Both Brian Mennecke and Martin Crossland have also been leaders in the investigation of spatial technologies and SDSS use in business and testing for their usefulness as decision support tools using scientific methods (Crossland et al. 1994, 1995; Crossland 2005; Mennecke and Crossland 1996, Mennecke et al. 2000). Many other researchers have also played an important role in the development of SDSS, but those mentioned specifically here have had a pronounced effect on SDSS becoming common and effective tools in many disciplines.

2.9 Summary

Spatial decision support systems have evolved greatly over the last few decades based on advances in underlying technologies such as computer hardware and software, networking, and communication technologies. After early development in the 1970s and 1980s, the concept of SDSS gained traction in the 1990s. The development of SDSS became much more common in the late 1990s when ever greater amounts of digital spatial data were becoming available and personal computers were becoming widely used. The growth has continued into the 2000s with diversification based on technological developments. The development of SDSS generally followed developments in Geographic Information Systems, with many concepts and techniques of the science taken from decision support systems research and advances. The first SDSS were developed for workstations using command-line-driven GIS. These technologies were the domain of experts with high-end computing resources. In the 1980s and 1990s, the cost of computing consistently fell, leading to the development of the personal computer and software with graphical user interfaces. Due to these advances, practitioners from a wider range of domains began utilizing computers and GIS software. The 1990s saw tremendous growth in the applications of SDSS to a variety of problem domains, including urban, transportation, environmental, natural resources, business, agricultural, emergency planning, and others. In addition, in the 1990s and into the 2000s, the development of object-oriented programming languages and component-based software provided developers much more power in developing coupled and customized software. Knowledge-based and artificial intelligence techniques were also introduced into SDSS in the 1990s. The great advances in networking technology and the use of the Web led to the increased use of Web-based technologies in SDSS architectures in the last decade. In addition, component-based software development, based on software and data compatibility standards, is providing greater flexibility in combining techniques from a variety of disciplines and programs. With improving wireless communication technologies, the ubiquity of GPS-enabled devices, and distributed software techniques, SDSS that operate with mobile components are becoming feasible. The combination of all these technologies is leading to an increase in Web-based and mobile-based software components in SDSS, which provide greater flexibility for use of real-time data as well as the inclusion of non-expert users in participatory systems. The rapid growth in SDSS development that began in the 1990s and has lasted until now should be expected to continue with a greater number of Web- and mobile-based SDSS being developed for a variety of disciplines.

Suggested Readings

DSS

Sprague, Ralph H., Jr., and Hugh J. Watson. *Decision Support Systems: Putting Theory into Practice* (3rd ed.). Englewood Cliffs: Prentice Hall, 1993.

Sprague, Ralph H., and Eric D. Carlson. *Building Effective Decision Support Systems.* Englewood Cliffs, NJ: Prentice Hall, 1982.

Keen, Peter G. W. *Decision Support Systems: A Research Perspective.* Cambridge: Center for Information Systems Research, Alfred P. Sloan School of Management, 1980.

Fick, G., and Ralph H. Sprague. *Decision Support Systems: Issues and Challenges.* Oxford; New York: Pergamon Press, 1980.

Alter, Steven L. *Decision Support Systems: Current Practice and Continuing Challenges.* Reading, MA: Addison-Wesley, 1980.

Turban, Efraim. *Decision Support and Expert Systems: Management Support Systems* (4th ed.). Englewood Cliffs, NJ: Prentice Hall, 1995.

Power, Daniel J. *Decision Support Systems: Concepts and Resources for Managers.* Westport, CT: Quorum Books, 2002.

Marakas, George M. *Decision Support Systems* (2nd ed.). Englewood Cliffs, NJ: Prentice Hall, 2002.

Mora, Manuel, Guisseppi A. Forgionne, and Jatinder N. D. Gupta. *Decision Making Support Systems: Achievements, Trends and Challenges for the New Decade.* Hershey, PA: Idea Group, 2002.

GIS

Longley, P. *Geographic Information Systems and Science.* New York: John Wiley & Sons, 2005.

Heywood, Ian, Sarah Cornelius, and Steve Carver. *An Introduction to Geographical Information Systems* (3rd ed.). Englewood Cliffs, NJ: Prentice Hall, 2006.

DeMers, Michael N. *Fundamentals of Geographical Information Systems* (4th ed.). New York: John Wiley & Sons, 2008.

Burrough, Peter A., and Rachael A. McDonnell. *Principles of Geographical Information Systems* (2nd ed.). Oxford; New York: Oxford University Press, 1998.

Chang, Kang-tsung. *Introduction to Geographic Information Systems with Data Files CD-ROM.* New York: McGraw-Hill Science/Engineering/Math, 2009.

Chrisman, Nicholas. *Exploring Geographical Information Systems, 2nd Edition.* New York: John Wiley & Sons, 2001.

Delaney, J., and Niel, K. V. *Geographical Information Systems: An Introduction* (2nd ed.). Oxford; New York: Oxford University Press, 2007.

Neteler, Markus, and Helena Mitasova. *Open Source GIS: A GRASS GIS Approach* (3rd ed.). New York: Springer, 2007.

Peng, Zhong-Ren, and Ming-Hsiang Tsou. *Internet GIS: Distributed Geographic Information Services for the Internet and Wireless Network.* Hoboken, NJ: John Wiley & Sons, 2003.

SDSS

Malczewski, J. *GIS and Multicriteria Decision Analysis*. New York: John Wiley & Sons, 1999.

Leung, Y. *Intelligent Spatial Decision Support Systems*. Berlin; New York: Springer, 1997.

Kersten, Gregory E., Zbigniew Mikolajuk, and Anthony Gar-On Yeh. *Decision Support Systems for Sustainable Development*. Boston: Kluwer Academic Publishers, 1999.

Mora, Manuel, Guisseppi. A. Forgionne, and Jatlinder N. D. Gupta. *Decision Making Support Systems. Hershey, PA:* Idea Group (IGI), 2003.

Burstein, Frada, and Clyde W. Holsapple. *Handbook on Decision Support Systems 2*. Netherlands: Springer, 2008.

Hilton, Brian N. *Emerging Spatial Information Systems and Applications*. Hershey, PA: Idea Group, 2006.

Geertman, Stan, and John Stillwell. *Planning Support Systems in Practice*. New York: Springer, 2003.

Yeung, Albert K. W., and G. Brent Hall. *Spatial Database Systems*. New York: Springer, 2007.

Halls, Peter J. *Spatial Information and the Environment*. New York: Taylor & Francis, 2001.

Mora, Manuel, Guisseppi A. Forgionne, and Jatinder N. D. Gupta. *Decision Making Support Systems: Achievements, Trends and Challenges for the New Decade*. Hershey, PA: Idea Group, 2002.

References

Alter, S. L. 1980. *Decision support systems: Current practice and continuing challenges*. Reading, MA: Addison-Wesley.

Armstrong, M. P. 1988. Distance imprecision and error in spatial decision support systems. Paper presented at the International Geographic Information Systems (IGIS) Symposium, Washington.

Armstrong, M. P. 1993. Perspectives on the development of group decision support systems for locational problem solving. *Geographical Systems* 1:69–81.

Armstrong, M. P. 1994. Requirements for the development of GIS-based group decision-support systems. *Journal of the American Society for Information Science* 45(9):669–677.

Armstrong, M. P., and P. J. Densham. 1990. Database organization strategies for spatial decision support systems. *International Journal of Geographical Information Science* 4(1):3–20.

Armstrong, M. P., P. J. Densham, and G. Rushton. 1986. Architecture for a microcomputer based spatial decision support system. Paper presented at Second International Symposium on Spatial Data Handling, Seattle, Washington.

Armstrong, M. P., and P. J. Densham. 1995. Cartographic support for collaborative spatial decision-making. Paper presented at the Twelfth International Symposium on Computer-Assisted Cartography, Charlotte, North Carolina.

Armstrong, M. P., and P. Lolonis. 1989. Interactive analytical displays for spatial decision support systems. Paper presented at the Ninth International Symposium on Computer-Assisted Cartography, Baltimore, Maryland.

Armstrong, M. P., G. Rushton, R. Honey, B. Dalziel, P. Lolonis, S. De, and P. J. Densham. 1991. Decision support for regionalization: A spatial decision support system for regionalizing service delivery systems. *Computers, Environment and Urban Systems* 15:37–53.

Ascough, J., II, H. Rector, D. Hoag, G. McMaster, B. Vandenberg, M. Shaffer, M. Weltz, and L. Ahuj. 2002. Multicriteria spatial decision support systems for agriculture: Overview, applications, and future research directions. Paper presented at the Integrated Assessment and Decision Support (iEMSs 2002), Lugano, Switzerland.

Barbosa, L. C., and R. G. Hirko. 1980. Integration of algorithmic aids into decision support systems. *Management Information Systems Quarterly* 4(1):1–12.

Barlishen, K. D., and B. W. Baetz. 1996. Development of a decision support system for municipal solid waste management systems planning. *Waste Management & Research* 14:71–86.

Baskent, E. Z., and S. Keles. 2005. Spatial forest planning: A review. *Ecological Modelling* 188(2–4):145–173.

Bellamy, J. A., and D. Lowes. 1999. Modelling change in state of complex ecological systems in space and time: An application to sustainable grazing management. *Environment International* 25(6–7):701–712.

Bertolotto, M., J. D. Carswell, L. McGeown, and J. McMahon. 2001. E-spatial technology for spatial analysis and decision making in web-based land information management systems. *Journal of Geographic Information and Decision Analysis* 5(2):95–114.

Bhargava, H. K., Andrew S. King, and D. S. McQuay. 1995. DecisionNet: An architecture for modeling and decision support over the world wide web. Paper presented at the Third International Conference on Decision Support Systems, Hong Kong.

Bhargava, H. K., D. J. Power, and D. Sun. 2007. Progress in web based decision support technologies. Decision Support Systems 43(4):1083–1095.

Bhargava, H. K., and C. G. Tettelbach. 1997. A web-based DSS for waste disposal and recycling. *Computer, Environment and Urban Systems* 21(1):47–65.

Bolstad, Paul V. 2005. *GIS fundamentals: A first text on geographic information systems. Second Edition*. White Bear Lake: Eider Press, Inc.

Bonazountas, M., D. Kallidromitou, P. Kassomenos, and N. Passas. 2007. A decision support system for managing forest fire casualties. *Journal of Environmental Management* 84(4):412–418.

Carlson, E. D., B. F. Grace, and J. A. Sutton. 1977. Case studies of end user requirements for interactive problem-solving systems. *MIS Quarterly* 1(1):51–63.

Carver, S., A. Evans, R. Kingston, and I. Turton. 2001. Public participation, GIS, and cyberdemocracy: Evaluating on-line spatial decision support systems. *Environment and Planning B: Planning and Design* 28(6):907–921.

Chang, N-B., Y. L. Wei, C. C. Tseng, and C. Y. J. Kao. 1997. The design of a GIS-based decision support system for chemical emergency preparedness and response in an urban environment. *Computers, Environment and Urban Systems* 21(1):67–94.

Chen, Y. and L. Daoliang. 2008. Spatial decision support system for reclamation in opencast coal mine dump. *WSEAS Transactions on Computers* 5(7):519–531.

Colby, M. M., and Y. J. Johnson. 2002. Potential uses for geographic information system-based planning and decision support technology in intensive food animal production. *Animal Health Research Reviews* 3(1):31–42.

Compas, E., and R. Sugumaran. 2004. Urban growth modeling on the web: A decision support tool for community planners. Paper presented at the 27th Annual Applied Geography Conference, St. Louis, Missouri.

Courtney, J. F., D. B. Paradice, and N. A. Mohammed. 1987. A knowledge-based DSS for managerial problem diagnosis. *Decision Sciences* 18(3):373–399.

Crossland, M. D. 1990. HydroLOGIC—A prototype geographic information expert system for examining an artificial intelligence application in a GIS environment. Paper presented at the Annual GIS/LIS Conference, Anaheim, California.

Crossland, M. D., and B. E. Wynne. 1994. Measuring and testing the effectiveness of a spatial decision support system. Paper presented at the 27th Annual International Conference on System Sciences, Honolulu, Hawaii.

Crossland, M. D., B. E. Wynne, and W. C. Perkins. 1995. Spatial decision support systems: An overview of technology and a test of efficacy. *Decision Support Systems* 14(3):219–235.

Crossland, M. D. 2005. Geographic information systems as decision tools. In *Encyclopedia of information science and technology (II)*, ed. M. Khosrow-Pour, 1274–1277. Hershey, PA: Idea Group.

Delen, D., R. Sharda, and P. Kumar. 2007. Movie forecast guru: A Web-based DSS for Hollywood managers. *Decision Support Systems* 43(4):1151–1170.

Densham, P. J. 1991. Spatial decision support systems. In *Geographical information systems: Principles and applications*, ed. J. Maguire, M. S. Goodchild, and D. W. Rhind. London: Longman Publishing Group.

Densham, P. J., and M. P. Armstrong. 1987. A spatial decision support system for locational planning: design, implementation and operation. Paper presented at the Eighth International Symposium on Computer-Assisted Cartography, Baltimore, Maryland.

Densham, P. J., and M. F. Goodchild. 1989. Spatial decision support systems: A research agenda. Paper presented at the Annual GIS/LIS Conference, Orlando, Florida.

Deveson, T., and D. Hunter. 2008. The operation of a GIS-based decision support system for Australian locust management. *Insect Science* 9(4):1–12.

Di, L. 2005. A framework for construction of web-services based intelligent geospatial knowledge systems. *Journal of Geographic Information Science* 11(1):24–28.

Dickson, G. W., J. E. L. Partridge, and L. H. Robinson. 1993. Exploring modes of facilitative support for GDSS technology. *MIS Quarterly* 17(2):173–194.

Dobson, M. W. 1983. A high resolution microcomputer based color system for examining the human factors aspects of cartographic displays in a real-time user environment. Paper presented at the Sixth International Symposium on Computer Assisted Cartography, Toronto, Canada.

Dragicevic, S., S. Balram, and J. Lewis. 2000. The role of web GIS tools in the environmental modeling and decision-making process. Paper presented at the 4th International Conference on Integrating GIS and Environmental Modeling, Banff, Canada.

Dutta, A. 1996. Integrating AI and optimization for decision support: A survey. *Decision Support Systems* 18:217–226.

Earle, N. and Keen, P. 2000. From .com to .profit: inventing business models that deliver value and profit. New York: John Wiley & Sons.

Enache, M. 1994. Integrating GIS with DSS: A research agenda. Paper presented at the Urban and Regional Information Association, Milwaukee, Wisconsin.

Engel, B. A., R. Srinivasan, and C. Rewerts. 1993. A spatial decision support system for modeling and managing agricultural non-point source pollution. In *Environmental modeling with GIS*, ed. M. F. Goodchild, B. O. Parks, and L. T. Steyaert, 231–237. New York: Oxford University Press.

Ferrand, N. 1996. Modeling and supporting multi-actor spatial planning using multi-agents systems. Paper presented at Third International Conference Integrating GIS and Environmental Modeling, Santa Fe, New Mexico.

Gheorge, A. V., and D. Vamanu. 1995. A pilot decision support system for nuclear power emergency management. *Safety Science* 20(1):13–16.

Girvetz, E., and F. Shilling. 2003. Decision support for road system analysis and modification on the Tahoe National Forest. *Environmental Management* 32(2):218–233.

Gorry, G. A., and M. S. Scott-Morton. 1971. A framework for management information systems. *Sloan Management Review* 13(1):55–70.

Gould, M. D. 1989. The value of spatial decision support systems for oil and hazardous chemical spill response. Paper presented at the 12th Applied Geography Conference. Binghamton, New York.

Grayman, W. M., J. P. Heath, and R. M. Males. 1992. Spatial decision support system for toxic spill modeling in the Ohio River. Paper presented at 1992 National Conference on Water Resources Planning and Management—Water Forum '92, Baltimore.

Guarnieri, F., and J. L. Wybo. 1995. Spatial decision support and information management application to wildland fire prevention—The WILFRIED System. *Safety Science* 20(1):3–12.

Holsapple, C. W., and A. B. Whinston. 1976. A decision support system for area-wide water quality planning. *Socio-Economic Sciences* 10:265–273.

Holsapple, C. W., and A. B. Whinston. 1996. *Decision support systems. A knowledge-based approach.* Minneapolis/St. Paul, MN: West Publishing.

Hopkins, L., and M. P. Armstrong. 1985. Analytic and cartographic data storage: A two-tiered approach to spatial decision support systems. Paper presented at the 7th International Symposium on Automated Cartography, Washington, DC.

Jain, D. K., U. S. Tim, and R. Jolly. 1995. A spatial decision support system for livestock production planning and environmental management. *Applied Engineering in Agriculture* 11(5):711–719.

Jankowski, P., G. L. Andrienko, and N. V. Andrienko. 2001. Map-centered exploratory approach to multiple criteria spatial decision making. *International Journal of Geographical Information Science* 15(2):101–127.

Jankowski, P., T. L. Nyerges, A. Smith, T. J. Moore, and E. Horvath. 1997. Spatial group choice: A SDSS tool for collaborative spatial decision-making. *International Journal of Geographical Information Science* 11(6):577–602.
Jankowski, P., and T. Nyerges. 2001. *Geographic information systems for group decision making*. London: Taylor & Francis.
Jankowski, P., T. Nyerges, S. Robischon, K. Ramsey, and D. Tuthill. 2006. Design considerations and evaluation of a collaborative, spatio-temporal decision support system. *Transactions in GIS* 10(3):335–354.
Jarupathirun, S., and F. M. Zahedi. 2005. GIS as spatial support systems. In *GIS in business*, ed. J. B. Pick, 151–174. Hershey, PA: Idea Group.
Jarupathirun, S., and F. M. Zahedi. 2007. Exploring the influence of perceptual factors in the success of web-based spatial DSS. *Decision Support Systems* 43(3):933–951.
Ji, W. 1993. Integrating a resource assessment model into arc/info GIS: A spatial decision support system development. Paper presented at the ACSM/ASPRS Annual Convention & Exposition Technical Papers, New Orleans, Louisiana.
Jung, C. T., and C. H. Sun. 2006. Development of a web-based spatial decision support system for business location choice in Taipei City. Paper presented at the ESRI 2006 User Conference, San Diego, California.
Keenan, P. B. 1996. Using a GIS as a DSS Generator. In *Perspectives on DSS*, ed. J. Darzentas, J. S. Darzentas, and T. Spyrou, 33–40. Mytilene, Greece: University of the Aegean Press.
Keenan, P. B. 1998. Spatial decision support systems: extending the technology to a broader user community. Paper presented at the IFIP TC8.3 in Bled, Slovenia.
Keenan, P. B. 2003. Spatial decision support systems. In *Decision making support systems: Achievements, trends and challenges for the new decade*, ed. M. Mora, G. Forgionne, and J. N. D. Gupta, 28–39. Hershey, PA: Idea Group.
Keenan, P. B. 2005. Concepts and theories of GIS in business. In *Geographic information systems in business*, ed. J. B. Pick, 1–19. Hershey, PA: Idea Group Publishing.
Keenan, P. B. 2006. Spatial decision support systems: A coming of age. *Control and Cybernetics* 35(1):9–27.
Keenan, P. B. 2008. Geographic information and analysis for decision support. In *Handbook on decision support system 2*, ed. F. Burstein and, C. W. Holsapple, 65–79. New York: Springer.
Kirkby, S. D. 1996. Integrating a GIS with an expert system to identify and manage dryland salinization. *Applied Geography* 16:289–303.
Lepreux S., C. Kolski, and M. Abed. 2003. Design method for component-based DSS. Paper presented at the 1st International Wokshop on Component-Based Business Information Systems Engineering, Geneva, Switzerland.
Leung, Y. 1997. *Intelligent spatial decision support systems*. Berlin: Springer.
Li, L., J. Wang, and C. Wang. 2005. Typhoon insurance pricing with spatial decision support tools. *International Journal of Geographical Information Science* 19(3):363–384.
Liou, Y., M. Chen, C. W. Wang, Y. W. Fan, and Y. P. J. Chi. 2007. Team-spirit: Design, implementation, and evaluation of a web-based group decision support system. *Decision Support Systems* 43(4):1186–1202.
Malczewski, J. 1997. Spatial decision support systems, NCGIA Core Curriculum in GIScience. Available on line at: http://www.ncgia.ucsb.edu/giscc/units/u127/u127.html.

Malczewski, J. 1999. *GIS and multicriteria decision analysis*. New York: John Wiley & Sons.

Malczewski, J. 2004. GIS-based land-use suitability analysis: A critical overview. *Progress in Planning* 62(1):3–65.

Malczewski, J. 2006. GIS-based multicriteria decision analysis: A survey of the literature. *International Journal of Geographical Information Science* 20(7):703–726.

Malczewski, J., T. Chapman, C. Flegel, D. Walters, D. Shrubsole, and M. A. Healy. 2003. GIS-multicriteria evaluation with ordered weighted averaging (OWA): Developing management strategies for rehabilitation and enhancement projects in the Cedar Creek watershed, Ontario, Canada. *Environment and Planning A* 35(10):1769–1784.

Mantey, P. E., and E. D. Carlson. 1980. Integrated geographic data bases: The GADS experience. In *Data base techniques for pictorial applications*, ed. A. Blaser, 173–198. New York: Springer.

Marble, D. F. 1999. Geographic information system technology and decision support systems. Paper presented at the 32nd Annual International Conference on System Sciences. Honolulu, Hawaii.

Matthews, K. B., A. R. Sibbald, and S. Craw. 1999. Implementation of a spatial decision support system for rural land use planning: Integrating geographic information system and environmental models with search and optimisation algorithms. *Computers and Electronics in Agriculture* 23(1):9–26.

Meaden, G. J. and J. M. Kapetsky. 1991. Geographical information systems and remote sensing in inland fisheries and aquaculture. FAO Fisheries Technical Paper No. 318.

Mennecke, B. E., and M. D. Crossland. 1996. Geographic information systems: Applications and research opportunities for information systems researchers. Paper presented at the 29th International Conference on System Science, Honolulu, Hawaii.

Mennecke, B. E., M. D. Crossland, and B. L. Killingsworth. 2000. Is a map more than a picture? The role of SDSS technology, subject characteristics, and problem complexity on map reading and problem solving. *MIS Quarterly* 24(4):601–629.

Molenaar, K. R. and A. D. Songer. 2001. Web-based decision support systems: case study in project delivery. *Journal of Computing in Civil Engineering* 15(4):259–267.

Moller-Jensen, L. 1997. Classification of urban land cover based on expert systems, object models and texture. *Computers, Environment and Urban Systems* 21:291–302.

Murphy, L. D. 1995. Geographic information systems: Are they decision support systems? Paper presented at the Twenty-Eighth Annual International Conference on System Sciences. Honolulu, Hawaii.

NASA (National Aeronautics and Space Administration). 2005. Earth Science REASoN (Research, Education and Applications Solution Network) project. http://geoinfo.sdsu.edu/reason/task.htm#top (accessed November 2009).

Nath, S. S., J. P. Bolte, L. G. Ross, and J. Aguilar-Manjarrez. 2000. Applications of geographical information systems (GIS) for spatial decision support in aquaculture. *Aquacultural Engineering* 23(1–3):233–278.

Negahban, B., C. Fonyo, W. G. Boggess, J. W. Jones, K. L. Campbell, G. Kiker, E. Flaig, and H. Lal. 1995. LOADSS: A GIS-based decision support system for regional environmental planning. *Ecological Engineering* 5(2–3):391–404.

Neto, S. L. R., and M. Rodrigues. 1999. A taxonomy of strategies for developing spatial decision support system. In *Systems development methods for databases, enterprise modeling, and workflow management*, ed. W. Wojtkowski, W. G. Wojtkowski, S. Wrycza, and J. Zupančič, 139–155. New York: Kluwer Academic/Plenum Publishers.

Nunamaker, J. F. 1989. Experience with and future challenges in GDSS (Group decision support systems). *Decision Support Systems* 5(2):115–118.

Nyerges, T., P. Jankowski, and C. Drew. 2002. Data gathering strategies for social-behaviour research about participatory geographic information system use. *International Journal of Geographical Information Science* 16(1):1–22.

Nyerges, T., P. Jankowski, D. Tuthill, and K. Ramsey. 2006. Collaborative water resource decision support: Results of a field experiment. *Annals of the Association of American Geographers* 96(4):699–725.

Özbayrak, M., and R. Bell. 2003. A knowledge-based decision support system for the management of parts and tools in FMS. *Decision Support Systems* 35:487–515.

Ozernoy, V. M., D. R. Smith, and A. Sicherman. 1981. Evaluating computerized geographic information systems using decision analysis. *Interfaces* 11(5):92–99.

Peterson, K. 1989. Toward the specification of a spatial decision support system for real estate investment analysis. Paper presented at the GIS/LIS '89 Conference, Orlando, Florida.

Peterson, K. 1998. Development of spatial decision support systems for residential real estate. *Journal of Housing Research* 9(1):135–56.

Pittman, R. H. 1990. Geographic information systems: An important new tool for economic development professionals. *Economic Development Review* 8(4):4–7.

Powell, R. 2001. DM Review: A 10 Year Journey. DM Review, URL:www.dmreview.com.

Power, D. J. 2008. Decision support systems: a historical overview. In *Handbook on decision support systems 1*, ed. F. Burnstein, and C. W. Holsapple, 121–140. Leipzig: Springer.

Power, D. J. 2002. Decision support systems: concepts and resources for managers. Westport, CT: Quorum Books.

Power, D. J., and S. Kaparthi. 2002. Building web-based decision support systems. *Studies in Informatics and Control* 11(4):291–302.

Qiu, Z., T. Prato, L. Godsey, and V. Benson. 2002. Integrated assessment of uses of woody draws in agricultural landscapes. *Journal of the American Water Resources Association* 38(5): 1255–1269.

Ray, J. J. 2007. A web-based spatial decision support system optimizes routes for oversize/overweight vehicles in Delaware. *Decision Support Systems* 43(4):1171–1185.

Rinner, C. 2003. Web-based spatial decision support: Status and research directions. *Journal of Geographic Information and Decision Analysis* 7(1):14–31.

Rinner, C., and P. Jankowski. 2002. Web-based spatial decision support—Technical foundations and applications. In *The encyclopedia of life support systems* (EOLSS), Theme 1.9—Advanced Geographic Information Systems (edited by Claudia Bauzer Medeiros). Oxford, U.K.: UNESCO/EOLSS Publishers.

Rodrigues, A., C. Grueau, J. Raper, and N. Neves. 1997. Environmental planning using spatial agents. In *Innovations in GIS 5—Selected papers from the papers presented at Proceedings GIS Research in the UK*, ed. S. Carver, and S. Openshaw, Bristol: Taylor & Francis.

Segrera, S., R. Ponce-Hernandez, and J. Arcia. 2003. Evolution of decision support system architectures: Applications for land planning and management in Cuba. *Journal of Computer Science & Technology* 3(1):40–46.

Shim, J. P., M. Warkentin, J. F. Courtney, D. J. Power, R. Sharda, and C. Carlsson. 2002. Past, present, and future of decision support technology. *Decision Support Systems* 33(2):111–126.

Sieber, R. 2006. Public participation geographic information systems: A literature review and framework. *Annals of the Association of American Geographers* 96(3):491–507.

Sikder, I., and A. Gangopadhyay. 2002. Design and implementation of a web-based collaborative spatial decision support system: Organizational and managerial implications. *Information Resources Management Journal* 15(4):33–47.

Sikder, I., and A. Gangopadhyay. 2004. Collaborative decision making in web-based GIS. In *Advanced topics in information resources management*, ed. M. Khosrow-Pour, 147–162. Hershey, PA: Idea Group.

Sikder, I. U. 2008. Knowledge-based spatial decision support systems: An assessment of environmental adaptability of crops. *Expert Systems with Applications* 36(3):5341–5347.

Sugumaran, R., S. Ilavajhala, and V. Sugumaran. 2007. Development of a web-based intelligent spatial decision support system WEBSDSS: A case study with snow removal operations. In *Emerging spatial information systems and applications*, ed. B. N. Hilton, 184–202. Hershey, PA: Idea Group.

Sugumaran, R., Meyer, J. and J. Davis. 2004. A web-based environmental decision support system (WEDSS) for environmental planning and watershed management. *Journal of Geographical Systems* 6:1–16.

Sugumaran, V., and L. Mobley. 2002. Integrating spatial regression into healthcare decision support systems. *International Journal of Healthcare Technology and Management* 4(1–2):132–147.

Sugumaran, R., M. K. Pavuluri, and D. Zerr. 2003. The use of high-resolution imagery for identification of urban climax forest species using traditional and rule-based classification approach. *IEEE Transactions on Geoscience and Remote Sensing* 41(9):1933–1939.

Sugumaran, V., and R. Sugumaran. 2003. Spatial decision support systems using intelligent agents and GIS web services. Paper presented at the Americas Conference on Information Systems, Tampa, Florida.

Sugumaran, V., and R. Sugumaran. 2007. Web-based spatial decision support systems (WebSDSS): Evolution, architecture, and challenges. *Communications of the Association for Information Systems* 19:844–875.

Tsou, M. H., and B. P. Buttenfield. 2002. A dynamic architecture for distributed geographic information services. *Transactions in GIS* 6(4):355–381.

Van der Vlugt, M. 1989. The use of a GIS-based decision support system in physical planning. Paper presented at the Annual GIS/LIS Conference. Orlando, Florida.

Wan, Q., J. Zhang, H. Lin, and C. Beijing. 1999. On-line group spatial decision support system for investment environment analysis. Paper presented at the International Conference on GeoInformatics, Ann Arbor, Michigan.

Wang, L., and Q. Cheng. 2006. Web-based collaborative decision support services: Concept, challenges and application. Paper presented at the ISPRS Technical Commission II Symposium, Vienna, Austria.

Wild, R. H., and K. A. Griggs. 2004. A web portal/decision support system architecture for collaborative intra-governmental planning. *Electronic Government* 1:61–76.

Witlox, F. 2005. Expert systems in land-use planning: An overview. *Expert Systems with Applications* 29(2):437–445.

Worrall, L. 1991. *Spatial analysis and spatial policy using geographic information systems.* London: Belhaven Press.

Wu, X., S. Zhang, and S. Goddard. 2004. Development of a component-based GIS using GRASS. Paper presented at the FOSS/GRASS Users Conference, Bangkok, Thailand.

Yan, Q., T. Chungsheng, X. Qiao, W. Jian, and Z. Yi. 1999. Spatial decision support system and its software tools. Paper presented at the Proceedings of Geoinformatics '99 Conference, Ann Arbor, Michigan.

Yeh, A. G. O., and J. J. Qiao. 2004. Component-based approach in the development of a knowledge-based planning support system (KBPSS). Part 1: The architecture of KBPSS. *Environment and Planning B: Planning and Design* 31(4):517–537.

Yue, P., L. Di, W. Yang, G. Yu, P. Zhao, and J. Gong. 2009. Semantic web services-based process planning for earth science applications. *International Journal of Geographical Information Science* 23(9):1139–1163.

Zhao, P., G. Yu, and L. Di. 2007. Geospatial web services. In *Emerging spatial information systems and applications*, ed. B. N. Hilton, 1–35. Hershey, PA: Idea Group.

Zhu, X., J. McCosker, A. P. Dale, and R. J. Bischof. 2001. Web-based decision support for regional vegetation management. *Computers, Environment and Urban Systems* 25(6):605–627.

3

Components of SDSS I: Geographic Information Systems

Learning Objectives

- Understand the basic components of a spatial decision support system.
- Learn about the basics of spatial data creation, storage, and management in GIS environments.
- Be introduced to spatial data exploration, visualization, processing, and analyses techniques in GIS.

3.1 Introduction

The difficulties presented by unstructured decision situations, the need for computer support systems, and the concept of spatial decision support systems were discussed in Chapter 1. In Chapter 2, the progression of SDSS was presented in order to give an historical perspective and also to provide an idea of where the techniques and technology are heading in the future. Spatial decision support systems are characterized by a wide variety of approaches, application domains, development techniques, technologies used, and the complexity of the software configurations. However, to be considered spatial decision support systems (SDSS), they must contain certain components. These common components as well as how they are specifically included in SDSS and how interaction among the components is facilitated will be discussed in this and the next chapter.

The overall purpose of SDSS is to provide an integrated set of flexible capabilities for decision making for tackling semi- or ill-structured spatial problems. Spatial decision support systems should be designed for ease of use, to provide solutions through presentation of a series of alternatives, for flexibility of use and easy adaptation, and to support analytical methods. In order to achieve these attributes, there are several common components that every SDSS should possess. These include a database, spatially explicit models, user interfaces, visualization and reporting capabilities, and alternatively, application domain knowledge. This chapter will provide an overview of the different components that compose an SDSS and how they interact. As spatial data management and analysis, usually with Geographic Information Systems (GIS), are often a focal component of many SDSS, a broad overview of GIS will be presented in detail. Chapter 4 will investigate the other SDSS components and how to integrate these components to successfully build an SDSS.

3.2 Components of Traditional DSS and GIS

As discussed earlier in Chapter 2, SDSS have evolved from DSS and GIS. Figure 3.1 depicts the combined components of GIS and DSS. A traditional DSS has three primary components: a database, a model base, and a user interface (Sprague and Carlson 1982). On the other hand, GIS can also be considered to be composed of three major components: a database, a user interface, and spatial data creation, analysis, and presentation capabilities.

The database component within a decision support system (DSS) mostly deals with nonspatial data collection, retrieval, management, and analysis.

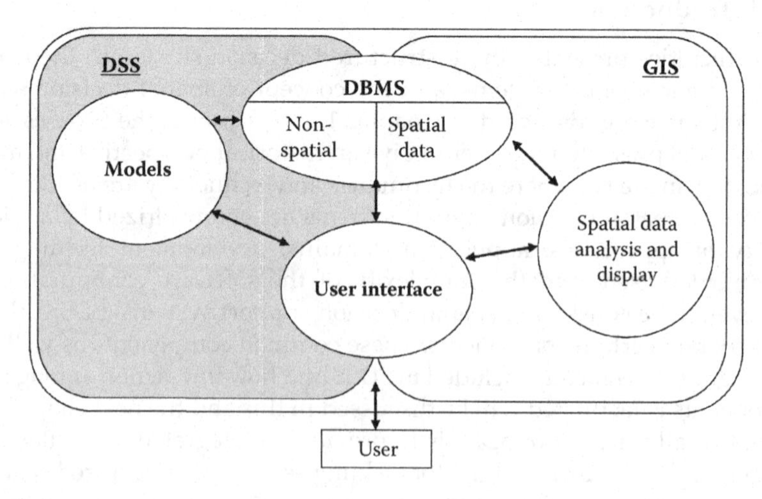

FIGURE 3.1
Traditional DSS and GIS components.

This component usually does not support cartographic presentation or mapping functionality, which are essential to spatial decision making. On the other hand, GIS provides spatial and nonspatial data collection, storage, management, and cartographic display functionalities. The database component in both systems feeds appropriate information to the other components as needed.

The model base component provides decision makers with access to a variety of models to help them in the decision-making process. Some of the examples of models include statistical, process based, mathematical, and multi-criteria evaluation. Traditional DSS have been built to use various specific modeling techniques., However, GIS software, while containing generic spatial analytical functions, does not contain specific analytical modeling capabilities (Densham 1991; Keenan 2006). The model base component operates on the data available in the database by either directly accessing data or through the use of intermediate data translation routines.

The user interface component in both GIS and DSS facilitates the interaction between the user and the computer system. This component is important as unintuitive or awkward user interfaces can frustrate users and render an otherwise sound system inoperable. Thus, it is vital for this component to be considered carefully in conjunction with users of the system.

As seen in Figure 3.1, GIS lacks the necessary modeling capabilities, whereas DSS do not support spatial data analysis and cartographic display functionalities. The development of SDSS has evolved to utilize components from both DSS and GIS. The description of different SDSS components and their roles follows.

3.3 Components of SDSS

As mentioned earlier in Chapter 1, the spatial decision-making process involves (1) identifying the issue, (2) collecting the necessary data, (3) defining the problem, including objectives, assumptions, and constraints, (4) finding appropriate solution procedures, and (5) solving the problem by finding an optimal solution (Keller 1997). At the most basic level, there are three major components: database, model, and user interface. However, the number and exact description of components mentioned in the SDSS literature varies. For example, Lolonis (1990) and Malczewski (1999) mentioned three components—database management, model base management, and dialog management, while Densham and Goodchild (1989) mentioned four components—a database management system, analysis procedures, display and report generators, and user interfaces. Armstrong and Densham (1990) mentioned the database management system (DBMS), model base, screen generator, report generator, and user interface as the five components, while Gao et al. (2004) reported six components: data, model, solvers, visualization, scenario, and knowledge.

Though the number of components varies, in this book we identify four core components and one optional component. Core components of SDSS include (1) the database management component (DBMC), (2) the model management component (MMC), (3) the dialog management component (DMC), and (4) the stakeholder component (SC) (Figure 3.2). The knowledge component (KC) is a common but not essential component in an SDSS. Figure 3.2 provides an overview of the components in an SDSS.

This chapter will focus on the capabilities of GIS as these types of software are frequently at the core of SDSS. Indeed, GIS software often fulfils the role of database management and dialog management components. Our research into the SDSS literature, as discussed in Chapter 2, demonstrated that the great majority of SDSS had GIS as the key software in database management and dialog management functions. In this chapter, we also want to provide a GIS overview as some readers of this book come from disciplines in which GIS education and use have not been common. The following sections detail GIS definitions, history, data collection, data management, data analysis, and data display. Chapter 4 provides a detailed description of the remaining SDSS components (i.e., model, dialog, and the knowledge management, and stakeholder components).

3.4 Geographical Information Systems (GIS) Overview

In the following sections we will provide a brief history of using spatial data and GIS software. We will also investigate some key concepts such as the definition of GIS, coordinate systems, spatial data models, spatial data collection methods, spatial data management, spatial data analysis and processing, and data visualization and cartography. Finally, we will provide an overview of commercial and open source GIS software.

3.4.1 History of Spatial Information and Data Use

Maps have played an important historical role for thousands of years while spatial analytical techniques have developed substantially over the past two centuries. Ancient civilizations such as the Egyptians and Greeks were some of the earliest mapmakers. The Romans developed extensive cadastral (map-based property register) mapping to help manage their empire and to levy taxes on citizens (Bernhardsen 2002). Exploration and military

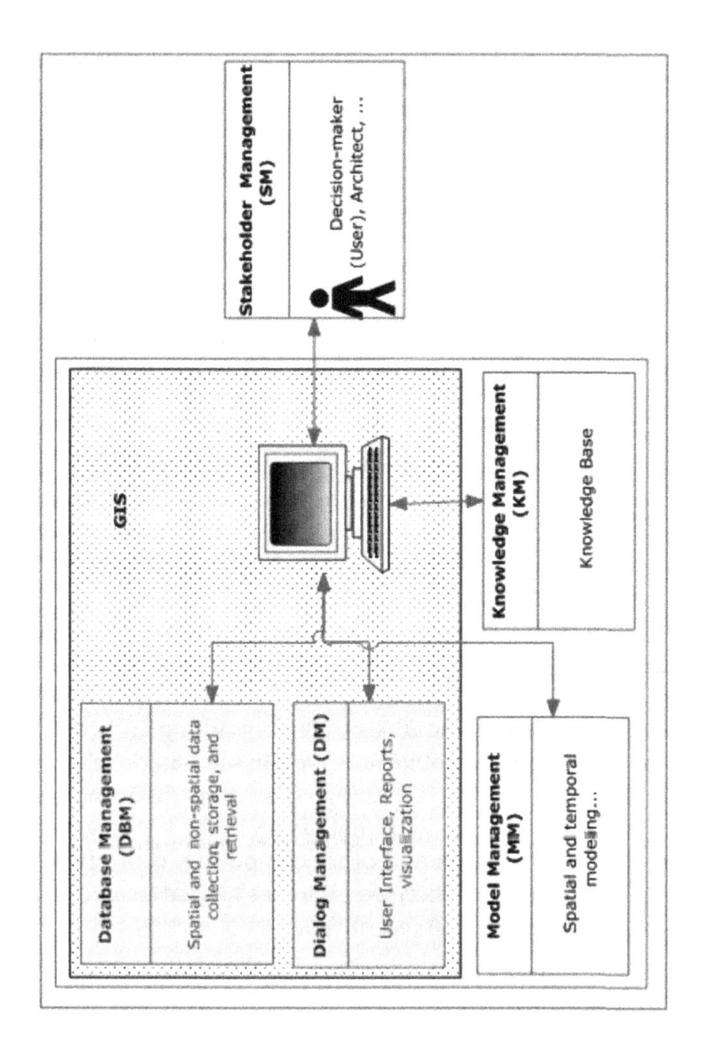

FIGURE 3.2
Components of SDSS.

purposes continued to drive mapmaking techniques and production. By the nineteenth century, more systematic attempts at planning drove more sophisticated mapping techniques. In 1838, the Irish government compiled a series of maps for railway planning, an effort some consider the first manual geographical information system (Bernhardsen, 2002). The use of aerial photography and the development of photogrammetrical techniques advanced mapmaking after World War I (Bernhardsen 2002).

The move from paper-based map use to automated digital spatial information systems began in earnest in the second half of the twentieth century. In the 1960s, the Canadian government embarked on an ambitious project to develop a multilayer land use/planning map for analyzing areas in or available for forestry, agriculture, or recreation (Keenan 2006) using mainframe computers that by today's standards were very limited. This system was called the Canada Geographic Information System. By the 1970s, computer processing was aided by the development of microprocessors, and by the 1980s, computers were of three types: mainframes for major data processing and computational tasks, personal computers (PCs) or desktop computers, and workstation or minicomputers, which were larger and more powerful than PCs but much smaller than mainframes (Bernhardsen 2002). In the 1980s, powerful workstation computers were the common hardware systems for which GIS development was most commonly taking place. However, with cheaper and more powerful computing power and memory resources, GIS development for PCs took off in the 1990s. The development of graphical user interfaces in the 1990s led to immense growth in the use of GIS in the 1990s, with Environmental Systems Research Institute (ESRI)'s ArcView being an example of a desktop-based GIS with user-friendly interfaces. The uptake and use of GIS technology has continued into the 2000s with networking and communication technologies leading to mobile and Web-based GIS developments. The growth in other geospatial technologies, such as GPS and remote sensing, has been commensurate with that of GIS in the last few decades. Indeed, the combination of these and other computing, communication, and networking and communication technologies has led to an explosion of spatially explicit applications in recent years.

3.4.2 Definitions of GIS

There is not a single agreed-upon definition of GIS. However, a review of several GIS textbooks shows some commonality in definitions of GIS:

- "A geographic information system (GIS) is a computer system for capturing, storing, querying, analyzing, and displaying geospatial data" (Chang 2009, p. 1).

- "[G]eographic information systems are systems designed to input, store, edit, retrieve, analyze, and output geographic data and information" (DeMers 2009, p. 19).
- "A GIS is designed for the collection, storage, and analysis of objects and phenomena where geographic location is an important characteristic or critical to the analysis" (Aronoff 1995, p. 1).
- "A computer-based system to aid in the collection, maintenance, storage, analysis, output, and distribution of spatial data and information" (Bolstad 2005, p. 1).

What makes GIS very powerful is the ability to explicitly handle spatial data as well as nonspatial data. Spatial data can take up a considerable amount of hard-disk space and require a considerable amount of computer processing power and memory. Over the last several decades, the great growth in computing power and accessibility to this computing power has aided the diffusion of GIS technology to a greater number of organizations. Spatial data have become an integral portion of many organizational databases being combined with extensive nonspatial datasets. The way spatial data are stored, managed, and analyzed will be discussed in the following sections.

3.4.3 Coordinate Systems

Spatial data hold explicit locational information based on a defined reference system. Locations in the context of cartography and GIS are referenced to either a projected or geographical coordinate system, which is used to reference the surface of the Earth. A geographical coordinate system is one that represents locations on a model of the Earth with latitude and longitude coordinates. These geographical coordinate systems often use a spherical model (Figure 3.3) to represent the Earth. A geographical coordinate system has the center of the Earth as an origin and measures latitude as degrees north or south of the equatorial plane and longitude measured in degrees east or west of the prime meridian (Chang 2009). Calculating areas and distances is complicated when using geographic coordinate systems since the distance corresponding to one degree of latitude or longitude varies depending on the location on the Earth's surface. Also, in relation to paper maps, traditionally it has been impractical to carry around a spherical model of the Earth conveniently. Due to these complications, projected coordinate systems were developed by cartographers. Projected coordinate systems use map projections, which are techniques for representing the surface of the Earth on a flat, two-dimensional surface such as a paper map. The science and mathematics behind projecting data is beyond the scope of this book. However, the basic concept can

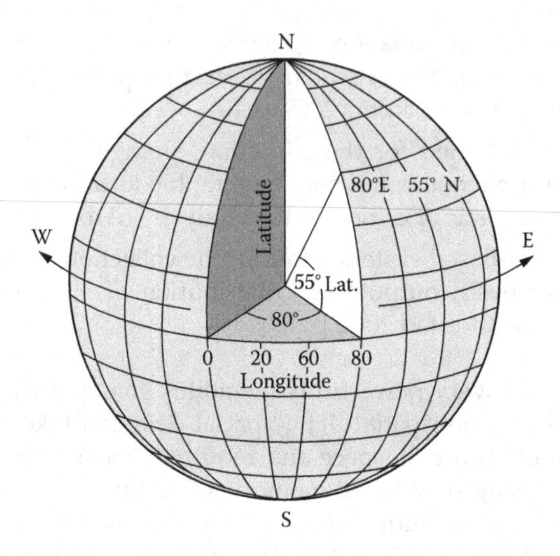

FIGURE 3.3
A spherical representation of the Earth. (http://publib.boulder.ibm.com/infocenter/db2luw/v8/index.jsp?topic=/com.ibm.db2.udb.doc/opt/csb3022b.htm)

be thought of as putting a light source inside a transparent globe and projecting the lines onto a two-dimensional surface surrounding the globe. There are different kinds (Figure 3.4) of projections, each with its own advantages and disadvantages. Each projection emphasizes the ability to accurately represent size, shape, distance, or direction at the expense of accurately representing at least one of these parameters. The ideal projection varies depending on the part of the world (i.e., at the poles or around the equator). When representing data in a small area such as a state, county, or small administrative unit, the distortions for a specific projection are minimal. However, when storing information at the global scale, it is often more useful to use unprojected data, as any individual projection would incur large distortions. There are great advantages to using projected coordinate systems, including the ease of calculating distance and areas, the portability of the projected information (e.g., paper maps), and representation in a computing environment. Map projections allow any point to be defined in a Cartesian x,y coordinate system. There are many different projections used throughout the world, and GIS programs often contain functionality for handling transformations between different projections and coordinate systems.

3.4.4 Data Models

In order to represent the real world in a digital environment, models must be created that can effectively characterize real-world phenomena

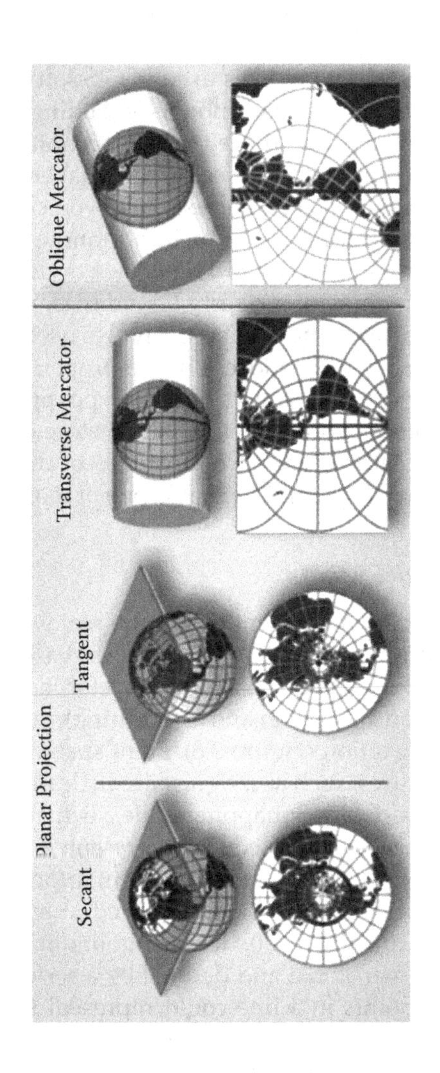

FIGURE 3.4
Examples of map projections. (Taken from http://nationalatlas.gov/articles/mapping/a_projections.html)

and reduce them to the binary (0 and 1) representation necessary for computer storage. As objects in the real world are complex, this is not a trivial issue. Various digital models created to represent real-world objects in GIS and other spatial software applications exist. The choice of data models impacts efficiency, the operations that can be carried out, the output, and the analytical power experienced (Demers 2009). Although there are many individual ways to represent spatial data in a digital environment, there are two main types of data models used in GIS software: vector and raster. Details on vector and raster data models, including advantages and disadvantages, will be examined in the coming sections. The ability to represent the spatial attributes of an object in a digital environment does not capture all aspects of the object. For example, you could represent a stand of trees as a polygon, but it might be important to also store information on the kind of trees or the approximate age of the trees. Another example would be the mapping of property boundaries. While it is useful to store the exact boundaries of the property, it is also useful to store information on the owner's name, the address of the property, and the value of the land and buildings on the property. These types of information are called *attributes*. The combination of being able to store data on the geographic coordinates of features and also attributes of those features are what makes GIS a powerful tool and the backbone of most SDSS.

3.4.4.1 Vector Data Model

When using the vector data model, entities in the real world are divided into clearly defined features represented by geometry based on point, line, or polygon geometry (Figure 3.5). The simplest vector features are points that are represented by an x (easting) and y (northing) value and possibly a z value to represent elevation (Figure 3.6). Point spatial data can represent a wide range of features, such as a sign, a fire hydrant, a tree, a house, or even a town. At a less-detailed mapping scale, such as continentwide, features such as towns can be collected and represented as points. At the time of data collection, there might not be sufficient information to collect detailed boundaries of cities, and when displayed at small scale (large area—e.g., global or continental level), detailed boundaries would not be visible. Lines are one-dimensional and defined by a series of x, y coordinates (Figure 3.7). The points in a line could represent the beginning, a break in, or the end point of a line. The lines, as stored spatially, do not have any width. Again, depending on the scale of mapping and intended use, a road or river feature could be defined as a line or a polygon. At a continental scale, a large river would be defined as a line, but at a city level it might be defined as a polygon. A polygon is a two-dimensional representation of area defined by a series of points and line segments connecting the points with the start and ending point being at the same x, y

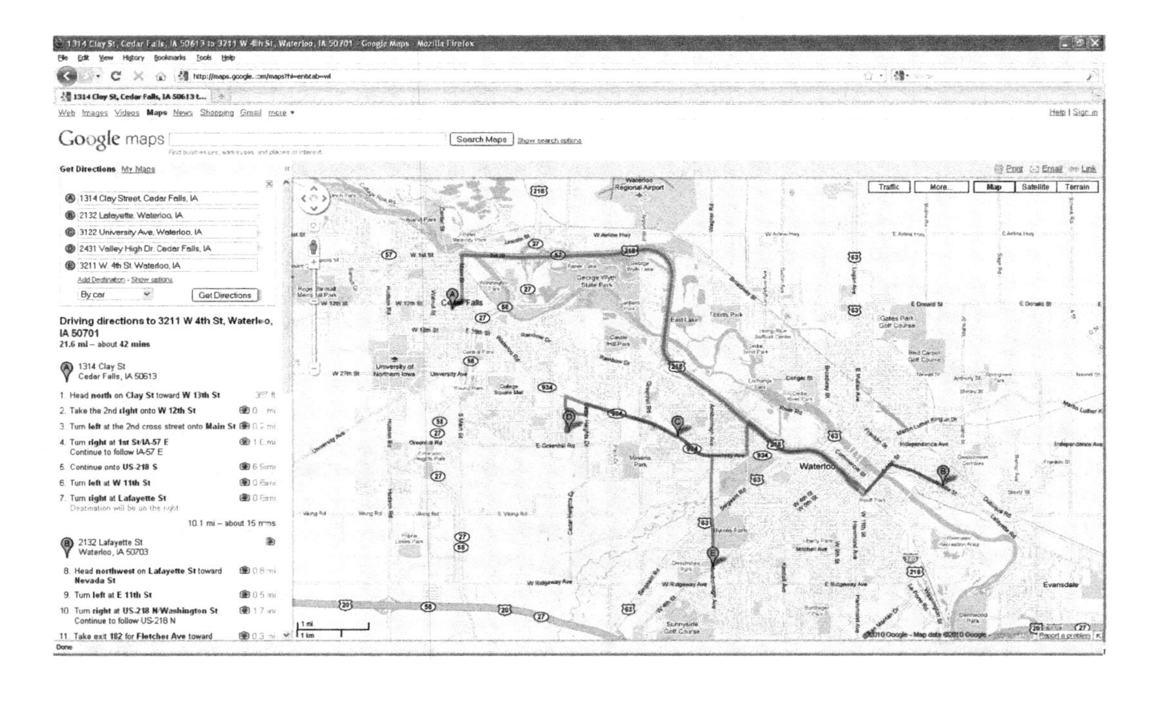

FIGURE 1.1
An example of routing in Google Maps.

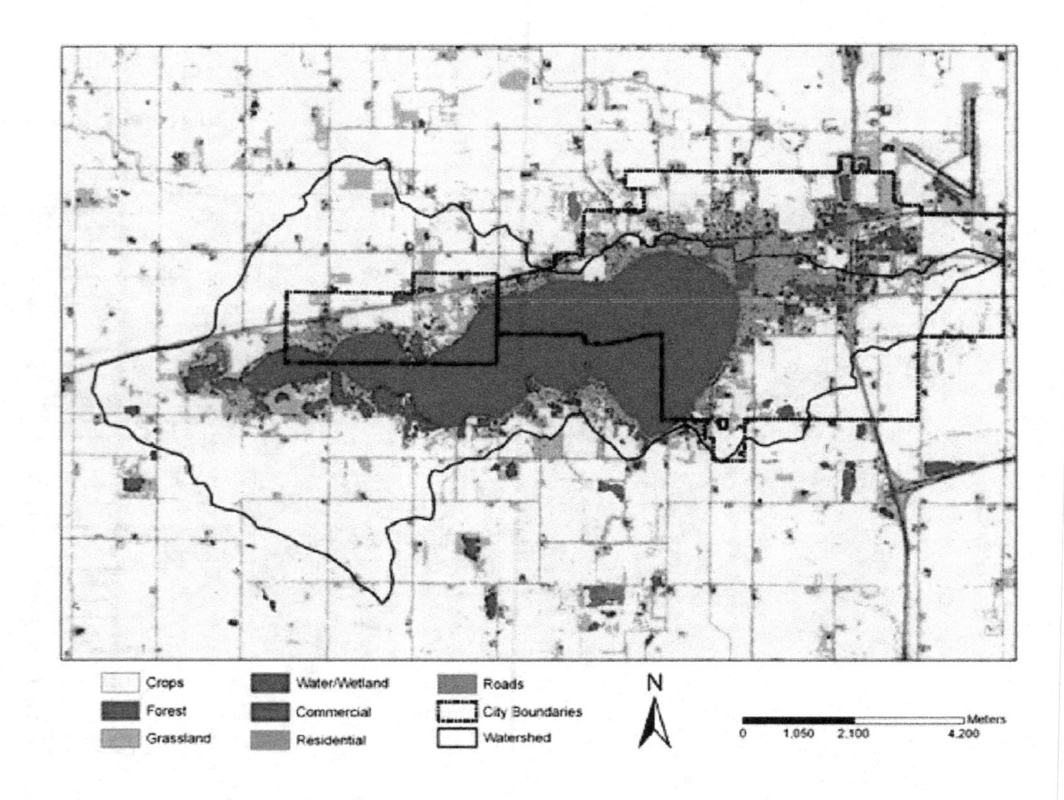

FIGURE 1.2
Land use patterns and watershed and city boundaries for a watershed with lake water quality problems.

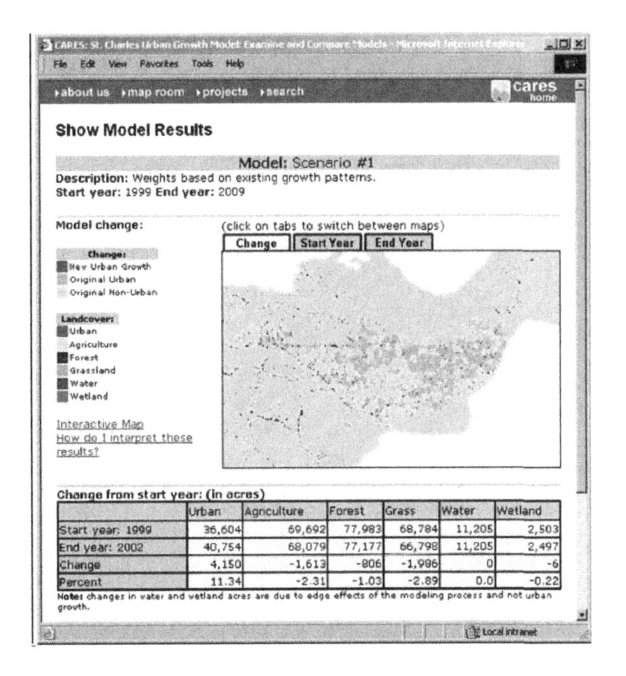

FIGURE 2.9
Web-based SDSS developed for urban growth prediction. (Compas, E. and R. Sugumaran. 2004. Urban growth modeling on the web: A decision support tool for community planners. Paper presented at the 27th Annual Applied Geography Conference, St. Louis.)

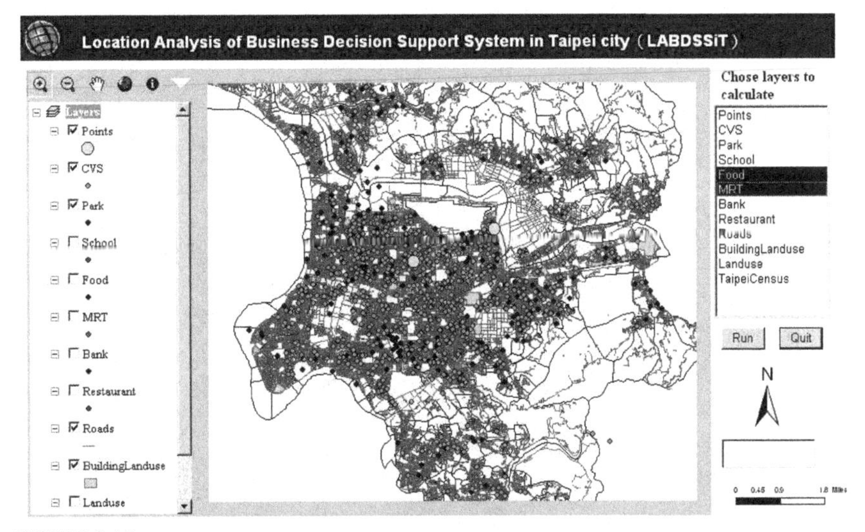

FIGURE 2.10
Service-based SDSS example. (Jung, C. T. and C. H. Sun. 2006. Development of a web-based spatial decision support system for business location choice in Taipei City. Paper presented at the ESRI 2006 User Conference, San Diego.)

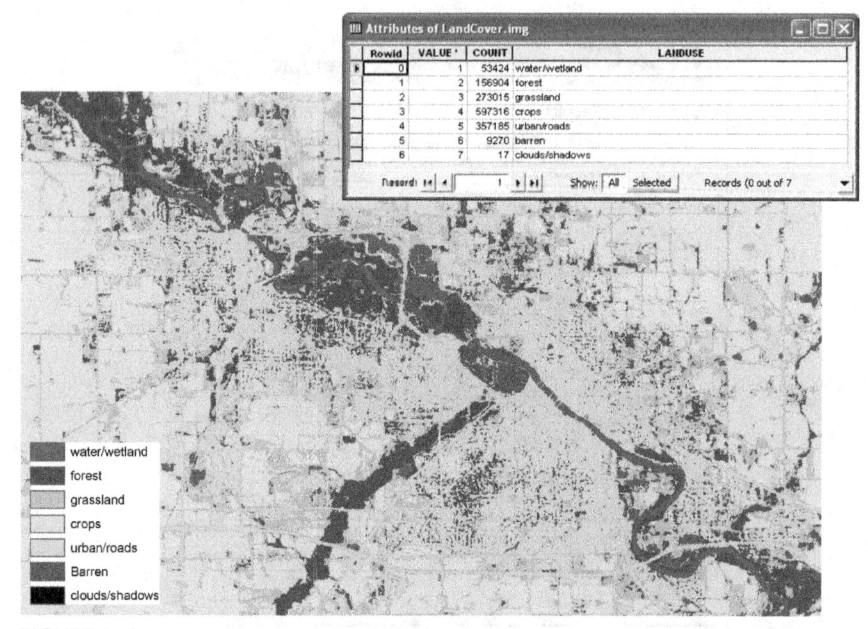

FIGURE 3.14
An ESRI GRID format integer raster representing land cover classes with attribute table shown.

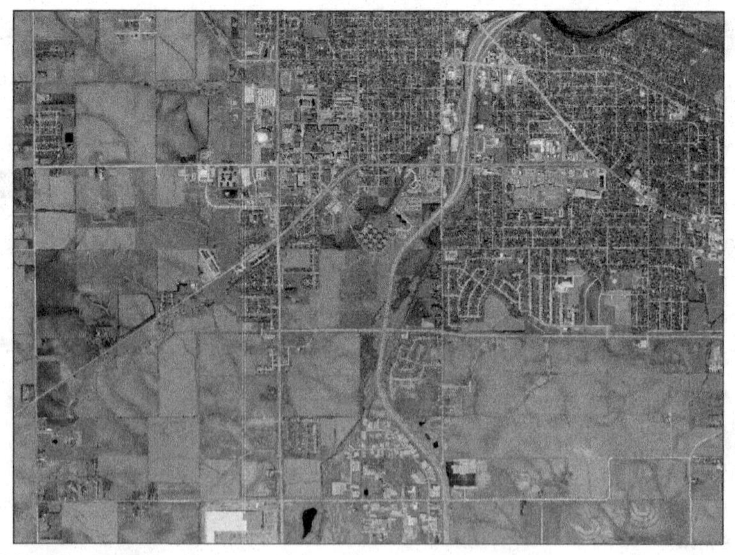

FIGURE 3.16
An example of a digital orthophotograph.

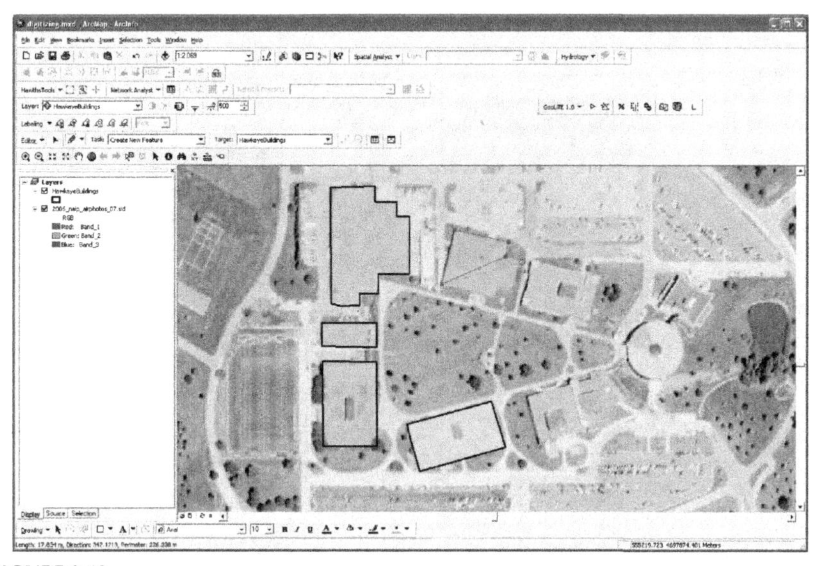

FIGURE 3.19
On-screen digitizing of polygon features in ArcGIS software using a digital orthophotograph as reference.

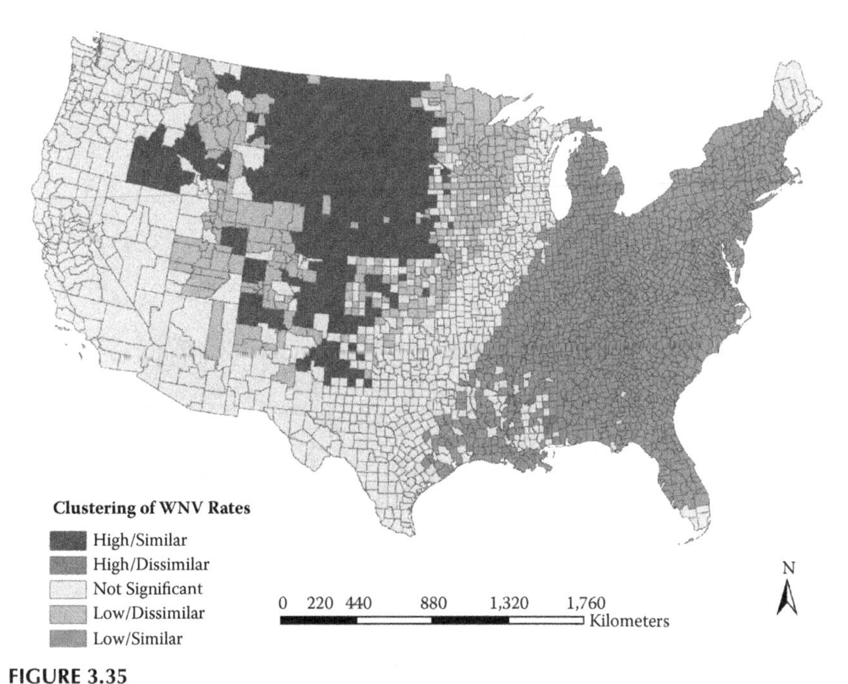

FIGURE 3.35
Results of running spatial autocorrelation tool from ArcGIS on the rate of WNV occurrence by county in the contiguous United States from 2003–2008 (From Sugumaran et al. 2009).

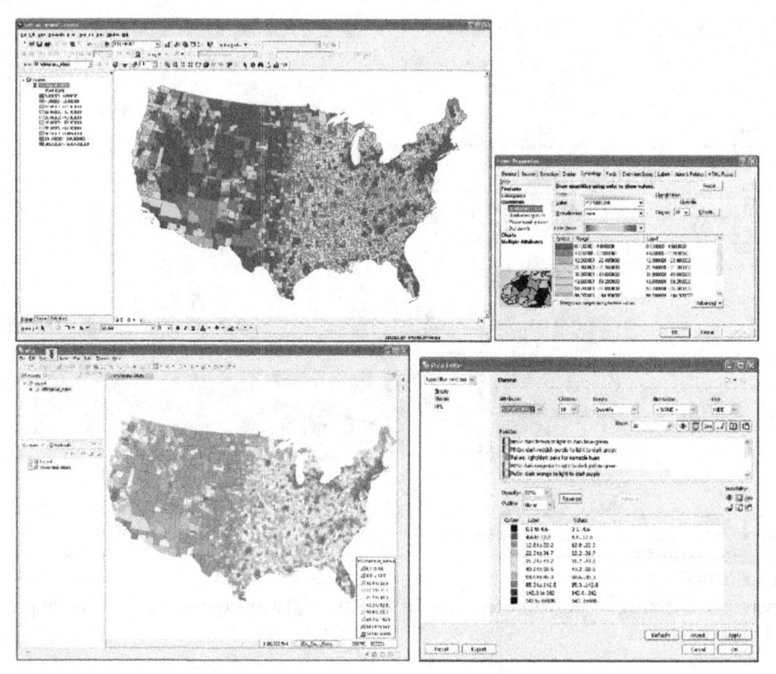

FIGURE 3.46
Both maps represent population densities by county in the contiguous United States using a quantile method. The top images show the map and dialog for setting classification properties in ArcGIS 9.3, while the bottom two show the same for the uDig software.

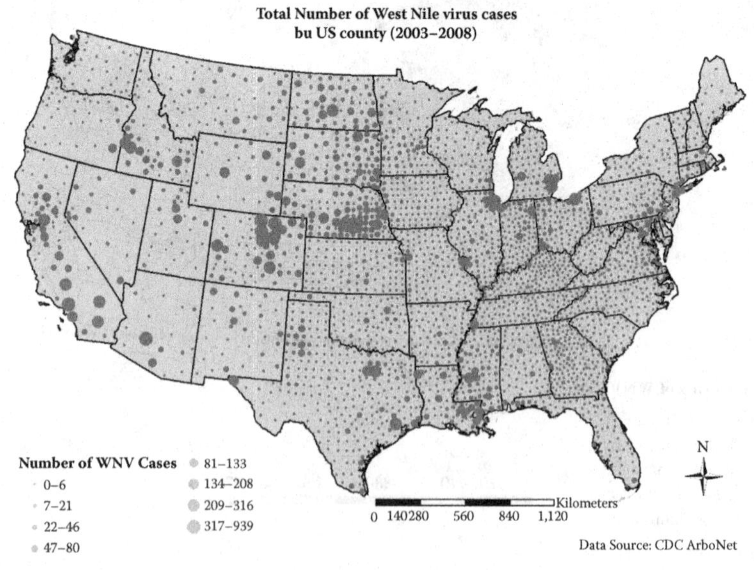

FIGURE 3.48
An example map composition created using ArcGIS 9.3.

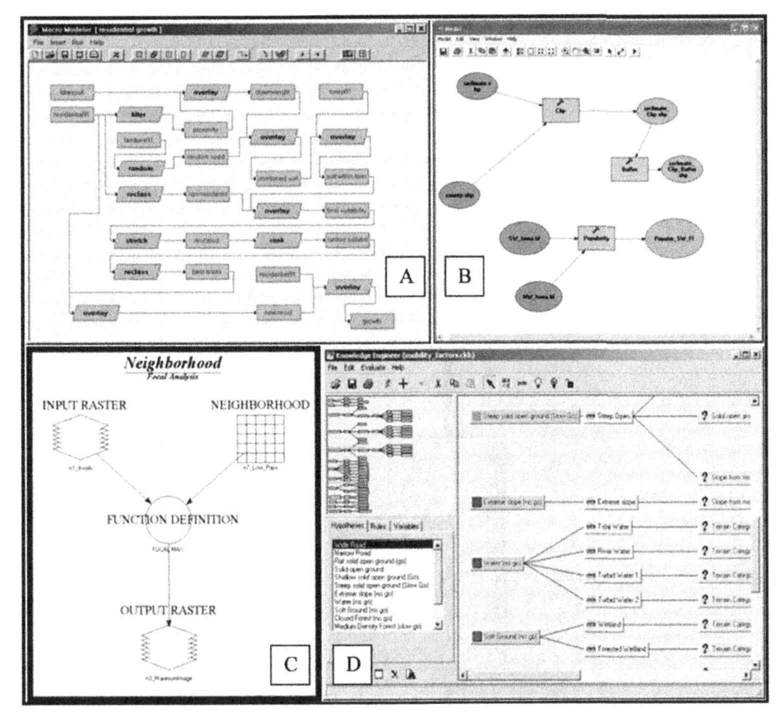

FIGURE 4.1
Modeling frameworks available in different GIS software. (A) IDRISI Macro modeler, (B) ESRI ModelBuilder, (C) ERDAS Imagine Model Maker, and (D) and ERDAS Imagine Knowledge Engineer.

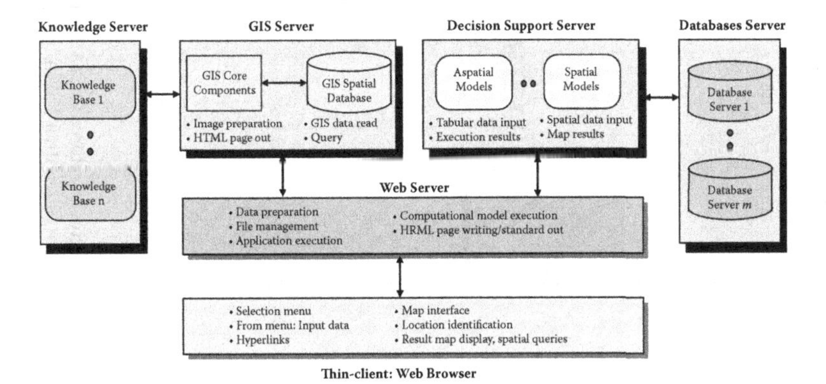

FIGURE 6.9
Schmatic representation of Web-based SDSS components. (Sugumaran, V. and R. Sugumaran. 2007. Web-based spatial decision support systems (WebSDSS): evolution, architecture, and challenges. *Communications of the Association for Information Systems* 19:844-875)

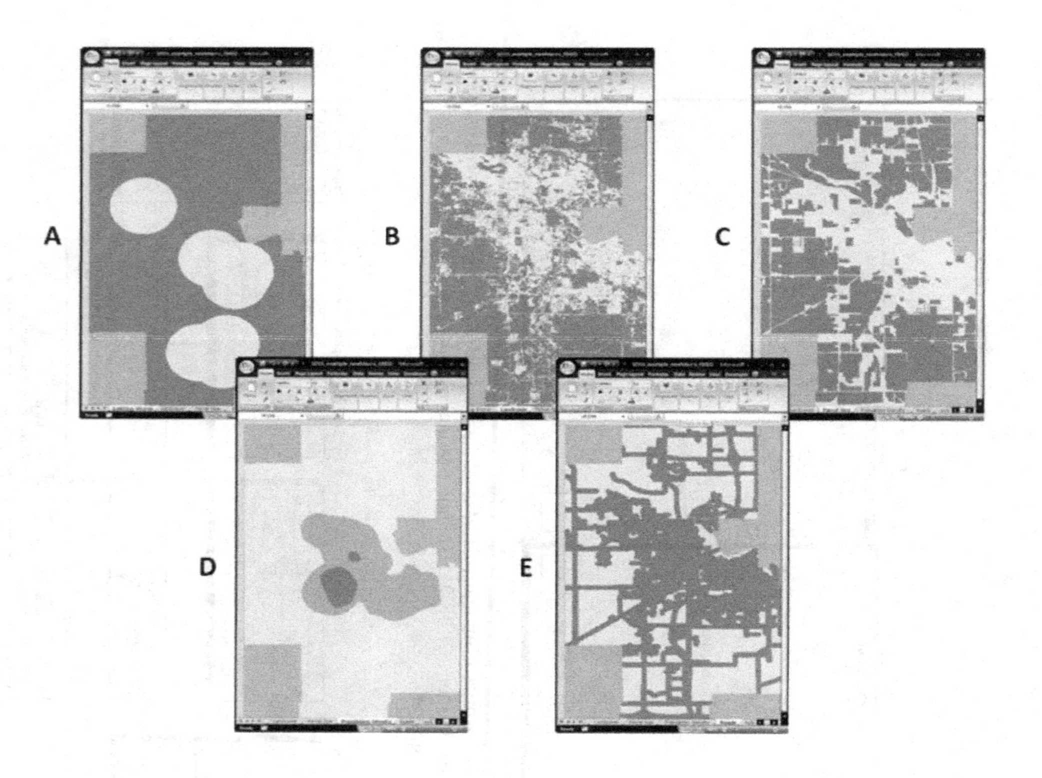

FIGURE 7.10
Representations of (A) existing grocery stores with a 1.5-km buffer, (B) suitable landcover classes, (C) areas with suitable parcel size, (D) population density classes, and (E) existing roads with a 100-m buffer.

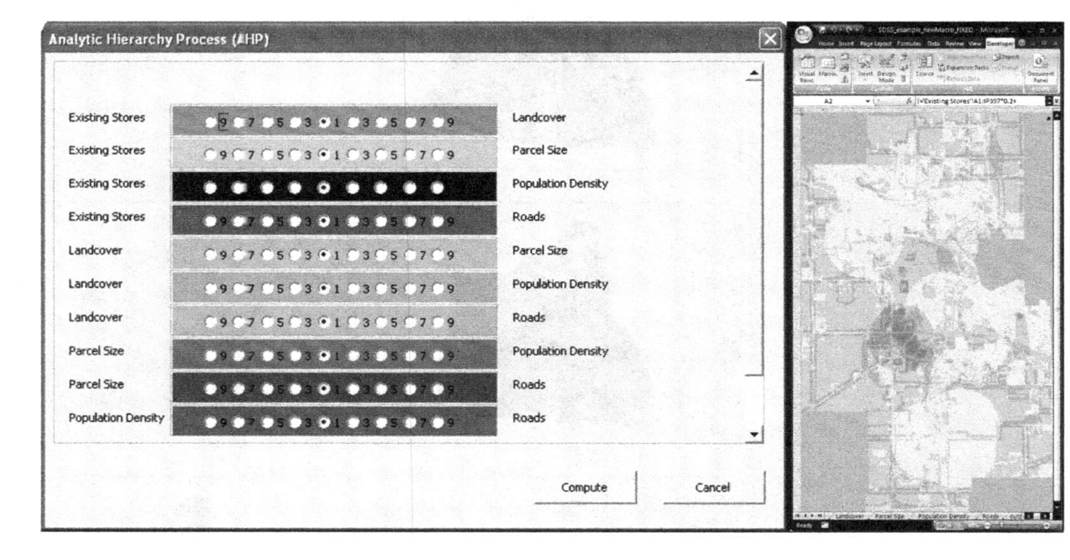

FIGURE 7.11
AHP set at default with all layers set as equally important to each other.

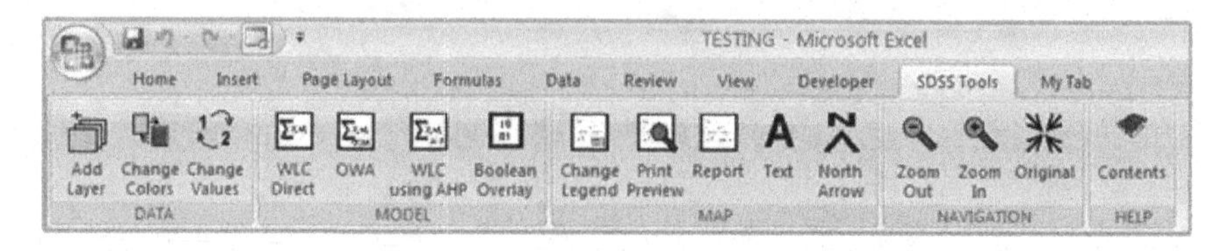

FIGURE 7.14
The main menu of the Spreadsheet SDSS in the Microsoft Excel interface.

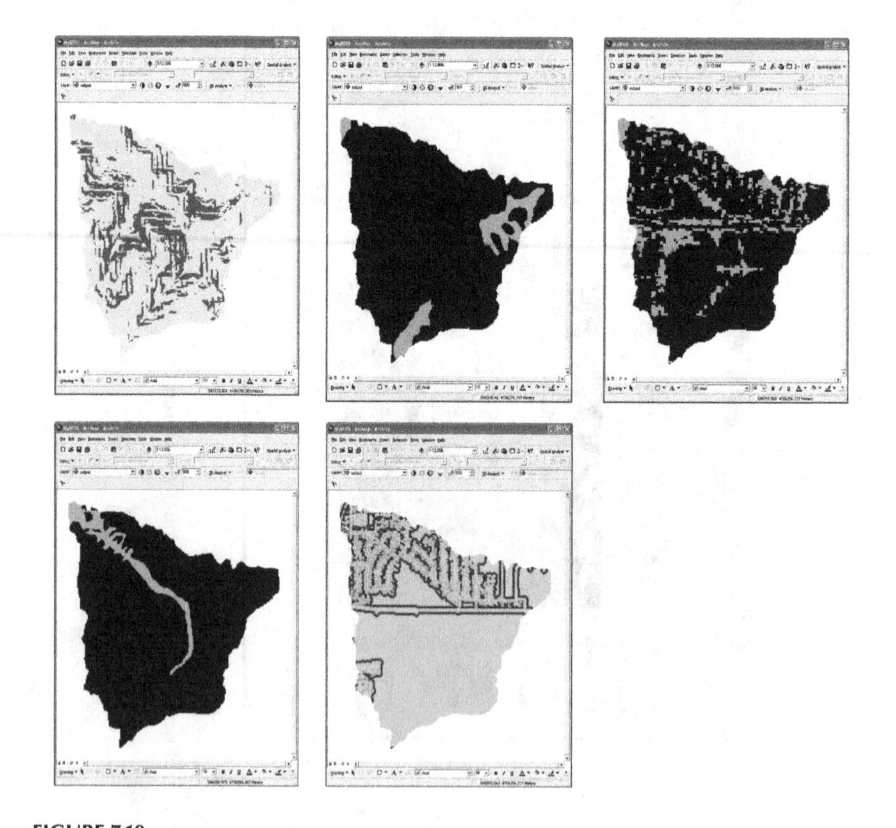

FIGURE 7.19
Original layers included in the WLC analysis of environmentally sensitive areas. From top left to bottom right: slope, hydrological soil group, green space, FEMA 100-year floodplain, and buffered impervious areas.

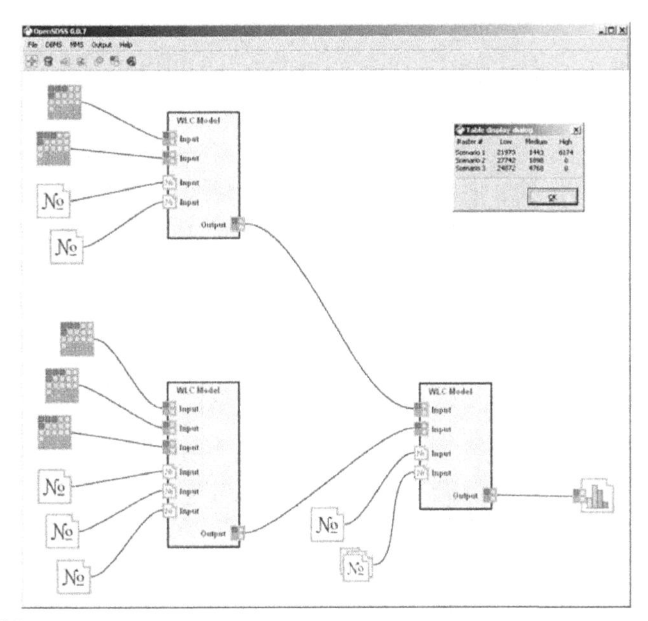

FIGURE 7.26
An example of a more complex model in OpenSDSS.

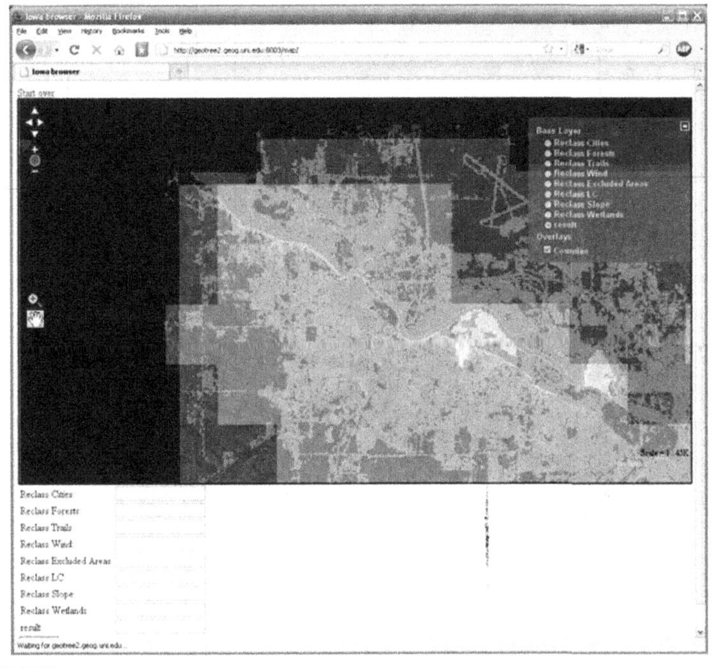

FIGURE 8.17
Results of the WLC calculation on the study area. Dark areas suggest high suitability and light areas highlight areas of low suitability for the placement of wind turbines.

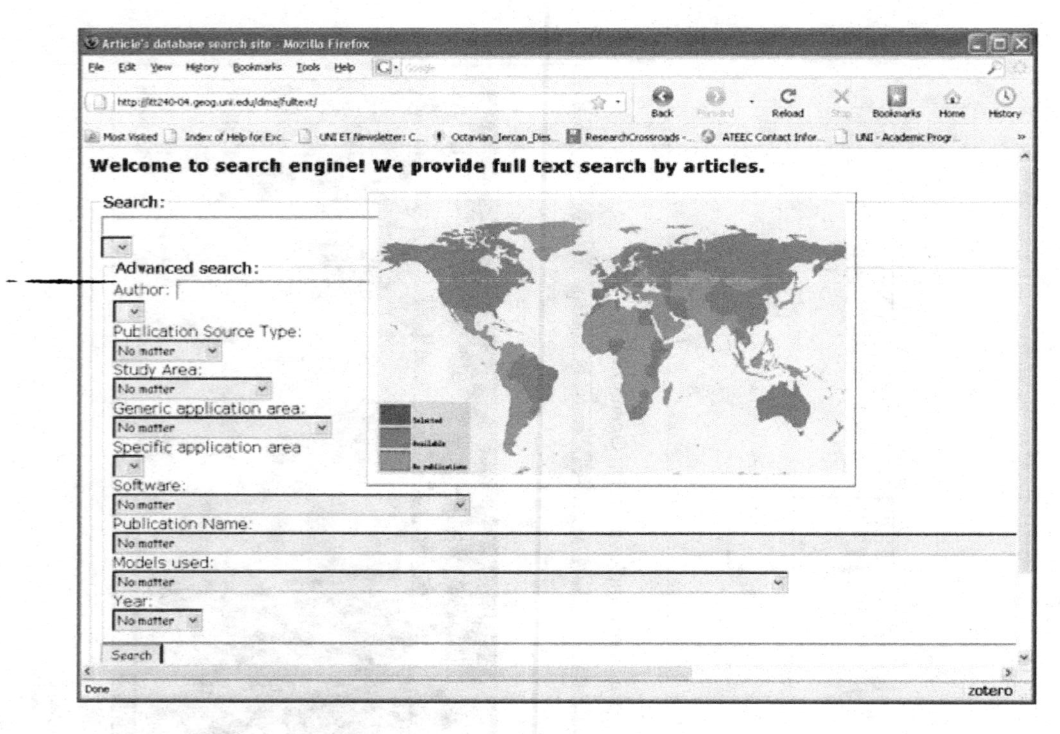

FIGURE 9.1
SDSS Web portal interface.

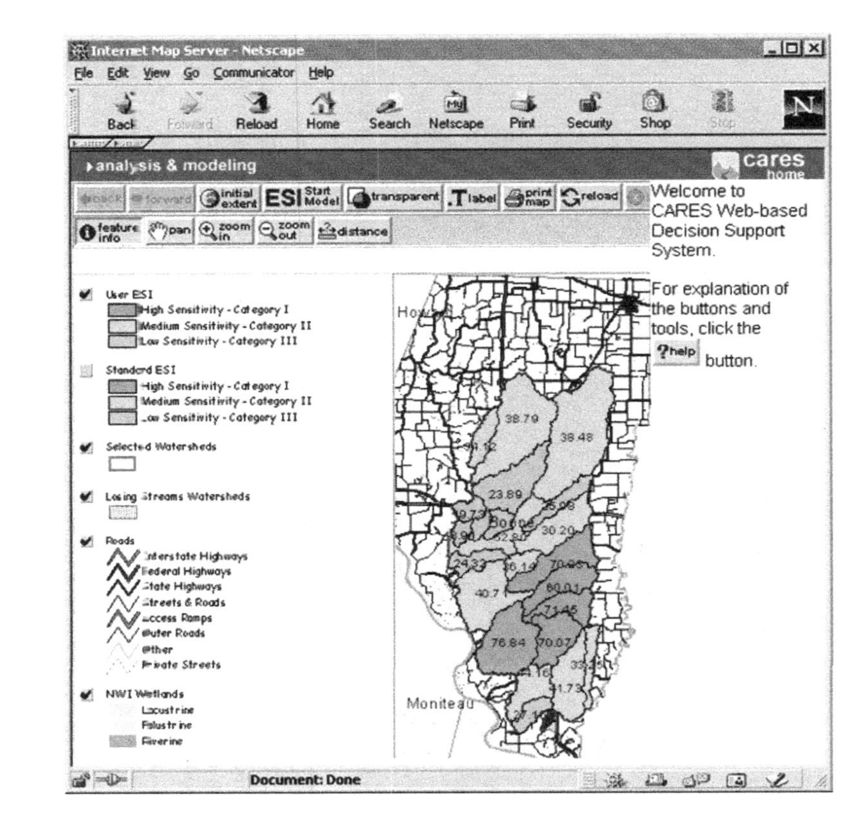

FIGURE 9.8b

The WEDSS user interface. (Sugumaran, F., J. C. Meyer, and J. Davis. 2004. A Web-based environmental decision support system (WEDSS) for environmental planning and watershed management. *Journal of Geographical Systems* 6(3):307–322.)

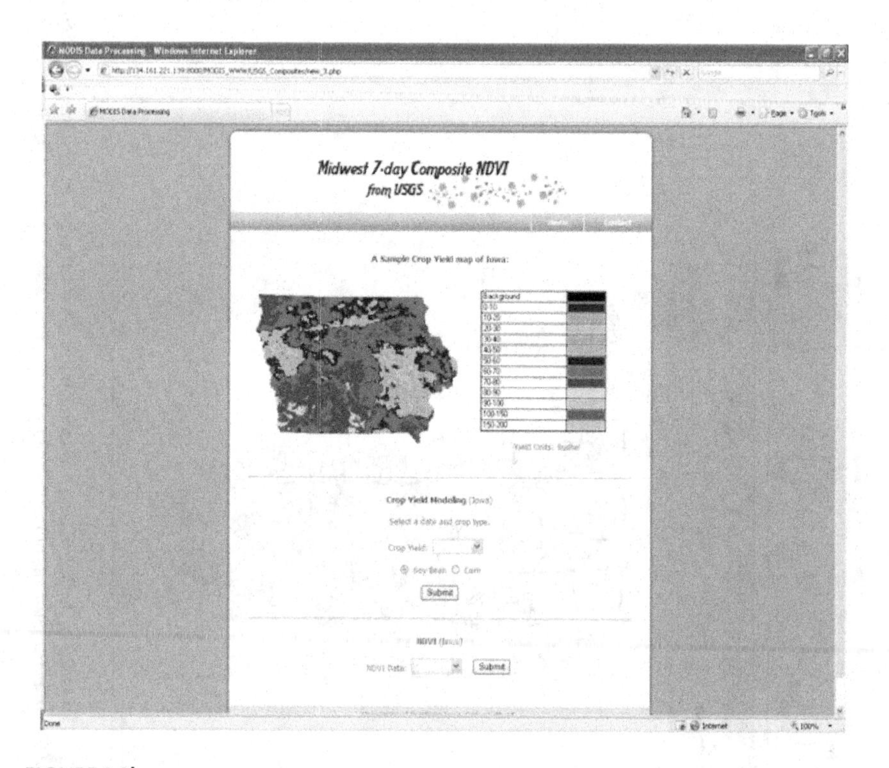

FIGURE 9.9b
Screenshot of the user interface developed for the crop yield SDSS. (Kaparthi, P., and R. Sugumaran. 2009. A Web-based agricultural crop condition and yield prediction modeling system using real-time data. Paper presented at the Iowa Geographical Information Council Conference, Waterloo, Iowa.)

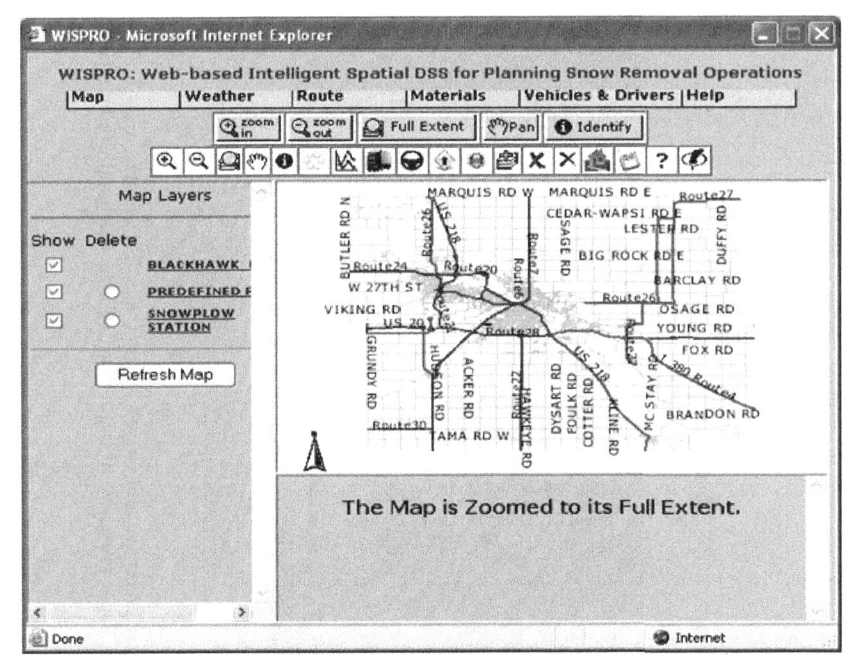

FIGURE 9.10b
Web interface of the WebISDSS.

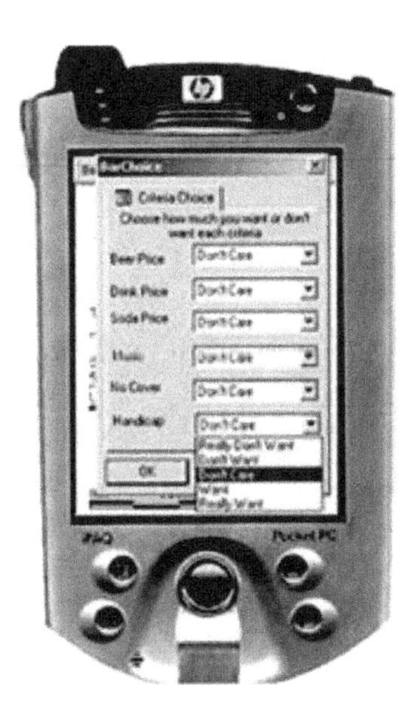

FIGURE 9.11
User interface from the Bar Finder application. (Rinner, C., M. Raubal, and B. Spigel. 2005. User interface design for location-based decision services. Paper presented at 13th International Conference on GeoInformatics, Toronto.).

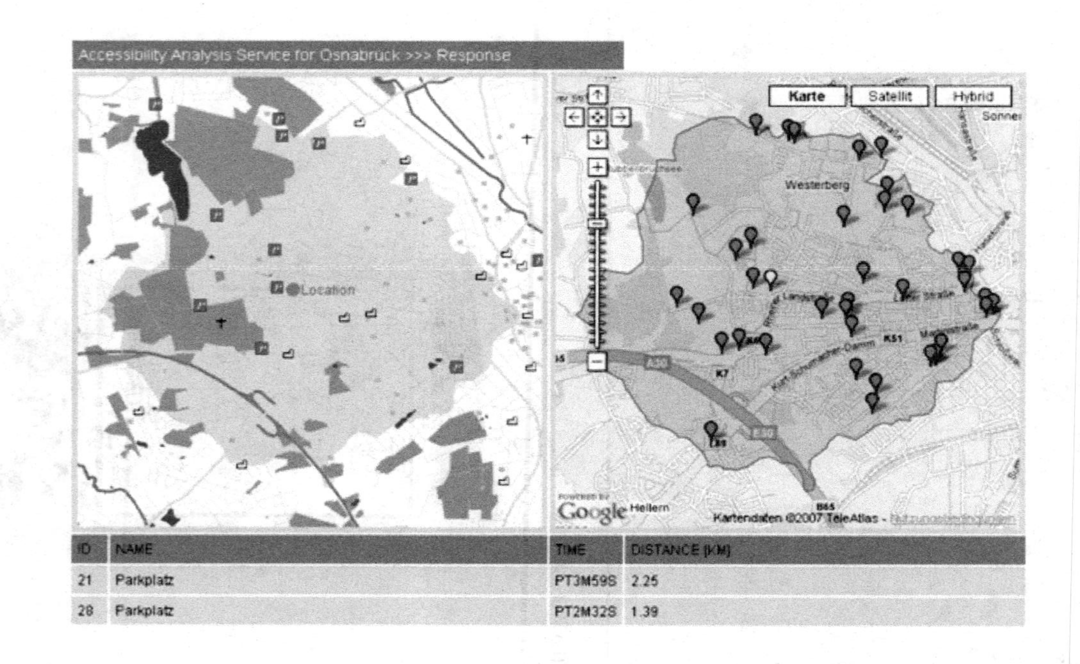

FIGURE 9.12
Example of the analysis results from the accessibility analyses displayed in their Web page and Google Maps. (Neis, P., L. Dietze, and A. Zipf. 2007. A Web accessibility analysis service based on the OpenLS Route Service. Paper presented at the 10th AGILE International Conference on Geographic Information Science, Aalborg University, Denmark.)

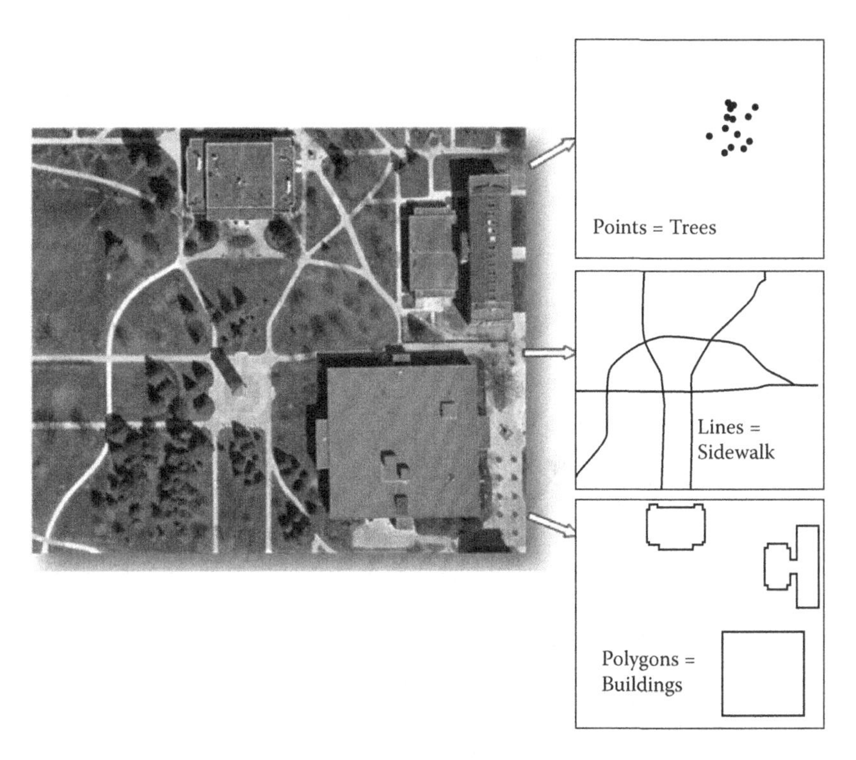

FIGURE 3.5
Point, line, and polygon features using the vector data model.

location (Figure 3.8). A polygon is made up of connected, closed, and non-intersecting line segments (Chang 2009). Polygon features have area and perimeter values, and are used to represent a wide range of physical (soil types, forest stands, water bodies), anthropogenic (land parcels, administrative boundaries), and other (cellular phone coverage) features.

Although the vector model, at the most basic level, is made up of point, line, and polygon features, there are more specific methods for implementing this type of data model within a digital environment. One of the most basic and earliest methods was called the *spaghetti model* (Bernhardsen 2002). In this model, each feature is stored without any reference to its relation to other features. The spaghetti label implies a series of lines lying on top of each other with no defined relationship. This is an inefficient way for encoding data as the same feature can be stored multiple times, often with slightly differing coordinate points causing errors of overlap or gaps between polygons (Figure 3.9). This model also makes spatial calculations computationally intensive (Demers 2009). This led, historically, to the storage of geographic features in spatial data models in which the geometric relationships between features are explicitly defined within the data

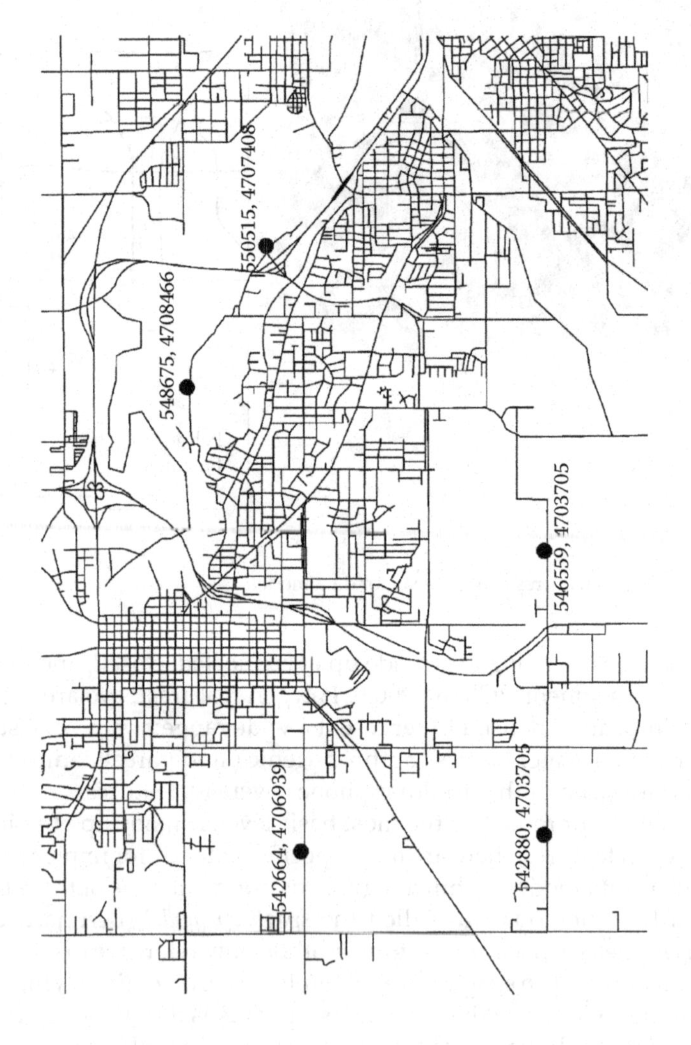

FIGURE 3.6
An example of point representation with x, y coordinates displayed.

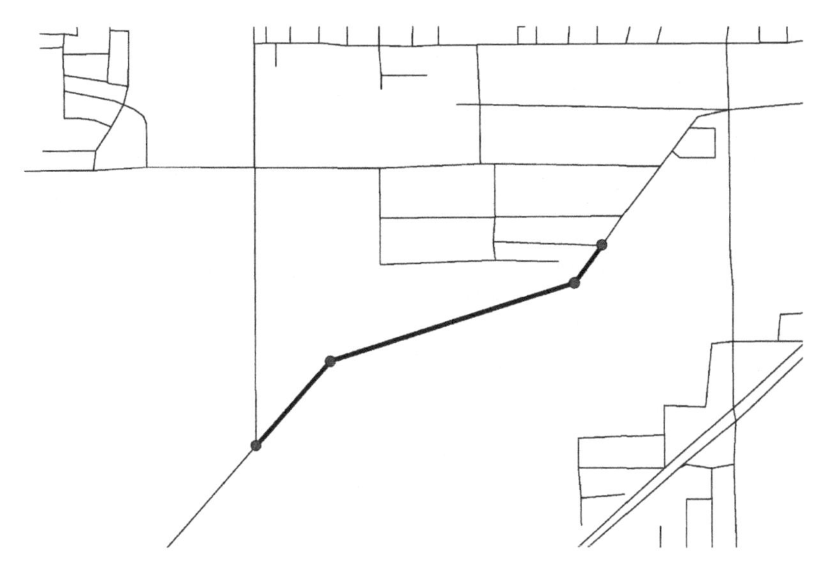

FIGURE 3.7
An example of a road line representation. This road segment is made up of a series of four points and connected line segments.

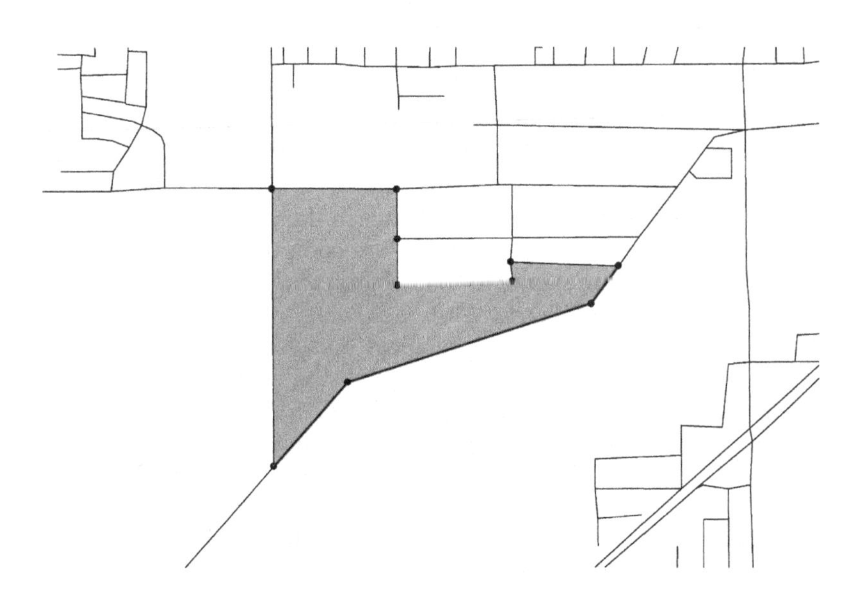

FIGURE 3.8
An example of a polygon feature representation. This polygon is made up of a series of ten points and connected line segments.

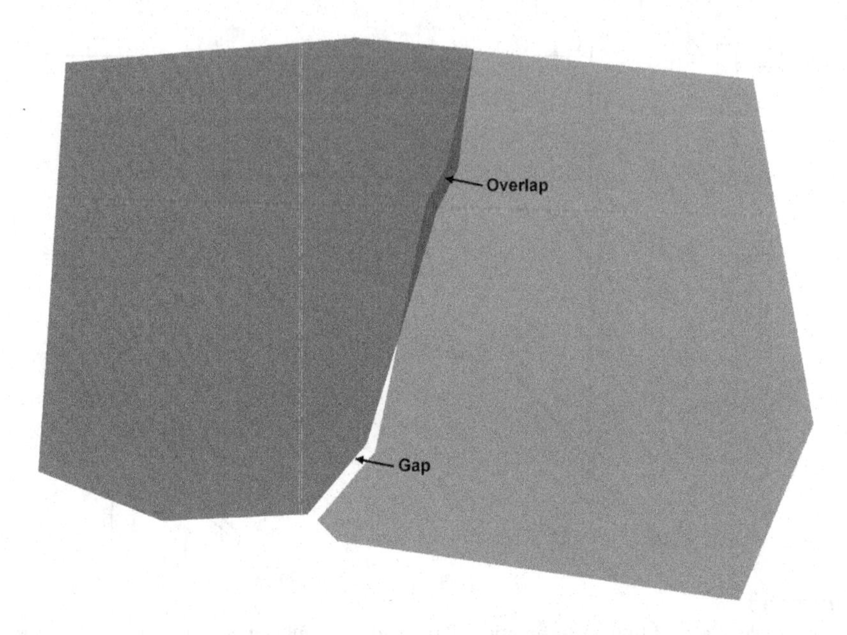

FIGURE 3.9
An example in which there are two polygons representing features with a common boundary. In the spaghetti model, mistakes can often be introduced such as overlaps and gaps because there is no enforcement of spatial relationships as in topological data models.

model. This type of data model is called a *topological data model*. *Topology* refers to the continuity of space and spatial properties, unaffected by distortion (Burrough and McDonnell 1998). In a topological data model, a line segment is defined when it either contacts or intersects another line or when the line changes direction (Demers 2009). The line segment is the basic entity in the topological data model. Each line segment has a beginning and ending node held explicitly in the data model along with a unique identifying number. Polygons are composed of these line segments, and for each line segment, the polygon to the left and right of it is stored in the data model (Figure 3.10).

There have been numerous topological data models developed over time. The U.S. Census Bureau developed the geographic base file/dual independent map encoding (GBF/DIME) format in the 1960s and improved upon it with the Topologically Integrated Geographic Encoding and Reference system (TIGER) for the 1990 census (Demers 2009). ESRI Inc. introduced the coverage topological data model in the 1980s to separate GIS from computer-aided design (CAD) software, which used nontopological data structures (Chang 2009). The coverage was a very common data model in the 1980s and 1990s because it was used in ESRI's popular ArcInfo GIS software. The coverage data model stored points with the x,y coordinate and a unique identifying number. Lines or *arcs* (ESRI's term)

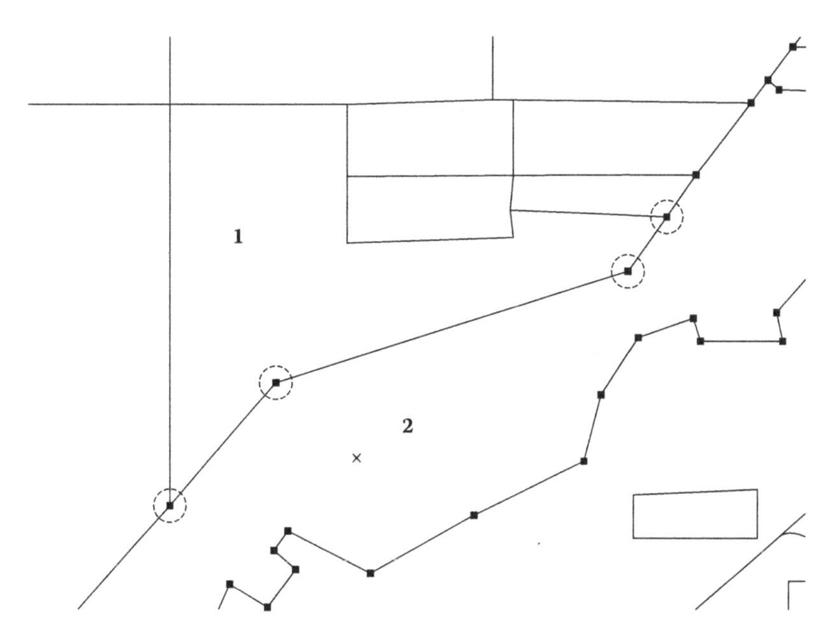

FIGURE 3.10
Adjacent polygons that share an edge. Polygons 1 and 2 share the four points highlighted in the circles.

were defined by both a *from* (or beginning) and a *to* (or ending) node and a list of x,y coordinates that made up the entire arc (Figure 3.11). A polygon was defined by a unique identifier, a list of arcs defining the polygon, and a list of points making up those arcs (Figure 3.12). Topological spatial data models explicitly hold information on the spatial relationships between features, and thus reduce computational requirements when carrying out spatial analysis. They also help minimize mismatches in spatial data such as seen in Figure 3.9. The elimination of these mistakes can remove incorrect interpretations of spatial analysis results carried out on mismatched or incorrect data.

Although topological data structures have clear advantages, they were developed in a time when computers were slower, had less memory, and were less powerful. Faster computer processing speeds and greater computing power have made some concerns of the topological data structure insignificant (e.g., file size due to coordinates being stored multiple times as compared to the spaghetti model). One of the most common spatial data formats used in GIS is ESRI's *shapefile*. This format was introduced by ESRI in the 1990s for faster display and easier data sharing (many other spatial software programs can read shapefiles). The shapefile was introduced with the ArcView software, which was originally meant to be more of a data-viewing program but grew into a true GIS software

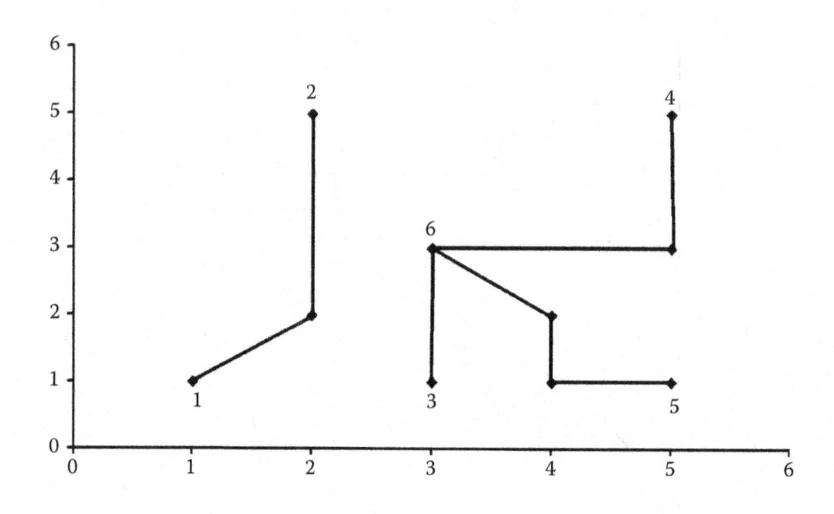

Arc#	FromNode	ToNode
1	1	2
2	3	4
3	5	6

Arc#	x,y coordinates
1	(1,1)(2,2)(2,5)
2	(3,1)(3,3)(5,3)(5,5)
3	(5,1)(4,1)(4,2)(3,3)

FIGURE 3.11
Storage of arcs in coverage data model.

package during the 1990s. Even though shapefiles use a nontopological data model, the format does accommodate more sophisticated features than basic spaghetti models, such as the use of rings (a closed non-self-intersecting loop), which allows the storage of multipart polygon (e.g., multiple islands in the state of Hawaii) features. Topological data models were developed at a time when the digital spatial data creation process required a rigorous, automated method to clean up data entry errors from digitizing paper maps. The shapefile was developed as a feature-centric perspective, and ArcView software provided tools for creating polygons in which boundaries matched exactly, for dissecting polygons and maintaining a perfectly matched boundary, and for splitting lines and having the shared node stored exactly in both lines (ESRI, 2001). The disadvantages of calculating spatial relationships on the fly were ameliorated by the improvements in computing power over time, making the shapefile format more feasible. The shapefile continues to be one of the most common spatial data formats in use in GIS and SDSS.

More sophisticated data structures have been developed recently based on object-based data models. The object-based model could use topological or nontopological data modes, but treats different features as objects that can have specific properties associated with them. For example, pipes

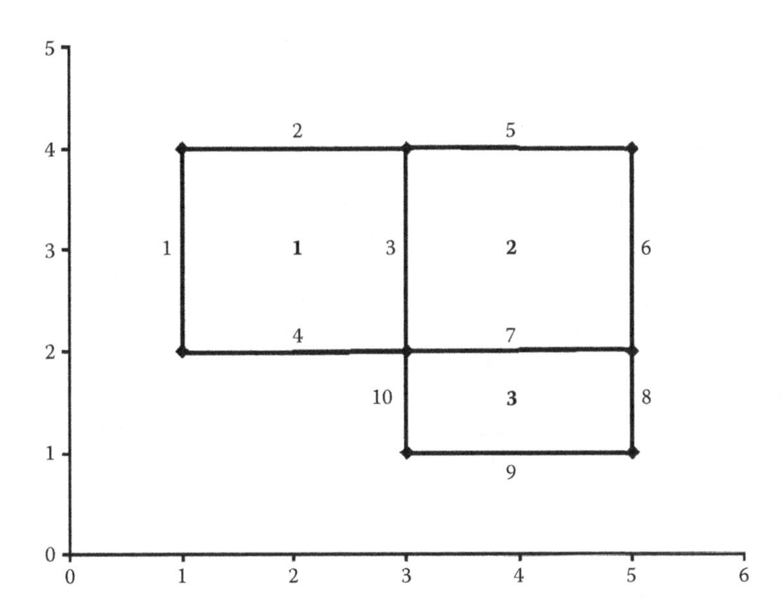

FromNode	ToNode
1	2
2	3
3	4
4	1
3	5
5	6
6	4
6	7
7	8
8	4

Arc#	x,y coordinates
1	(1,2)(1,4)
2	(1,4)(3,4)
3	(3,4)(3,2)
4	(3,2)(1,2)
5	(3,4)(5,5)
6	(5,5)(5,2)
7	(5,2)(3,2)
8	(5,2)(5,1)
9	(5,1)(3,1)
10	(3,1)(3,2)

Polygon#	Arc#
1	4,1,2,3
2	5,6,7,3
3	7,8,9,10

FIGURE 3.12
Storage of polygons in coverage data model. Polygons are labeled in bold while arcs are labeled in italics in the image.

are stored as a specific kind of line that has specific defined attributes, such as width, material, use, and so on. Also, specific relationships and rules can be built into the object-based data model, such as a requirement that pipes used for a specific task must have a certain capacity (Demers 2009). Topological rules and relationships can be built into these object-based model structures. For example, in ESRI's Geodatabase model, there are many generic topological rules (e.g., must not overlap, must not have gaps, must not intersect) that can be applied to a specific feature class or between different feature classes in different ways. A feature class is a set of geographic features that must have the same type of geometry and the same set of attributes. For example, in a Geodatabase, there could be two separate feature classes of different detailed census boundaries, which have a rule that they must have coincident boundaries. For example, five census blocks (more detailed) will fit exactly into one census tract (less detailed) (Chang 2009). While the development of object-based spatial databases requires early investments of time, they can lead to significant improvements in efficiency in the long run, including ease in data sharing. Numerous governmental organizations distribute large spatial datasets using ESRI's Geodatabase, including the New York State Department of Transportation and the U.S. National Hydrography Dataset (Chang 2009).

There are many vector data formats presently in use, and it is impossible to cover them all. The most commonly used GIS program presently is ESRI's ArcGIS, which primarily works with shapefiles and geodatabases, but can also still work with coverages and read other formats such as CAD data. There are many other formats including vendor-specific formats, such as MapInfo's TAB, government-created data formats such as the U.K. Ordnance Survey's National Transfer Format (NTF), and open standard data storage mechanisms such as Geography Markup Language (GML). The Open Geospatial Consortium (OGC), an international industry consortium, was formed in 1994 and has worked toward developing international standards for geospatial interoperability, including data formats such as GML and Keyhole Markup Language (KML).

3.4.4.2 Raster Data Model

The second major data model category used in GIS is the raster data model. The *raster data model* is usually based on a regular two-dimensional grid that is used to represent real-world phenomena, with each cell in the grid representing an individual value for the characteristic being represented. The raster data model is useful for representing phenomena that vary continuously over space, such as precipitation, elevation, or soil erosion (Chang 2009). Figure 3.13a demonstrates a raster dataset representing elevation while Figure 3.13b shows the same data but at a scale close enough to see individual cells. Each cell holds one elevation value, in this case stored in feet units. Raster data are handled differently depending on whether they hold integer or floating-point values. Integer rasters usually represent categorical data (for example, in a land cover raster 1 might represent water, 2—urban, 3—coniferous forest, etc.). Floating-point cell values represent continuous numeric data such as the average amount of precipitation that falls or elevation values with precision to one or more decimal places. Floating-point rasters require more memory for storage and generally will not have an associated attribute table for viewing and querying values from the data. This is because there are often too many

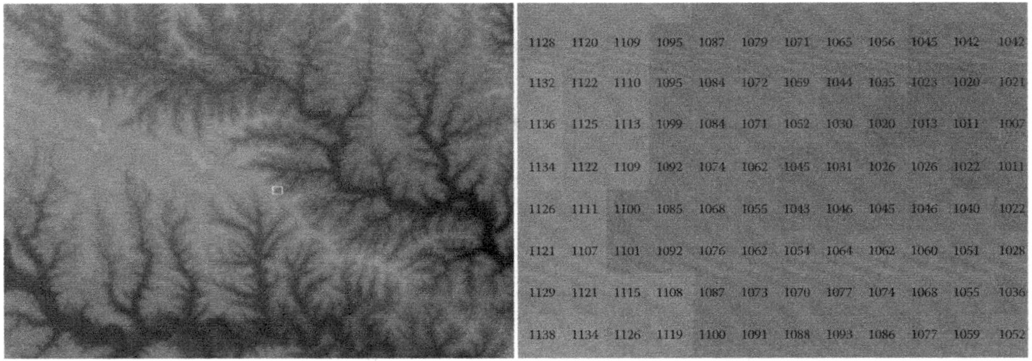

1128 1120 1109 1095 1087 1079 1071 1065 1056 1045 1042 1042
1132 1122 1110 1095 1084 1072 1059 1044 1035 1023 1020 1021
1136 1125 1113 1099 1084 1071 1052 1030 1020 1013 1011 1007
1134 1122 1109 1092 1074 1062 1045 1031 1026 1026 1022 1011
1126 1111 1100 1085 1068 1055 1043 1046 1045 1046 1040 1022
1121 1107 1101 1092 1076 1062 1054 1064 1062 1060 1051 1028
1129 1121 1115 1108 1087 1073 1070 1077 1074 1068 1055 1036
1138 1134 1126 1119 1100 1091 1088 1093 1086 1077 1059 1052

FIGURE 3.13
(a) A raster representation of an elevation with darker areas representing lower river valleys or stream channels; lighter colors represent higher elevations. The white rectangle represents the area shown in (b). (b) A zoomed-in view of elevation raster representation with the elevation value of each cell labeled.

unique values in a floating-point raster (for example, in a raster that is 1000 × 1000 cells, if each cell had a unique value, there would be one million values). An integer raster representing categorical data might only have a few unique values but many cells that have that value (Figure 3.14). In ESRI's GRID format (Figure 3.14), the Value field in the attribute table holds the integer value, while the Count field holds the number of cells in the raster that hold that unique value.

Although the raster data model is conceptually simple, there are intricacies in making the storage of different kinds of surface data efficient. The raster data must hold information on the size of the cell that defines the resolution of the data. A cell size of 10 meters (10 × 10) means that each cell represents 100 m^2 in the real world, while a 100-m cell size represents 10,000 m^2, or a hectare. Raster datasets with smaller cell sizes can capture more detail across shorter distances, but also require greater computer memory and processing resources. For example, a small raster dataset with only a single integer value for each cell, but with resolutions of 10 m and 100 m, take up approximately 3.2 mB and 32 kB on disk, respectively. Various methods have been developed for efficiently storing raster data, but are beyond the scope of this book.

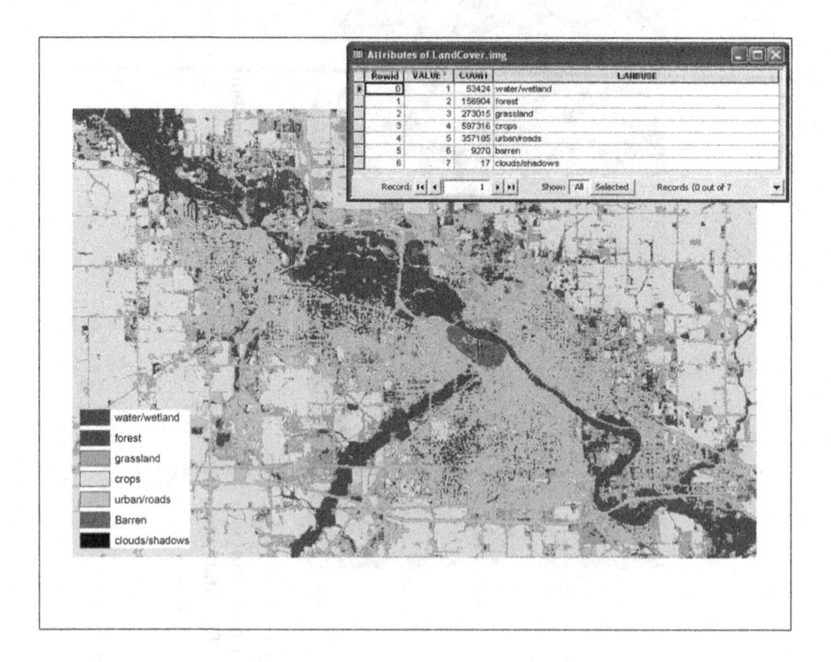

Rowid	VALUE	COUNT	LANDUSE
0	1	53424	water/wetland
1	2	156904	forest
2	3	273015	grassland
3	4	597316	crops
4	5	357185	urban/roads
5	6	9270	barren
6	7	17	clouds/shadows

FIGURE 3.14
(See color insert following page 74.) An ESRI GRID format integer raster representing land cover classes with attribute table shown.

One of the most common types of datasets using the raster data model is the digital elevation model (DEM). As can be seen in Figure 3.13, DEMs represent elevation by storing a single elevation value for each cell in the raster. There are many different sources of DEMs. During an 11-day mission in 2000, the NASA Space Shuttle Endeavor collected data on elevation for the majority of the globe. The Shuttle Radar Topography Mission (SRTM) resulted in the development of an almost complete global coverage of DEMs at an approximate resolution or cell size of 90 m. Other nations have developed more detailed DEMs. For example, in the United States, DEMs are available for the majority of the country at a resolution of 10 or 30 m. In recent years, advanced technologies such as Light Detection and Ranging (LiDAR) have led to the development of very high-resolution (e.g., 1 m) DEMs. These high-resolution DEMs are being collected for local or regional areas for detailed applications. Figure 3.15 represents four levels of DEM resolutions. The 90-m resolution is from the STRM data, the 10- and 30-m resolutions are from the U.S. Geological Survey (USGS) DEMs,

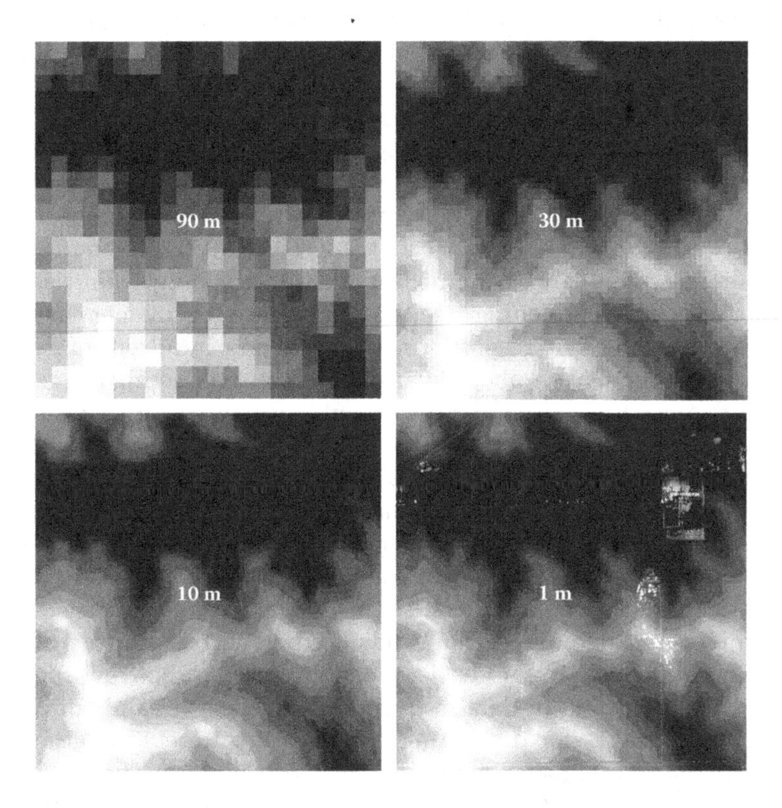

FIGURE 3.15
Digital elevation models at 90-, 30-, 10-, and 1-m resolutions. The area represented in each of the images is a 2 × 2 km tile.

and the 1-m resolution is derived from LiDAR data. At the 90-m resolution, few features can be picked out, but as the resolution increases, more and more landscape features can be seen. The choice of which resolution to use depends on data availability and the purpose of the application. The 90-m resolution data would be insufficient for a spatial decision-making process about where to locate a new road, but the 1-m resolution would be impractical in looking at countrywide land use suitability. The size of the raster files shown in Figure 3.15 goes from approximately 1 kB for the 90-m raster to approximately 15 mB for the 1-m resolution data. Thus, trade-offs between the detail provided by higher resolution and the higher computing costs must be weighed when deciding which resolution data to use in a given spatial decision-making process.

Imagery of the Earth's surface taken from various platforms including satellites and airplanes are stored using the raster data model. The use of aerial photography as a means of gathering information has been common since World War I. With the advent of spatial processing software, aerial photography has become a very common digital spatial data source. Digital orthophotographs are photos in which the displacement caused by camera tilt and the natural terrain have been removed (Demers 2009). Many organizations, especially local, state, and federal governments, regularly collect aerial imagery and produce digital orthophotography that can be used in GIS and spatial image processing software (Figure 3.16). Imagery of the Earth's surface has been collected by satellites for several decades (e.g., Figure 3.17). The number of satellites has grown greatly in the last decade with many countries and private companies launching satellites that can collect imagery at varying spatial and spectral resolutions. There are now hundreds of private and government sensors aboard various satellites collecting imagery of the Earth's surface. The spatial resolution of satellites ranges from less than 1 m (e.g., GeoEye-1) to greater than 1 km (Advanced Very High Resolution Radiometer [AVHRR]). Different sensors collect data in different wavelengths of the electromagnetic spectrum (e.g., ultraviolet, visible, infrared). Satellites often carry sensors that produce multispectral imagery or images stored separately for different ranges of wavelengths. For example, the Landsat Thematic Mapper instrument collects data in seven narrow spectral bands. Hyperspectral sensors collect data simultaneously in hundreds of contiguous, very narrow spectral bands. The collection of data in different wavelength bands provides various advantages for discerning patterns of different materials on the Earth's surface. Image file formats can store numerous bands in one file. In this case, each file holds several raster datasets (bands) within one file. Each of these files can be viewed separately or as a combined image. Different spatial software, such as image processing and GIS,

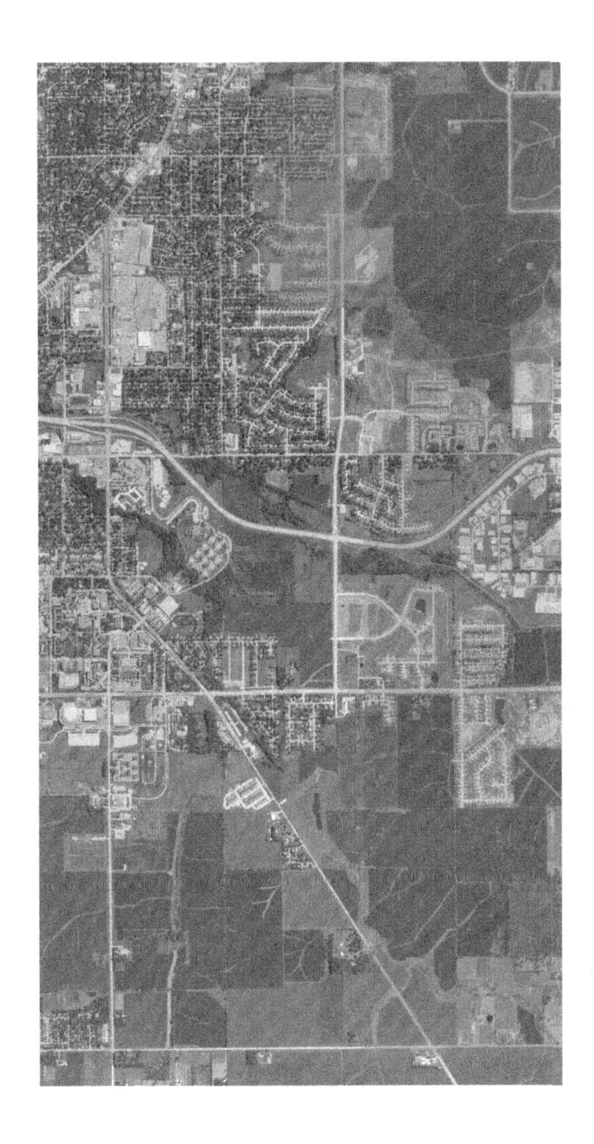

FIGURE 3.16
(See color insert following page 74.) An example of a digital orthophotograph.

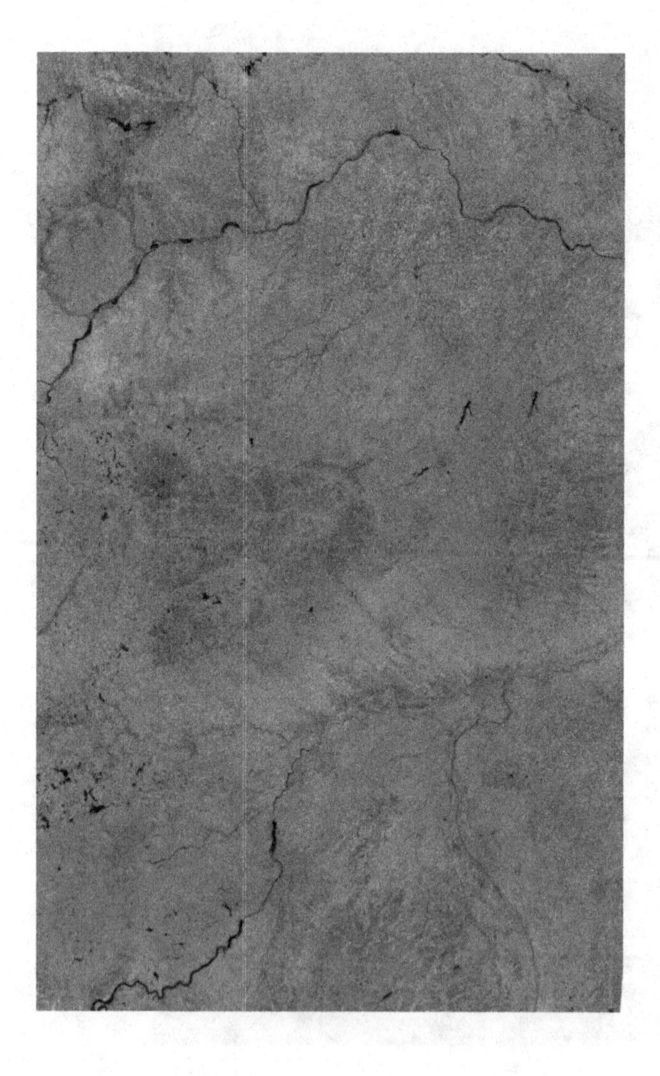

FIGURE 3.17
A satellite image of Iowa and parts of surrounding states compiled from multiple Landsat scenes.

provide different mechanisms for viewing these data in different band combinations.

3.4.4.3 **Raster versus Vector**

Due to the variety of real-world phenomena, both the raster and vector data models have advantages and disadvantages in their representation. The vector data model clearly matches some discrete natural and anthropogenic features in the real world better than raster data. For example, features with clearly defined geometry, such as roads, buildings, lakes, and so on, are more efficiently and accurately represented in the vector data structure. On the other hand, phenomena that vary continuously across space, such as elevation, climatic characteristics, air or water quality concentrations, etc., are better represented in the raster data model. If these phenomena were represented with the vector model, the number of individual features would be great, leading to large file sizes. Other real-world phenomena are somewhere in between discrete and continuous patterns. For example, depending on how real-world patterns are interpreted, land cover could be defined as discrete generalized polygons using the vector model or with a raster data model, which can capture small-scale variation in land cover through automated image processing techniques. The decision of which data model is to be used depends to a degree on the method for collecting or developing the spatial data. For land cover mapping over a large geographic area, usually some image processing techniques used on raster satellite imagery are used to delineate land cover classes, as manually digitizing would be too time-consuming. However, manual digitizing of vector features based on aerial photography might be used at a more local level to define discrete land cover classes (Figure 3.18). Most GIS and other spatial programs allow at least viewing of both vector and raster datasets, and most provide functionality for converting between the two data models.

3.4.5 Spatial Data Collection

There are many different methods for acquiring or collecting digital spatial data, and new methods have evolved with technological advances. The amount of available spatial data has grown tremendously, especially over the last two decades. When developing a spatial database for a given project, both primary and secondary sources might be used (Longely et al., 2005). A primary data source would be considered one that is created by the database developer while secondary data would be considered data that was created by a separate organization (primary data for them) but can be incorporated into a given project or database. The data development process is one of the most time-consuming aspects of any spatial

FIGURE 3.18a
A raster data model representation of land cover derived from automated image processing of satellite imagery.

decision-making process and requires significant resources. A common mistake is made when underestimating the time and resources necessary to build acceptable spatial databases to support the spatial decision-making process.

There are many different methods for creating primary data. A common method for translating hard-copy maps into digital data was manual digitizing. A digitizing tablet could be used by attaching a map to the tablet, registering the corners of the map (matching digitizing tablet coordinates to real-world coordinates of map), and then tracing features on the map with a mouselike device. This method has been and continues to be used by private and governmental organizations to convert many hard-copy maps to digital data. In developed countries, much of this work was done in the 1980s and 1990s. Another method for converting hard-copy maps or photos to digital data is using a scanner to save the map or photo as a digital image. After the map or image is stored as digital data, it can be referenced to real-world coordinates using GIS or other spatial software. This data can then be visualized in GIS and other software and be used to create derivative spatial datasets. One common operation is to take a scanned map and carry out manual on-screen digitizing for the creation of vector features. For example, after the scanning and registration process is carried out on an aerial photo and using functionality from GIS software, all of the building footprints could be traced on screen to produce a polygon

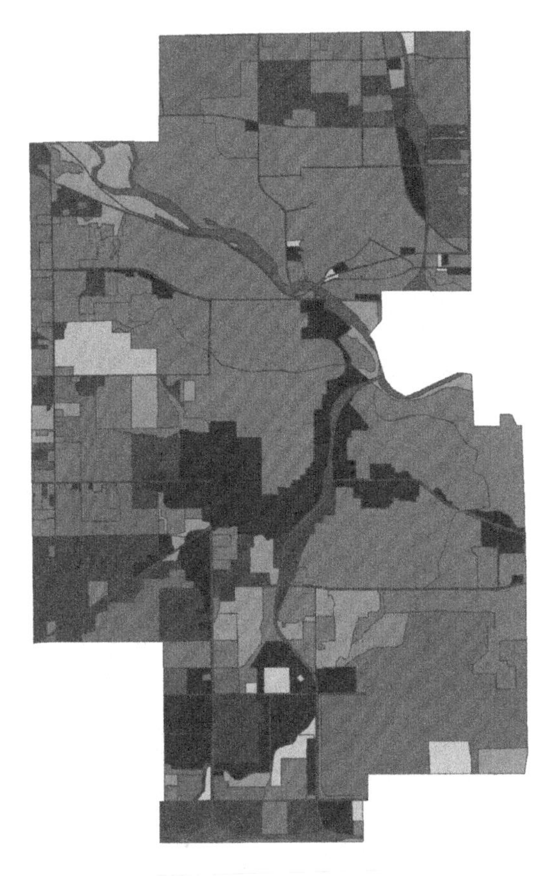

FIGURE 3.18b
Manually digitized vector model representation of land cover.

vector theme (Figure 3.19). The entire process of moving from a hard-copy data source to digital spatial data requires careful attention to detail and is described in more detail in texts dedicated to GIS.

Another method of creating digital spatial data is by carrying out field surveys using Global Positioning Systems (GPS). The development of GPS by the United States military originally was for defense applications but has grown exponentially over time in myriad private and government sectors. GPS uses a constellation of satellites in space that broadcasts information that allows a handheld GPS receiver on the ground to accurately calculate real-world positions. Sophisticated GPS hardware and software have been developed to facilitate collection of spatial data in the field and for incorporating that data easily back into GIS software. New handheld computers allow users to display GIS base layers while in the field using submeter-accuracy GPS to record spatial data (as well as related attribute data) on

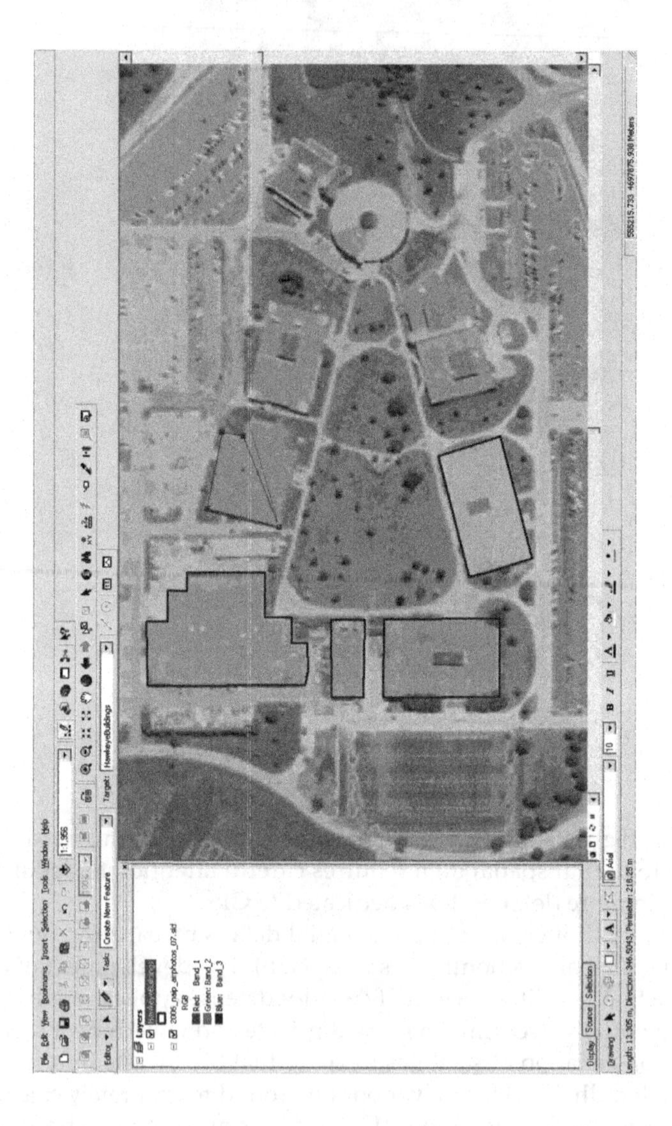

FIGURE 3.19
(See color insert following page 74.) On-screen digitizing of polygon features in ArcGIS software using a digital orthophotograph as reference.

real-world features. The accuracy of GPS units has improved greatly and, depending on the sophistication of the unit, can be accurate to the centimeter level. Point, line, and polygon features can be collected with GPS units and easily incorporated into GIS databases. Advances in GPS and communication technology are allowing the tracking of vehicle movement in real time. GPS are also used for applications such as precision farming, in which the rate at which fertilizer, water, or pesticides are applied is controlled through a system using GIS and field maps in real time.

A final common method for creating spatial data is automated processing of remotely sensed imagery. This imagery is most commonly collected by satellite or airplane-based sensors that record reflected or emitted energy along the electromagnetic spectrum. There are many providers of remotely sensed imagery led by national governments (e.g., NASA) with some commercial providers (e.g., GeoEye). The digital imagery provided by these organizations is useful for visualization purposes but also can be used for the derivation of useful spatial products. A remote sensing analyst is a person who uses various tools to derive useful information from the digital numbers representing reflectance in the electromagnetic spectrum stored in remotely sensed imagery. Many automated image processing applications have been developed specifically for aiding remote sensing analysts. Software such as IDRISI, Erdas Imagine, ENVI, ER Mapper, PCI, and others can be used to process imagery and derive spatial information on vegetation, geology, soils, settlement patterns, air quality, water quality, climate, land use, disasters, and many other phenomena. The process of moving from a raw satellite-derived image to a purpose-specific derived spatial product requires considerable effort and expertise, even with the use of image processing software. Spatial and radiometric (characteristics of the electromagnetic spectrum recorded) resolution, revisit frequency, cost, and other attributes of different remotely sensed imagery dictate which type of imagery would be useful for addressing a certain spatial problem.

There are many distributors of spatial data in various formats. Some spatial datasets are provided free of charge while others are sold commercially. Appendix A provides links to useful Web sites for each state in the United States where spatial data can be acquired or with instructions on how to acquire spatial data. Appendix B lists some links for global datasets.

3.4.6 Database Management

The majority of discussion up to now has dealt solely with the way spatial information is collected and stored in GIS and related software. However, one of the main strengths of GIS is the ability to store large amounts of nonspatial information that are either directly or indirectly related to spatial features. Pieces of information directly related to spatial features are called *attributes*. A wide variety of characteristics of any vector feature can be recorded in the attribute table of the vector feature class. For example, a line representation of pipe network attributes could include the material it is made of, diameter, date installed, manufacturer, and so on. There are a variety of ways to store spatial and attribute data, including the storage of both in the same file, in separate files or databases, and in a single relational database (Bernhardsen 2002). The advantage of the first method is rapid search capabilities, but with increasing amounts of data this method becomes less efficient. The second method stores spatial data in one file while attribute data are stored in a separate file with a feature identifier allowing them to be linked. The ESRI shapefile format (which is made up of anywhere from three to seven files) is an example, with the attribute data being stored in a .dbf (dBase format) and the primitive geometry data stored in the .shp file. Similarly, in the MapInfo TAB format, the .DAT file holds the attribute data while the .MAP file holds the geometry data. The relational database method stores the geometrical information in a table with related tables holding attribute data. An advantage is that these methods are based on standard technologies, allowing easy transfer and ease in using technologies such as the Structured Query Language (SQL) (Bernhardsen 2002).

Regardless of the storage technique, GIS software can carry out a variety of operations based on the geometry and attributes of a feature. Figure 3.20 shows a polygon representation of census tracts for the state of Iowa in the United States with the attribute table also shown. Attributes include a unique identifier for the tract, an area field storing the number of square kilometers, the population, and population density. Many of these attributes were recorded at the time of data creation. However, the values in the AREA field can be calculated at any time using built-in functions in the GIS software. In general, GIS software contains functionality for adding new fields and populating them with new values using mathematical, text, or date functions. For example, in ArcGIS software, numerical (long integer, short integer, float, double), text, date, and binary large object (BLOB; used to store objects such as images, audio, or multimedia objects) attribute field types can be added to a feature class attribute table. In addition to handling attribute tables, most GIS software can read and manipulate stand-alone tables that do not hold spatial data. The format of these tables varies by software but includes dBase, Excel, text, DAT, and other files. These are considered flat files in that they are a simple collection of records for which fields are used to store identical data in each row (Malczewski 1999). Advantages to using flat files include simplicity and speed in retrieval.

As spatial technologies evolved and their use grew tremendously in various disciplines, more sophisticated database technologies were developed. In contrast to flat file structures, relational databases organize

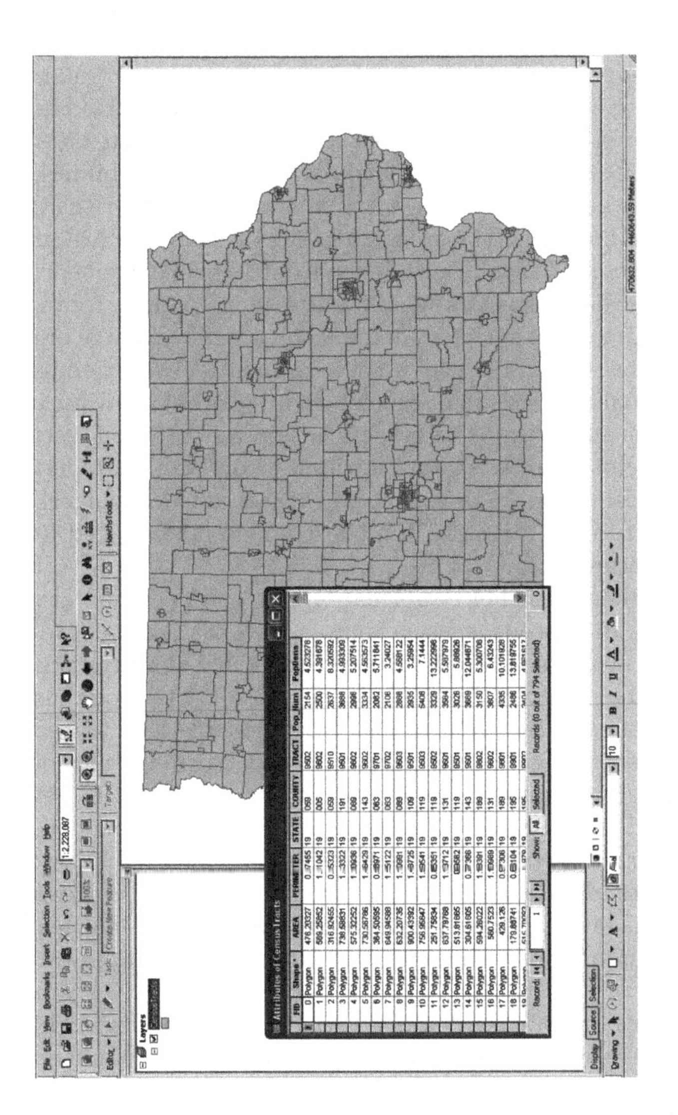

FIGURE 3.20
A polygon shapefile with associated attribute table in ArcGIS 9.3.

information in tables that hold keys used to define relationships between various tables (Chang 2009). Companies such as Oracle developed spatial capabilities in their relational database structure. Oracle Spatial is software that incorporates functionality for handling spatial data in its Oracle Standard and Enterprise database editions. This platform provides, according to Oracle, support for a wide range of applications, from automated mapping/facilities management and geographic information systems (GIS) to wireless location services and location-enabled business services. Other enterprise-level database management systems, such as Microsoft SQL Server, also have spatial data handling capabilities. In the open source arena, PostGIS software adds functionality for spatial data handling to the PostGRE database management system. ArcSDE software by ESRI manages spatial data in a relational database management system and enables it to be accessed by ArcGIS software. All of these technologies allow the integration of spatial data into enterprise-level (large organizational) databases. Object-oriented databases, such as ESRI's Geodatabase, rely on relational database management techniques but allow for specialized rules and relationships to be defined for and between features or objects. This helps to protect against attribute errors being introduced into databases. These types of relational and object-oriented databases can require greater initial investment in construction but can lead to much greater efficiencies later.

Nonspatial data often hold information that can be joined or related to spatial data. In the census tract example, the U.S. Census Bureau will have a whole range of demographic and economic characteristics held in tables that have a tract unique identifier. The tract unique identifier matches the value in the feature class attribute table. The two separate tables can be joined together for carrying out spatial and nonspatial queries and for subsequent map visualization. For example, we could find census tracts that fall within 1 km of major highways and then query the joined demographic/economic data to see what percentage of people in those tracts are of a certain ethnic group. In this example, if available, data on asthma rates by census tract could be used to investigate if there is any correlation among ethnicity, proximity to major highways, and asthma rates.

Attribute data can be recorded in a variety of ways. In the digitization process, a database developer would define the attributes of interest for a given type of feature, and when the features were created, the attribute data could be added at the same time. It is important that a common identifier is added for future relation of spatial data to associated information from separate tables. In the case of field surveys using GPS, it is often possible to collect and enter attribute information simultaneously while recording positional information. The creation of rules and relationships in relational or object-oriented databases can help protect against the introduction of errors in the data creation process.

3.4.7 Data Considerations

The collection of spatial data and the development of spatial databases should always follow a careful planning process. There are many considerations for developing suitable spatial and nonspatial datasets in a comprehensive database for a specific application. These considerations include the cost of data acquisition and management, the scale or resolution necessary, and the accuracy and precision levels required. Scale and resolution imply the amount of detail that is represented in the spatial data. The level of detail needed depends on the spatial decision problem. For example, a field-based precision farming operation would require highly detailed information on terrain and soil conditions. However, for a statewide assessment of land suitable for residential development, lower resolution spatial data on soil types or terrain would be necessary. The spatial precision and accuracy are dependent on the method of data creation. Database developers must decide on the amount of resources to be invested in order to meet accuracy and precision requirements.

Even with careful database development, errors are almost always present. Demers (2009) mentioned two primary types of errors, the first of which is mainly associated with vector data and is called an *entity error.* Entity errors deal with incorrect spatial data, such as missing features, extraneous features, misplaced or misshaped entities, or misconnected features. These errors can be introduced during the data creation process and should be checked while creating data either through manual checks or using topological tools of the GIS software. Indeed, there should be ongoing quality control checks at the time of data creation, completion, and use. Topological data models and topological functions in GIS software can be used to discover and correct many of these errors without having to carry out manual inspection. The second primary error type—*attribute errors*—can occur in both raster and vector data. If spatial analyses are carried out with flawed spatial data, then any resulting data will contain errors.

3.4.8 Spatial Data Exploration, Processing, and Analysis

A tremendous number of spatial operations can be carried out on vector and raster data. These operations can be combined to create a sequence of operations that can be considered as a model or a modeling operation. There are a range of GIS packages available, and the amount of functionality and the exact algorithms they use vary. However, some basic functions are included in most software. In addition, specialized functions are developed as extensions to the basic GIS software package either by the original GIS developer or as third-party add-ons. The specific discipline or research area in which the GIS software will be applied often dictates the type of operations needed. For example, in watershed modeling, raster operations are important as the movement of water through the landscape is modeled more efficiently in a raster environment. In this GIS overview chapter, it is impossible to cover the full breadth of spatial analysis and processing operations. However, some broad and important categories of spatial analysis will be covered with some specific examples given. In the following section, we provide an overview of some of the most common vector and raster analyses. For more detailed coverage of spatial processing and analyses, please refer to Chrisman (2001), Langley et al., (2005), Heywood et al., (2006), or Demers (2009).

3.4.9 Map Data Exploration

Numerous tools built into GIS software exist for exploring spatial data and associated attributes to uncover patterns that are not readily evident. In all GIS software, there are tools for dynamically viewing spatial data that allow a user to navigate across the represented space in a map view. There is usually functionality that allows for zooming in or zooming out based on either dynamic drawing or by set amounts, panning the map, zooming to previous extents, or zooming to the extent of all datasets in the map. Figure 3.21 shows the standard toolbar in ArcGIS ArcMap software. The first eight buttons on the toolbar represent tools used for dynamic navigation. There are also functionalities for zooming to a given layer's extent and to zoom to selected features in a layer. The ability to navigate dynamically provides the user great opportunity to investigate spatial data. As GIS allows dynamic navigation through spatial data, the scale of the data viewed changes dynamically. Although the amount of detail visible in the map display changes depending on the level a user zooms to, the amount of detail that is present in a given map layer does not change. That amount of detail is determined by the source scale of the data. For example, in the left panel of Figure 3.22, it is difficult to see differences in the two state borders. However, when zoomed in to a smaller area along the boundary, significantly more detail is seen in the polygon boundary that was created from source data of a larger scale. The user also has the ability to set the scale at which a given layer will be viewed in some GIS

FIGURE 3.21
Standard toolbar for data exploration in ArcMap.

FIGURE 3.22
Varying levels of detail visible as viewed in GIS at two separate scales. A polygon representing the area of Iowa is depicted in two datasets that were produced from data sources of varying resolution.

software. For example, when zoomed out to a state level, a layer representing land parcels would not be displayed, but when zoomed in to the city level, this layer would become visible. Most GIS also have functionality for measuring distances (and sometimes area) in the map view (last button on right in Figure 3.21).

3.4.10 Data Identification, Examination, and Query

There are a wide variety of tools available in GIS software for spatially identifying, selecting, and querying features and for forming queries based on attributes held in associated attribute tables. The ninth button from the left in Figure 3.21 is a spatial select tool. This tool allows a user to draw a rectangle or a point. All features that the point or rectangle intersects will be selected. This is useful as the user can export the selected features to a new dataset for specific analysis purposes. The button with the "i" icon (Figure 3.21) allows the user to click on any spatial feature, and a dialog (Figure 3.23a) will open that shows all of the associated attribute information coupled with that feature. In the ArcMap software, there is also search functionality (third button from right in Figure 3.21) in which the user can search for features in the feature class, places (e.g., cities, rivers) from an ESRI map service (i.e., from ESRI servers), address locations, and routes. Spatial selection of features from a vector layer can be carried out based on features from another vector layer. There are a variety of spatial selection techniques available, such as intersect, within a distance of, contain, are completely within, etc. (Figure 3.23b). An example of a spatial selection involves choosing schools or daycares that fall within 1000 meters of federally permitted facilities monitored for air quality (Figure 3.24). This functionality allows for very useful data exploration, for selecting features for specific further analysis, or for mapping purposes.

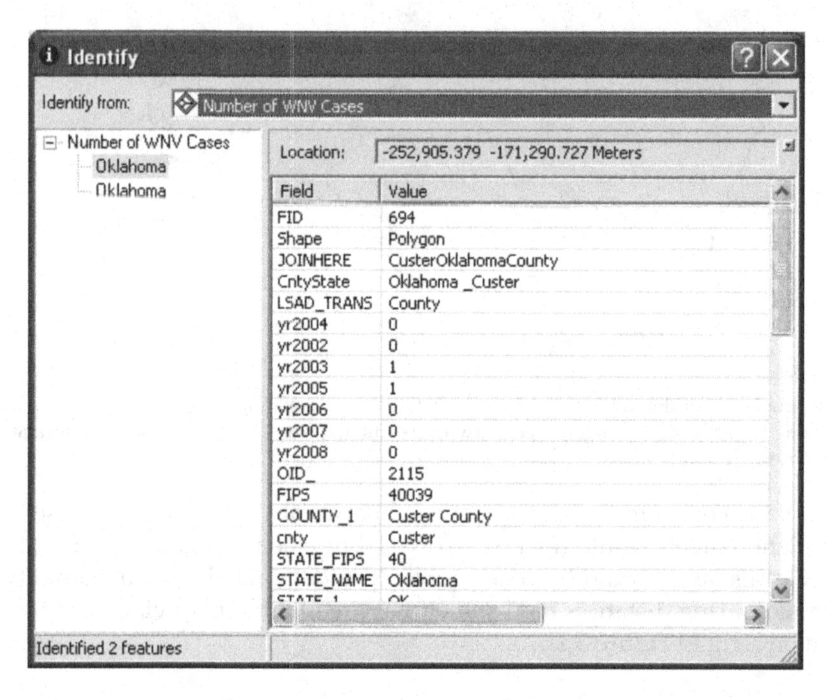

FIGURE 3.23a
The information dialog that appears when a user clicks on a feature with the Identify tool chosen in ArcGIS software.

The ability to select features from spatial data based on attributes is an important way to examine vector data. The user can select a set of features that match a given set of criteria. This is called *querying the database*. Queries are common operations in database management systems. A query is the selection of a subset of records based on values of specified attributes (Bolstad 2005). Queries can be simple and based on a single attribute or complicated and based on many attributes. Rows or records in the attribute table are selected when a certain condition is met through a query. For example, when looking at a county polygon spatial feature class with many demographic variables, various queries can be created to see the spatial distribution of certain specific properties of the population. Imagine that a user wants to visualize where there are counties with a high percentage of elderly persons in the United States. Figure 3.25a shows the Select By Attributes dialog from ArcGIS for creating this query using county data and also the attribute table (Figure 3.25b) and map (Figure 3.25c) in which the selected features are highlighted. A GIS user can construct very complex queries using various operators. In ArcGIS software, the Structured Query Language (SQL), a very common database management query language developed in the 1970s, is used for

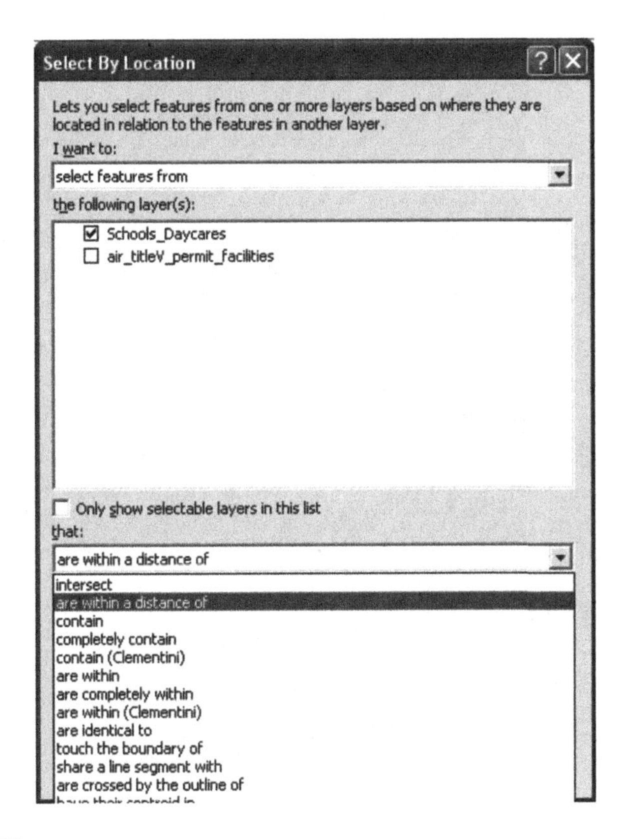

FIGURE 3.23b
(Continued.)

constructing queries. The Select By Attributes dialog builds the first part of the SQL statement as it knows which table it is making the selection from (in Figure 3.25a, SELECT * FROM Counties WHERE: field). The user then builds the remaining part of the query using the tools in the dialog. There are other functions available for working with tables in GIS that are useful for data exploration in combination with queries. For example, in ArcGIS, records can be sorted based on values in a single or multiple fields, statistics can be calculated for numeric values in a field, and new fields can be created and values calculated based on other field values. Users can build multiple queries in order to ask and answer questions of the data. An example based on county census data would be asking what percentage of the elderly population in the United States lives in counties with low population density. In this case, users could check the total elderly population in the country by using the statistics tool with no records in the table selected. They could decide on a population density threshold by sorting the records in the table or by using the statistics tool

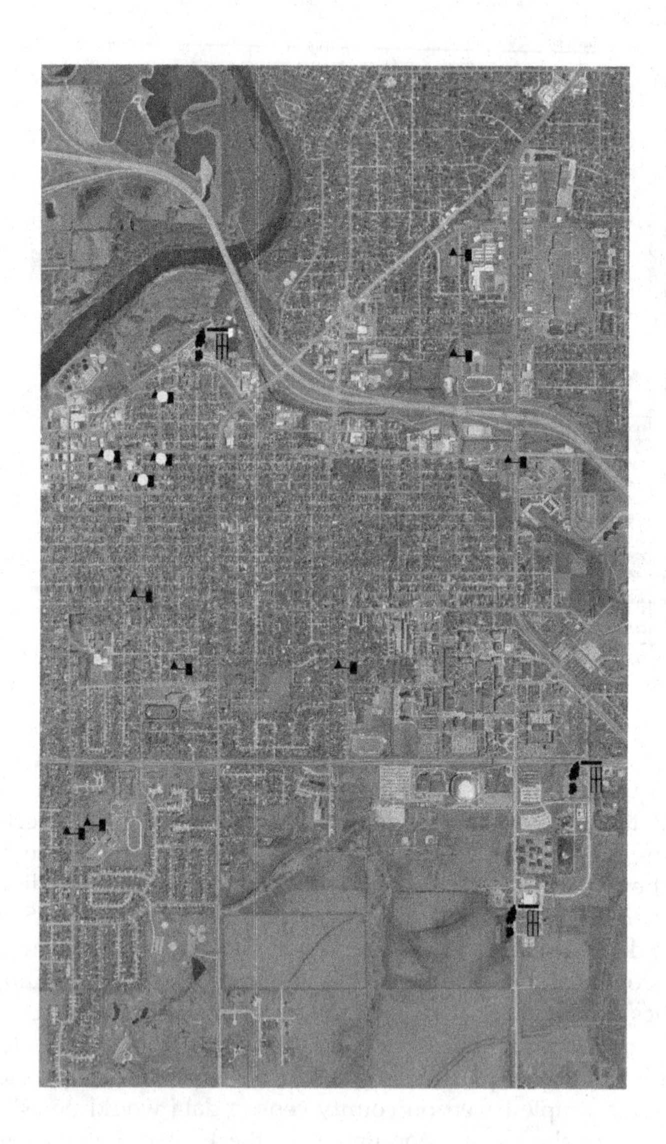

FIGURE 3.24
The Select By Location dialog in ArcGIS software and an example of a spatial selection result with highlighted points (highlighted dots on school icons) showing schools or daycares within 1000 meters of a federally permitted facility monitored for air quality.

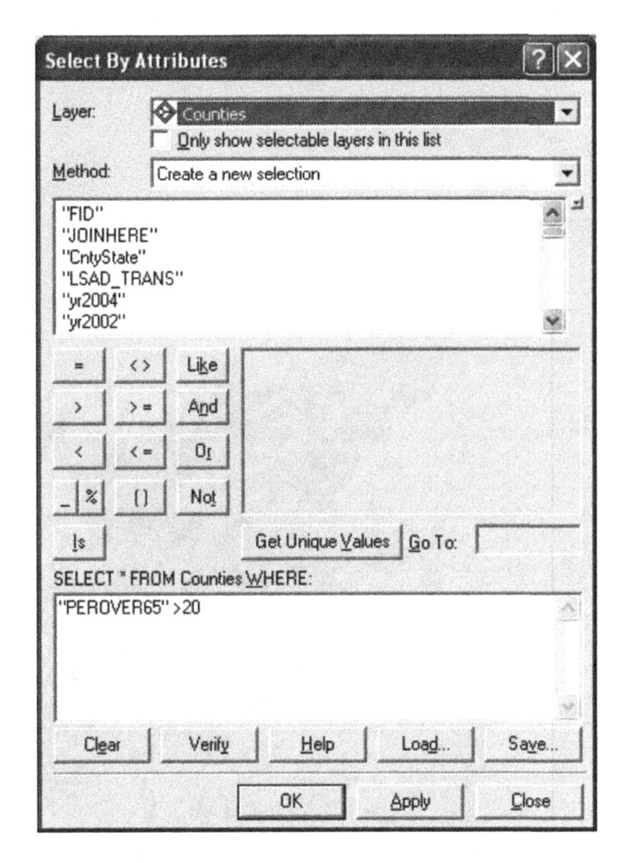

FIGURE 3.25a
The Select by Attributes dialog with a query for selecting those counties with the percentage of the elderly being greater than 20%.

on the population density field to determine the minimum, maximum, mean, and standard deviation of the population densities. They then could create a query in which they select sparsely populated counties according to their criteria. They could then recalculate the statistics for only the selected records to see the total number of elderly people living in sparsely populated areas. With both of these values, they could calculate a percentage of the elderly population living in sparsely populated areas (rural) as compared to urban areas. These data query and exploration functions are very useful in understanding underlying spatial patterns and where users or stakeholders need to gain a better understanding of the real-world situation that is represented by spatial data.

Some GIS software packages contain reporting and graph-producing capabilities useful for data exploration and for summarizing information. These capabilities are not usually as sophisticated as the

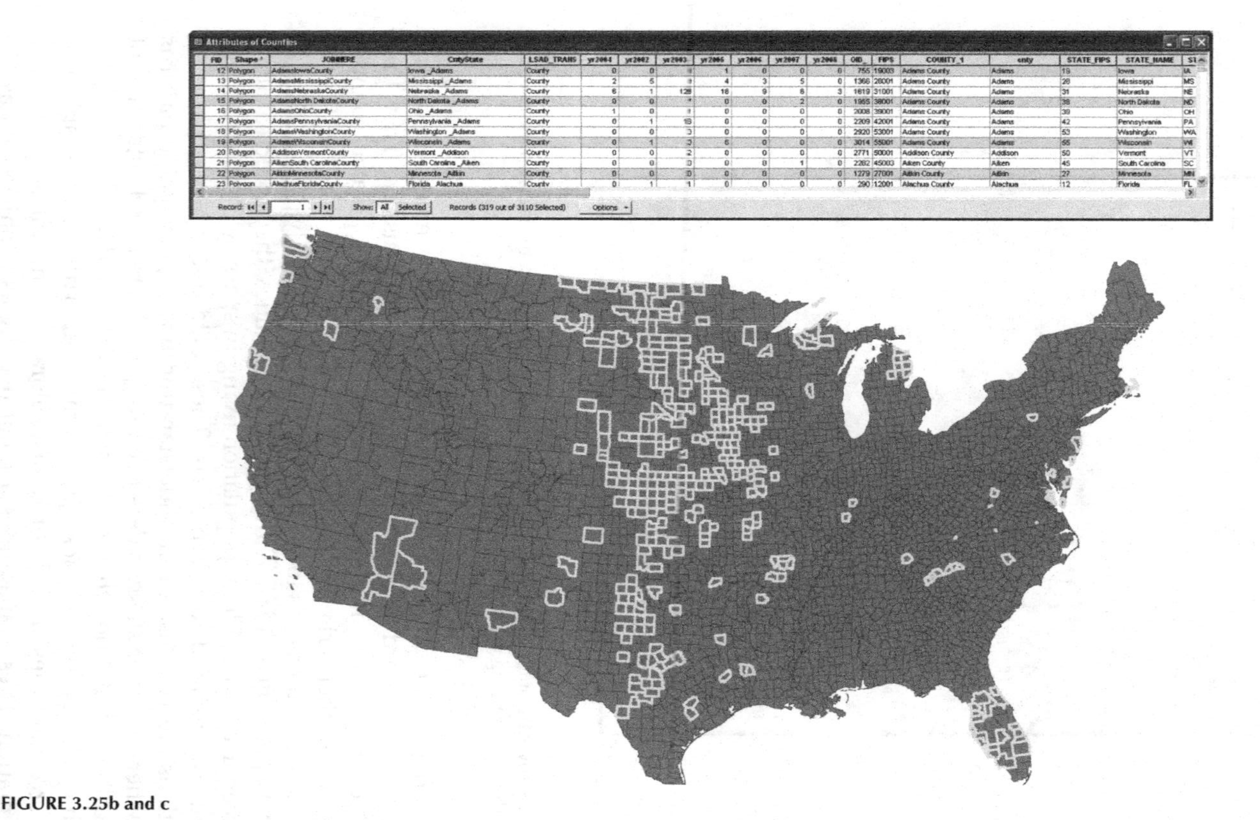

FIGURE 3.25b and c
The top panel shows the attribute table showing that 319 of 3010 counties meet the criterion. The map highlights that most of the counties that meet the criterion are in the middle sections of the U.S. and in Florida and Arizona.

cartographic functionality seen in the GIS software but are capable of producing basic graphs and reports. The reports or graphs can be produced from the attribute tables of spatial data and also from standalone tables. More recent software technology allows the incorporation of specialized software components for making graphs and reports. ArcGIS software has a Create Graph Wizard (Figure 3.26), which can be used to create a variety of graphs, including bar charts, line graphs, scatter plots, box plots, pie charts, and others. Similarly, ArcGIS has a report-creating tool built into the software as seen in Figure 3.27. Both the report and charts built with ArcGIS tools can be inserted into map compositions created in ArcMap.

3.4.11 Vector Processing and Analysis

Many real-world elements are more suitably or efficiently represented as vector features. There are also many specific spatial processing and analysis techniques that have been developed specifically for point, line, and polygon vector data. These types of operations are detailed in the following sections.

3.4.11.1 Buffering

One of the most basic vector processing functions is *buffering*. In a buffering operation, a user will indicate a distance at which to create a buffer around features in a point, line, or polygon feature class. The GIS software uses the x and y locations that define the given feature and a buffering algorithm to create polygons around each of the features (Figure 3.28). An individual buffer polygon created for each feature or overlapping buffer polygons can be dissolved into a single polygon. The distance used for buffering does not have to be uniform but can also be varied based on values from a field in the attribute table of the feature class. In addition, multiple buffers can be produced. For example, an analyst interested in seeing the population falling in a 10- and 50-km radius around a nuclear power plant could produce multiple buffers. There are many potential applications in which buffering would be useful. A few examples include eliminating areas within 1000 m of schools from the opening of restricted businesses such as liquor stores, removing land from consideration for the development of a new landfill if it is beyond a certain distance from a major road, or restriction of certain activities around streams to protect stream water quality. The spatial data created from a buffering operation is often subsequently combined with other data for further analysis.

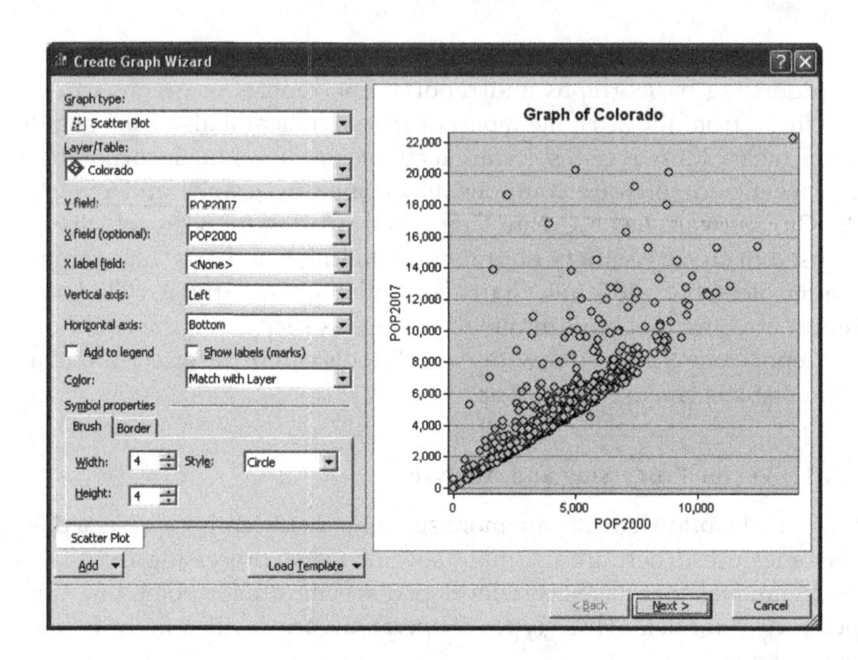

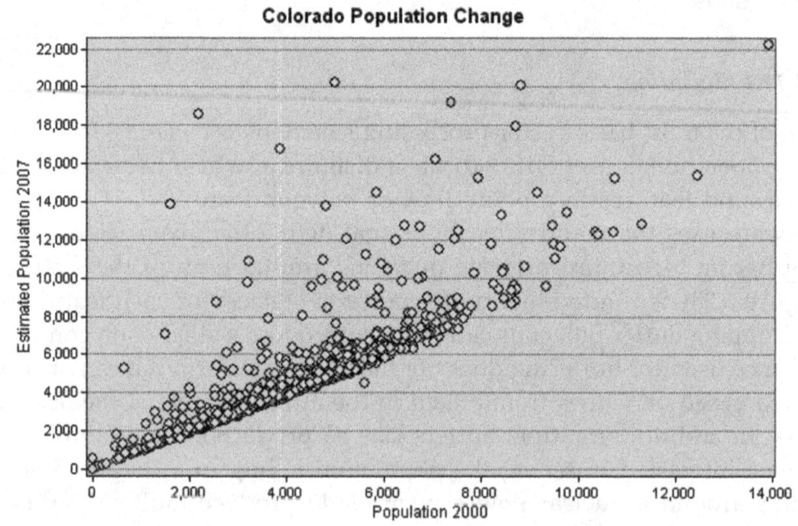

FIGURE 3.26
The ArcGIS graph wizard and an example scatter plot graph created from that wizard.

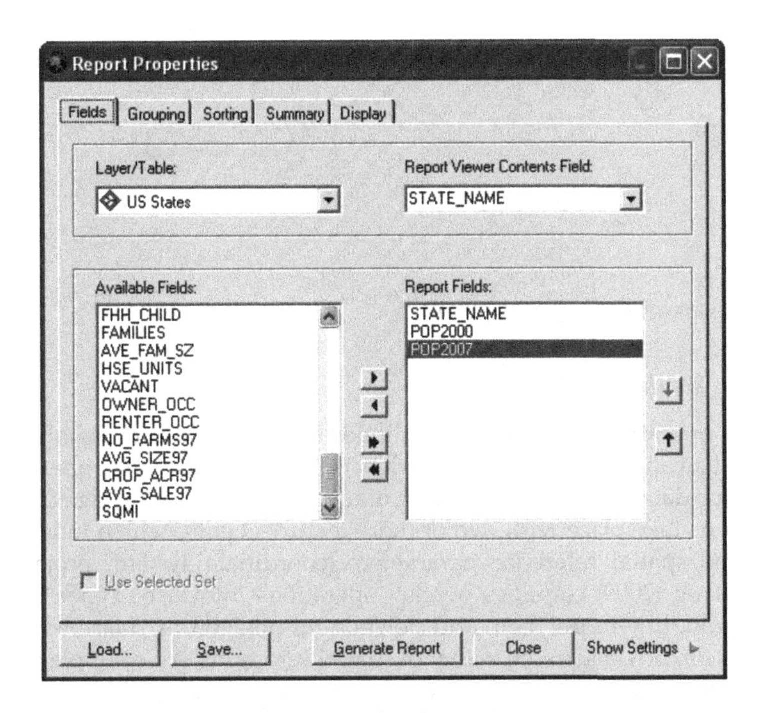

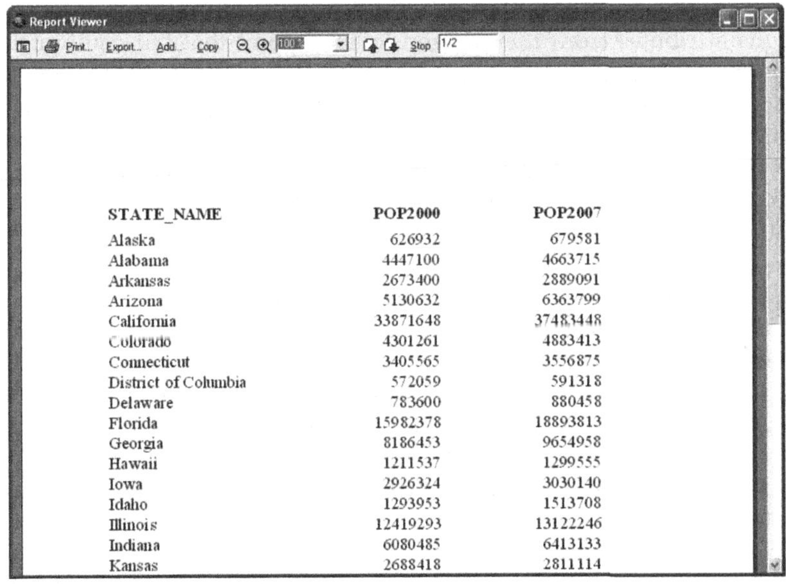

STATE_NAME	POP2000	POP2007
Alaska	626932	679581
Alabama	4447100	4663715
Arkansas	2673400	2889091
Arizona	5130632	6363799
California	33871648	37483448
Colorado	4301261	4883413
Connecticut	3405565	3556875
District of Columbia	572059	591318
Delaware	783600	880458
Florida	15982378	18893813
Georgia	8186453	9654958
Hawaii	1211537	1299555
Iowa	2926324	3030140
Idaho	1293953	1513708
Illinois	12419293	13122246
Indiana	6080485	6413133
Kansas	2688418	2811114

FIGURE 3.27
Report Properties dialog in ArcGIS software and an example report.

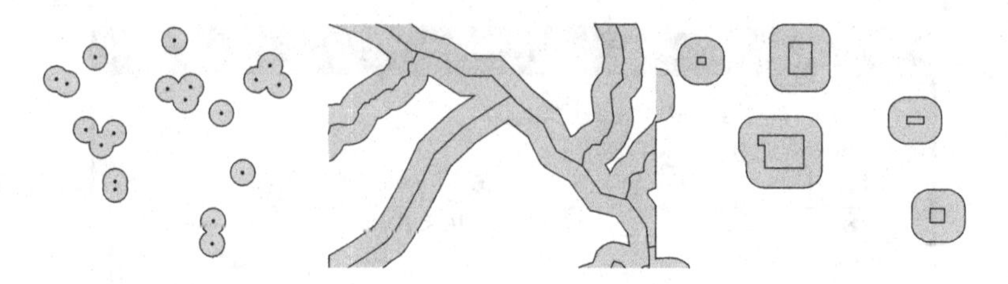

FIGURE 3.28
Buffering of point, line, and polygon features.

3.4.11.2 Spatial Overlay

Overlay operations are one of the unique and important functionalities in spatial data analysis. There are a variety of geometrical overlay operations for vector data that are available in many GIS software packages. These operations take place with two or more feature classes, which must have the same spatial reference parameters (coordinate system, projection, etc.). Chang (2009) classifies overlay operations into three types: point-in-polygon, line-in-polygon, and polygon-on-polygon. A *point-in-polygon* analysis already discussed is that of spatial selections. An example of this operation is finding how many confined animal feeding operations fall in a given watershed. Another point-in-polygon analysis is when the GIS assigns attributes from the polygon to each point that falls within that given polygon. This is called a *spatial join*. This operation could be useful for investigating relationships between different features. An example would be medical entomologists collecting ticks, which are responsible for spreading Lyme disease, and recording tick locations as points using GPS. Attributes from soils and vegetation data could be associated with the tick data with a point-in-polygon analysis to see if these ticks occur more commonly in areas with wetter soils and specific vegetation types. In a *line-in-polygon* analysis, such as an intersection operation, the line segments are altered based on the polygons that overlay them. Figure 3.29 demonstrates the result of intersecting a river with a town boundary, which could be useful for the city government. Depending on the specific operation carried out in the GIS, intersections mean that the resulting line feature class gains the attributes of the polygon. When the features are clipped, the attributes from the polygon are not incorporated in the new line features attribute table.

There are many *polygon-on-polygon* spatial operations available in GIS software. These include clip, union, intersect, identity, split, erase, and others. Graphical representations of these operations are shown in Figures 3.30 and 3.31. The difference between the *intersect* and *union* operations is that the latter preserves all features from the inputs while the former preserves only those that overlap from the two sets of features. There are many

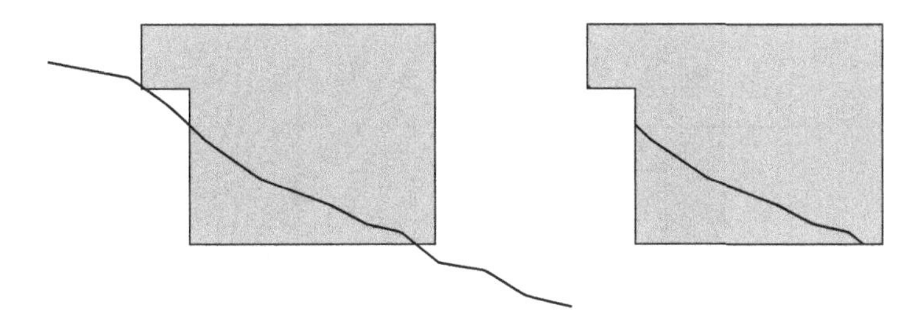

FIGURE 3.29
Line-in-polygon intersect operation in which the result is only the part of the line that intersects the polygon.

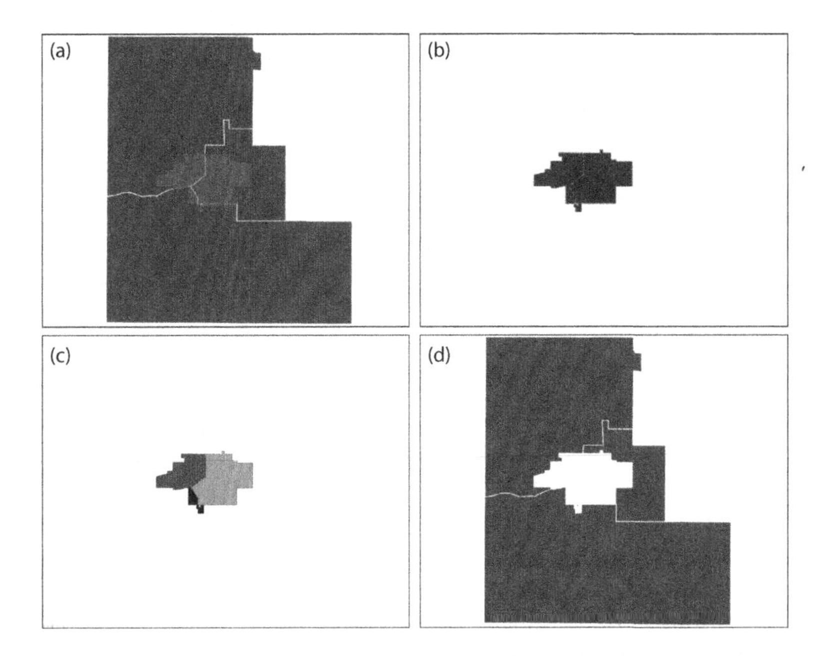

FIGURE 3.30
(a) Separate input polygon features classes and results of (b) clip, (c) split, and (d) erase operations.

applications for each of these operations. Intersect operations could be used with a buffered river polygon to investigate land uses that fall within that buffer area. The *clip* operation is often used to limit a spatial dataset to a specific area of interest, such as clipping a land cover feature class to a single administrative boundary. The *split* operation is similar to the clip function, except the clip is based on each individual feature, producing a new feature class for each splitting feature (Figure 3.30c). Thus, a land cover feature class could be clipped into many separate feature classes

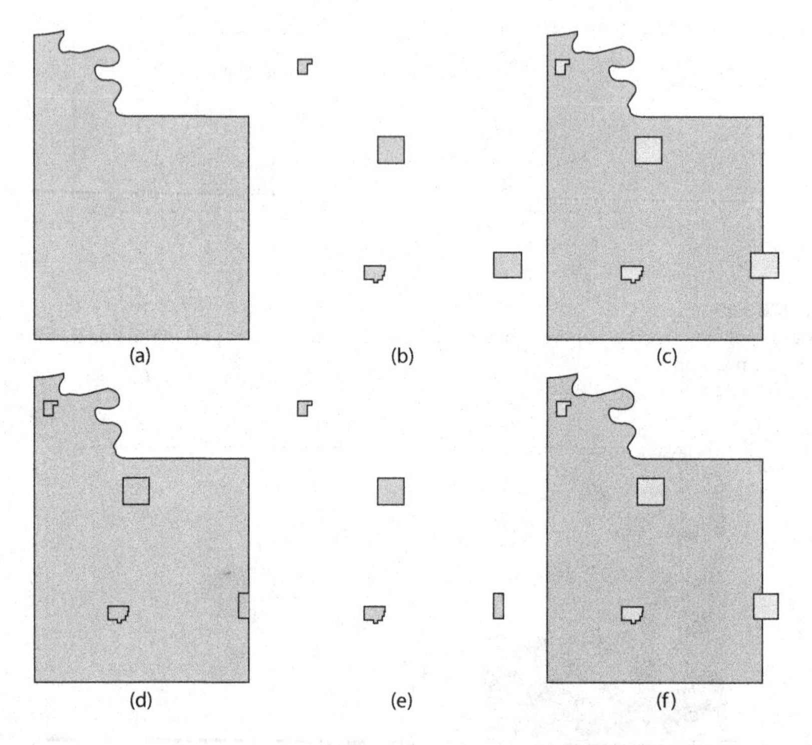

FIGURE 3.31
(a) Input features (a and b) and overlain on right (c) and results of (d) identity, (e) intersect, and (f) union.

using multiple administrative boundaries. An *erase* operation could be used to identify areas in a county that are not part of incorporated city boundaries within that county. The *symmetrical difference* operation might be used to examine areas that fall outside certain zoning and soil types as a way of looking at potentially developable land.

There are other spatial operations that can be carried out on point, line, and polygon vector data in which features between feature classes might not directly overlap. There are also spatial operations that can be carried out on a single feature class. *Merge* is a common operation in which two separate feature classes are combined into one resulting feature class. An example of this would be to merge two adjoining county river feature classes into a single feature class (Figure 3.32). The *dissolve* operation aggregates features based on a common attribute. In ArcGIS, these spatial operations are organized as tools in toolboxes (Figure 3.33). These tools can easily be incorporated into spatial processing models developed with ModelBuilder in ArcGIS. ModelBuilder can be used as a generic SDSS generator and is discussed in more detail in Chapter 5.

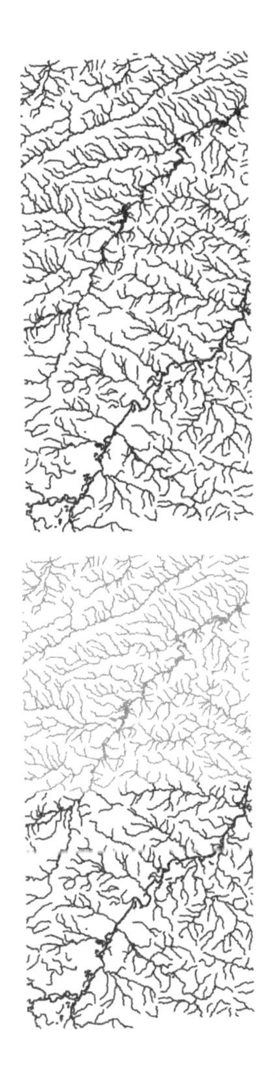

FIGURE 3.32
Left: Two separate river feature classes. Right: River feature classes merged into a single feature class.

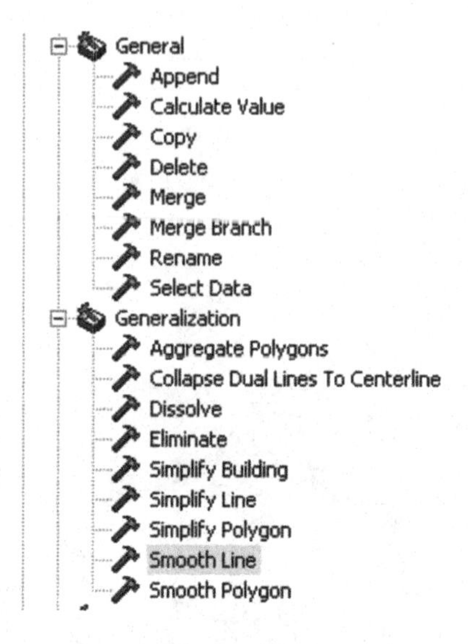

FIGURE 3.33
Some of the spatial operation tools in ArcGIS 9.3 General and Generalization toolboxes.

3.4.11.3 Pattern Analysis and Spatial Statistics

Functions in GIS for pattern analysis and spatial statistics provide great insight into spatial relationships from the real world. *Pattern analysis* is the use of quantitative methods for describing and analyzing the distribution of spatial features (Chang 2009). Pattern analysis can be considered a data exploratory activity that leads into more formal analysis carried out using spatial statistics. Given a set of geographic features, such as points representing crime locations, an analyst might want to know if these points show a random, dispersed, or clustered distribution. There are numerous algorithms useful for examining spatial patterns, some of which have been incorporated into GIS software. Example output from ArcGIS (Figure 3.34) demonstrates a measure of spatial autocorrelation (or whether nearby features are similar) for a given feature class.

Although it is very useful to know that there is some clustering, it is also useful to demonstrate these patterns by mapping the clusters themselves. In the last decade, spatial statistics tools moved from specialized standalone packages into GIS software. Various algorithms have been incorporated into GIS tools that help to identify spatially where hot spots or clusters exist. An example of output for one of these algorithms in ArcGIS is shown in Figure 3.35. The tool actually operates on the centroids of the polygons and finds whether there is similarity, no pattern, or dissimilarity between polygons close to each other. These types of tools are especially useful in such disciplines as public health, epidemiology, crime analysis,

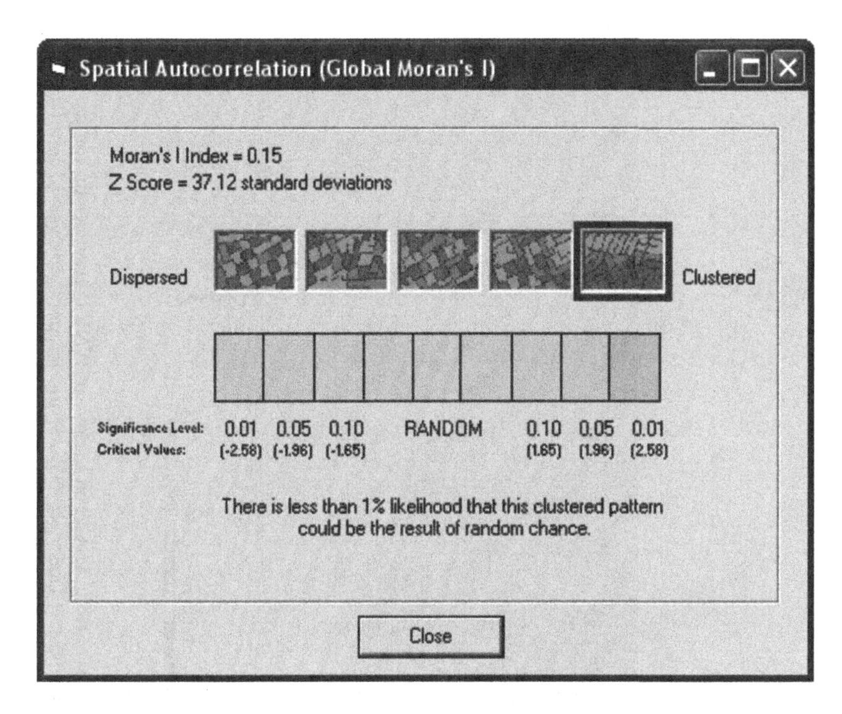

FIGURE 3.34
Graphic demonstrating level of spatial autocorrelation (ArcGIS 9.3).

business, and ecology. Li et al. (2005) incorporated spatial statistics tools in a typhoon insurance pricing SDSS.

3.4.11.4 Routing and Network Analysis

Most computer- and Internet-savvy users today are quite familiar with routing and network applications, as they show up in many different Web sites such as Google Maps, Yahoo! Local Maps, Microsoft Bing Maps, and many others as well as in vehicle-mounted GPS units. Network routing algorithms similar to those in these applications are also present in GIS. These algorithms find the shortest path between any two points along a network. The algorithms use attribute information of the line segments in the network, such as speed limits, as well as the geometrical information on the line features themselves to calculate the most efficient route. Important to these algorithms is how they handle information at nodes. Nodes are points at the intersections of two or more lines that denote either the beginning or end of a line (Demers 2009). These are important in network traffic applications because they can be used to represent traffic signs and signals in a networking algorithm. Network analysis in GIS is also used in modeling utility networks. Network techniques and data

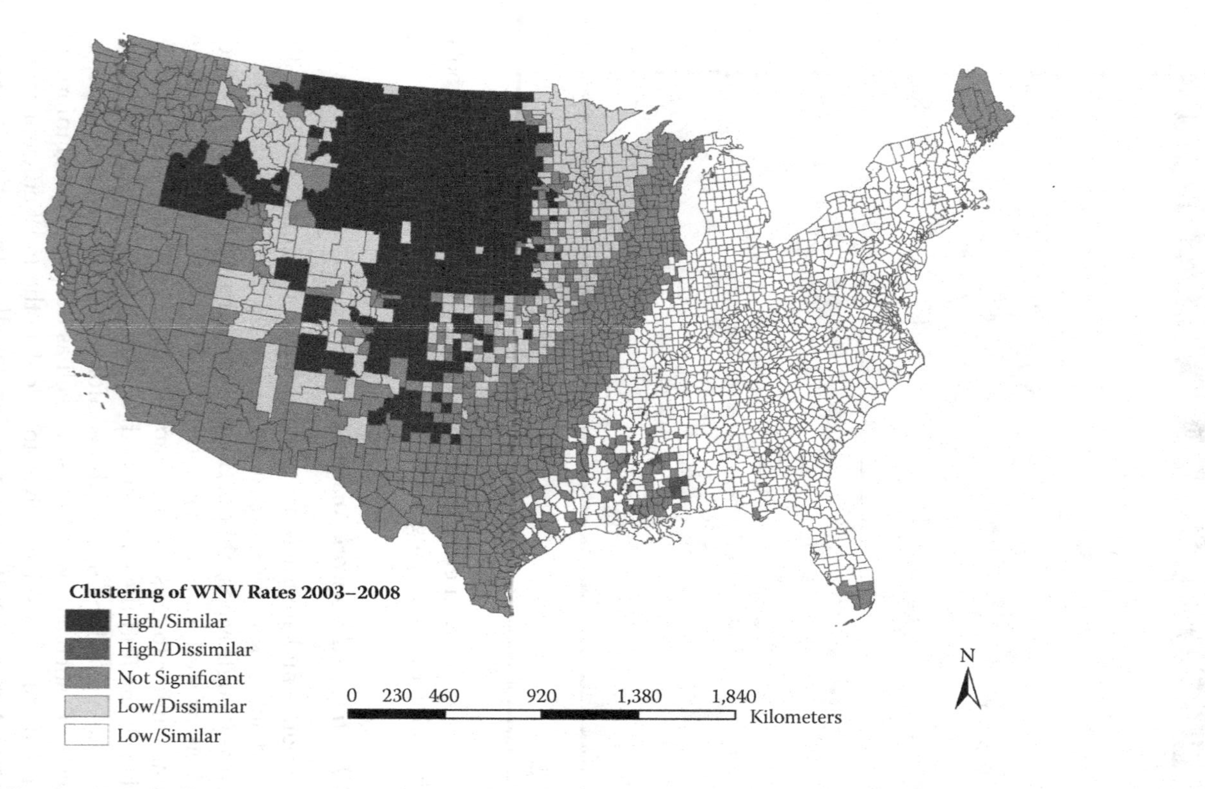

FIGURE 3.35
(See color insert following page 74.) Results of running spatial autocorrelation tool from ArcGIS on the rate of WNV occurrence by county in the contiguous United States from 2003–2008 (From Sugumaran R., S. R. Larson,, and J. P. DeGroote, 2009. Spatio-temporal analysis of county-based human West Nile virus incidence in the continental United States. *International Journal of Health Geographics* 8:43.).

structures can also be used for allocation problems such as deciding the location of a fire station and the population that would be served from any given location based on a certain travel time. This type of operation could be applied in many ways, such as investigating hospital service areas or finding locations of businesses when considering travel times of potential customers. Kang et al. (2008) used the ArcGIS Network Analyst for modeling optimal transportation solutions in their poultry litter management SDSS.

3.4.12 Raster Data Analysis

A wide range of specialized data analysis and processing techniques operate on raster data. The simplicity of the raster data model allows a wide variety of operations to be carried out in a computationally efficient manner (Chang 2009). Operations involving raster data can be carried out on a single raster, on multiple raster datasets, or even with a combination of vector and raster data. In the latter case, the vector data is usually converted to raster data behind the scenes or unseen to the user before the desired operation is carried out. Generally, these raster operations can be classified into local, neighborhood, or zonal operations. The basics of each of these types are discussed below and illustrated with practical examples.

3.4.12.1 Local Operations

Local operations are those that operate on a cell-by-cell basis either on a single raster or multiple rasters. There are a variety of local numeric operations that can be used, including arithmetic, logarithmic, trigonometric, and power functions (Chang 2009). Figure 3.36 represents a local raster operation in which all cell values are multiplied by two to derive a new raster. This basic idea can be applied with a wide variety of mathematical techniques and complex formulas. Many standard local operations can be used for single rasters, including data conversion operations such as converting floating point rasters to integers and a variety of logical operators. A common local raster operation used for generalizing data is reclassification. There are many instances in which a user would like to go from many classes to fewer classes or to reclassify a range of numeric values to a more easily understandable set of classes. An example of reclassifying a range of topographic slope values into three classes of flat, moderate, or steep is shown in Figure 3.37. The user indicates a range for each new class, and then the program assigns the new class value on a cell-by-cell basis. Local mathematical and logical operations can be carried out on multiple rasters also. These types of operations are often called *map* or *raster algebra*, and in ArcGIS these operations (as well as those on single rasters) are formulated in the Raster Calculator (Figure 3.38). The user can

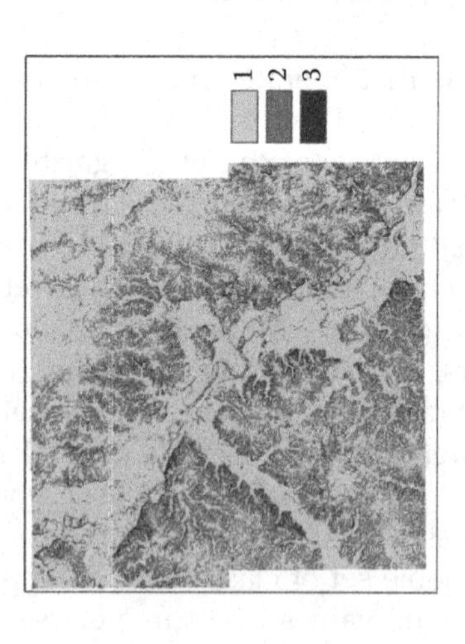

× 2 =

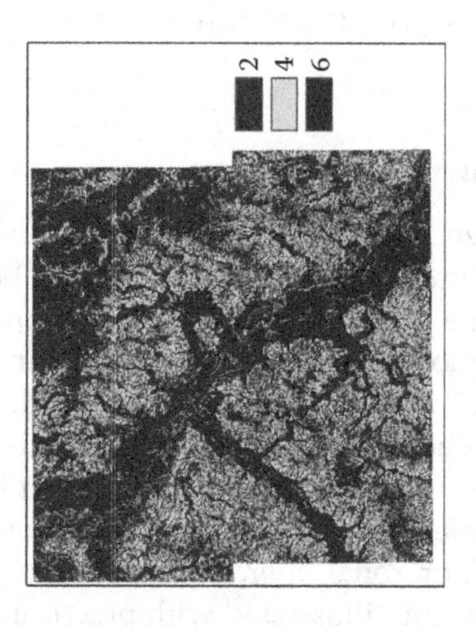

FIGURE 3.36
A local multiplication operation on a single raster to produce a new raster.

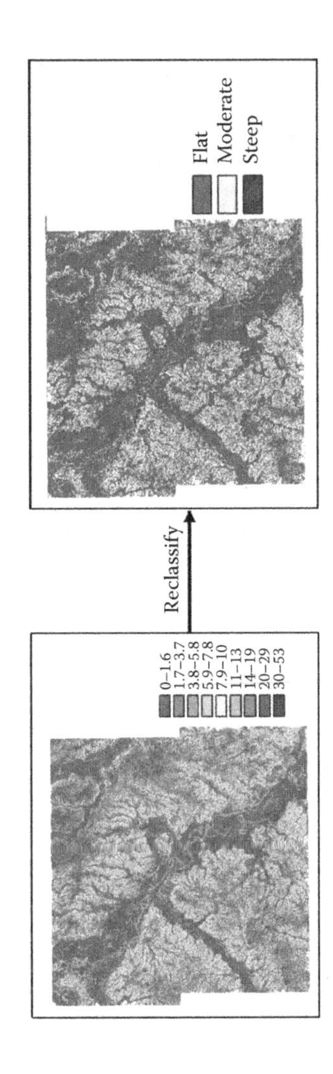

FIGURE 3.37
An example of a reclassification operation on a single floating point raster to produce a new raster with three classes.

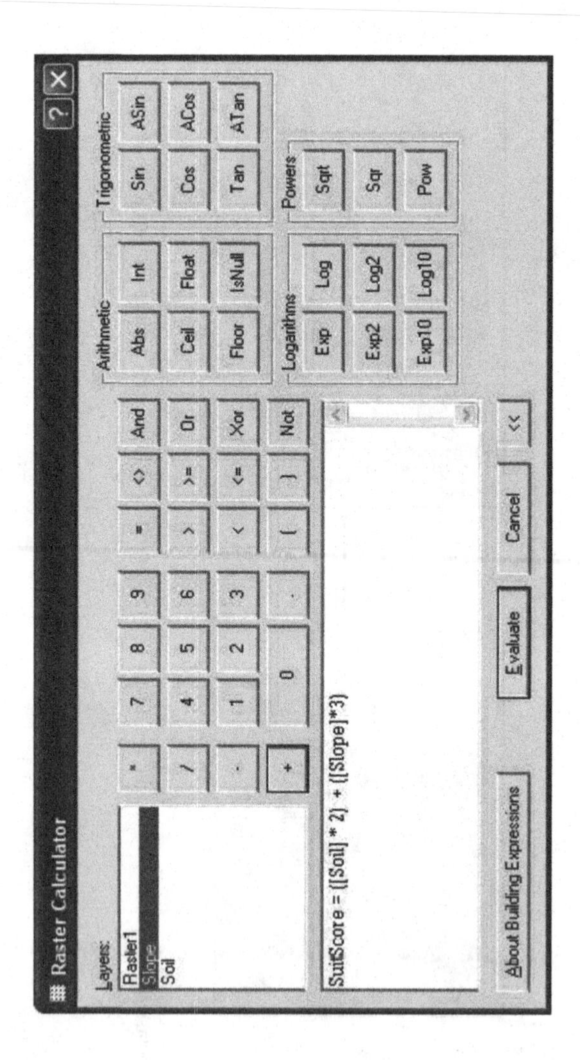

FIGURE 3.38
The Raster Calculator dialog in ArcGIS software.

select the raster from the upper left panel and then choose functions from those shown or from many other functions that can be accessed. The user can string together multiple rasters and functions into one final equation to produce a new output raster. Thus, a tremendous variety of local raster operations can be carried out for a variety of applications. An example would be a land suitability analysis in which soil and slope rasters could be reclassified into ranks of favorability (e.g., 1–5 with 5 being most suitable). Each raster could be multiplied by a weight, and these would be summed together to get a final suitability score (Figure 3.38).

3.4.12.2 Neighborhood Operations

In contrast to local operations, neighborhood operations examine not only the cell of interest but also surrounding cells. The focal cell is the one for which the calculation is carried out, but the resulting value is dependent on values from cells in some defined neighborhood. The neighborhoods can be of varying sizes and shapes (e.g., rectangle, circle). Generally, if the center of any cell falls within the neighborhood shape, then the cell is considered in the analysis (Chang 2009). The neighborhood functions use cell values within the neighborhood to compute a new value for the focal cell. The algorithms consider each cell in the neighborhood for the focal cell and then move the neighborhood to the next focal cell. Neighborhood statistical calculations include minimum, maximum, range, sum, mean, standard deviation, variety, majority, minority, or median. One reason for carrying out these types of calculations is data simplification for a smoothing out of the variation in the data (Figure 3.39). These types of operations are often used in image processing of remotely sensed imagery and are sometimes called *filtering, convolution,* or *moving window* operations (Chang 2009).

There is a whole class of terrain or topographic operations, carried out on digital elevation models (DEM), which use neighborhood-based algorithms. These operations include slope, aspect, hillshade (used for topographical visualization), filling sinks, and flow direction. Slope and aspect calculations are commonly used in site selection, environmental, and ecological applications. Sink-filling algorithms correct errors in DEMs. These algorithms and flow direction calculations are necessary steps for defining watershed boundaries, which are often used in hydrological and water quality modeling SDSS applications.

3.4.12.3 Zonal Operations

Zonal operations calculate summary statistics from an input raster based on a set of cells with common values in another raster or based on a feature vector dataset. In ArcGIS, the statistics calculated are minimum,

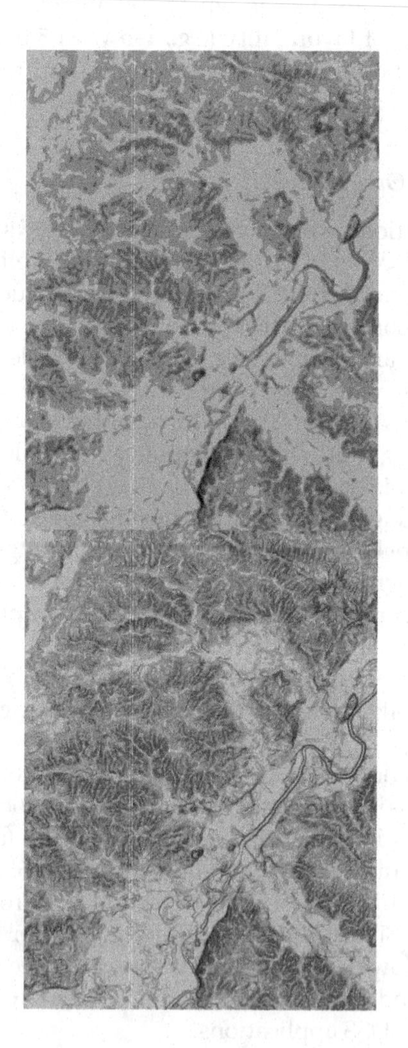

FIGURE 3.39
An example of a neighborhood mean statistic calculated with a 5 × 5 rectangular neighborhood on a slope raster. The left side represents the original raster, while the right represents the result of running the mean neighborhood function.

maximum, range, mean, standard deviation, and sum. An example of a zonal statistics operation could be to calculate the average slope from a slope raster based on vegetation or land use classes from a separate raster. Using this technique, a user could investigate what type of topographic condition is associated with a certain vegetation class or land use. Another example might be using a series of watershed boundaries to calculate summary statistics on the slope conditions in each of the watersheds. This type of example is illustrated in Figure 3.40, in which zonal statistics of a slope raster have been calculated in ArcGIS for six watershed boundaries. The output of the zonal statistics operation is a table storing the summary statistics and a chart representing a chosen statistic (in this case the mean).

3.4.13 Data Visualization and Cartography

Maps have been used for many centuries as a way of conveying information about space and the relationships between real-world objects. Geographic information systems software has revolutionized and facilitated the ease of visualizing spatial data in a digital environment and producing hard-copy maps. Cartography can be considered the making and study of maps in all their aspects (Robinson et al. 1995). Facets of cartography, such as projections and coordinate systems, were covered previously and form important components of cartography. The art of making maps is tremendously important and has been greatly democratized in the last several decades by the introduction of digital techniques. By the 1970s there were considerable investments already made into computer-assisted cartography (Burrough and McDonnell 1998). However, it was not until later, with the great growth of GIS, that computer-assisted map-making became widely used by nonprofessional cartographers. One of the main areas of emphasis for GIS software development has been in the development of effective cartographic tools. The cartographic functionality in modern GIS software provides users with great ability to produce a wide variety of maps. However, it is still the ultimate responsibility of the mapmaker to produce maps that efficiently and effectively convey spatial information. Due to GIS capabilities for easily making maps, there are greater risks in unqualified individuals making maps that are in error or are misleading.

There are many techniques for visualizing features (i.e., points, lines, polygons, and raster) in a variety of ways to convey certain information. Features can be symbolized based on a combination of color, size, texture, shape, pattern, and so on. For example, points could be represented by many different symbol types and colors, as seen in the ArcGIS 9.3 software dialogs for selecting symbols and colors (Figure 3.41). There are thousands of different symbols available, such as generic shapes, but also

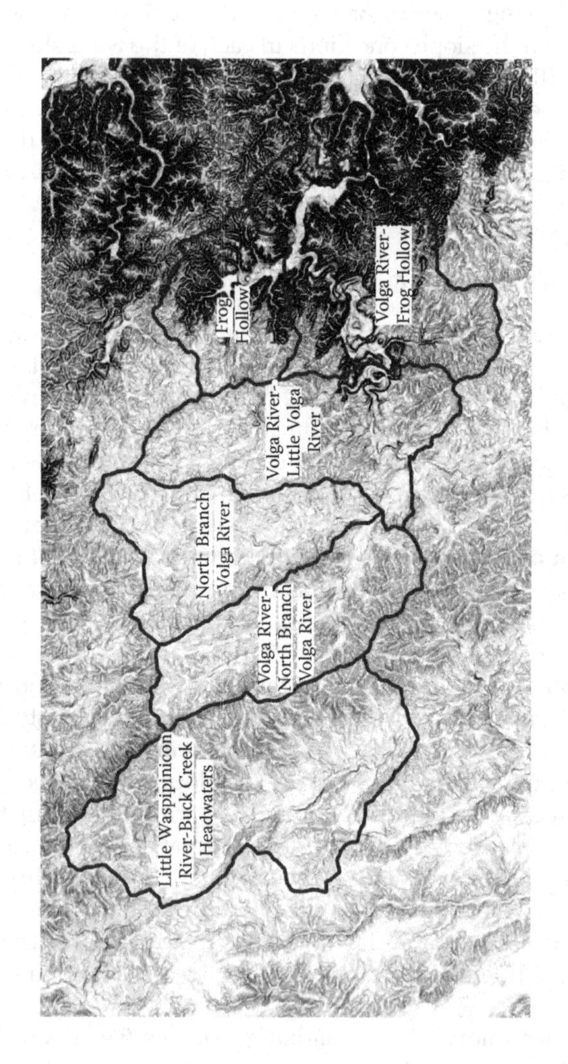

FIGURE 3.40a
The original raster dataset (slope raster) and watershed polygon boundaries used in an ArcGIS Spatial Analyst zonal statistics operation.

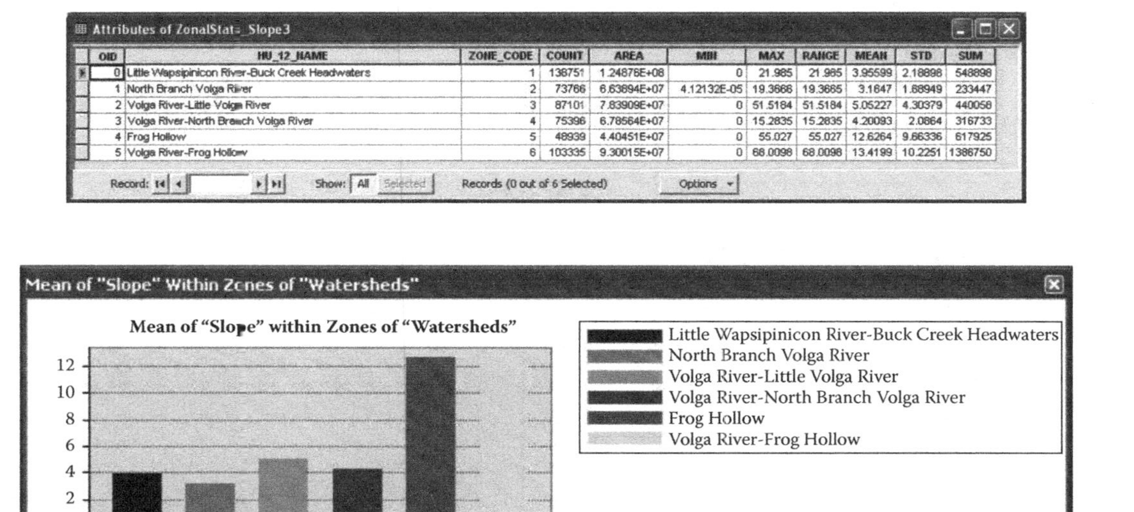

Attributes of ZonalStats_Slope3

OID	HU_12_NAME	ZONE_CODE	COUNT	AREA	MIN	MAX	RANGE	MEAN	STD	SUM
0	Little Wapsipinicon River-Buck Creek Headwaters	1	138751	1.24876E+08	0	21.985	21.985	3.95599	2.18898	548898
1	North Branch Volga River	2	73766	6.63894E+07	4.12132E-05	19.3666	19.3665	3.1647	1.68949	233447
2	Volga River-Little Volga River	3	87101	7.83909E+07	0	51.5184	51.5184	5.05227	4.30379	440058
3	Volga River-North Branch Volga River	4	75396	6.78564E+07	0	15.2835	15.2835	4.20093	2.0864	316733
4	Frog Hollow	5	48939	4.40451E+07	0	55.027	55.027	12.6264	9.66336	617925
5	Volga River-Frog Hollow	6	103335	9.30015E+07	0	68.0098	68.0098	13.4199	10.2251	1386750

FIGURE 3.40b and c
Top: Output table of zonal statistics operation with summary statistics from slope raster stored for each individual watershed. Bottom: Chart of mean slope value for each watershed.

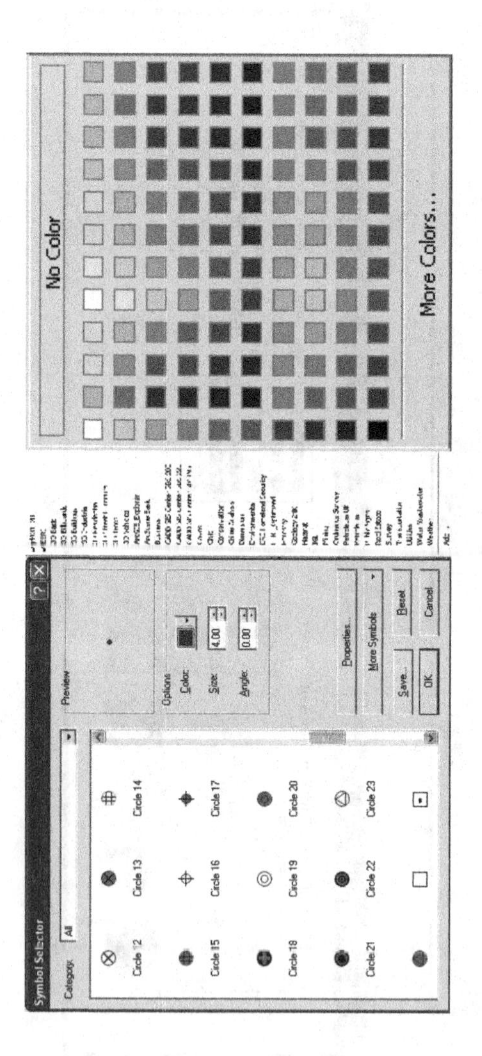

FIGURE 3.41
Point symbol selector, list of extra categories of symbols, and color palette available in ArcGIS 9.3.

those specific to certain disciplines such as crime, business, transportation, utilities, and so on. In Figure 3.41, it is seen that the user can also adjust the size of the symbol and the angle at which it is represented. Similarly, Figure 3.42 demonstrates many options for representing line and polygon features. Figure 3.43 demonstrates the Style Editor used for symbolization purposes in the free GIS software called uDig (User-friendly Desktop Internet GIS). Most GIS software applications also allow the importing of custom symbology allowing users to develop highly specialized representations of spatial features in their maps.

One of the most powerful aspects of GIS is the ability to store not only geometrical properties of geographic features but also attribute information associated with those features. GIS software has specialized functionality for varying the representation of individual spatial features in a given layer based on attributes held in the layer's attribute table. Feature attributes can be represented categorically or quantitatively depending on the type of attribute. There are many types of vector features which have categorical attributes that allow meaningful symbolization. For example, a roads network (Figure 3.44) could be symbolized based on the type of road (e.g., highway, city street) or land ownership parcels could be symbolized based on ownership type (e.g., commercial, residential, industrial). Raster data can also be symbolized based on categories as seen in Figure 3.45, which shows a land cover raster dataset. Numeric data can be represented categorically if there are a limited number of values. However, if there are many unique numeric values, it is better to use various classification methods that are available in GIS software for meaningful segmentation of the data into a set of classes. For example, the color of individual counties could be varied based on the population density of the county as seen in Figure 3.46. In this figure, population density for the contiguous United States is shown in ArcGIS 9.3 and the uDig software. There are different classification methods within GIS software. The user can set class breaks or the GIS software can do it automatically using statistical techniques. Point, line, and polygon layers can all be symbolized using these techniques if they contain numeric attributes. The color or size of symbols can be varied in the classification as seen in Figure 3.47, which shows the number of animals at confined animal feeding operations in the vicinity of rivers ranked by the Strahler stream order classification.

Cartographic functionality in GIS software can be used to produce simple or complex maps, which can then be exported to various digital image formats for easy distribution or printed as hard-copy maps of various sizes depending on the user's hardware. Map components, such as scale bars, titles, north arrows, legends, charts, tables, images, and other features, can be added to map compositions in GIS software (Figure 3.48). Some GIS software provides map templates intended to make the map-making process easier for the user. These programs allow the user to save

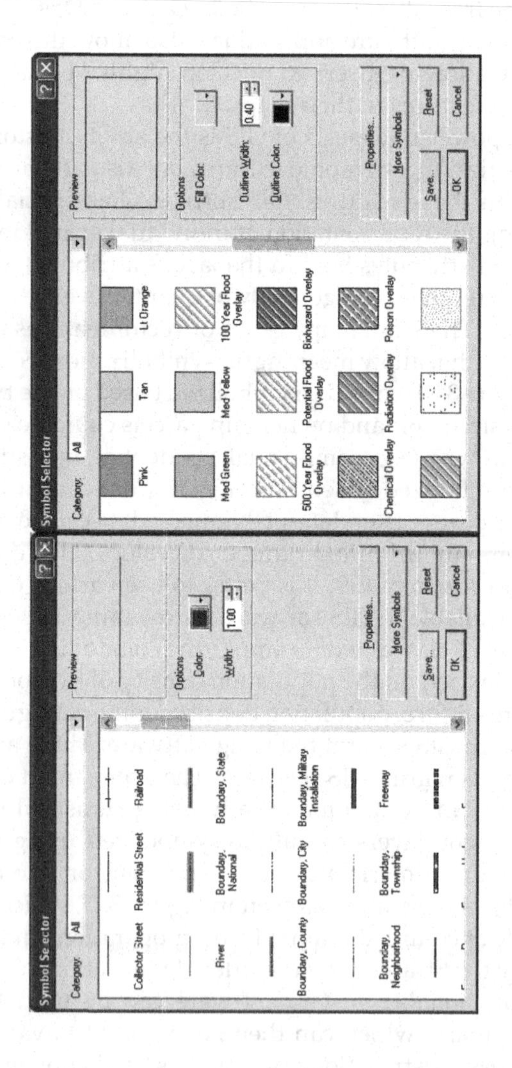

FIGURE 3.42
Line and polygon symbol selector dialogs from ArcGIS 9.3 software.

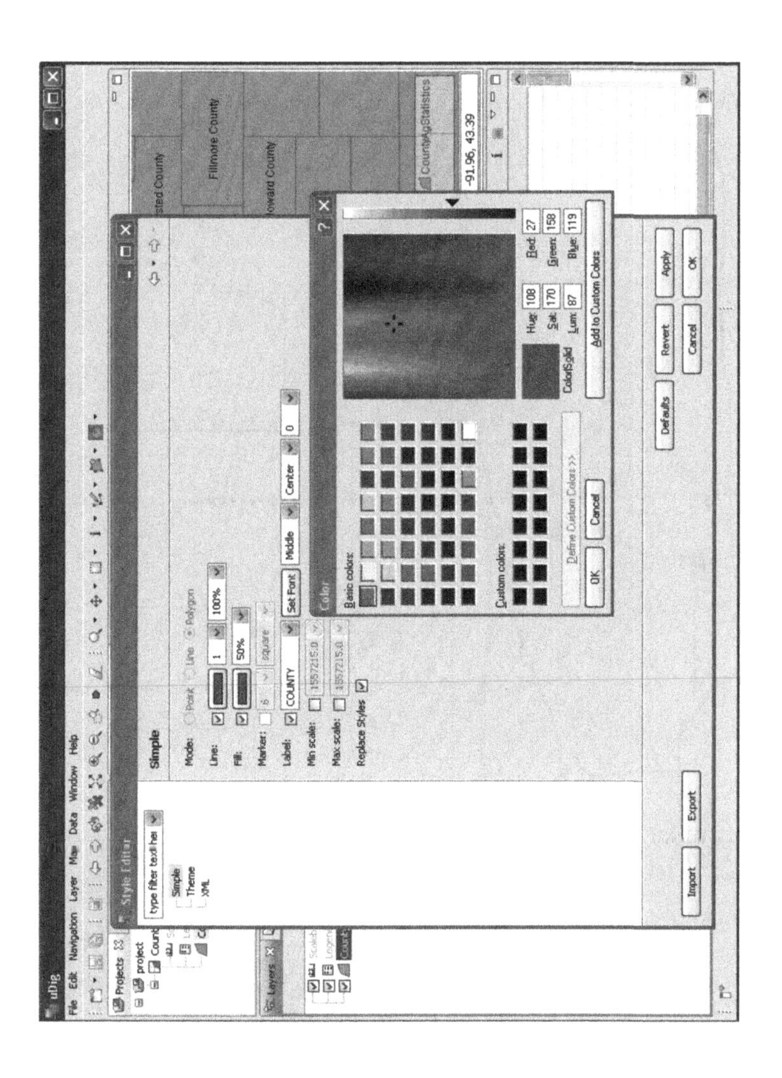

FIGURE 3.43
The Style Editor used for symbolizing spatial data in the uDig software.

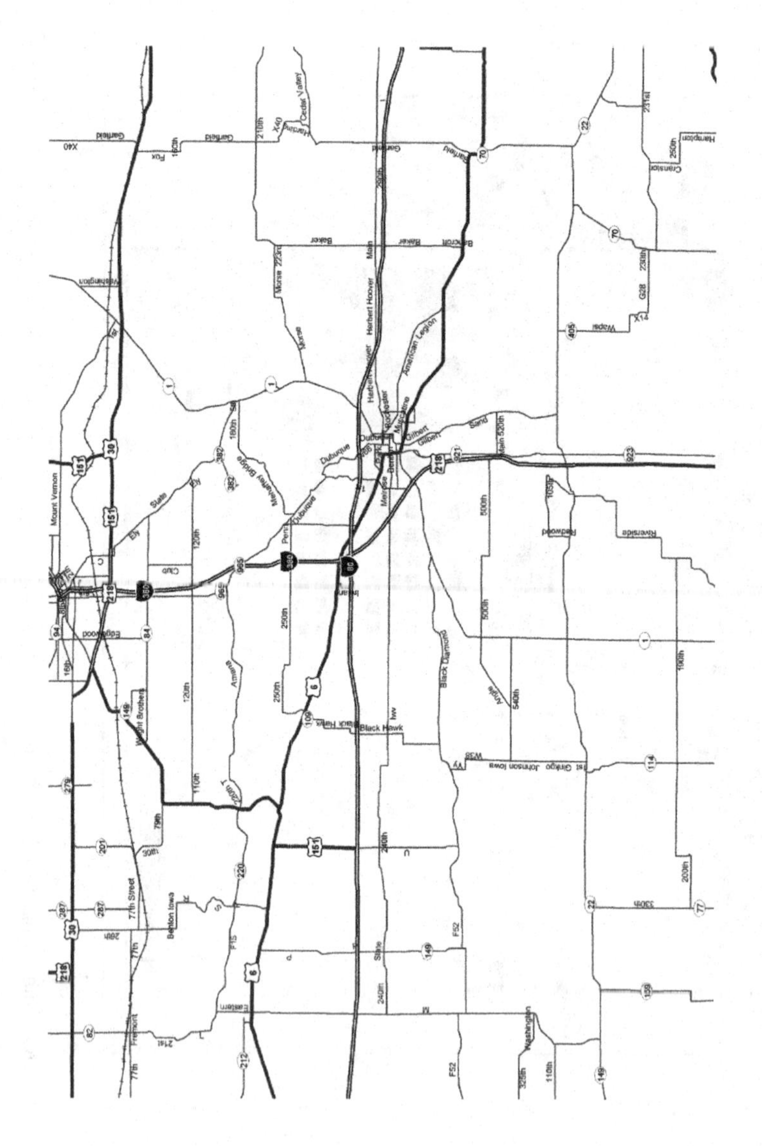

FIGURE 3.44
A map depicting a road network with different symbology based on type of road.

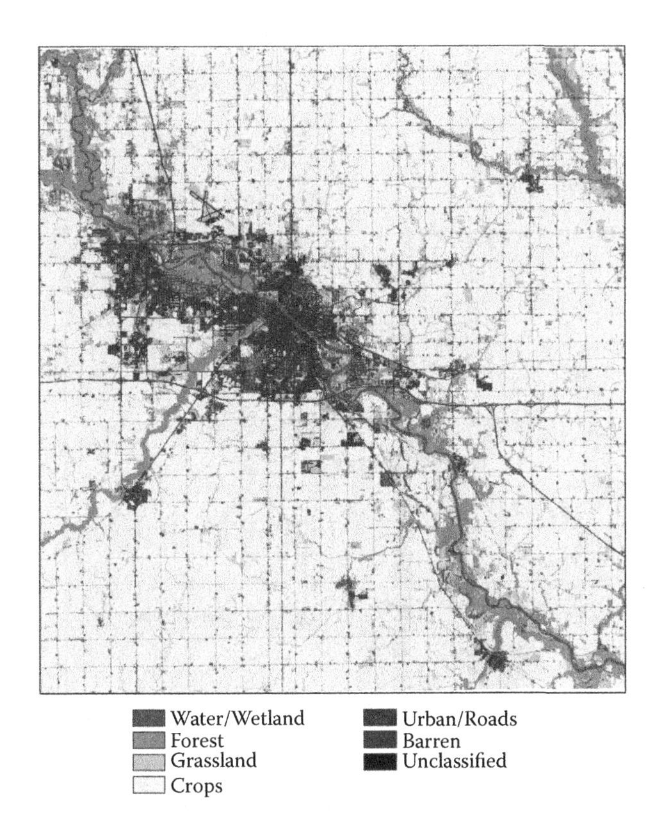

FIGURE 3.45
Categorical land cover raster map (ArcGIS 9.3).

custom templates in order to make it easier to produce a similar series of maps. There are other functionalities present in GIS software for producing effective maps. Labeling (such as seen in Figure 3.44) is often important for conveying information in a map, and a set of tools is often available for labeling features in a GIS program. The cartographic flexibility and functionality available in most modern GIS software provide a significant amount of functionality for a user to produce maps of varying degrees of quality. It is incumbent on GIS users to learn at least the basics of cartographic production in order to produce useful maps.

As greater hardware capability has been developed, more sophisticated spatial data visualization methods have been developed, including 3D techniques. A variety of spatial data are useful for viewing in 3D perspectives for various purposes. The use of DEM data for representing the shape of the surface of the Earth is common in conjunction with other spatial data, including vector features as well as remotely sensed imagery. In addition, 3D models of buildings, trees, and other landscape items can

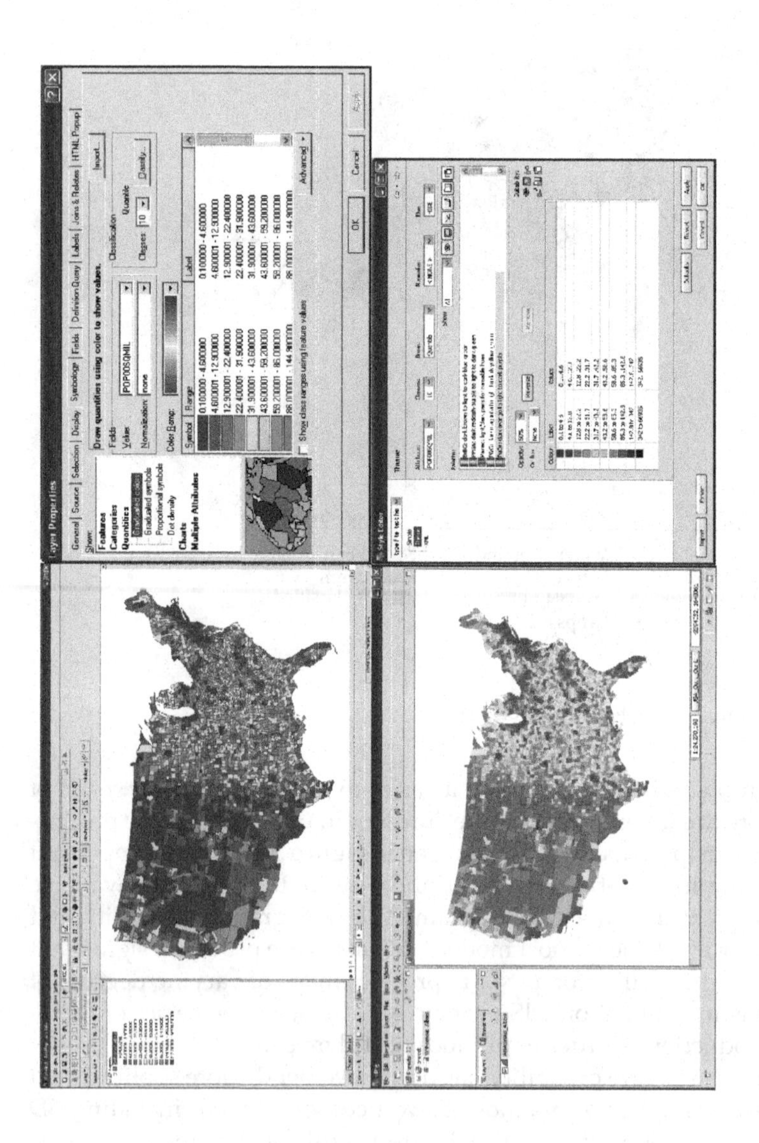

FIGURE 3.46
(See color insert following page 74.) Both maps represent population densities by county in the contiguous United States using a quantile method. The top images show the map and dialog for setting classification properties in ArcGIS 9.3, while the bottom two show the same for the uDig software.

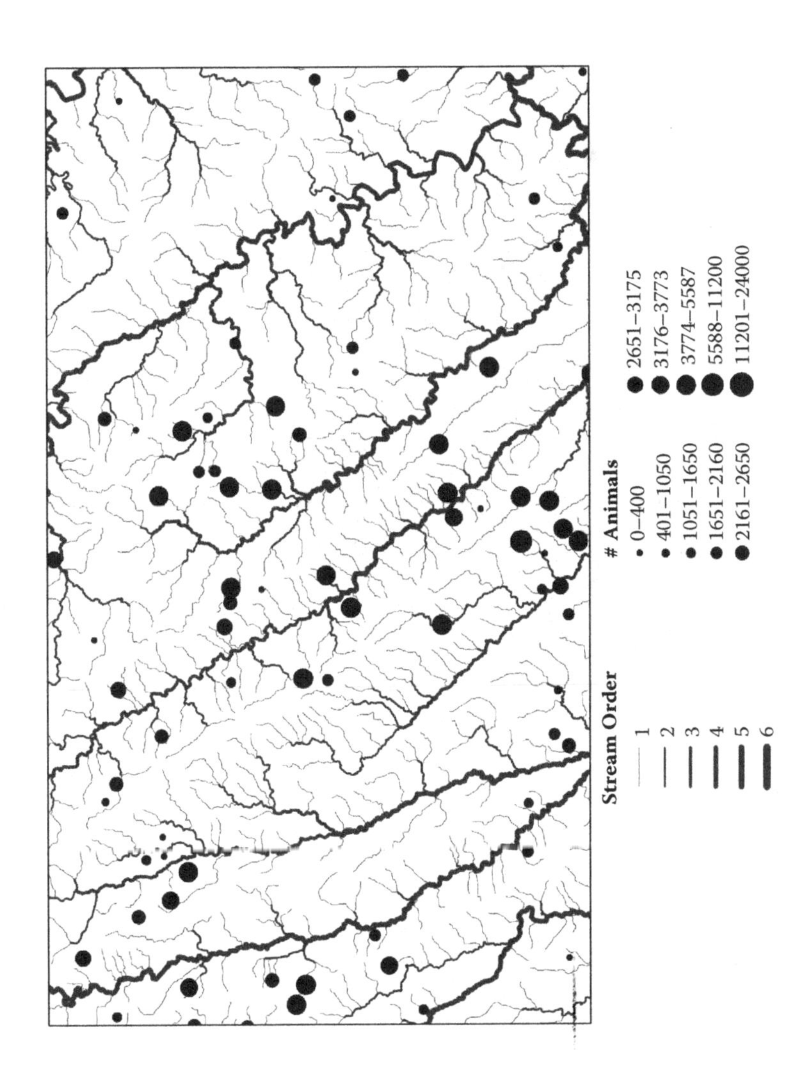

FIGURE 3.47
Graduated symbology for point and line features.

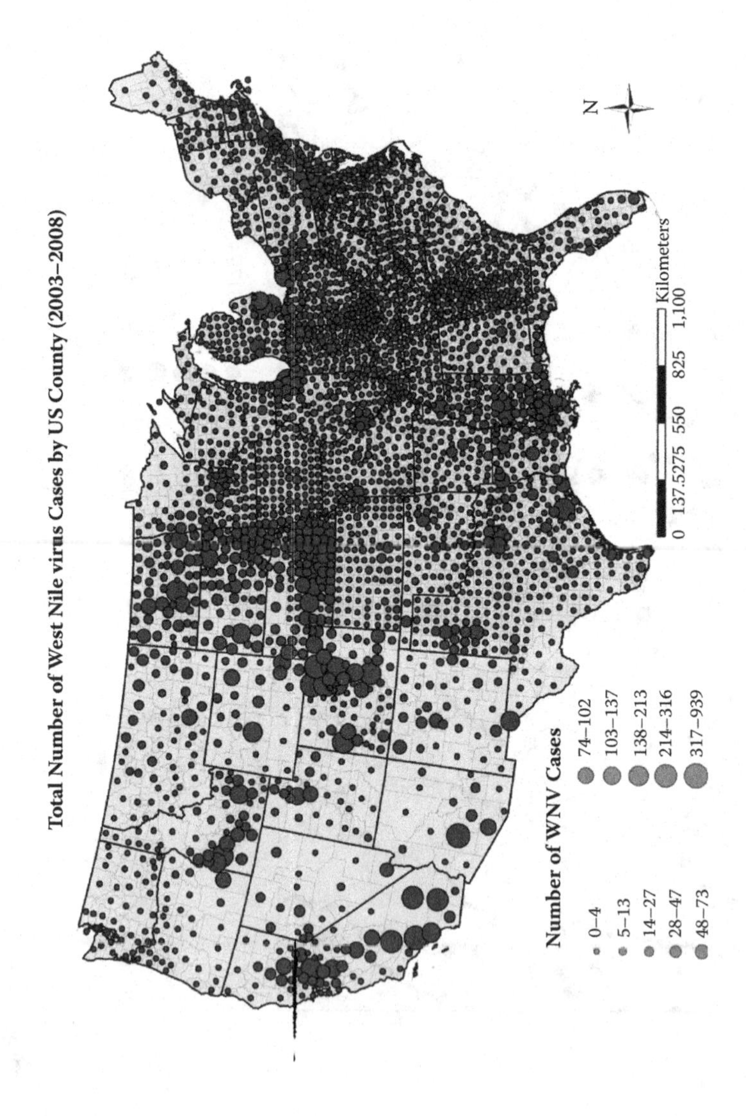

FIGURE 3.48
(See color insert following page 74.) An example map composition created using ArcGIS 9.3.

be placed on the landscape to visualize pseudo-realistic views of reality or potential future developments. The most common example of this type of 3D visualization is freely available in the form of virtual globes such as Google Earth, NASA World Wind, and Microsoft Bing Maps 3D. With these globes, users can move around the globe virtually with navigation tools that allow them to zoom in to a place on the Earth and then to rotate their view. They can also type in place names (e.g., cities, addresses, landmarks, national parks), and the application will zoom in to that location if it is stored in the database and a match is found. An example of zooming into the Grand Canyon in Google Earth is shown in Figure 3.49. In Figure 3.50, an oblique view of Manhattan in New York City is shown with 3D models of buildings as shown in Google Earth. There are many software packages for building and viewing 3D visualizations, some of which are extensions of GIS software such as ArcScene and 3D Analyst which are extensions to ESRI's ArcGIS software. These types of software allow a user to add a variety of data such as DEMs, aerial or satellite imagery, vector features, and 3D models of buildings or other objects. Users can dynamically visualize data in a variety of ways and also produce fly-throughs and animations of the landscape in the form of movie files. Figure 3.51 shows a screenshot of a 3D animation created in ArcScene to dynamically represent the Parkersburg, Iowa, tornado path that occurred on May 25, 2008. The visualization was created using a DEM, aerial photography, vector points representing building locations, 3D models to represent those buildings, a series of vector points with time stamps to simulate movement of tornado in time and space, and a 3D model of a funnel cloud. These types of animations are often used for visualizing potential future developments such as residential housing, wind farms, or new businesses. The use of 3D visualization techniques can provide constructive aids in the planning process.

3.4.14 GIS Software

There is a wide variety of both commercial and open source GIS programs available with varying range of functionalities. All of these software packages cannot be covered individually here. Rather, we will mention some of the most important and widely used software and provide a set of links in Appendix C for sites that provide coverage of the wide variety of software available. Some of the most widely used commercial GIS packages that have a significant amount of functionality include ESRI's ArcGIS, Intergraph's GeoMedia, Pitney Bowes MapInfo, Clark Lab's IDRISI, Manifold System GIS, and General Electric's Smallworld. ESRI has generally held the largest market share with its GIS software. ESRI's ArcGIS software is the product of evolution over several decades as ESRI has been producing commercial software since the early 1980s (Bolstad 2005). This software provides

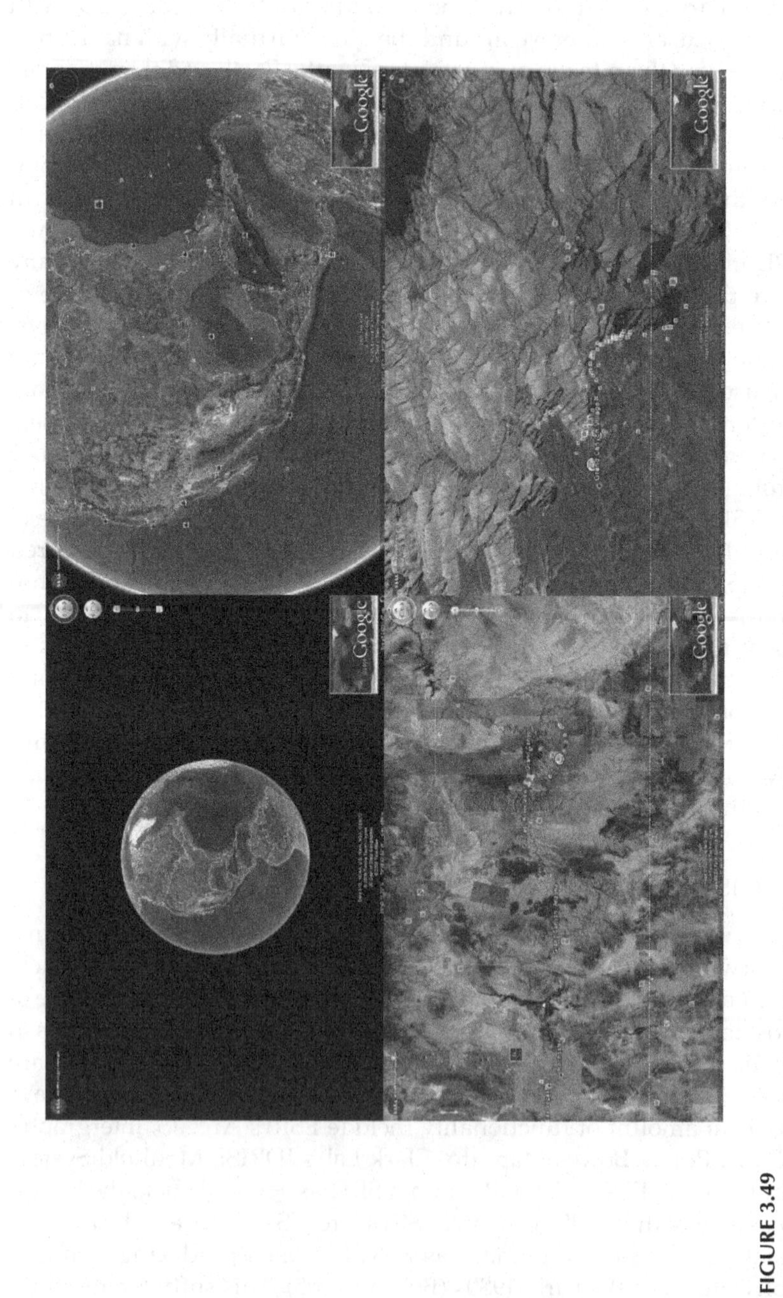

FIGURE 3.49
A series of views using Google Earth software zooming from (top left) globe level to (bottom right) zoomed-in view of the Grand Canyon from an oblique view.

FIGURE 3.50
An oblique view in Google Earth of the Manhattan borough of New York City with 3D models of buildings displayed.

FIGURE 3.51
A screenshot from a movie file produced in ArcScene software dynamically modeling the movement and simulated damage of the Parkersburg, Iowa, tornado that occurred on May 25, 2008.

the broadest and most complete set of methods for spatial data storage, management, and processing (Bolstad 2005). GeoMedia by Intergraph is another large player in the GIS market and contains significant functionality for data creation, management, and analysis. MapInfo is not as widely used as these others, and its use is concentrated in business and municipal applications (Bolstad 2005). The IDRISI software suite was developed in the Graduate School of Geography at Clark University and has undergone considerable evolution over the last two decades. IDRISI was developed to be an affordable GIS and image processing software. For this reason, it has been used widely in academic institutions and in many developing countries (Bolstad 2005). The Manifold System GIS is a lower-cost GIS that has significant functionality but a smaller market share. The Smallworld GIS software is commonly used by the utility industry.

Open source GIS software is in a phase of great growth presently. There are no open source GIS packages that have the amount of functionality that is available in some of the full commercial packages mentioned above. However, some open source software packages have an impressive array of functionalities. One of the oldest and most extensive open source GIS software is the Geographic Resource Analysis Support System (GRASS). The GRASS software was originally developed by the U.S. Army Construction Engineering Research Laboratories, but is currently maintained and developed by a community of developers as open source software (Chang 2009). Many new open source GIS products are under continuous development with improvements and increased functionality being added. Some of the most popular include Quantum GIS (or QGIS), SAGA, uDIG, and others. A list of Web sites that detail GIS software is provided in Appendix C.

3.5 Summary

This chapter introduced the idea that SDSS are built from components found both in decision support systems and geographic information systems. Both DSS and GIS have user interfaces and database management components, while DSS have modeling components, and GIS have spatial data analyses and presentation components. In addition, the GIS software has special functionality for handling spatial database management. The architectures of SDSS have generally adapted from both of these disciplines and are composed of user interfaces or dialog management, model management, and database management components (with spatial database management capabilities).

As GIS software is so important in SDSS development, and due to the potential lack of knowledge in the subject, this chapter has focused on GIS software characteristics and functionality as well as spatial data management issues. An overview of the capabilities for GIS in spatial data creation, management, processing, analysis, and visualization was provided in this chapter. The two most basic spatial data models (vector and raster) were introduced and compared. The vector data model uses points, lines, and polygons to represent real-world features in a GIS environment. The raster data model uses a two-dimensional grid in which each cell in the grid stores a value representing some real-world characteristic. The basic methods of spatial data creation (digitizing from existing maps or aerial imagery, image processing of remotely sensed data, and using GPS) were discussed. The evolution of database management technology in GIS, with a movement toward relational and objects-oriented systems, was discussed. Useful vector spatial processing and analysis techniques such as buffering, spatial overlay, pattern analysis, spatial statistics, and routing were touched upon in this chapter. Three main classes of raster analysis techniques were discussed. Local functions operate on a cell-by-cell basis and can include mathematical and logical operations. Neighborhood functions are used to calculate new values for focal cells based on neighboring cells from the input raster. Zonal functions use zones from a raster or vector dataset to calculate statistics from a given raster. Finally, the data visualization and cartographic capabilities of GIS software were presented with an emphasis placed on the responsibility of the GIS analyst to handle these capabilities responsibly.

References

Aronoff, Stan. 1995. *Geographic information systems: A management perspective.* Ottawa, Canada: WDL Publications.

Armstrong, M. P., and P. J. Densham. 1990. Database organization strategies for spatial decision support systems. *International Journal of Geographical Information Science* 4(1):3–20.

Bernhardsen, Tor. 2002. *Geographical information systems: An introduction.* New York: John Wiley & Sons, Inc.

Bolstad, Paul V. 2005. *GIS fundamentals: A first text on geographic information systems. Second Edition.* White Bear Lake: Eider Press, Inc.

Burrough, Peter A., and Rachael A. McDonnell. 1998. *Principles of geographical information systems.* New York: Oxford University Press.

Chang, K. 2009. *Introduction to geographic information systems with data files CD-ROM.* New York: McGraw-Hill Science/Engineering/Math.

Chrisman, Nicholas. 2001. *Exploring geographical information systems.* New York: John Wiley & Sons, Inc.
DeMers, Michael N. 2009. *Fundamentals of geographical information systems.* New York: John Wiley & Sons, Inc.
Densham, P. J. 1991. Spatial decision support systems. In *Geographical information systems: Principles and applications*, ed. J. Maguire, M. S. Goodchild, and D. W. Rhind, p. 403–412. London: Longman Publishing Group.
Densham, P. J., and M. F. Goodchild. 1989. Spatial decision support systems: A research agenda. Paper presented at the Annual GIS/LIS Conference, Orlando, Florida.
ESRI, 2001. Understanding topology and shapefiles. http://www.esri.com/news/arcuser/0401/topo.html (accessed November 11, 2009).
Gao, S., D. Sundaram, and J. Paynter. 2004. Flexible support for spatial decision-making. Paper presented at the 37th Annual International Conference on System Sciences, Honolulu, Hawaii.
Heywood, Ian, Sarah Cornelius, and Steve Carver. 2006. *An introduction to geographical information systems.* Harlow: Prentice Hall.
Kang, M. S., P. Srivastava, T. Tyson, J. P. Fulton, W. F. Owsley, and K. H. Yoo. 2008. A comprehensive GIS-based poultry litter management system for nutrient management planning and litter transportation. *Computers and Electronics in Agriculture* 64:212–224.
Keenan, P. B. 2006. Spatial decision support systems: A coming of age. *Control and Cybernetics* 35(1):9–27.
Keller, C. P. 1997. Unit 57—Decision-making using multiple criteria. http://www.geog.ubc.ca/courses/klink/gis.notes/ncgia/u57.html (accessed November 10, 2009).
Lolonis, P. 1990. Methodologies for supporting location decision-making: State of the art and research directions. Working paper, Department of Geography, the University of Iowa.
Longley, Paul. A., Michael F. Goodchild, David J. Maguire, and David W. Rhind. 2005. *Geographic information systems and science.* New York: John Wiley & Sons, Inc.
Malczewski, J. 1999. *GIS and multicriteria decision analysis.* New York: John Wiley & Sons, Inc.
Robinson, Arthur H., Joel L. Morrison, Phillip C. Muehrcke, A. Jon Kimmerling, and Stephen C. Guptill. 1995. *Elements of cartography.* New York: John Wiley & Sons, Inc.
Sprague, R. H., and Carlson, E. D. 1982. *Building effective decision support systems.* Englewood Cliffs, NJ: Prentice-Hall, Inc.
Sugumaran, R., S. R. Larson, and J. P. DeGroote. 2009. Spatio-temporal cluster analysis of county-based human West Nile virus incidence in the continental United States. *International Journal of Health Geographics* 8:43.

Appendix A: Spatial Data Sources for the United States

State	URL for Downloadable Spatial Data
Alabama	http://www.fws.gov/data/statdata/aldata.html
Alabama	http://www.aces.edu/waterquality/gis_data/
Alaska	http://www.asgdc.state.ak.us/
Alaska	http://gina.uas.alaska.edu/joomla/index.php?option=com_content&task=section&id=6&Itemid=138
Arizona	http://agic.az.gov/portal/dataList.do?sort=theme&dataset=0
Arizona	http://atlas.library.arizona.edu/atlas/index.jsp?theme=EnvironmentandPopulation
Arkansas	http://www.geostor.arkansas.gov/G6/Home.html
Arkansas	http://www.arkansassiteselection.com/gis-data-download.aspx
California	http://www.atlas.ca.gov/download.html
California	http://www.dot.ca.gov/hq/tsip/gis/datalibrary/gisdatalibrary.html#soil
Colorado	http://emaps.dphe.state.co.us/gis/maps.asp
Colorado	http://ucblibraries.colorado.edu/map/links/gis.htm#co
Connecticut	http://www.ct.gov/dep/cwp/view.asp?a=2698&q=322898&depNav_GID=1707
Connecticut	http://magic.lib.uconn.edu/connecticut_data.html
Delaware	http://datamil.delaware.gov/geonetwork/srv/en/main.home
Delaware	http://www.udel.edu/FREC/spatlab/
Florida	http://www.fgdl.org/metadataexplorer/explorer.jsp
Florida	http://data.labins.org/2003/MappingData/drg/drg_utm27.cfm
Georgia	http://data.georgiaspatial.org/index.asp
Georgia	http://csat.er.usgs.gov/statewide/downloads.html
Hawaii	http://hawaii.gov/dbedt/gis/download.htm
Hawaii	http://hawaii.wr.usgs.gov
Idaho	http://inside.uidaho.edu/asp/GeoData.asp
Idaho	http://www.idwr.idaho.gov/GeographicInfo/GISdata/gis_data.htm
Illinois	http://www.isgs.uiuc.edu/nsdihome/webdocs/browse.html
Illinois	http://www.isgs.uiuc.edu/nsdihome/ISGSindex.html
Indiana	http://inmap.indiana.edu/download.html
Indiana	http://in.gisinventory.net/
Iowa	http://www.igsb.uiowa.edu/nrgislibx/gishome.htm
Iowa	http://ortho.gis.iastate.edu/
Kansas	http://www.kansasgis.org/catalog/catalog.cfm
Kansas	http://www.kars.ku.edu/products/ksid/index.shtml
Kentucky	http://www.uky.edu/KGS/gis/kgs_gis.html
Kentucky	http://kygeonet.ky.gov/geographicexplorer/explorer.jsf
Louisiana	http://atlas.lsu.edu/rasterdown.htm
Louisiana	http://lagic.lsu.edu/datacatalog/theme_form.asp

Maine http://megis.maine.gov/catalog/
Maine http://www.maine.gov/dep/gis/datamaps/
Maryland http://dnrweb.dnr.state.md.us/gis/data/data.asp
Maryland http://www.mgs.md.gov/coastal/data/index.html
Massachusetts http://www.mass.gov/mgis/laylist.htm
Massachusetts http://maps.massgis.state.ma.us/massgis_viewer/index.htm
Michigan http://www.mcgi.state.mi.us/mgdl/?action=thm
Michigan http://www.michigan.gov/dnr/0,1607,7-153-10371_14546---,00.html
Minnesota http://www.lmic.state.mn.us/chouse/clipnship.html
Minnesota http://www.gis.leg.mn/html/download.html
Minnesota http://geogateway.state.mn.us/documents/index.html
Mississippi http://www.maris.state.ms.us/HTM/DownloadData/Statewide-Theme.htm
Mississippi http://www.gis.ms.gov/Portal/dataDownload.do
Missouri http://www.msdis.missouri.edu/datasearch/ThemeList.jsp
Missouri http://ims.missouri.edu/moims2008/
Montana http://nris.mt.gov/gis/gisdatalib/gisDataList.aspx?datagroup=statewide&searchTerms=
Montana http://bogc.dnrc.mt.gov/web_mapper.asp
Nebraska http://www.dnr.ne.gov/databank/spat.html
Nebraska http://snr.unl.edu/data/geographygis/NebrGISdata.asp
New Hampshire http://www.granit.unh.edu/data/downloadfreedata/category/databycategory.html
New Jersey https://njgin.state.nj.us/NJ_NJGINExplorer/BrowseByTheme.jsp
New Jersey http://www.state.nj.us/dep/gis/lists.html
New Mexico http://rgis.unm.edu/data_entry.cfm
New Mexico http://sar.lanl.gov/maps_by_name.html
New York http://www.nysgis.state.ny.us/gisdata/
New York http://cugir.mannlib.cornell.edu/browse.jsp
North Carolina http://www.nconemap.com/Default.aspx?tabid=286
North Carolina http://www.ncdot.org/it/gis/DataDistribution/
North Dakota http://web.apps.state.nd.us/metadataexplorer/explorer.jsf
North Dakota http://wms-sites.com/catalog/North-Dakota-GIS-Hub-Web-Map-Service---NDWMS_GeneralInfo
Ohio http://metadataexplorer.gis.state.oh.us/metadataexplorer/explorer.jsp
Ohio http://www.dnr.state.oh.us/gims/category/tabid/10528/Default.aspx
Oklahoma http://geo.ou.edu/DataFrame.htm
Oklahoma http://www.ocgi.okstate.edu/zipped/
Oregon http://www.oregon.gov/DAS/EISPD/GEO/alphalist.shtml
Oregon http://libweb.uoregon.edu/map/map_section/map_librarydata.html
Pennsylvania http://www.pasda.psu.edu/uci/SearchPage.aspx
Pennsylvania http://www.gis.state.pa.us/portal/server.pt/community/imagery_ftp/1615

Rhode Island	http://www.edc.uri.edu/rigis/data/default.html
Rhode Island	http://ortho.edc.uri.edu/
South Carolina	http://www.dnr.sc.gov/GIS/gisdownload.html
South Carolina	http://www.cas.sc.edu/gis/dataindex.html
South Dakota	http://arcgis.sd.gov/IMS/sdgis/Data.aspx
South Dakota	http://www.sdgs.usd.edu/other/db.html
Tennessee	http://www.tngis.org/data.html
Tennessee	http://tnmap.state.tn.us/portal/Default.aspx
Texas	http://www.tnris.state.tx.us/datadownload/download.jsp
Texas	http://www.glo.state.tx.us/gisdata/gisdata.html
Utah	http://gis.utah.gov/download
Utah	http://www.emrl.byu.edu/gsda/
Vermont	http://www.vcgi.org/dataware/
Vermont	http://maps.vermont.gov/imf/sites/VCGI_basemap/jsp/launch.jsp
Virginia	http://gisdata.virginia.gov/Portal/ptk?command=openchannel&channel=22
Virginia	http://fisher.lib.virginia.edu/collections/gis/virginia_gis_data.html
Washington	http://fortress.wa.gov/dnr/app1/dataweb/dmmatrix.html
Washington	http://wagic.wa.gov/washdat.htm
West Virginia	http://wvgis.wvu.edu/data/data.php
West Virginia	http://gis.wvdep.org/
Wisconsin	ftp://dnrftp01.wi.gov/geodata/
Wisconsin	http://www.sco.wisc.edu/wisclinc/findgeodata.php
Wyoming	http://partners.wygisc.uwyo.edu/website/dataserver/viewer.htm
Wyoming	http://partners.wygisc.uwyo.edu/wygeolibrary/explorer.jsf

Appendix B: Global Spatial Data Sources

DIVA-GIS	http://www.diva-gis.org/Data.htm
40 Libraries	http://libweb.uoregon.edu/map/map_section/map_globaldatasets.html
Arc GIS	http://resources.esri.com/arcgisdesktop/index.cfm?fa=content
Atmospheric Science Data Center	http://eosweb.larc.nasa.gov/HPDOCS/proj_sup.html
Global Land Cover Facility	http://glcf.umiacs.umd.edu/data/
World Clim	http://www.worldclim.org/
UN GEO Data Portal	http://geodata.grid.unep.ch/
GIS @ University of Chicago	http://gis.uchicago.edu/data.htm
NOAA Class	http://www.nsof.class.noaa.gov/saa/products/catSearch
USGS EROS Center	http://edc.usgs.gov/
HYDRO1k	http://edc.usgs.gov/Find_Data/Products_and_Data_Available/gtopo30/hydro
SEDAC Dataset Catalog	http://sedac.ciesin.columbia.edu/gateway/databygis-nongis.html
World Database on Protected Areas	http://www.wdpa.org/Download.aspx
UN Spatial Data	http://geomatics.nlr.nl/unsdi/srv/en/main.search?category=datasets
FAO-UNESCO Soil Maps	http://www.lib.berkeley.edu/EART/fao.html
Climate Grids	http://www.cru.uea.ac.uk/~timm/grid/index.html

Appendix C: Links for Lists of Commercial and Open Source GIS Software

Wikipedia	http://en.wikipedia.org/wiki/List_of_GIS_software
Geoplan	http://www.geoplan.com/Mapping_Solutions/GIS_Mapping_Software
Open Source GIS	http://opensourcegis.org/
FreeGIS	http://www.freegis.org/

4

Components of SDSS II

Learning Objectives

- To understand the remaining components of SDSS:
  - Model management
  - Dialog management
  - Knowledge management
  - Stakeholder

4.1 Introduction

In the first two chapters, we provided an overview of the spatial decision-making process, an introduction to spatial decision support systems (SDSS), and how the evolution of geographic information systems (GIS) and decision support systems (DSS) have combined to lead to the development of SDSS. In the last chapter, we began discussing the different components of SDSS and provided a detailed description of GIS and their spatial database management components. The purpose of this chapter is to discuss in detail the remaining components, including the model management, dialog management, and knowledge management components, as well as the important component of the people involved in SDSS development and use—the stakeholders.

4.2 Model Management Component

The model management component (MMC) of SDSS specifically helps to manage, execute, and integrate different models (Chakhar and Martel

2004). Spatial models provide analytical capabilities to the SDSS and help in examining the locations, attributes, and relationships of features in spatial data through various overlay and analytical methods. As mentioned in Chapters 2 and 3, most existing GIS provide overlay functions but lack advanced analytical modeling capability. In recent years, a few GIS software programs have incorporated analytical spatial models. Examples of analytical modeling capabilities within GIS programs include a location-allocation model in ArcInfo, ideal point analysis in CommonGIS, and the analytic hierarchy process (AHP) and ordered weighted averaging (OWA) capabilities in IDRISI. In other cases, basic modeling frameworks, with specific user interfaces for developing spatial modeling processes, have been introduced in GIS and spatial analysis software. Modeling management frameworks built into GIS and other spatial analysis software include Spatial Modeler and Knowledge Engineer from ERDAS Imagine, Macro Modeler from IDIRISI, and ModelBuilder from Environmental Systems Research Institute (ESRI) (Figure 4.1). However, as these modeling frameworks rely on existing functions within the GIS, most do not provide users with the range of spatially explicit modeling capabilities necessary for complex spatial decision making. These modeling frameworks are growing in sophistication. However, SDSS have traditionally and often still do require the development of a specific model management component that manages a set of models that interact with the spatial database and GIS functionality to produce new information relevant for the decision-making process. The following sections provide a summary of different spatially explicit models often used in spatial decision-making processes.

4.3 Modeling Techniques in SDSS

A wide variety of spatial modeling techniques have been utilized in spatial decision-making processes and in SDSS. Many of these models are not inherently spatially explicit but are adapted for spatially explicit use in an SDSS. Some examples include mathematical models, statistical models, simulation models, prediction models, spatiotemporal models, land suitability models, and dynamic models. The classification of models used in SDSS is difficult because there is such a wide variety falling into different disciplines. Among these models, land suitability models are one of the most widely used in SDSS. Land suitability models estimate the ability of a given type of land to support a defined use. Collins et al. (2001) and Malczewski (2004) classified land suitability models within GIS into three major groups: (1) computer-assisted overlay, (2) multi-criteria

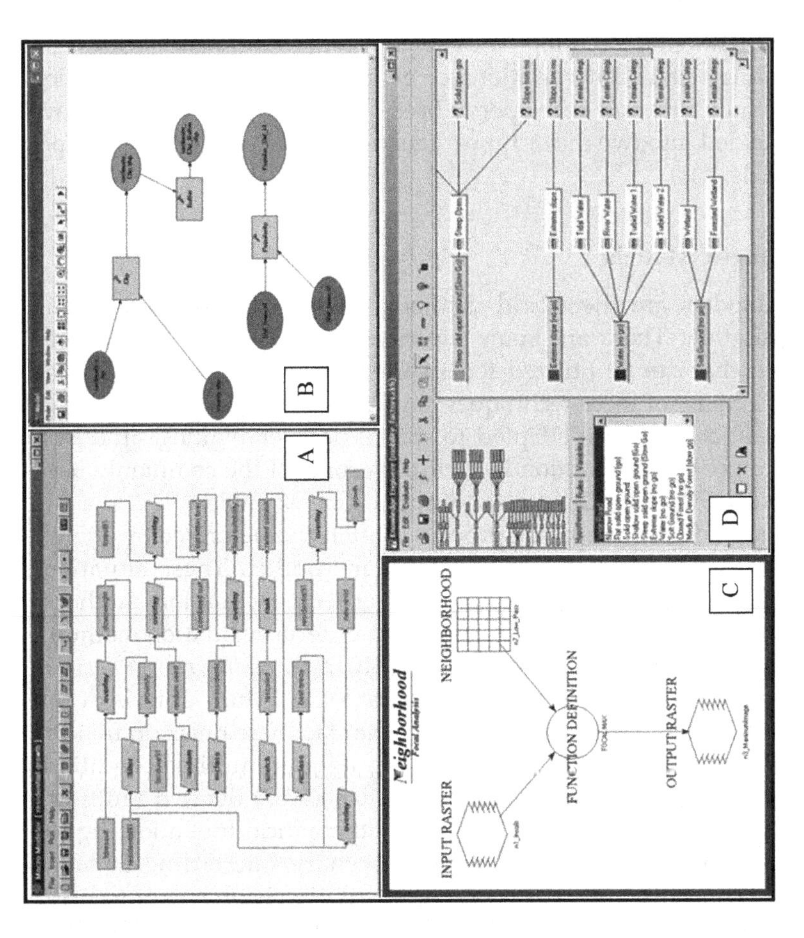

FIGURE 4.1
(See color insert following page 74.) Modeling frameworks available in different GIS software. (A) IDRISI Macro modeler, (B) ESRI ModelBuilder, (C) ERDAS Imagine Model Maker, and (D) ard ERDAS Imagine Knowledge Engineer.

decision-making methods, and (3) artificial intelligence (AI) methods. They also described different models within each group. For example, weighted linear combination (WLC), analytical hierarchy process (AHP), and ordered weighted averaging (OWA) are examples of multi-criteria decision-making methods. It is not possible to cover in detail the full range of modeling techniques in this book. Thus, the goal here is to provide readers with an overview of the range of models commonly utilized in SDSS based largely on the SDSS applications database described in Chapter 2. This chapter will provide several application examples instead of providing detailed descriptions of each model. To provide decision makers (end users) and developers a broad overview, all the models have been classified into two major types: generic models and application-specific models.

4.3.1 Generic Models

Generic models are theoretical methods that can be implemented in any application. There are many modeling techniques that have been developed that can be utilized for a variety of spatial decision-making situations. Many of these techniques have been developed for nonspatial problems and have been adapted to spatial decision-making situations. The major goal of this section is to discuss some of the commonly used models in SDSS within the context of application examples.

The spatial decision-making process is often characterized by a range of attributes and criteria that are often in conflict. These situations have often been addressed with multi-criteria evaluation methods. Numerous terms are used to discuss multi-criteria evaluation methods, including *multi-criteria decision-making* (MCDM), *multi-criteria decision analysis* (MCDA), and *multi-criteria analysis* (MCA). Multi-criteria evaluation methods offer a set of procedures that facilitate decision making by examining a number of alternatives in light of multiple conditions and conflicting objectives (Voogd 1983). Roy (1985) defined multi-criteria analysis as "a decision aid and a mathematical tool allowing the comparison of different alternatives or scenarios according to many criteria, often conflicting, in order to guide the decision maker towards a judicious choice." Malczewski (1999, 81) characterized MCDM problems as those that "involve a set of alternatives that are evaluated on the basis of conflicting and incommensurate criteria." For problems with a spatial dimension, GIS provide a natural complement to multi-criteria methods. Malczewski (1999 and 2004) covered MCDM in spatial decision making in much greater depth than we will here. This section attempts to provide a broad overview of some of the multi-criteria and other models and applications particularly relevant to SDSS.

4.3.1.1 Boolean Overlays

A simple method of combining spatial data that represent attributes in a multi-criteria decision-making process is the use of map overlays, and specifically Boolean overlays. Map overlay occurs in a GIS environment when two or more separately stored datasets, which are representations of some real-world phenomenon or human constructs related to the real world, are combined in a variety of ways to derive a new dataset. The use of Boolean logic in map overlay is quite common in GIS and usually is carried out using raster data. Operating on the raster data model, GIS software contains algorithms that allow for the calculation of many algebraic and logical (e.g., AND, OR, NOT) functions to be carried out on a cell-by-cell basis. This can be very useful for carrying out modeling operations such as land suitability analyses. Imagine a situation where a local authority is trying to site a landfill. They could set multiple criteria, such as the site must fall on nonsteep slopes, be at least 500 m from a river, and occur outside all city limits. Given that there is sufficient spatial data on these real-world conditions, a GIS analyst could set up a Boolean overlay model in which he or she identifies parcels of land that fit the criteria defined (Figure 4.2). In the final operation, Boolean logic is used to check on a cell-by-cell basis if each included raster layer is suitable (e.g., outside of city limits AND at least 500 m from rivers AND on gentle slopes). This is a classic raster overlay procedure that can be carried out with many combinations of data and logical Boolean operations. While this type of operation is easily understandable and useful for initial analysis, it often does not capture the complexities of spatial decision-making processes such as land use planning. Beedasy and Whyatt (1999) used Boolean logical models, in addition to other models, in an SDSS whose goal was to assist in locating a new hotel project in Mauritius using data on transportation features, communities, land use, elevation, existing tourist features, and the coastline. They decided that the Boolean combination technique was inferior to the weighted linear combination techniques they used.

4.3.1.2 Weighted Linear Combination

Weighted linear combination (WLC) models have been commonly used in SDSS. WLC is a method of map overlay that tries to capture, in more depth, some of the value judgments and expert opinions within the spatial analysis procedure. These types of procedures are also sometimes called *simple additive weighting*. The WLC approach involves the assigning of weights of relative importance to each map layer. Malczewski (1999) summarized the steps in this process as follows: (1) define the evaluation criteria or map layers, (2) standardize each criterion map layer, (3) define the criterion weights or weight of relative importance, (4) construct the weighted standardized

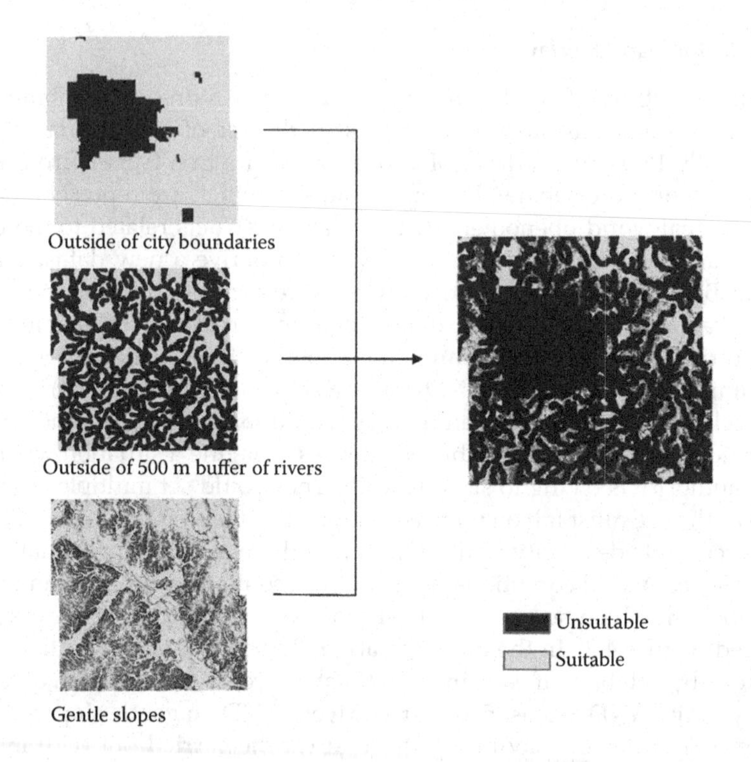

FIGURE 4.2
A Boolean overlay process for a hypothetical decision on lands potentially suitable for a new landfill.

map layers, (5) generate an overall score by adding the weighted standardized map layers, and (6) rank the alternatives according to the overall performance score. If we carry forward the previous example, we could use the city limits and buffers around rivers as absolute limitations (Figure 4.3). We could then use reclassified layers on soils and slopes as evaluation criteria where weighting is used. In Figure 4.3, data layers on soils and slopes are reclassified into values of 1 to 4, with 1 representing the least attractive and 4 representing the most attractive in relation to land suitability (Steps 1 and 2). These layers are then multiplied by weights of two and three, respectively, with the higher score for slope indicating that it is more important. The result is two new weighted raster layers (Steps 3 and 4). The weight-adjusted map layers are then added together to get a final score (Step 5). Finally, the scores can be given a ranking or classification (Step 6). This method adds greater sophistication in defining portions of the landscape that are desirable by producing a range of scores instead of just suitable or unsuitable classes. Although this technique has been commonly applied (e.g., Sugumaran et al. 2004), others (Lai and Hopkins 1989; Malczewski

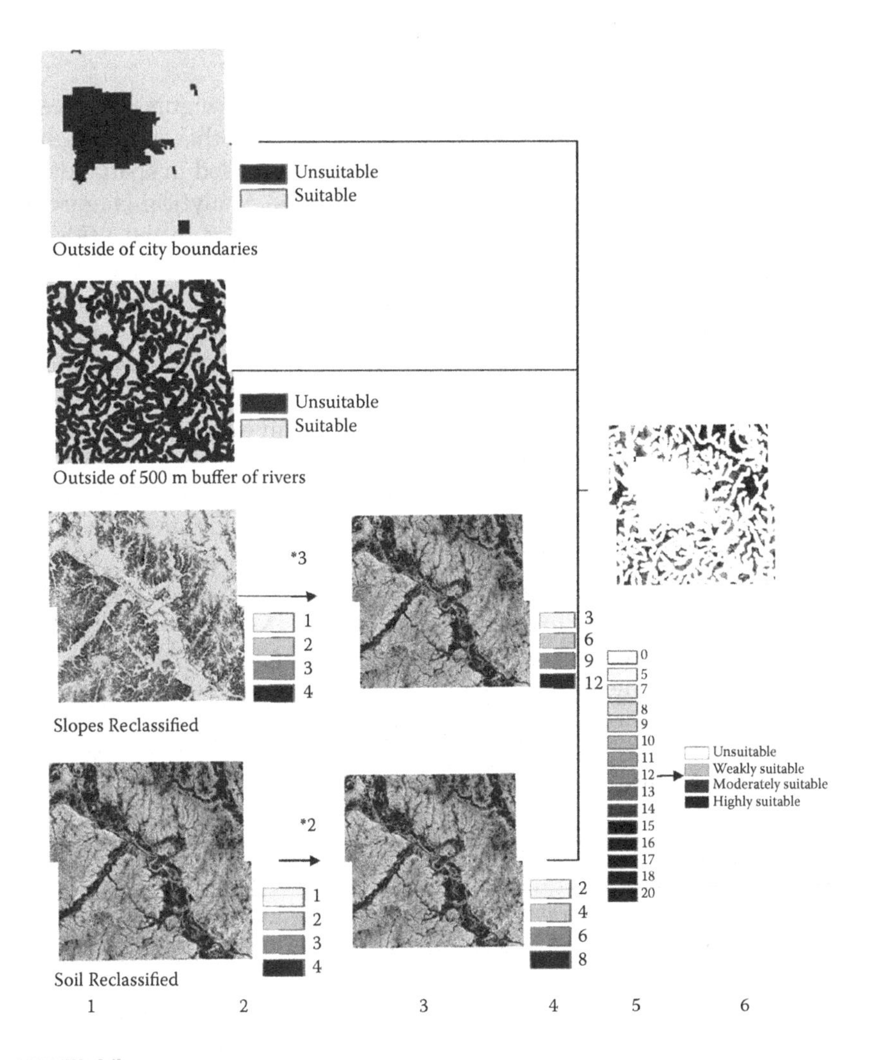

FIGURE 4.3
An example of the weighted linear combination method based on the six steps identified by Malczewski, Jacek. 1999. *GIS and Multicriteria Decision Analysis*. New York: John Wiley & Sons, Inc.

and Jackson 2000) have criticized the technique because of inappropriate methods of standardizing suitability maps and unverified assumptions of independence among suitability criteria (e.g., the soil attributes in our example might be correlated with the slope classes). Other critiques of this methodology include assigning weights in decision making that are purely dependent on the experience of decision makers, and that the weighting techniques do not capture the complexities of a real-world decision-making situation (Heywood et al. 1995).

4.3.1.3 Analytical Hierarchy Process

One of the weaknesses of WLC is in establishing the weights effectively and in a realistic fashion without user bias. To address this shortcoming, another multiattribute technique that has often been used in spatial decision-making processes and SDSS applications is the Analytical Hierarchy Analysis or Process (AHP). This method was developed in the 1970s by Saaty (1980) and is widely used in a variety of decision-making situations. The AHP technique is a priority-ranking technique that helps break down complex problems into component parts. These parts are arranged into a hierarchy using subjective judgment in order to assign numerical values representing that subjective judgment based on the relative importance of these factors. Those values are then combined in order to identify the highest score or priority (Ahmad et al. 2004). In a spatial context, AHP can be used to derive associated weighted suitability map layers, and then these weighted map layers can be combined in ways similar to the WLC methods (Malczewski 2004). This method is effective when a large number of alternatives are represented by means of raster data. However, the method is also applicable to the vector data model (Malczewski 2004). The general process of AHP is to define the unstructured problem, break the problem up into a hierarchical structure of general and detailed criteria and alternative elements, carry out pairwise comparisons based on comparison matrices, estimate the relative weights of decision elements, check consistency of comparison matrices, and finally, aggregate the relative weights of decision elements to obtain an overall rating (Lee et al. 2008). The power of AHP comes in its ability to carry out pairwise comparisons between each pair of general and detailed criteria and to calculate weights that are used to determine final scores for all potential alternatives. In a spatial context, this pairwise comparison leads to the development of weights for spatial layers (general criteria) and for cells or vector features (detailed criteria). These weights are then used in a WLC method to get a final raster cell-by-cell or vector feature-by-feature score. The AHP process provides a more systematic way to address complex multi-criteria decision analysis as compared to WLC, which relies heavily on the subjective decision making of expert users. Figure 4.4 displays an example where AHP was used to identify environmentally sensitive areas. In this example, the user develops the pairwise comparisons between layers representing slope, forested areas, wetland soils, impervious areas, and floodplain zones, and the AHP algorithm calculates the weights and uses this to determine the final scoring on a cell-by-cell basis representing the most environmentally sensitive areas. The AHP procedure has been utilized for a variety of spatial decision-making situations within SDSS, including those for land conservation (Strager and Rosenberger 2006), biodiversity conservation (Karnatak et al. 2007), urban land use (Taleai et al. 2006),

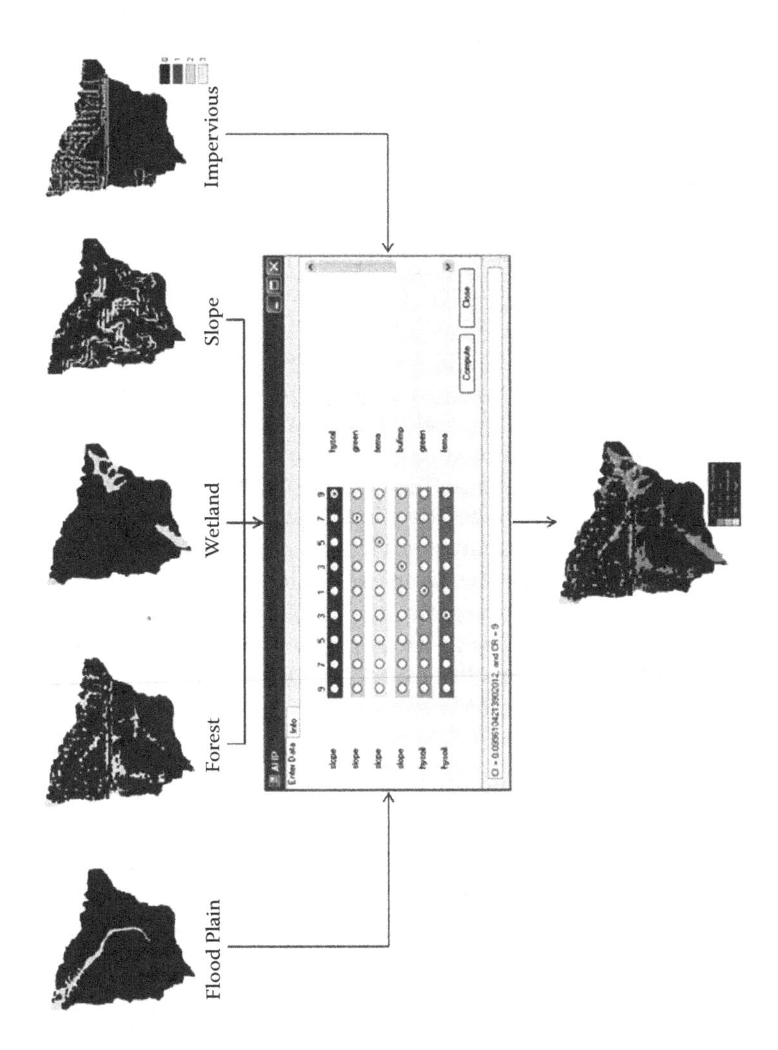

FIGURE 4.4
Example of an Analytical Hierarchy Process application.

and other applications. A more complete discussion of spatial AHP can be seen in Malczewski (1999).

Common problems with AHP include the following: (1) it allows the comparison of only two criteria at once and converts subjective assessments of relative importance into a linear set of weights, (2) it is based on discrete decisions that do not take into account the uncertainty associated with decision makers' judgment (Cheng 1996 and 1999; Lee et al., 2008), and (3) the subjective judgment, selection, and preference of decision makers have great influence on the AHP results (Tuzkaya et al. 2009).

4.3.1.4 Ordered Weighted Approach

The ordered weighted approach (OWA) technique is a multi-criteria method that was developed by Yager (1988). In an OWA approach, two types of weights are used: weights of importance like those used in the WLC method and ordered evaluation criteria (Rinner and Malczewski, 2002). A criterion importance weight is assigned to an attribute for all locations, while ordered weights vary on a location-by-location or feature-by-feature basis. The order weights allow for direct control over the levels of trade-off among criteria (Eastman and Jiang 1996; Eastman 1997; Jiang and Eastman 2000). Figure 4.5 (left side) is an interface developed within software called *SpreadsheetSDSS*, developed by the authors of this book, where OWA was used to identify environmentally sensitive areas. This OWA analysis used all the five layers presented in Figure 4.4, provided equal criterion weights to each, and used the following ordered weights: w = (0.24, 0.36, 0.2, 0.1, 0.1) as an example to run the model. There are several commercial programs (for example, IDRISI) that have also implemented the OWA in their system.

4.3.1.5 Artificial Neural Networks

Artificial neural networks (ANN) are meant to simulate the workings and learning capabilities of the human brain (Malczewski 2004). They can be used to model complex nonlinear relationships between inputs and outputs. Neural networks are composed of individual processing units (neurons) and joining interconnected weights. Artificial neural networks are adaptive systems that use training or learning phases. Based on these training or learning phases, the ANN can begin to identify relationships between inputs and output data. The ANN method tests created models by putting new input data through the network and analyzing output versus known conditions to get an accuracy score. Ideally, the ANN should learn which inputs are or are not influential and decipher what relationships exist (Gimblett et al. 1994). Gimblett et al. found ANN useful because of adaptive autonomous rule-generation capability, which handled a large

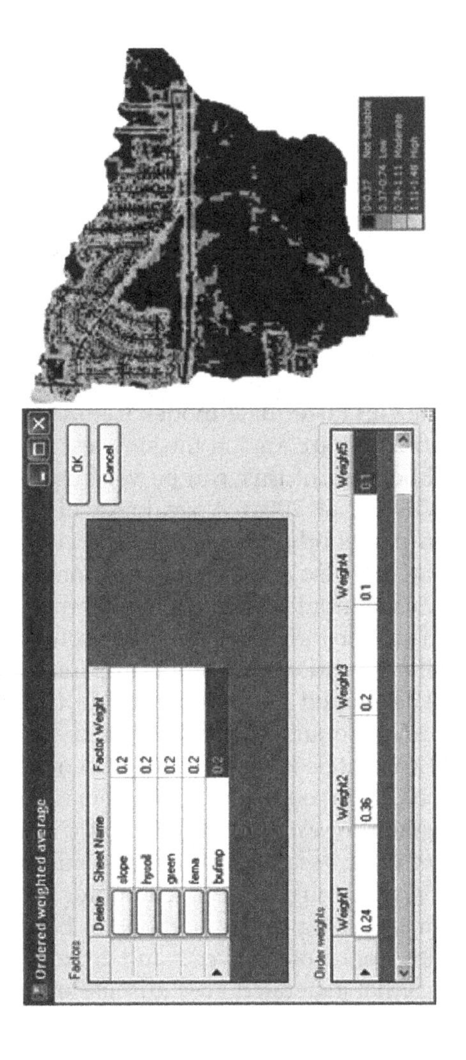

FIGURE 4.5
Ordered weighted averaging (OWA) analysis output for the identification of environmentally sensitive areas.

number of different combinations of interdependent land suitability factors. The ANN approaches are best suited for planning tasks when there is little understanding of the problem structure and when working with large datasets (Malczewski 1999). One of the problems with the ANN methods is that they are essentially black boxes in which the inner workings are hidden from the analysts. Artificial neural networks have been incorporated in SDSS for estimating snow water equivalents across watersheds (Kim 2003) and flood forecasting (Shim and Shim 2000).

4.3.1.6 Cellular Automata

Cellular automata (CA) use a discrete representation of a regular grid of cells, each of which has a finite number of states such as *on* or *off* or *alive* or *not alive*. The state of a given cell at any point in time is dependent on its and its neighbor's previous state and is decided by a set of deterministic or probabilistic rules (Malczewski 2004). Each of the cells is updated simultaneously according to a given time step. The time step and the neighborhood to be considered for each cell must be defined. The CA data structure is very analogous to the GIS raster data model. Cellular automata methods are inherently spatial and are among the simplest representations of dynamic systems, and because of this, can be very useful for modeling land use dynamics (White et al. 2004). A weakness of CA techniques is that they are based on neighborhood relationships and generally don't account for global effects that also affect spatial phenomena. For example, the housing state (e.g., residential or no development) of a given area (a cell in CA) can be influenced by global conditions such as economic conditions in addition to the condition of neighboring areas (nearby cells in CA). White et al. (2004) developed a system for the development and support of policies for improving quality of the populated environment in the Netherlands with an SDSS that used a constrained CA model representing land use dynamics. The CA linked to GIS data layers representing physical characteristics, accessibility, and zoning, while the CA was based on a regionalized model of macro-scale dynamics of demographic and economic activities. The CA modeling was carried out for the entire country with a cell size of 500 m and using a neighborhood of 196 cells (a radius of 8 cells). Various scenarios were modeled years and decades in advance and were used in workshops that generated great discussion and strong reactions. Luca (2007) used a CA technique in conjunction with agent-based modeling techniques for a dynamic urban and regional design program for Torino, Italy. Shah et al. (2008) use a CA technique in a system called the Freeway Incident Analysis System (FIAS) for traffic management in Korea. The FIAS successfully simulated traffic effects and could be used by traffic managers for decision making according to the authors. A GIS-embedded CA model called *iCity* was developed for predictive modeling of urban growth by Stevens et al. (2007). This implementation was unique in that it used irregularly shaped parcel polygons as the spatial unit.

4.3.1.7 Genetic Algorithms

Another set of AI modeling techniques utilized in SDSS are genetic algorithms (GA). Genetic algorithms are search methods that mimic biological evolution in that they involve a competitive selection that eliminates poor solutions. The most useful portions of successful solutions can be recombined with other solutions to find even more optimal solutions (Malczewski 2004). These types of algorithms have been used commonly in land use planning and suitability analyses (Malczewski 2004). In a land use planning example, a population would be made up of individual land use plans, which in turn are made up of genes that specify characteristics of any given land use plan. The GA can mutate by changing genes, or characteristics, of the individual or by swapping genes with another individual. For example, an individual land use plan could adopt genes or characteristics of another land use plan to find a more ideal solution. Each individual is assigned a value indicating its adequacy as a solution to the problem based on some fitness metrics such as financial returns or impact on the environment (Matthews et al. 1999). These techniques are useful when conventional multi-criteria methods are insufficient, usually due to the complex nature of the problem and the size of the potential solution space (Malczewski 2004). O'Sullivan and Unwin (2003) pointed out that it is difficult to apply GA to specific problems, and that if a good solution can be described, it is likely that one can be found without applying GA. Matthews et al. (1999) implemented a spatial land allocation decision support system called the Land Allocation Decision Support System (LADSS), which consisted of a GIS, spatial databases, knowledge base, graphical user interface, and land use planning tools that used GA techniques. The prototype system was demonstrated as a flexible tool for the exploration of land use planning scenarios. Odeyemi et al. (1998) used GA techniques for land allocation modeling used to support decision-making processes for budgeting for agricultural extension services and human resource allocation in Zimbabwe.

4.3.1.8 Agent-Based Models

Another AI-based approach is agent-based modeling, which is also sometimes called individual-oriented or distributed artificial intelligence-based modeling. These techniques are used for simulating complex phenomena in dynamic systems and are distinctive in their ability to simulate future unpredictable situations (Lampert 2002). The agents are capable of acting autonomously while interacting with their environment and other agents

(Sugumaran and Sugumaran, 2007). The rationale of this approach is that it is easier to model the behavior of individuals than it is to model a system as a whole. However, by modeling the behavior of individual agents, system-level lessons can be discovered.

Several research efforts have been reported that used agent technology in addressing spatial decision-making problems (Gimblett et al. 2000; Manson 2000; Sengupta et al. 2003). Ferrand (1996) reported on a system utilized to solve complex spatial optimization problems encountered in the search for environmental impact areas. Papadias and Egenhofer (1995) reported on using agents in qualitative collaborative planning. Agents represent topological, direction, and distance constraints applied to a spatial planning problem. The focus of this work was on using spatial access methods that can effectively process qualitative constraints represented by agents. Rodrigues et al. (1997a) described a multiagent-based system used for modeling geographic elements for environmental analysis in land use management. The system was aimed at establishing methodologies for evaluation and standard simulation in environmental quality description scenarios. In another study, Sengupta and Bennett (2003) developed an agent-oriented modeling framework to overcome some of the limitations of traditional SDSS. Their SDSS, called Distributed Intelligent Geographical Modelling Environment (DIGME), could be used to evaluate the ecological and economic impacts of agricultural policies. Recently, Ligtenberg et al. (2009) proposed a multiagent system to simulate potential locations for new urban development.

4.3.1.9 Fuzzy Modeling Techniques

Many phenomena in the real world are difficult to classify into discrete or crisp categories, in spatial or nonspatial data as well as in preferences and criteria used in modeling activities. The inclusion of crisp boundaries in spatial and nonspatial data produces similar definitive or crisp results in any modeling or SDSS activity. The outputs from these activities can possibly give an elevated level of confidence not truly warranted. In order to address uncertainty that is often implicit in the inputs and modeling processes, fuzzy techniques have been developed. These techniques utilize fuzzy set theory or fuzzy logic. The fuzzy set theory allows objects or locations to belong partially to multiple sets instead of being completely discrete (Malczewski 1999). An example would be in land use planning in which a constraint of 1 km is used to exclude areas surrounding a river. There are generally few real-world attributes that would indicate a location 0.99 km away is acceptable while a location 1.01 km is unacceptable (Malczewski 2004). Fuzzy set theory can be used to address this kind of situation by characterizing membership not as absolute but as partial. Fuzzy set theory can also be used for linguistic statements that are used to provide an ordering of criteria such as low, moderate, and high (Malczewski 2004). In fuzzy sets, the degree of membership in a class is expressed on a continuous scale between 0 and 1, with 0 meaning not a member and 1 being a full member (Burrough and McDonnell 1998). In a spatial context, fuzzy sets can be used to represent geographical entities with imprecisely defined boundaries. For example, in soil mapping, a distinct boundary might be drawn between two soil types even though in reality there is a gradient between the types. Using fuzzy sets, the values would provide a value between 0 and 1 for membership in either of the soil groups (Figure 4.6). The methodologies for developing membership values are beyond the scope of this textbook. More detailed examination can be found in Burrough and McDonnell (1998), Malczewski (1999), and many other publications.

Spatial data developed based on fuzzy sets can be utilized in various spatial operations and SDSS, including multi-criteria analysis techniques including AHP and ANN. Klungboonkrong and Taylor (1998) used fuzzy set theory and fuzzy multi-attribute decision making in an SDSS for evaluating environmental impacts of urban road networks. Girvetz and Shilling (2003) used fuzzy logic rules in a raster-based system to evaluate degree of truth for assertions about the environmental impact of forest road development. Li et al. (2005) used the fuzzy comprehensive evaluation technique, based on the determination of fuzzy weights by experts, in an SDSS for typhoon insurance pricing. Boroushaki and Malczewski (2008) used fuzzy linguistic qualifiers in a customized ArcGIS-based AHP-OWA system. Chang et al. (2008) used fuzzy multi-criteria decision making for landfill siting based on the translation of linguistic ratings (e.g., good, fair, poor, etc.) into fuzzy spatial datasets regarding environmental impacts, transportation issues, economic impacts, and public nuisance.

4.3.2 Application-Specific Models

Many SDSS applications have adopted modeling methodologies designed for specific problem situations or to represent specific anthropogenic, chemical, biological, or physical processes occurring in the real world. Specific modeling methods have been developed for many domain-specific spatial decision-making situations including environmental, natural resources management, agricultural, emergency planning, public health, transportation, urban, utilities, and others. Although each of these modeling techniques cannot be examined in depth, some common types of modeling techniques utilized in SDSS will be touched upon in this section.

The evolution of SDSS coincided somewhat with increased concern about environmental issues such as air and water quality. The Clean Water Act of the 1970s and other legislation prompted greater interest in understanding and controlling water pollution. The complexity of

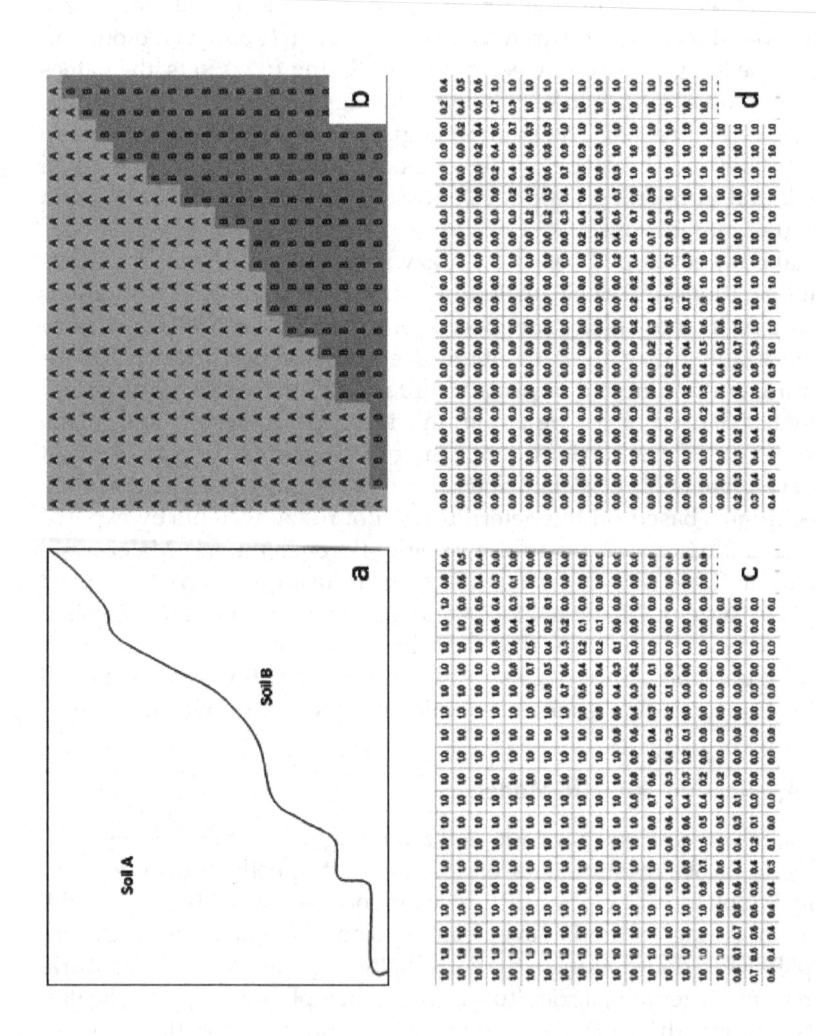

FIGURE 4.6
(a) Vector representation of crisp boundaries between two soil types and (b) a rasterized version of these crisp boundaries, (c) fuzzy representation with values for membership in the Soil A category, and (d) fuzzy representation with values for membership in Soil B category.

hydrological systems and limited resources for monitoring in-situ water quality and point source dischargers led to the development of modeling techniques to assist in the management of water resources. Numerous SDSS applications over the last few decades have utilized hydrological or water quality modeling techniques. By the time the first SDSS were developed, hydrological and water quality simulation models already existed, usually without explicit spatial representation. However, one of the first systems considered an SDSS included the MULQAL water quality simulation model within its framework (Holsapple and Whinston 1976). Since that time, numerous hydrological and water quality modeling routines have been included in SDSS (Table 4.1). A few examples of these types of models incorporated within SDSS include the Agricultural Non-Point Source Pollution Model (AGNPS) to predict nonpoint source pollutant loadings within agricultural watersheds, the Hydrological Simulation Program-Fortran (HSPF) for simulation of watershed hydrology and water quality for both conventional and toxic organic pollutants, the Long-Term Hydrological Assessment (L-THIA) tool, and the Soil and Water Assessment Tool (SWAT). These models, with varying degrees of input parameterization and complexity of algorithms, mathematically approximate physical and chemical processes governing water quantity and quality either at a watershed or stream level. The AGNPS model uses raster spatial data on land cover/use, soils, topographic conditions, and rainfall to model hydrology, erosion, and the transport of sediment and chemicals (Figure 4.7) through a watershed (Bennett and Vitale 2001). Engel et al. (1993) were the first to utilize AGNPS in an SDSS when they developed a system integrating the AGNPS model with GRASS GIS to assist in the management of runoff, erosion, and nutrient movement in agricultural watersheds. The AGNPS model has subsequently been implemented numerous times (Table 4.1) in SDSS. The HSPF model has a more complex set of algorithms that compute the movement of water (and associated chemical and physical constituents) through a complete hydrological cycle (rainfall, evapotranspiration, runoff, infiltration, and groundwater flow) (Wilkerson et al. 2007). The SWAT model partitions a watershed into a series of subbasins or subwatersheds (Figure 4.8) and uses a variety of algorithms that incorporate information on weather, soils, topography, vegetation, and land management practices to model water movement, sediment movement, crop growth, and nutrient cycling (Neitsch et al. 2005). The SWAT model was first integrated with GIS using ArcView GIS,and more recently was integrated with ArcGIS software and has been used in numerous SDSS applications (Rao et al. 2007; Volk et al. 2008). The L-THIA model estimates long-term average runoff in a watershed using data on climate, soils, and land use and was used in an SDSS for evaluating land use changes in a watershed by Engel et al. (2003).

TABLE 4.1

A Sampling of Application-Specific Modeling Examples Used in SDSS

Model(s) Used	Model Type	Application	Author and Year
AGNPS	Hydrologic, erosion, and water quality simulation	Watershed management	Srinivasan and Engel, 1994 Sun, 1995 Lovejoy et al. 1997
SWAT			Bennett and Vitale, 2001 Sengupta et al. 2003 Lant et al. 2005 Rao et al. 2007 Volk et al. 2008
L-THIA			Engel et al. 2003 Choi et al. 2005
HSPF	Hydrologic simulation	Watershed management	Taylor et al. 1999 Wilkerson et al. 2005
Water Evaluation and Planning (WEAP)	Water resources simulation	Management of sewage practices and investigation of effect on water quality	Assaf and Saadeh, 2008
CAMEO	Gaussian plume model	Evaluation of toxic spill events	Chang et al. 1997
AgroClimatic Change and European Soil Suitability (EURO-Access 2)	Crop forecast model	Crop yield estimation at regional scale	Lagacherie et al. 2000
Dymex	Population modeling software	Locust pest modeling	Deveson, 2008
Adaptive ecosystem model (AEM)	Biophysical and social carrying capacity model	Modeling carrying capacity for national park	Prato, 2001
GAYA	Forest stand simulator	Long-term forest management planning	Næsset, 1997
Land Use Evolution and Impact Assessment Modeling (LEAM)	Land use change description	Assessing the costs of urban sprawl	Deal and Schunk, 2004

International Food Policy Simulation Model (IFPSIM)	Economic model	Prediction of crop demand and crop market price variation to alternative policies	Tan et al. 2004
Unnamed	Location allocation	Defining regions and service locations within regions	Armstrong et al. 1991
Unnamed	3D network model	Emergency response (ground transportation and multilevel buildings)	Kwan and Lee, 2005
Flight Leg Allocation Problem	Shortest path problem solver	Identify optimal configurations of migratory bird stopovers	Downs and Horner, 2008
DUO	Dynamic network flow modeling	Route choice in congested urban road networks	Wu et al. 2001
PossPOP	Tuberculosis simulation model in possums	Simulation of ecological and infection dynamics	McKenzie et al. 1997
SIS/SIR	Epidemic disease spread	Modeling of AIDS spread	Yang et al. 2007

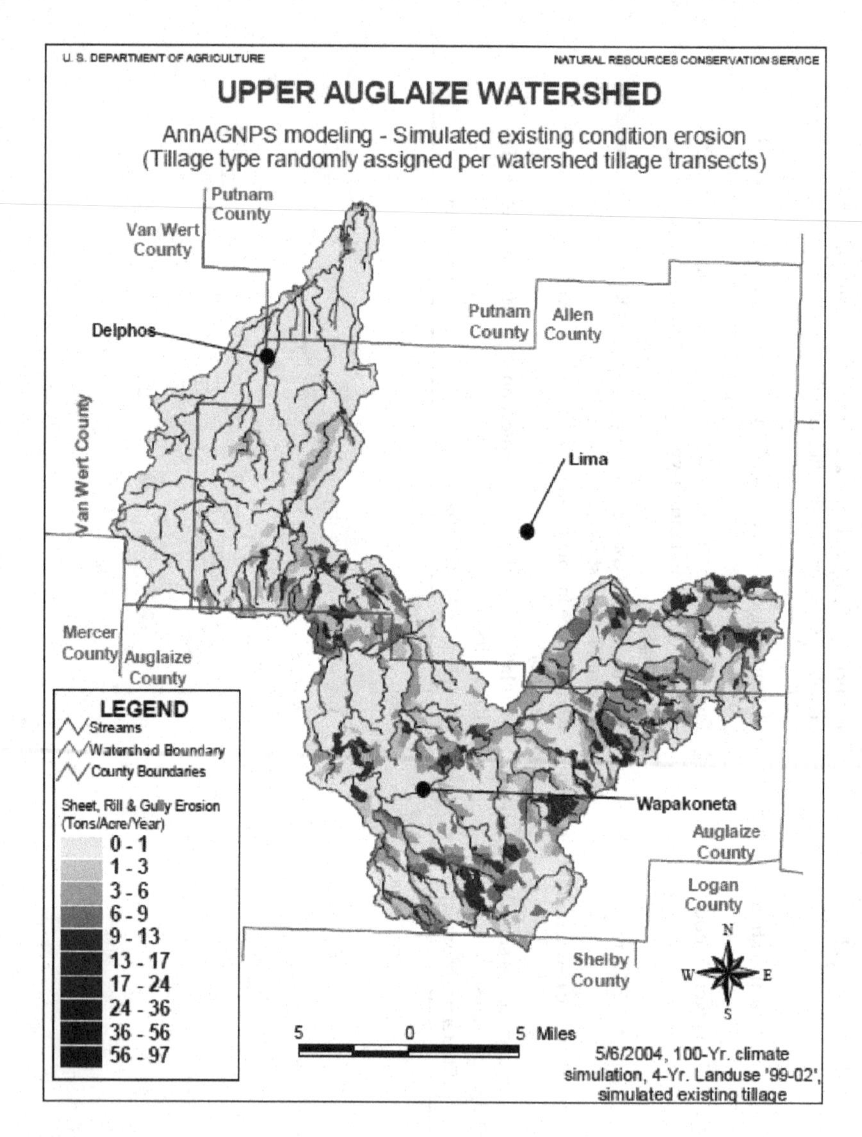

FIGURE 4.7
An example of erosion estimate output from the AGNPS model. (Taken from ftp://ftp-fc.sc.egov.usda.gov/OH/pub/Programs/agnps/Final_UA_Master_Report_with_Appendix_02.14.05.pdf)

A variety of other models have been integrated within SDSS for various spatial decision-making situations, with several examples provided in Table 4.1 and in this paragraph. The models have been utilized in a variety of disciplines. A crop yield estimation model was developed and included in an SDSS for the development of decision maps (Lagacherie et

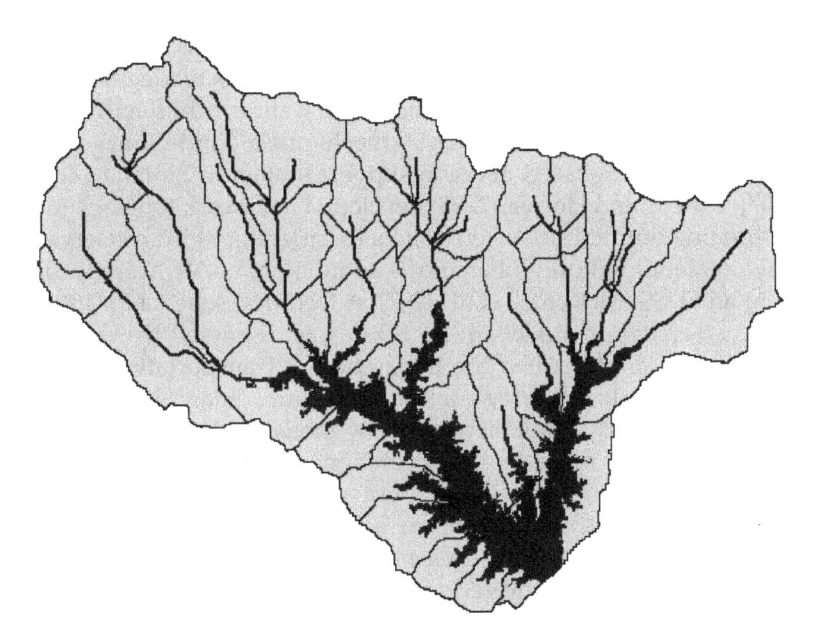

FIGURE 4.8
The SWAT model uses subwatersheds or subbasins as a unit of modeling. (Taken from SWAT Version 2005 user manual, ftp://ftp.brc.tamus.edu/pub/swat/doc/swat2005/SWAT%202005%20theory%20final.pdf)

al. 2000). In Australia, a locust development model was incorporated in an SDSS for agricultural pest control (Deveson 2008). The International Food and Agricultural Policy Simulation Model (IFPSIM) was used for predicting crop demands and crop market prices that change under alternative policy scenarios in an SDSS examining land suitability and use (Tan et al. 2004). Modeling delivery services for home food aid delivery were developed in an SDSS by Gorr et al. (2001). An ecosystem management model was used to examine carrying capacities of national parks in the United States (Prato 2001). Two separate wildfire modeling routines were included in an SDSS for examining fire hazards in California (Radke 1995). Mathematical models used for representing potential movement of radioactive pollution were included in an SDSS for nuclear power emergency management (Gheorghe and Vamanu 1995). Toxic spill modeling functionality was built into a GIS-based SDSS for emergency planning by Chang et al. (1997) in Taiwan. A 3D network data model was utilized in an SDSS intended for emergency response purposes (Kwan and Lee 2005). A forest stand simulator called GAYA was used for projecting forest stand development as part of an SDSS for long-term forest management planning (Næsset 1997). A statistical principal components analysis approach used in a prototype SDSS called SimilarAreas supported landscape planning (Bryan 2003). In France, a biodecision model was incorporated in an SDSS for estimating irrigation water demand. This biodecision model consisted of a plant development model as well as a human irrigation behavior model (Leenhardt 2004). A mechanistic wind-damage model was built into a GIS-based decision support system in Finland (Zeng et al. 2007). Downs and Horner (2008) developed a network topology model for estimating optimal stopover habitats for migratory bird conservation. In New Zealand, wildlife tuberculosis simulation models were incorporated in an SDSS (McKenzie et al. 1997). A technical schema for integrating disease epidemic models in an SDSS was discussed by Yang et al. (2007). Choi (1996) described the integration of transportation planning models and optimized route-selecting procedures in an SDSS. Wu et al. (2001) interfaced a dynamic network flow model with a GIS to form an SDSS for analysis of route choices. Evolutionary routing algorithms were combined with GIS in a decision support system by Mendoza et al. (2009). Armstrong et al. (1991) used an heuristic location allocation model in an SDSS identifying regions and service locations within regions.

In this chapter we have covered some of the generic modeling strategies, and in this section tried to give examples of application specific modeling techniques. There are many other specific models or modeling routines that have been used which we were not able to cover explicitly here. This section has attempted only to give an introduction to the variety of methods that have integrated into SDSS for a wide variety of applications.

4.4 Dialog Management Component

A key to any successful SDSS is the development of effective mechanisms for user interaction with software components. These mechanisms are termed the dialog management component (DMC). The DMC provides the interface between the user and the rest of the components of any SDSS. It provides mechanisms whereby data and information are input to the system from the user and output from the system to the user. In Chapter 3, the cartographic and tabular display capabilities of GIS were detailed. The ability to represent outputs cartographically in maps and also as 3D models was described. In addition, the capabilities of GIS software to produce effective reports, tables, and charts were investigated. The remaining portion of this section will focus on the importance of effective user interfaces.

As mentioned in Chapter 1, spatial decision-making processes involve iterative, interactive, and participative involvement of a decision maker or end users. The user interface components of an SDSS provide these functionalities and act as a channel through which the user connects to the computer system to generate and compare different solutions to a problem and to view potential outcomes from decision alternatives. The importance of user interfaces has gained much attention in the past two decades, mainly because there has been a realization that usability is a key for the success of any software product. One can build an advanced SDSS that might solve complex problems, but if the user interfaces do not allow easy use, there is a high possibility for failure of the system. Some of the following characteristics should be considered during user interface design.

In the design of an effective user interface, Malczewski (1999) summarized five issues that need to be considered: (1) accessibility, (2) flexibility, (3) interactivity, (4) ergonomic layout, and (5) processing-driven functionality. By accessibility, he meant that the user interfaces should be intuitive, facilitating new users' applications. The ability to recover from unintended or mistaken actions would constitute a flexible system. An interactive system would allow efficient information flow back and forth between the user and the system itself. An ergonomic layout implies efficient communication between the user and the system. A processing-driven interface allows the user to understand the upcoming and completed tasks clearly. All of these characteristics can be met by careful planning of the system in advance with input from users, thoroughly documented and commented software code, and significant software testing by potential users.

As discussed in Chapter 2, the evolution of user interfaces in GIS and SDSS has followed that of computer software in general. This progression broadly included a move from command-line driven applications to more and more sophisticated graphical user interfaces. Earlier hardware and software configurations of SDSS generally were characterized by fairly complicated command-line-driven GIS and modeling software. These systems required someone with significant experience in the spatial sciences and often in computer programming as well. In many of the SDSS from the 1980s and 1990s, ArcInfo Workstation software was used. This software generally required significant expertise and operated with various modules and commands typed into a command line window requiring strict syntax compliance (Figure 4.9). The development of customized user interfaces for SDSS applications was possible but often required the users to type in commands to run different modules of the system. In addition, the display of outputs, such as maps, was generally basic with limited or unintuitive interactive capabilities for nonexpert users (Figure 4.10). In the 1990s, the introduction of Windows-based computing (Figure 4.11) and the development of GIS and modeling software with more intuitive graphical user interfaces greatly increased the number of organizations using these types of software. In addition, the number of individuals able to use and understand at least the basic functionality of GIS rose greatly during this time period. The original desktop personal computer GIS was

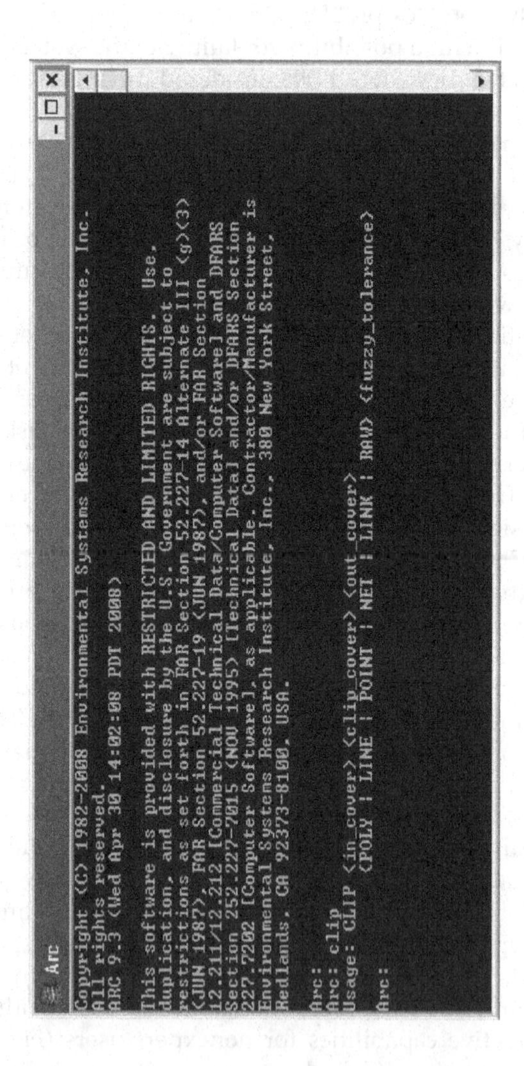

FIGURE 4.9
The Arc command line window from ArcInfo Workstation.

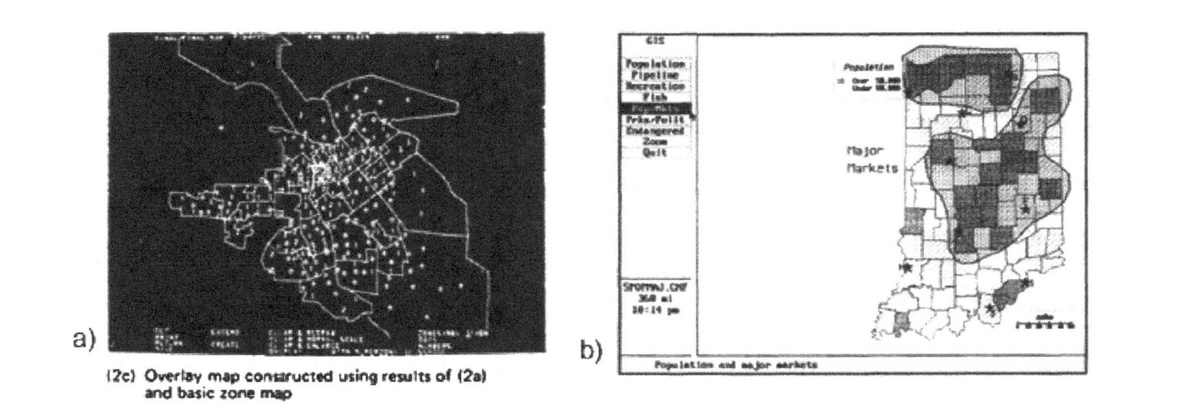

FIGURE 4.10
Two example map interfaces taken from papers describing the (a) Geo-data Analysis and Display System (taken from Carlson, E. D., B. F. Grace, and J. A. Sutton. 1977. Case studies of end user requirements for interactive problem-solving systems. *MIS Quarterly* 1(1):51–63) and (b) MapInfo software (taken from Crossland, M. D., and B. E. Wynne. 1994. Measuring and testing the effectiveness of a spatial decision support system. Paper presented at the 27th Annual International Conference on System Sciences, Honolulu, Hawaii.).

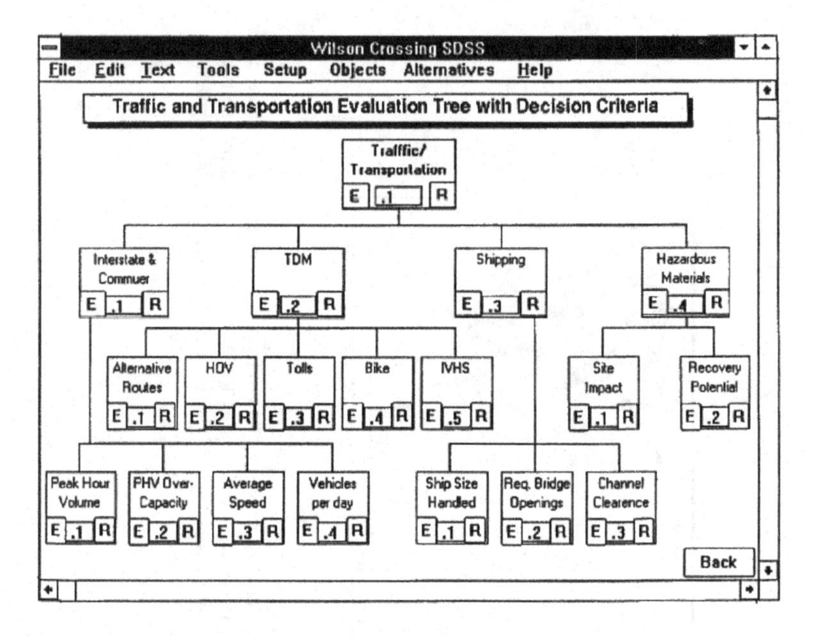

FIGURE 4.11
An example user interface for an SDSS from 1992. (Taken from Armour, F. J. 1992. Utilizing hypermedia in a MAU model-based spatial decision support system (SDSS). Paper presented at URISA 92 Annual Conference Proceedings, Washington, D.C.)

developed by MapInfo (Figure 4.10) in the latter half of the 1980s. In the late 1990s and the early 2000s, the GIS software that was most commonly used in SDSS applications became ESRI's ArcView (Figure 4.12). ArcView was originally designed to be a GIS data viewer, but after successive versions, became a more full-fledged GIS. This software became popular because of its user-friendly interfaces and also because the software came with a development language and environment called Avenue and a user-interface development environment, both of which provided great capabilities for developing customized functionality, tools, and interfaces (Figure 4.13). Similar software with user interfaces and development environments were developed by other companies including MapInfo and Intergraph's GeoMedia, and these were also utilized in SDSS. Further evolution in user interfaces has taken place with new generations of commercial and open source GIS software. ArcGIS, from ESRI, has evolved from ArcInfo Workstation and ArcView software and has become the most commonly used GIS software in recent years in SDSS applications. The ArcGIS software comes packaged with Visual Basic for Applications (VBA), which allows the development of graphical user interfaces and customized GIS applications. ArcGIS provides the extensive functionality that evolved over decades in the ArcInfo workstation software and also

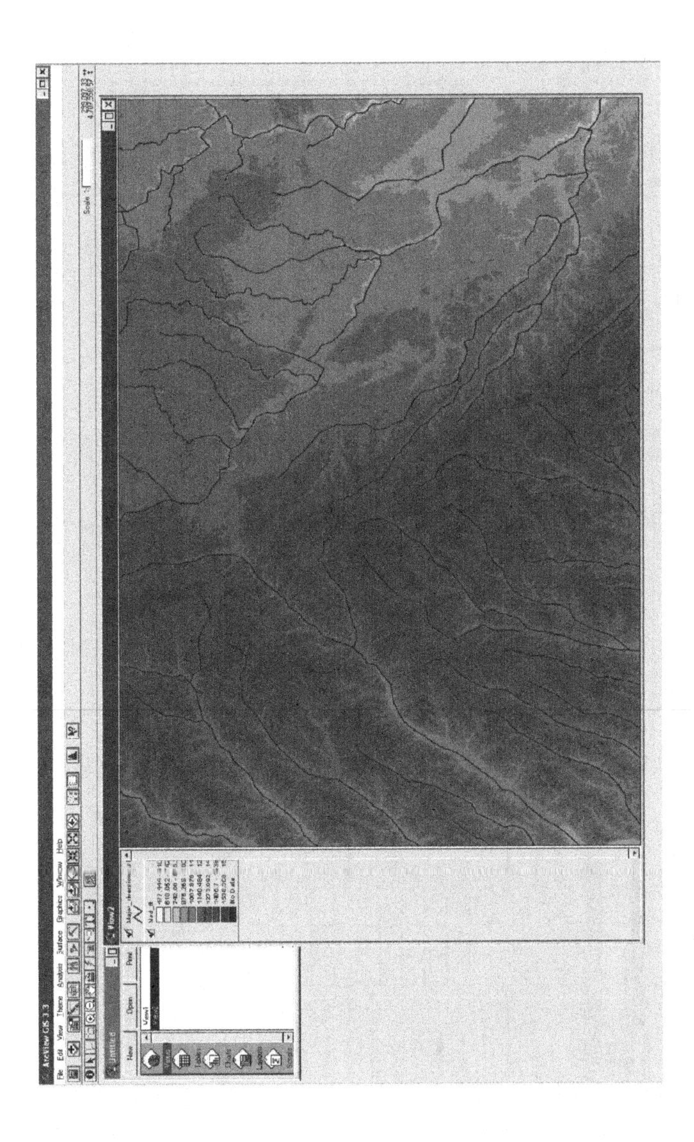

FIGURE 4.12
ArcView GIS 3.3 user interface.

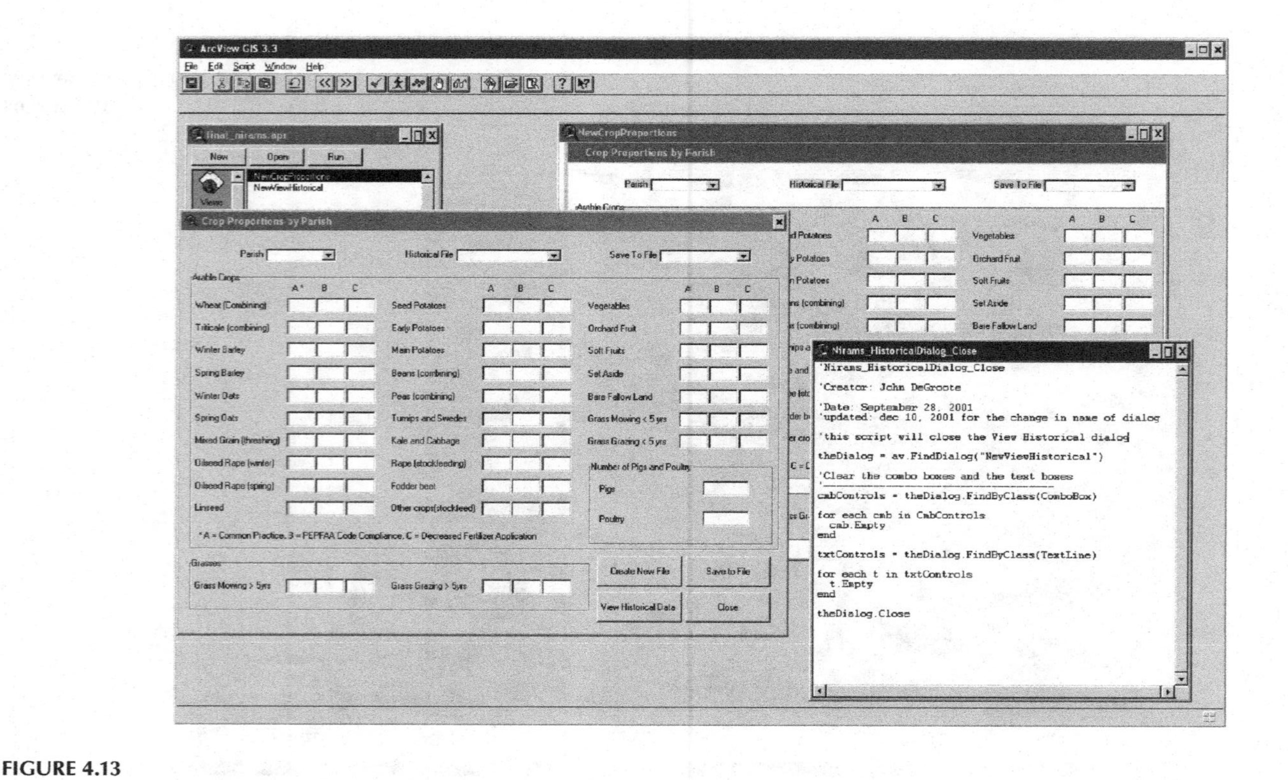

FIGURE 4.13
The Avenue scripting interface and the Dialog Designer extension interface in ArcView 3.3.

modern graphical user interfaces. The ArcGIS software is also built on a set of objects called ArcObjects that can be accessed by users through a wide variety of application programming interfaces or development environments, including VBA. These capabilities allow for the development of customized GIS tools, algorithms, and user interfaces, thus providing greater context for SDSS development. The ArcGIS desktop environment was used in 15 out of 28 research articles published in 2008 detailing an application of SDSS.

The development of more intuitive and user-friendly interfaces across various platforms (desktop, Web, etc.) has led to the inclusion of more users and decision makers in hands-on applications of the software components of SDSS. An SDSS-related research area since the 1990s has been that of public participatory GIS, which was originally defined as "a variety of approaches to make GIS and other spatial decision-making tools available and accessible to all those with a stake in official decisions" (Schroeder 1996). Stasik and Jankowski (1997) discussed the development of an Internet-based decision support system using ESRI's Map Objects. The system was called Spatial Understanding and Decision Support System (SUDSS). The system was designed to facilitate group collaboration and was used in an experimental deliberation for a realistic land use planning problem. Beedasy and Ramloll (2000) discussed the development of collaborative SDSS, stressing that simple interface considerations can have large impacts. The interface design of their system (CoSpaME, Figure 4.14) was based on MacroMedia Flash and was built with a number of software widgets that allowed interaction with maps by multiple users and also interaction between users. Carver et al. (2000 and 2001) discussed the potentials presented by developing SDSS on the World Wide Web (WWW) for improving environmental decision making. In the examples they provided, the complexity of the spatial operations taking place was hidden from the user behind user-friendly graphical interfaces. They argued at the time that providing open access to decision-making problems over the WWW using GIS functionality would become more common and important. Barton et al. (2005) argued that the technology necessary for enabling the public to participate online in a geographic context was available and mature, but there is tremendous potential for misuse or abuse, arguing for careful design of systems including user interfaces. Jankowski et al. (2006) described the development of a participatory SDSS called WaterGroup that used ArcGIS software and technology for multiple computers to communicate with each other. They argued for the necessity of developers receiving feedback from end users both when defining system requirements and when usability of the software is evaluated. In a 2006 review of public participatory GIS, Sieber (2006) mentioned that improvements in software and user interfaces allow policies to be determined remotely and actually lead to less local public participation. Jarupathirun and Zahedi (2007)

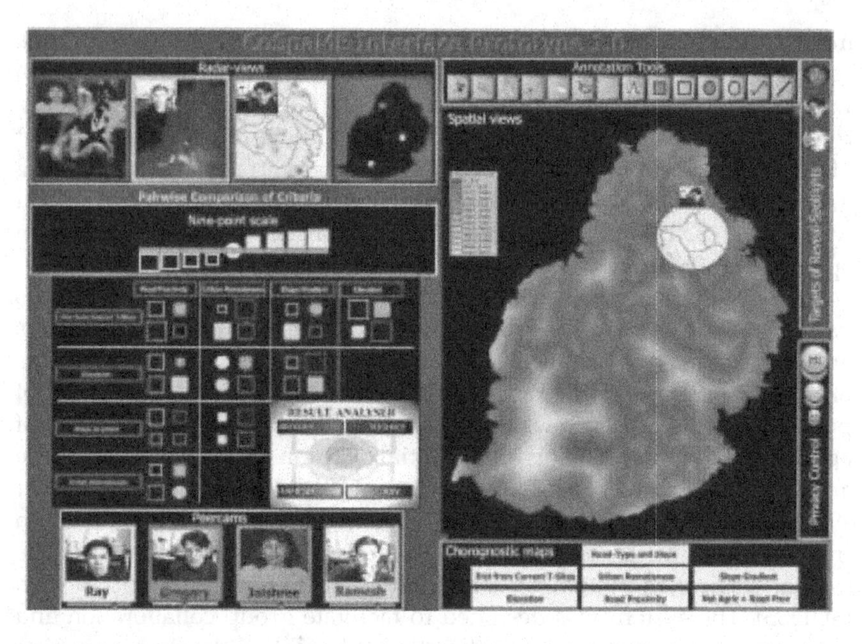

FIGURE 4.14
User interface for the CoSpaME system.

argued for intelligent interfaces that could help guide decision makers in using SDSS.

There has been a large push in development of cartographic and spatial analysis capabilities over the Web. There have been numerous commercial and free open source Web-based geospatial software packages developed that have been used at least as a component of SDSS. Web mapping software became popular in the latter half of the 1990s and early 2000s, with software such as ESRI's ArcIMS being commonly used for presenting spatial data online. Over the last decade, the ArcIMS software has been used in a multitude of SDSS applications (e.g., Dymond et al. 2004; Johnson 2005; Karnatak et al. 2007). ArcIMS has been used for presenting cartographic visualization capabilities in SDSS applications including participatory applications (Wan et al. 1999). Such applications as Google Maps and Google Earth are being utilized by many researchers for presenting results in a geographic context. However, as Andrienko et al. (2007) pointed out, these technologies are insufficient to provide effective decision support as they lack many necessary capabilities such as discerning information from various data sources, scenario development, data analysis, and visualization. The free open source software MapServer was used by Engel et al. (2003) in a watershed SDSS for presenting cartographic results. Taweepworadej et al. (2006) similarly used MapServer in an SDSS for point source pollution analysis in Thailand. Best et al. (2007)

developed a multicomponent SDSS that includes map visualizations using MapServer and Google Earth. Park et al. (2008) discussed the usefulness of using an online geospatial processing and visualization software program such as ArcGIS server for public participatory GIS.

Other technologies will also likely contribute to the adaption of SDSS for spatial decision-making situations. Interactive technologies such as Microsoft Surface or GeoWall will provide tools for spatial investigation and collaboration. Microsoft Surface is a multitouch computer that responds to natural hand gestures. This type of technology allows multiple users to collaboratively and simultaneously interact with data and each other (Microsoft 2009). The GeoWall mission is to expand scientific visualization tools for research and educational purposes (GeoWall 2009). Continuing advances and improvements in user interface development across hardware and software settings will continue with Web-based applications becoming even more prevalent.

The successful uptake of SDSS depends on the user's ability to successfully and efficiently negotiate the software interfaces. Beedsay and Ramloll (2000) stressed that significant design effort, including that for user interfaces, is necessary to develop successful collaborative SDSS. Rinner et al. (2005) mentioned that user interface design is somewhat dependent on the complexity of the application, and they provided some general recommendations for design, including providing results in tabular and map formats and the use of radio buttons for defining user weights with general categories such as high/medium/low instead of using percent slider-bar features.

4.5 Stakeholder Component (SC)

An important aspect of decision systems that is traditionally not explicitly discussed in DSS literature, and SDSS literature specifically, is the role of the stakeholders and decision makers. In a spatial decision-making situation, there are a wide variety of individuals and organizations that might have a stake in the potential outcomes. The successful application of an SDSS to a spatially dependent problem is dependent upon the effective involvement of a wide array of potential players. The different stakeholders function in various roles in the overall design, development, implementation, and usage of an SDSS. The general categories of stakeholders in situations where SDSS are applied include the decision maker or end user, the analyst, the developer or builder, and the expert (Figure 4.15).

The expert is often, but not always, the person who is a proponent for the SDSS, as he or she sees the potential value in the development of decision

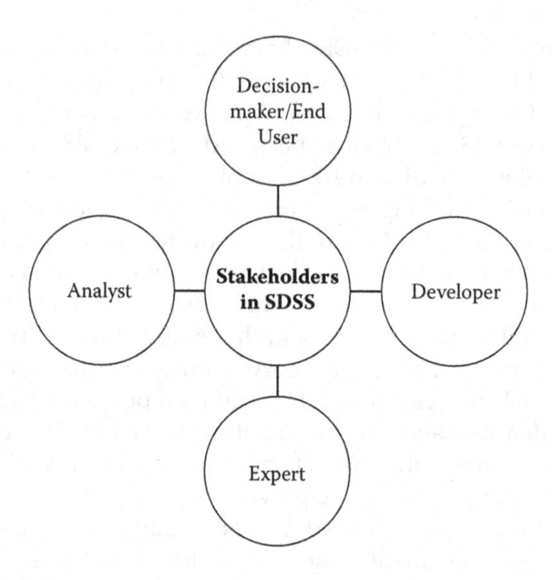

FIGURE 4.15
Stakeholders involved in SDSS.

support mechanisms. The expert has detailed knowledge in some crucial aspect of the spatial decision problem at hand and is familiar with the variety of techniques (e.g., software, algorithms, monitoring) that are available or possible for the development of tools that will help to address the spatial decision problem. At other times, experts play a smaller, more specific role in that they are experts in one aspect of the spatial decision problem. In either instance, the expert usually can provide unique knowledge and insight into the problem situation that can be incorporated into the SDSS. He or she might work with the analyst, decision makers, or architect in developing the knowledge or modeling components or by providing expert advice in the parameterization of the modeling component. However, an expert might not be aware of location-specific conditions in a given area, which are important to local spatial decision-making situations. Indeed, Strager and Rosenberger (2006) demonstrated differences when comparing groups representing outside experts and local stakeholders in a spatial multi-criteria analysis.

The developer of an SDSS collects requirements from end users, designs system architectures, develops user interfaces, and programs the functionality of the system. The design phase of SDSS development is crucial and requires input from the full spectrum of potential stakeholders and users. Jankowski et al. (2006) pointed out that knowledge of users' information needs can lead to the development of an SDSS that can fit the decision process. Systematic investigation of potential user and stakeholder needs should thus precede software development. Sahota and Jeffrey (2006) stressed that limited involvement of users in the development phase was an important reason for the lack of decision support tool uptake. They cited other studies that identified reasons for lack of uptake, including the fact that tools are time consuming to use, too complex, and had too much uncertainty in outputs. Defining user needs in planning stages can help to ameliorate these problems.

The analyst is a person who is often involved in selecting models, carrying out simulations in the SDSS, analyzing data, producing outputs, and interpreting results, which are used to aid the decision makers. Given the often GIS-centric nature of SDSS, this person might be a GIS analyst or a person familiar with the modeling aspects of the SDSS. There often might be more than one analyst, as there might be a modeling specialist and a GIS analyst. As large ill-structured spatial decision problems often call for an emphasis on multidisciplinary team approaches (Ascough et al. 2002), these situations will likely have multiple analysts from the different disciplines.

The decision makers are the stakeholders at the end of the process who need to be presented with meaningful information regarding various scenarios that deal with the spatial problem at hand and that can be used to make decisions. The decision makers rely on the experts and analysts to provide useful information through the SDSS and meaningful interpretation in order to aid in their decision making. Although these four separate roles in SDSS are somewhat distinct, there are often situations in which a single individual may operate in more than one role depending on the nature and size of the spatial decision problem and also the user's level of expertise.

Given the complex and multidisciplinary nature of many of the spatial decision problems in which an SDSS is used, it is not surprising that the ability to successfully manage the interactions of various stakeholders plays a key role in the success of the SDSS application. At a most basic level, the logistics can be difficult. Jankowski et al. (1997) noted the difficulties in having a meeting where participants are attending at different locations and at the same time or where participants are attending at different locations and different times. The vast improvement in communication, Internet, and other technologies has helped to alleviate some of these problems. Indeed, Ascough et al. (2002) noted that the ability to quickly access and process large spatial datasets over networks would offer tremendous improvement in how multi-criteria SDSS would be developed and used. Jankowski et al. (2006) identified the key issue in developing a collaborative SDSS as the anticipation of users' needs. This anticipation requires the involvement of various stakeholders at early stages of the spatial decision-making process. They demonstrated the utility of this concept for a spatiotemporal decision problem involving conjunctive water administration in Idaho. Ascough et al. (2002) posited that many SDSS tools were too complex for clients or users. They also pointed out that users frequently complain that not all objectives of stakeholders are captured. As is evinced by numerous authors (Ascough et al. 2002; Hirschfeld et al. 2005; Jankowski et al. 2006; Miller et al. 2004; Nyerges et al. 2006), substantive involvement across a spectrum of stakeholders is necessary throughout the SDSS development process including planning, development, testing, and application. Nygerges et al. (2006) specifically stated that stakeholders' perceptions of problems must be considered early in the process. Miller et al. (2004) pointed out that many management decisions are moving to a bottom-up approach with more stakeholders without expertise being involved in the process. The success of bottom-up decision making depends on the education of stakeholders about the issues and processes, and SDSS can be important for this purpose. Many of the social and institutional aspects of decision making have been studied in other disciplines, while in SDSS research there has been more of a focus on technological issues. Nyerges et al. (2006) argued for crossover research in deliberative democracy with participatory GIS as many public problems have geospatial underpinnings. They also point out that there are three main constructs to consider at the beginning of a complex decision process: (1) institutional influences such as mandates in laws or regulations, (2) group participant influences such as who is involved, and (3) information technology considerations. Much of the SDSS literature has focused on the third aspect while the first two can be crucial to the success of using information technology in ways such as SDSS. Uran and Janssen (2003) examined why many SDSS are underused and found that one reason is that support for evaluating outputs generated by SDSS was limited, leaving users unsatisfied. They traced this back to vague specifications defined during planning stages and stressed that the need for a closer link between developers and users is the most important lesson to be learned. Thus, more research is necessary into the social and institutional aspects of SDSS applications, and significant attention needs to be paid in carefully planning the development of SDSS based on communication between developers, analysts, experts, and decision makers or users.

4.6 Knowledge Management Component

A knowledge management system (KMS) is not an essential component of an SDSS but has been included in many SDSS. The purpose of a knowledge management component (KMC) is to provide expert knowledge that can aid users in finding a solution to the specific problem or to provide guidance to novice users in the overall decision-making process and also

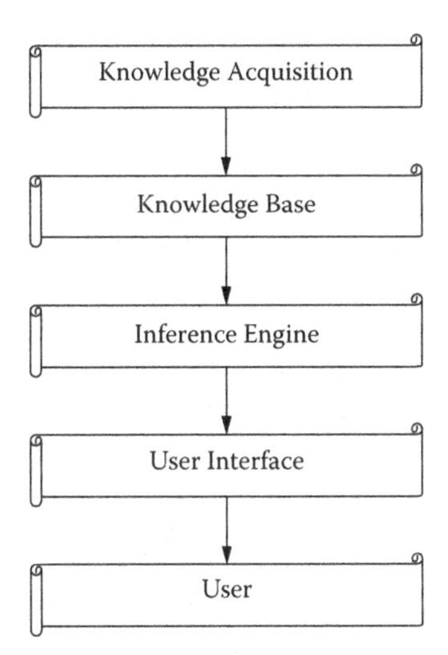

FIGURE 4.16
Typical system architecture used in the knowledge management component.

in selection of analytical models. Knowledge management systems are computer programs that manipulate a knowledge base to solve problems. The knowledge base is generally coded into a set of rules that are used to replicate decision-making processes by humans (Ehler 1994). The knowledge component software is usually composed of a knowledge base, inference engine, and user interface (Figure 4.16). The knowledge base (KB) is composed of domain-specific facts and knowledge-based rules, which are used by the inference engine. The inference engine uses some programmed logic to make decisions based on those rules and facts (Ehler 1994). Armstrong et al. (1990) argued that decision support systems should contain a repository of expert knowledge to help guide users and alleviate the need for experts to be called upon too frequently. Zhu et al. (1996) broke knowledge bases into five general categories: domain knowledge, model knowledge, utility program knowledge, metadata, and process knowledge. The model knowledge, in their example, provided descriptions of models, rules for selection of appropriate models, and assignment of relevant parameter values. The utility program knowledge provided information to users about tangential tools. Metadata provided information about data utilized and process knowledge was used to guide successful applications of steps in running SDSS. Peterson (1998) defined the knowledge base as "the collection of facts, definitions, rules of thumb, and computational procedures that apply to the domain." In order to develop a knowledge base, knowledge has to be acquired from experts and transformed into a set of rules and facts (Figure 4.16). The knowledge base is then analyzed by the inference engine to reach conclusions in the decision process. The user interface provides the link between the user, the knowledge base, and the inference engine.

There are numerous commercial, free, and open source knowledge base system shells for the development of knowledge-based systems. For example, CLIPS (C Language Interface Production System), JLisa, Mandarex, and TyRuBa are some of the free KB shells. Some of the commercial shells include Jess, EXSYS, Teknowledge, OpenRules, and Gemsym.

Many SDSS have included a knowledge base component. In the development of an SDSS for ameliorating gypsy moth damage to deciduous forests, Elmes and Cai (1992) built in a knowledge base component and a graphical GIS tutorial that provided users with guidance in selecting inputs, operations, and data characteristics required for spatial analysis. Their knowledge management system had a user needs analysis subsystem, project planning subsystem, GIS processing subsystem, product evaluation, and refinement system. Yeh and Qiao (2004) used a meta-model knowledge base to provide users with intelligent support in choosing appropriate models in a planning support SDSS. Witlox (2005) stressed that certain conditions need to be met for using knowledge-based systems, including that the knowledge is specialized and based on true expert input and that the tasks to be addressed with KBS are not trivial or overly complicated. Scheibe et al. (2006) utilized a KBS for an SDSS addressing the location of wireless broadband communication systems. The system had built-in knowledge such as information on curvature of the Earth, degradation of signals over distances, and characteristics of radio frequency ranges. Sugumaran et al. (2007) used an expert system for the development of an intelligent SDSS for snow plow routing in Iowa, USA.

4.7 Summary

As defined in Chapter 3, decision support systems (DSS) traditionally have been considered to require at least three key components: a database management system, a model base management system, and a dialog or user interface management system. Spatial decision support systems are unique in that they generally rely on spatial modeling and analysis functionality, usually from GIS software. Due to the GIS-centric nature of SDSS, Chapter 3 focused on covering all the basic aspects of GIS software that are utilized in SDSS. This chapter has attempted to give a thorough introduction to the remaining components of SDSS: model management, dialog management, stakeholders, and knowledge management systems (not in all SDSS). Numerous generic modeling techniques commonly used within SDSS frameworks were reviewed. These methodologies included fairly straightforward methodologies easily implemented in GIS and SDSS frameworks including Boolean overlay and weighted linear combination. Although these techniques have been utilized for SDSS applications such as evaluating ecological sensitivity by watershed and tourism planning, they are generally considered rudimentary and not capable of capturing the complexities of spatial decision problems and the subjectivity in choosing weights for the latter technique. More sophisticated multi-criteria techniques such as AHP and OWA have been introduced into SDSS to provide more robust handling of the complexity of spatial decision-making problems. These techniques have proven useful in a wide variety of application areas as they provide a level of sophistication often necessary while being amenable to implementation in GIS software.

In more recent years, artificial intelligence methods have been incorporated into spatial analysis systems, including SDSS. Artificial neural networks imitate the workings of a human brain and are useful for situations with great uncertainty or for large datasets. Genetic algorithms are methods that search large datasets and eliminate parts that lead to poor solutions. This technique is again useful when datasets are large and the spatial problem is complex. These techniques are less transparent and thus difficult for users to understand. These methods (WLC, AHP, OWA, ANN, GA, etc.) are generic in nature and can be applied to a variety of spatial decision problems. There are many SDSS applications that use modeling techniques that are specific to certain phenomena. This chapter highlighted some commonly used types of specific models used in SDSS, including watershed and hydrological models, which approximate physical and chemical processes, and biological models, which estimate biological and ecological processes. Other commonly used types of models included in SDSS applications were economic, location allocation, network/transportation, and public health techniques.

The evolution over the last few decades of software user interfaces has proved important in the evolution and increased application of SDSS. There has been considerable evolution in the development and sophistication of user interfaces in SDSS. The evolution toward friendlier user interfaces has led to the democratization of the use of SDSS to a greater number of stakeholders and potential users. In the 1980s and first half of the 1990s, many SDSS applications relied on GIS and other software components that operated from command-line-driven interfaces that required significant expertise. The movement toward graphical user interfaces in GIS and other software led to the greater uptake of use in GIS software and subsequently SDSS. In addition, easier-to-use development environments within (e.g., Avenue and Dialog Designer in ArcView 3.x, VBA in ArcGIS) and outside (e.g., Visual Basic) GIS software allowed the development of customized interfaces and applications. The ability to develop user-friendly customized applications opened up the development of SDSS to a wider range of practitioners, which opened up the use of SDSS to a greater number of users and decision makers. In recent years, Web-based geospatial applications have become more prominent and are likely in the future to become more common in SDSS applications and allow for more effective public participation or collaborative applications.

The successful application of SDSS depends on the stakeholders that are involved in their design, development, and use. We have identified four major stakeholders in SDSS applications including developers, analysts, experts, and users. These stakeholder positions are not mutually exclusive, and a single individual may qualify as, for example, a developer and an analyst or an expert and an analyst depending on the level of complexity of the spatial decision. Past research has indicated that the success of an SDSS requires involvement of users from an early stage. There should be continual interaction between all of the stakeholders throughout the development, testing, and application phases. This is often difficult to do based on logistical constraints. However, with greater network and communication abilities, these logistical constraints can be more easily overcome. Thus, one of the most important lessons learned in SDSS development is the early and frequent involvement of the end user in conception, development, testing, and final use of any SDSS.

References

Ahmad, I., S. Azhar, and P. Lukauskis. 2004. Development of a decision support system using data warehousing to assist builders/developers in site selection. *Automation in Construction* 13(4):525–542.

Andrienko, G., N. Andrienko, P. Jankowski, D. Keim, M-J. Kraak, A. MacEachren, and S. Wrobel. 2007. Geovisual analytics for spatial decision support: Setting the research agenda. Special issue of the *International Journal of Geographical Information Science* 21(8):839–857.

Armour, F. J. 1992. Utilizing hypermedia in a MAU model-based spatial decision support system (SDSS). Paper presented at URISA 92 Annual Conference Proceedings, Washington, D.C.

Armstrong, M. P., S. De, P. J. Densham, P. Lolonis, G. Rushton, and V. Tewari. 1990. A knowledge-based approach for supporting locational decision-making. *Environment and Planning B: Planning and Design* 17:341–364.

Armstrong, M. P., G. Rushton, R. Honey, B. Dalziel, P. Lolonis, S. De, and P. J. Densham. 1991. A spatial decision support system for regionalizing service delivery systems. *Computers, Environment and Urban Systems* 15:37–53.

Ascough, J., II, H. Rector, D. Hoag, G. McMaster, B. Vandenberg, M. Shaffer, M. Weltz, and L. Ahuj. 2002. Multicriteria spatial decision support systems for agriculture: Overview, applications, and future research directions. Paper presented at the Integrated Assessment and Decision Support (iEMSs 2002), Lugano, Switzerland.

Assaf, H., and M. Saadeh. 2008. Assessing water quality management options in the Upper Litani Basin, Lebanon, using an integrated GIS-based decision support system. *Environmental Modelling & Software* 23(10–11):1327–1337.

Barton, J., J. Plume, and B. Parolin. 2005. Public participation in a spatial decision support system for public housing. *Computers, Environment and Urban Systems* 29(6):630–652.

Beedasy, J., and D. Whyatt. 1999. Diverting the tourists: A spatial decision-support system for tourism planning on a developing island. *International Journal of Applied Earth Observation and Geoinformation* 1(3):163–174.

Beedasy, J., and R. Ramloll. 2000. User interface considerations to facilitate negotiations in collaborative spatial decision support systems. Paper presented at the GIS Research U.K. National Conference, York, U.K.

Bennett, D., and A. J. Vitale. 2001. Evaluating nonpoint pollution policy using a tightly coupled spatial decision support system. *Environmental Management* 27(6):825–836.

Best, B. D., P. N. Halpin, E. Fujioka, A. J. Read, S. S. Qian, L. J. Hazen, and R. S. Schick. 2007. Geospatial web services within a scientific workflow: Predicting marine mammal habitats in a dynamic environment. *Ecological Informatics* 2(3):210–223.

Boroushaki, S., and J. Malczewski. 2008. Implementing an extension of the analytical hierarchy process using ordered weighted averaging operators with fuzzy quantifiers in ArcGIS. *Computers & Geosciences* 34:399–410.

Bryan, B. A. 2003. Physical environmental modeling, visualization and query for supporting landscape planning decisions. *Landscape and Urban Planning* 65(4):237–259.

Burrough, P. A., and R. A. McDonnell. 1998. *Principles of geographical information systems.* New York: Oxford University Press.

Carlson, E. D., B. F. Grace, and J. A. Sutton. 1977. Case studies of end user requirements for interactive problem-solving systems. *MIS Quarterly* 1(1):51–63.

Carver, S., A. Evans, R. Kingston, and I. Turton. 2000. Accessing geographical information systems over the World Wide Web: Improving public participation in environmental decision-making. *Information Infrastructure and Policy* 6:157–170.

Carver, S., A. Evans, R. Kingston, and I. Turton. 2001. Public participation, GIS, and cyberdemocracy: Evaluating on-line spatial decision support systems. *Environment and Planning B: Planning and Design* 28:907–921.

Chakhar, S., and J. M. Martel. 2004. Towards a spatial decision support system: Multi-criteria evaluation functions inside geographical information systems. *Annales du LAMSADE* 2(2):97–123.

Chang, N-B., G. Parvathinathan, and J. B. Breeden. 2008. Combining GIS with fuzzy multicriteria decision-making for landfill siting in a fast-growing urban region. *Journal of Environmental Management* 87:139–153.

Chang, N-B., Y. L. Wei, C. C. Tseng, and C. Y. J. Kao. 1997. The design of a GIS-based decision support system for chemical emergency preparedness and response in an urban environment. *Computers, Environment and Urban Systems* 21(1):67–94.

Cheng, C. H. 1996. Evaluating naval tactical missile systems by fuzzy AHP based on the grade value of membership function. *European Journal of Operational Research* 96:343–350.

Cheng, C. H. 1999. Evaluating weapon systems using ranking fuzzy numbers. *Fuzzy Sets and Systems* 107:25–35.

Choi, K. 1996. A development of expert spatial decision support system for optimum route planning and economic feasibility. Paper presented at the Geographic Information Systems for Transportation (GIS-T) Symposium, Kansas City.

Choi, J-Y., B. A. Engel, and R. L. Farnsworth. 2005. Web-based GIS and spatial decision support system for watershed management. *Journal of HydroInformatics* 7:165–174.

Collins M. G., F. R. Steiner, and M. J. Rushman. 2001. Land-use suitability analysis in the United States: Historical development and promising technological achievements. *Environmental Management* 28:611–621.

Crossland, M. D., and B. E. Wynne. 1994. Measuring and testing the effectiveness of a spatial decision support system. Paper presented at the 27th Annual International Conference on System Sciences, Honolulu, Hawaii.

Deal, B., and D. Schunk. 2004. Spatial dynamic modeling and urban land use transformation: A simulation approach to assessing the costs of urban sprawl. *Ecological Economics* 51 (1–2):79–95.

Deveson, T., and D. Hunter. 2008. The operation of a GIS-based decision support system for Australian locust management. *Insect Science* 9(4):1–12.

Downs, J. A., and M. W. Horner. 2008. Spatially modelling pathways of migratory birds for nature reserve site selection. *International Journal of Geographical Information Science* 22(6):687–702.

Dymond, R. L., B. Regmi, V. K. Lohani, and R. Dietz. 2004. Interdisciplinary web-enabled spatial decision support system for watershed management. *Journal of Water Resources Planning and Management* 130(4):290–300.

Eastman J. R., and H. Jiang. 1996. Fuzzy measures in multi-criteria evaluation. Paper presented at the International Symposium on Spatial Accuracy Assessment in Natural Resources and Environmental Studies, Fort Collins, Colorado.

Eastman, J. R. 1997. IDRISI for Windows, Version 2.0: Tutorial Exercises Graduate School of Geography, Clark University, Worcester, MA.

Ehler, G. 1994. Expertise: A knowledge-based spatial decision support system for industrial site evaluation. Paper presented at the Fourteenth Annual ESRI User Conference, Palm Springs, California.

Elmes, G. A., and G. Cai. 1992. Data quality issues in user interface design for a knowledge-based decision support system. Paper presented at the Fifth International Symposium on Spatial Data Handling, Charleston, South Carolina.

Engel, B. A., J-Y. Choi, J. Harbor, and S. Pandey. 2003. Web-based DSS for hydrologic impact evaluation of small watershed land use changes. *Computers and Electronics in Agriculture* 39(3):241–249.

Engel, B. A., R. Srinivasan, and C. Rewerts. 1993. A spatial decision support system for modeling and managing agricultural non-point source pollution. In *Environmental modeling with GIS*, ed. M. F. Goodchild, B. O. Parks, and L. T. Steyaert, 231–237. New York: Oxford University Press.

Ferrand, N. 1996. Modelling and supporting multi-actor spatial planning using multi-agents systems. Paper presented at the Third NCGIA Conference on GIS and Environmental Modelling, Santa Fe, New Mexico.

GeoWall. 2009. An introduction to the GeoWall. http://www.geowall.org/intro.html (accessed November 11, 2009).

Gheorghe, A. V., and D. Vamanu. 1995. A pilot decision support system for nuclear power emergency management. *Safety Science* 20(1):13–26.

Gimblett, H. R., C. Roberts, and T. C. Daniel. 2000. Intelligent agent modeling for simulating and evaluating river trip scheduling scenarios for the Grand Canyon National Park. In *Integrating geographic information systems and agent-based modeling techniques for understanding social and ecological processes*, ed. H. R. Gimblett, 245–276. New York: Oxford University Press.

Gimblett, R. H., G. L. Ball, and A. W. Guise. 1994. Autonomous rule generation and assessment for complex spatial modeling. *Landscape and Urban Planning* 30:13–26.

Girvetz, E., and F. Shilling. 2003. Decision support for road system analysis and modification on the Tahoe National Forest. *Environmental Management* 32(2):218–233.

Gorr, W., M. Johnson, and S. Roehrig. 2001. Spatial decision support system for home-delivered services. *Journal of Geographical Systems* 3(2):181–197.

Heywood, I., J. Oliver, and S. Tomlinson, 1995. Building an exploratory multi-criteria modeling environment for spatial decision support, In *Innovations in GIS 2*, ed. P. Fisher, 127–136. London: Taylor & Francis.

Hirschfeld, J., A. Dehnhardt, and J. Dietrich. 2005. Socioeconomic analysis within an interdisciplinary spatial decision support system for an integrated management of the Werra River Basin. *Limnologica—Ecology and Management of Inland Waters* 35(3):234–244.

Holsapple, C. W., and A. B. Whinston. 1976. A decision support system for area-wide water quality planning. *Socio-Economic Planning Sciences* 10:265–273.

Jankowski, P., T. Nyerges, S. Robischon, K. Ramsey, and D. Tuthill. 2006. Design considerations and evaluation of a collaborative, spatio-temporal decision support system. *Transactions in GIS* 10(3):335–354.

Jankowski, P., T. L. Nyerges, A. Smith, T. J. Moore, and E. Horvath. 1997. Spatial group choice: A SDSS tool for collaborative spatial decision-making. *International Journal of Geographical Information Science* 11(6):577–602.

Jarupathirun, S., and F. M. Zahedi. 2007. Exploring the influence of perceptual factors in the success of web-based spatial DSS. *Decision Support Systems* 43(3):933–951.

Jiang, H., and J. R. Eastman. 2000. Application of fuzzy measures in multi-criteria evaluation in GIS. *International Journal of Geographical Information Systems* 14:173–184.

Johnson, M. P. 2005. Spatial decision support for assisted housing mobility counseling. *Decision Support Systems* 41 (1):296–312.

Karnatak, H. C., S. Saran, K. Bhatia, and P. S. Roy. 2007. Multicriteria spatial decision analysis in Web GIS environment. *Geoinformatica* 11(4):407–429.

Kim, Y. S. 2003. Temporal-spatial decision support system. Paper presented at the Annual Meeting of the Decision Sciences Institute, Washington, D.C.

Klungboonkrong, P., and M. A. P. Taylor. 1998. A microcomputer-based-system for multicriteria environmental impacts evaluation of urban road networks. *Computers, Environment and Urban Systems* 22(5):425–446.

Kwan, M-P, and J. Lee. 2005. Emergency response after 9/11: The potential of real-time 3D GIS for quick emergency response in micro-spatial environments. *Computers, Environment and Urban Systems* 29(2):93–113.

Lagacherie, P., D. R. Cazemier, R. Martin-Clouaire, and T. Wassenaar. 2000. A spatial approach using imprecise soil data for modelling crop yields over vast areas. *Agriculture, Ecosystems & Environment* 81(1):5–16.

Lai, S. K., and L. D. Hopkins. 1989. The meanings of trade-offs in multiattribute evaluation methods: A comparison. *Environment and Planning B: Planning and Design* 16:155–170.

Lant, C. L., S. E. Kraft, J. Beaulieu, D. Bennett, T. Loftus, and J. Nicklow. Using GIS-based ecological-economic modeling to evaluate policies affecting agricultural watersheds. *Ecological Economics* 55:467–484.

Lee, A. H. I., W. C. Chen, and C. J. Chang. 2008. A fuzzy AHP and BSC approach for evaluating performance of IT department in the manufacturing industry in Taiwan. *Expert Systems with Applications* 34:96–107.

Leenhardt, D., J-L. Trouvat, G. Gonzales, V. Perarnaud, S. Prats, and J-E. Bergez. 2004. Estimating irrigation demand for water management on a regional scale II. Validation of ADEAUMIS. *Agricultural Water Management* 68:233–250.

Lempert, R. 2002. Agent-based modeling as organizational and public policy simulators. *Proceedings of the National Academy of Sciences of the United States of America* 99:7195–196.

Li, L., J. Wang, and C. Wang. 2005. Typhoon insurance pricing with spatial decision support tools. *International Journal of Geographical Information Science* 19(3):363–384.

Ligtenberg, A., A. Beulens, D. Kettenis, A. K. Bregt, and M. Wachowicz. 2009. Simulating knowledge sharing in spatial planning: an agent-based approach. *Environment and Planning B: Planning and Design* 36:644–663.

Lovejoy, S. B., J. G. Lee, T. O. Randhir, and B. A. Engel. 1997. Research needs for water quality management in the 21st century: A spatial decision support system. *Journal of Soil and Water Conservation* 52(1):18–22.

Luca, C. 2007. Generative platform for urban and regional design. *Automation in Construction* 16(1):70–77.

Malczewski, J. 2004. GIS-based land-use suitability analysis: a critical overview. *Progress in Planning* 62(1):3–65.

Malczewski, J., and M. Jackson. 2000. Multicriteria spatial allocation of educational resources: An overview. *Socio-Economic Planning Sciences* 34(3):219–235.

Malczewski, Jacek. 1999. *GIS and multicriteria decision analysis*. New York: John Wiley & Sons, Inc.

Manson, S. M. 2000. Agent-based dynamic spatial simulation of land-use/cover change in the Yucatan peninsula, Mexico. Paper presented at the 4th International Conference on Integrating GIS and Environmental Modeling (GIS/EM4): Problems, Prospects and Research, Banff, Canada.

Matthews, K. B., A. R. Sibbald, and S. Craw. 1999. Implementation of a spatial decision support system for rural land use planning: Integrating geographic information system and environmental models with search and optimisation algorithms. *Computers and Electronics in Agriculture* 23(1):9–26.

McKenzie, J. S., R. S. Morris, C. J. Tutty, and D. U. Pfeiffer. 1997. EpiMAN-TB, a decision support system using spatial information for the management of tuberculosis in cattle and deer in New Zealand. Paper presented at the second annual conference of Geocomputation'97, Dunedin, New Zealand.

Mendoza, J. E., A. L. Medaglia, and N. Velasco. 2009. An evolutionary-based decision support system for vehicle routing: the case of a public utility. *Decision Support Systems* 46(3):730-742.

Microsoft 2009. Microsoft surface. http://www.microsoft.com/surface/Pages/Product/WhatIs.aspx (accessed November 11, 2009).

Miller, R., P. Heilman, and D. Guertin. 2004. Information technology in watershed management decision making. *Journal of the American Water Resources Association* 40(2):347–357.

Næsset, E. 1997. A spatial decision support system for long-term forest management planning by means of linear programming and a geographic information system. *Scandinavian Journal of Forest Research* 12:77–88.

Neitsch, S. L., J. G. Arnold, J. R. Kiniry, and J. R. Williams. 2005. Soil and Water Assessment Tool (SWAT) User Manual. Version 2000. Temple, TX: USDA-Agricultural Research Service and Texas A&M Blackland Research Center.

Nyerges, T., P. Jankowski, D. Tuthill, and K. Ramsey. 2006. Collaborative water resource decision support: Results of a field experiment. *Annals of the Association of American Geographers* 96(4):699–725.

Odeyemi, I. A. O., D. C. Finnegan, N. B. Lilwall, R. M. Wilson, and R. L. Hodgart. 1998. Managing agricultural services delivery in less favoured areas: A role for geospatial models. Paper presented at the First International Conference on Geospatial Information in Agriculture and Forestry, Lake Buena Vista, Florida.

O'Sullivan, D., and D. J. Unwin. 2003. *Geographic information analysis.* Hoboken, NJ: Wiley.

Papadias, D., and M. Egenhofer. 1995. Qualitative collaborative planning in geographical space: Some computational issues. NCGIA Initiative, September 16–19, 1995.

Park, S., J. Lee, Y. Choi, and J. Y. Nam. 2008. To improve a public participation decision support system based on Web 2.0 technology. In *Proceedings of the Twenty-Eighth Annual ESRI User Conference*, San Diego, California.

Peterson, K. 1998. Development of spatial decision support systems for residential real estate. *Journal of Housing Research* 9(1):135–56.

Prato, T. 2001. Modeling carrying capacity for national parks. *Ecological Economics* 39:321–331.

Radke, J. 1995. A spatial decision support system for urban/wildland interface fire hazards. Paper presented at the ESRI International User Conference, Palm Springs, California.

Rao, M., G. Fan, J. Thomas, G. Cherian, V. Chudiwale, and M. Awawdeh. 2007. A web-based GIS decision support system for managing and planning USDA's conservation reserve program (CRP). *Environmental Modelling and Software* 22(9):1270–1280.

Rinner, C, and J. Malczewski. 2002. Web-enabled spatial decision analysis using ordered weighted averaging (OWA). *Journal of Geographical Systems* 4(4):385–403.

Rinner, C., Raubal, M. and B. Spigel. 2005. User interface design for location-based decision services. Paper presented at the 13th International Conference on GeoInformatics, Toronto, Canada.

Rodrigues, A., C. Grueau, J. Raper, and N. Neves. 1997. Environmental planning using spatial agents, Paper presented at Proceedings GIS Research in the U.K. 1997 (GISRUK '97), School of Geography, University of Leeds, U.K., April 9–11, 1997.

Roy, B. 1996. *Multicriteria methodology for decision aiding*. Dordrecht: Kluwer Academic Publishers.

Saaty, T. L. 1980. *The analytic hierarchy process: Planning, priority setting, resource allocation*. New York: McGraw-Hill International Book Co.

Scheibe, K. P., L. W. Carstensen Jr., T. R. Rakes, and L. P. Rees. 2006. Going the last mile: A spatial decision support system for wireless broadband communications. *Decision Support Systems* 42(2):557–570.

Schroeder, P. 1996. Report on public participation GIS workshop. In *GIS and Society: The Social Implications of How People, Space and Environment are Represented in GIS*, ed. T. Harris and D. Weiner. NCGIA Technical Report 96-97, Scientific Report for Initiative 19 Specialist Meeting. South Have, MN, March 2-5, 1996.

Sengupta, R., J. Beaulieu, and S. Kraft. 2003. Assisting decision-makers manage compensatory payments to preserve water quality in agricultural watersheds through modeling and simulation. In *Proceedings of the XIth World Water Congress,* Madrid, Spain. Carbondale, Illinois: International Water Resources Association.

Sengupta, R. R., and D. A. Bennett. 2003. Agent-based modeling environment for spatial decision support. *International Journal of Geographical Information Science* 17(2):157–180.

Shah, S.A.A., H. Kim, S. Baek, H. Chang, and B. H. Ahn. 2008. System architecture of decision support system for freeway incident management in Republic of Korea. *Transportation Research Part A* 42:799–810.

Shim, K. C., and S. B. Shim. 2000. A development of spatial decision support system for integrated river basin flood control. Paper presented at the 4th International Conference on Hydro-Science and Engineering, Seoul.

Sieber, R. 2006. Public participation geographic information systems: A literature review and framework. *Annals of the Association of American Geographers* 96(3):491–507.

Srinivasan, R., and B. A. Engel. 1994. A spatial decision support system for assessing agricultural nonpoint source pollution. *Water Resources Bulletin* 30(2):441–452.

Strager, M. P., and R. S. Rosenberger. 2006. Incorporating stakeholder preferences for land conservation: Weights and measures in spatial MCA. *Ecological Economics* 58(1):79–92.

Stasik, M., and P. Jankowski. 1997. Using MapObjects in the Internet-based decision support system paper text. Paper presented at the Seventeenth Annual ESRI User Conference, San Diego, California.

Stevens, D., S. Dragicevic, and K. Rothley. 2007. iCity: A GIS-CA modelling tool for urban planning and decision making. *Environmental Modelling and Software* 22(6):761–773.

Sugumaran, R., J. C. Meyer, and J. Davis. 2004. A web-based environmental decision support system (WEDSS) for environmental planning and watershed management. *Journal of Geographical Systems* 6(3):307–322.

Sugumaran, R., S. Ilavajhala, and V. Sugumaran. 2007. Development of a web-based intelligent spatial decision support system WEBSDSS: A case study with snow removal operations. In *Emerging spatial information systems and applications*, ed. B. N. Hilton, 184–202. Hershey, PA: Idea group.

Sugumaran, V. and R. Sugumaran. 2003. Spatial decision support systems using intelligent agents and GIS Web services. In *Proceedings of Americas Conference on Information Systems*, 2481–2486. Tampa, Florida, August 4–6, 2481–2486.

Sugumaran, V. and R. Sugumaran. 2007. Web-based spatial decision support systems (WebSDSS): evolution, architecture, and challenges. *Communications of the Association for Information Systems* 19:844-875.

Sun, C-H. 1995. A spatial decision support system for slope land management. Paper presented at the International Conference of GeoInformatics, Hong Kong.

Taleai, M., A. Sharifi, R. Sliuzas, and M. Mesgari. 2006. Evaluating the compatibility of multi-functional and intensive urban land uses. *International Journal of Applied Earth Observation and Geoinformation* 9(4):375–391.

Tan, G., R. Shibasaki, and K. Matsumura. 2004. Development of a GIS-based decision support system for assessing land use status. *Geo-Spatial Information Science* 7(1):72–78.

Taweepworadej, W., W. Kanarkard, R. G. Adams, N. Davey, and D. Hormdee. 2006. Development of a spatial decision support system (DSS) for the point-source pollution. Paper presented at TENCON 2006 IEEE Region 10 Conference, Hong Kong.

Taylor, K., G. Walker, and D. Abel. 1999. A framework for model integration in spatial decision support systems. *International Journal of Geographical Information Science* 13(6):533–555.

Tuzkaya, G., B. Gülsün, C. Kahraman, and D. Özgen. 2009. An integrated fuzzy multi-criteria decision making methodology for material handling equipment selection problem and an application. *Expert Systems with Applications* 37(4):2853–2863.

Uran, O., and R. Janssen. 2003. Why are spatial decision support systems not used? Some experiences from the Netherlands. *Computers, Environment and Urban Systems* 27(5):511–526.

Volk, M., J. Hirschfeld, A. Dehnhardt, G. Schmidt, C. Bohn, S. Liersch, and P. W. Gassman. 2008. Integrated ecological-economic modelling of water pollution abatement management options in the Upper Ems River Basin. *Ecological Economics* 66(1):66–76.

Voogd, H. 1983. *Multi-criteria evaluation for urban and regional planning.* London: Pion, Ltd.

Wan, Q., J. Zhang, H. Lin, H. and C. Beijing. 1999. On-line group spatial decision support system for investment environment analysis. Paper presented at the International Conference of GeoInformatics, Ann Arbor, Michigan.

White, R., B. Straatman, and G. Engelen. 2004. Planning scenario visualization and assessment: A cellular automata based integrated spatial decision support system. In *Spatially integrated social science,* ed. M. F. Goodchild, and D. G. Janelle, 420–442. New York: Oxford University Press, Inc.

Wilkerson, G. W., McAnally, W. H., and J.A. Ballweber. 2005. Assessing low impact development strategies through the integration of spatial decision support systems and best management practices. *International Association of Landscape Ecology Annual Symposium*. Syracuse, NY. 192.

Wilkerson, G. W., W. H. McAnally, J. L. Martin, J. A. Ballweber, K. Collins, and G. Savant. 2007. LATIS: A spatial decision support system to assess low impact site development strategies. Paper presented at the Second National Low Impact Development Conference, Wilmington, North Carolina.

Witlox, F. 2005. Expert systems in land-use planning: An overview. *Expert Systems with Applications* 29(2):437–445.

Wu, Y. H., H. J. Miller, and M. C. Hung. 2001. A GIS-based decision support system for analysis of route choice in congested urban road networks. *Journal of Geographical Systems* 3(1):3–24.

Yager, R. R. 1988. On ordered weighted averaging aggregation operators in multicriteria decision making. *IEEE Transactions on Systems, Man and Cybernetics* 18(1):183–190.

Yang, K., S. Peng, Q. Xu, and Y. Cao. 2007. A study on spatial decision support systems for epidemic disease prevention based on ArcGIS. In *GIS for health and the environment,* ed. W. Cartwright, G. Gartner, L. Meng, and M. P. Peterson, 30–43. New York: Springer.

Yeh, A. G. O., and J. J. Qiao. 2004. Component-based approach in the development of a knowledge-based planning support system (KBPSS). Part 1: The architecture of KBPSS. *Environment and Planning B: Planning and Design* 31(4):517–537.

Zeng, H., A. Talkkari, H. Peltola, and S. Kellomäki. 2007. A GIS-based decision support system for risk assessment of wind damage in forest management. *Environmental Modelling and Software* 22(9):1240–1249.

Zhu, X., R. J. Aspinall, and R. G. Healey. 1996. ILUDSS: A knowledge-based spatial decision support system for strategic land-use planning. *Computers and Electronics in Agriculture* 15(4):279–301.

5

SDSS Software

Learning Objectives

- Develop a broad understanding of different spatial decision support systems (SDSS) software available in order to guide modelers, managers, and decision makers in adopting, using, learning, and selecting a particular product.
- Gain a conception of SDSS software types and classes based on end users' or decision makers' perspective.
- Be exposed to the range of SDSS software according to specific application domains as well as more generic SDSS software.

5.1 Introduction

In Chapters 3 and 4, different SDSS components were examined and their role in the use of SDSS in spatial decision making was documented. This chapter focuses on providing an overview of existing SDSS software for the purpose of assisting potential users and developers in the selection of a specific type of SDSS software. The SDSS packages are often made up of separate software components, and the configurations and dependent software will be described. In other words, this chapter attempts to provide potential SDSS users with a thorough overview of existing SDSS software. The next chapter will focus on what tools and techniques are available for successful SDSS development.

The importance of computer-based systems for supporting complex spatial decision-making situations was established in earlier chapters. This importance has been evidenced by the development of hundreds of SDSS in the past 30 years. As seen in Chapter 2, this development is ongoing. For any potential new SDSS user or spatial decision maker, selecting a suitable system for a particular application is a challenging task. Many factors contribute to the complexity of choosing a system, including cost (commercial versus free), user expertise (novice or expert), deployment platform (desktop, Web-based, or mobile), operating system (Windows, Mac, Unix), availability or suitability of a particular model (neural network, cellular automata, Agricultural Non-Point Source Pollution Model [AGNPS]), data type necessary for model support (raster or vector), user level (single versus group), and user interface (command-line versus graphical user interface [GUI]).

In order to inform and guide decision makers, or end users, in the possible adoption of decision support systems [DSS] or SDSS, several decision support software classifications have been described. For example, Power (2002) classified decision support systems as falling into five categories based on key functionalities: (1) *Model-driven* DSS are built around statistical, financial, optimization, or simulation models; (2) *Communication-driven* DSS are designed to boost group decision making, enhancing communication between the project participants; (3) *Data-driven* DSS have their strength in the analysis and manipulation of large volumes of time series data; (4) *Document-driven* DSS are designed to retrieve, combine, and manage unstructured information from a variety of sources and formats; and (5) *Knowledge-driven* DSS provide facilities for storage, accumulation and use of facts, rules, procedures, and similar structures. In another study, Haettenschwiler (1999) characterized DSS based on the nature of the user's interaction with the system, i.e., passive, active, or cooperative. A *passive DSS* is a system that provides some aid in the process of decision making, but it is up to the user to come up with solutions and make the final decision. A communication- or a document-driven DSS would fit well in this category, though there is no direct correspondence between these classifications. An *active DSS*, on the contrary, is designed to actually produce solutions. Such systems have a low degree of flexibility. A *cooperative DSS* combines the best features of the previous two types. Just like an active system would do, a cooperative DSS provides users with solutions and possible action scenarios. However, they are designed to accept whatever modifications the user might implement. In this way, automated decision making becomes an iterative process, where different scenarios can be brought up, refined, and validated both by the user and by the DSS.

Sugumaran and Sugumaran (2007) classified SDSS into four categories based on the use and evolution of information and communication technology: (1) desktop SDSS, (2) Web SDSS/distributed SDSS, (3) mobile SDSS, and (4) service-based SDSS. Rizzoli and Young (1997) stated that there had only been two main types of environmental decision support systems (EDSS) developed. These were problem-specific EDSS, which were specific to an environmental domain but applicable to many locations, and those that were specific to both an environmental domain and also restricted to use in a specific location.

In this chapter we are going to adopt a classification method based on the level of usability or adoptability for which they were designed. We classify SDSS into three types: (1) problem-specific SDSS, which can be adapted for only a small range of problems; (2) application- or domain-specific SDSS, which can be used for a variety of problems within a specific application sphere; and (3) generic SDSS or SDSS-generating software, which can be used for a wide range of potential issues.

This chapter focuses on a review of existing SDSS software, while the next chapter focuses on techniques and technologies that can be used for developing new SDSS (Figure 5.1). To use SDSS for any application, whether the user(s) is a novice or expert, one of the first choices to be made is selecting suitable software for the decision-making situation. It is likely that those who have little programming or software development experience or limited access to resources for extensive software development will choose to utilize existing software. Those with the necessary resources for software development might be more likely to develop their own applications. The most crucial test as to whether to develop a new SDSS or use an existing SDSS is whether the existing software will meet all of the criteria established for the spatial decision-making process. There are a large number of spatial processing and scenario investigation software packages that are generic in nature and can be applied as SDSS to specific

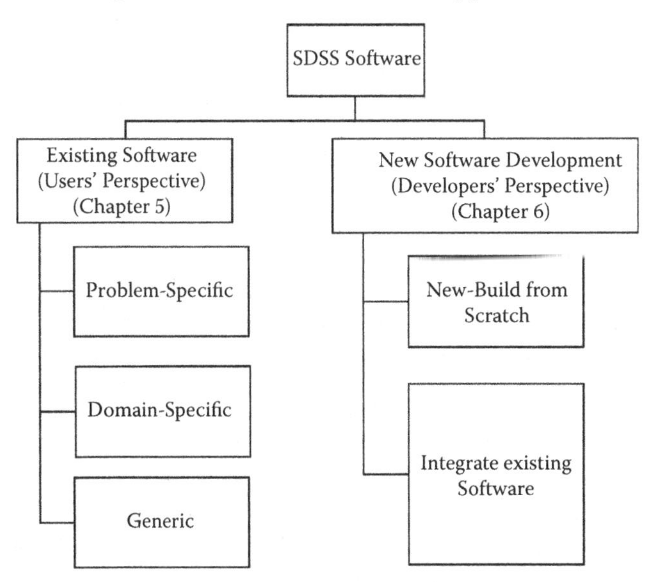

FIGURE 5.1
An overview of the contents of Chapters 5 and 6.

problems. There are also software systems that are considered SDSS that are more broadly applicable to problems within a certain domain or discipline, such as land use planning. There have also been many application-specific SDSS developed over the years that are applicable to a specific type of spatial decision problem. In this chapter, we are going to focus on the use of existing problem-specific, domain-specific, or generic SDSS software in which users do not have to carry out significant programming or software development. In Chapter 6, we will discuss techniques and technologies relevant to situations when there is a need to develop new functionality and software in order to address a spatial decision-making situation (Figure 5.1).

5.2 Existing SDSS Software

In this chapter, we propose a classification of SDSS software based on the level of specificity for problem areas for which they can be used. Our proposed classification scheme defines three distinct levels: problem-specific SDSS, domain-oriented SDSS, and generic SDSS. Geographic information systems (GIS) software often form a central and integral part of SDSS frameworks. A short synopsis of the most common GIS software used in SDSS configurations is given in the next section.

5.2.1 GIS Software Used in SDSS

The availability and familiarity of software to those responsible for SDSS development can often influence the nature of the SDSS. Based on the research conducted for this book, most common SDSS developments have been GIS-centric SDSS with the GIS software forming the core component of the systems. In examples of GIS-centric SDSS, products from Environmental Systems Research Institute (ESRI) (ArcInfo, ArcView, ArcGIS) have been by far the most commonly used (Figure 5.2). In general, ESRI has been a dominant player in the GIS market, and they are specifically well represented in academic environments in which many publications regarding SDSS have been developed. Often the potential users of these systems, such as federal, state, and local public agencies, businesses, or nonprofit organizations, might not have access to the GIS software or sufficient licenses to cover their potential use. For example, an academic institution might have a sitewide license to ESRI software, while a local agency might only have one or two licenses that are already dedicated to everyday use by GIS technicians, thus limiting potential uptake at that agency. These types of situations have limited the effective uptake

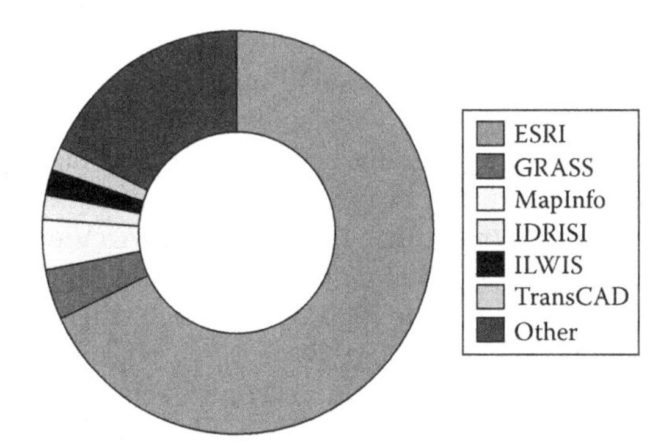

FIGURE 5.2
Graph representing proportion of GIS software used in GIS-centric SDSS development platforms. The ESRI category includes ArcInfo, ArcView, and ArcGIS software.

of SDSS. These types of problems and challenges will be discussed in more depth in the last chapter.

5.2.2 Problem-Specific SDSS

Problem-specific SDSS are tailored to address a specific issue, problem, or decision-making situation. The SDSS literature is dominated by examples of SDSS that were developed for specific problem situations. Indeed, Rizzoli and Young (1997) classified environmental decision support systems into two categories: (1) problem-specific systems and (2) situation- and problem-specific systems. The former are those that are applicable to a specific problem domain but could be applied to similar problem situations in different locations, while the latter represents those that are both specific to a problem situation and the geographic location for which the system can be used. As software development environments became more flexible, most SDSS developments were created with the idea of reusability in mind. Indeed, many SDSS, although developed for specific problems, had or have the potential for reuse in other geographic areas with the same problem or to slightly different problem situations. This reuse potential often goes unrealized for a variety of reasons. This section will focus on some of the specific problem areas for which SDSS have been developed, investigate some of the technological trends, and also investigate some of the limiting factors in successful reuse of problem-specific SDSS, with many examples provided from different disciplines.

While the use of commercial GIS software (such as ESRI products) has dominated SDSS development, there have been and continue to be alternatives. The use of freely available software such as GRASS GIS eliminates the need for users to purchase GIS software. Indeed, GRASS has been used in numerous SDSS applications. For example, Srinivasan and Engel (1994) developed an SDSS for assessing agricultural nonpoint source pollution using the AGNPS model and GRASS. Although there are clear advantages to using free GIS software, there can also be disadvantages. The commercial software products are generally easier to use and have more flexible or sophisticated customization or development environments. The availability of the Avenue programming language with ArcView GIS software promoted great third-party development including SDSS applications in the latter half of 1990s and into the 2000s. This is true still for ESRI's ArcGIS, which can be customized using Visual Basic for Applications or many other development environments. In addition, commercial software packages often come with greater resources to assist potential developers. Given all of this, however, it should be expected that the continued development of free GIS software will facilitate future SDSS development.

There are alternatives between free and expensive complete GIS packages. MapObjects, which provides a less-expensive licensing arrangement, has been used for the provision of necessary GIS functionality in numerous SDSS. MapObjects is an ESRI product that allows software designers to build lightweight applications with some GIS functionality. The users of these applications would not have to pay for a full GIS software license. For example, Johnson (2001) developed a lightweight distributable SDSS called the Housing Location Planner, which was used to aid housing mobility planning for a public housing authority. The system was developed with Microsoft Visual Basic 5.0 and ESRI MapObjects 1.2 (Figure 5.3). The MapObjects software has ostensibly been replaced by the ArcGIS Engine software. ArcGIS Engine follows a similar idea in which all of the objects (ArcObjects) on which the full ArcGIS software is built are exposed for development of applications. The developer buys a developer license, and the potential users would have to pay for a deployment license, which is much lower than the cost of the full ArcGIS software. A recent SDSS used ArcGIS Engine to provide tools for modeling water supplies in Western Africa (Laudien et al. 2009).

The delivery of spatial data display, management, and analysis capabilities through distributed or Web-based applications can also eliminate expensive licensing issues for the end user. The MapServer platform actually provides a free development environment for developing Web-based applications with spatial data publishing and interactive mapping capabilities. MapServer was used by Engel et al. (2003) in a Web-based SDSS for development of inputs to a hydrological model and for display of spatial data. There are also commercial software applications that require development licenses but allow for the provision of GIS capabilities through the Web. An example of this is ArcGIS Server technology. A prototype Web-based collaborative SDSS that utilizes weights of evidence modeling implemented in ArcGIS Server was developed by Wang and Cheng (2006).

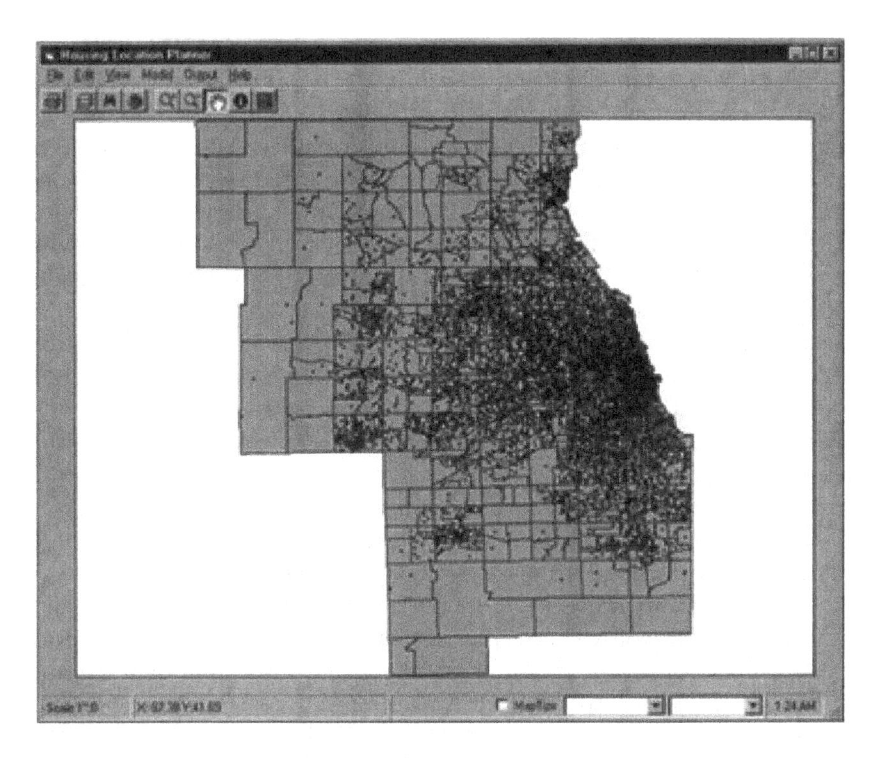

FIGURE 5.3
The MapObjects-based Housing Location Planner SDSS (Source: Johnson 2001).

In some cases, the development of an SDSS for a specific purpose and a specific location or organization makes sense due to potential efficiency gains by developing that SDSS. In Delaware, an ongoing, highly useful Web-based SDSS has been created for managing and processing the oversize/overweight vehicle permitting process (Ray 2007). The Delaware Department of Transportation (DOT) has used the system to replace previous manual processes, which has led to reduction in permit processing times, sometimes from weeks to seconds, and thus provided cost savings. The Web-based SDSS supports decisions made about which routes certain vehicles can travel based on characteristics of the roads and the vehicle. Ray (2007) points out that this type of spatial decision making (i.e., route to be taken by oversized vehicles) is an inherently spatial problem. The spatial information about addresses, street intersections, bridges, tollbooths, road signs, and road conditions can all be stored in spatial databases and can affect routing decisions. More generic SDSS with vehicle-routing applications were available; however, they were not able to meet the specific needs of the overweight/oversize vehicle permitting process in Delaware. The developed system, called the Oversize/Overweight Permitting System (OOPS), integrates with the Delaware DOT's Integrated Transportation

Management System. This is a good example of the use of Web technologies for the development of a much-utilized SDSS application that is used within a state agency's management activities.

In some cases, an SDSS is developed to use specific datasets that might not be available in other geographic locations (e.g., dataset available in one state or country but not in another). Salt and Dunsmore (2000) developed a GIS-based SDSS for long-term management of radioactively contaminated land resources. This SDSS was built specifically to work with spatial and nonspatial datasets that were specific to Scotland and thus could not be easily adapted for other countries. The SDSS was developed within ArcView GIS using the Avenue programming language. The application used spatial and nonspatial data on soils, land cover, contours, meteorological data, agricultural census, and land management data. These latter two data sources would likely not be available in the necessary formats in other countries. This SDSS was an example of an embedded GIS-based SDSS in which significant GIS and programming experience was necessary for development. To make systems that, even when designed to be application-specific, are generic in the sense that they can accept data from a variety of sources and in different formats (e.g., spatial data on land cover or soils from different countries) requires more detailed and careful development and programming efforts.

There have been several SDSS that utilize GIS for nonprofit or governmental assistance work. Gorr et al. (2001) provided an interesting example of the use of an SDSS for evaluating the efficiency of a nonprofit home food delivery system. The system also could be used to evaluate the location of the nonprofit facilities in relation to customers. They embedded their algorithms within the ArcView GIS-based SDSS, which they called Home Delivered Meals Decision Support System (HDM DSS). The HDM DSS made use of existing network algorithms in ArcView. The user interfaces were also developed for, and run in, ArcView GIS. This is an example of an embedded GIS-based SDSS that was delivered for a specific purpose but could easily be adapted for a broader set of analyses. Barton et al. (2005) used a much more complex software structure in an SDSS to assist in the planning, management, and evaluation of housing in densely populated areas. This SDSS was developed to address a wide set of issues, including the management of high-rise areas and crime mapping and analysis, while providing community interactivity functionality as well as secure internal and intergovernmental information sharing. The SDSS had a complicated structure that incorporated data and expert knowledge built into a knowledge base from the Department of Housing, intranet, and WWW servers, a 3D geometry processor, and a 3D map virtual reality presenter. The Web-based platform and interactive abilities of this system provided advantages in accessibility but required significant development expertise and relied on high bandwidth infrastructure for any users.

An example of a problem-specific SDSS that was applied in various locations was a system called the Geographic-Engineering Tool for Wireless: Evaluation of Broadband Systems (GETWEBS), which was developed for the specific problem of analyzing wireless systems and has been used in several applications (Bostian et al. 2002, Scheibe 2003, Scheibe et al. 2006). Scheibe et al. (2006) described the use of the system for determining optimal locations for wireless broadband equipment. In their application, using GETWEBS, information on topography, demography, and propensity to pay for services was used with the built-in expert knowledge base of the system. The software components were described by Bostian et al. (2002) and included spatial analysis and database management functionality through ESRI's MapObjects and modeling functionality through a financial modeling package for Microsoft Excel. The MapObjects components and linkages to Excel were accessed through Visual Basic programming. Scheibe et al. (2006, p. 561) noted that this application-specific SDSS has been extensively utilized, saying "this tool has proven to be invaluable in planning specific wireless networks." They discussed how the system was used to investigate numerous location models and scenarios for wireless broadband development.

An SDSS developed for assisting in the management of tuberculosis in cattle and deer was first described by McKenzie et al. (1997) and again discussed by McKenzie et al. (2002) as a potentially useful tool to assist field managers developing and implementing a risk-based approach to possum control for TB management in New Zealand. The EpiMAN-TB system was comprised of a database, map display tools, a simulation model of TB in possums, and decision aids based on expert systems. Included in the database was the Agribase database, which contained property ownership and land use information. Some of the spatial operations in EpiMAN-TB were developed externally so commercial GIS or database management software would not have to be relied upon. However, there seemed to be some need for preprocessing of spatial data with GIS software in order to have proper inputs into their SDSS. This is an example of a specific SDSS that was built without the reliance upon commercial software. Although designed specifically for TB in wildlife, the authors suggest that it could be adapted to manage other endemic diseases (McKenzie et al. 1997). Indeed, a similar system was adapted for examining foot and mouth disease in Great Britain (Morris et al. 2002).

Although there have been a large number and wide range of problem-specific SDSS applications developed, many have had limited reuse. There are many reasons for the lack of reuse. The development of these systems for a specific problem indicates that they might be too specific for reuse in other geographic locations. Reusability is dependent on how flexible the SDSS is in terms of data formats accepted and software dependencies. If the application allows some flexibility in types and formats of data accepted, it would more likely be adopted in other geographic locations. When developing a problem-specific SDSS, however, it is likely that the developers do not take the extra time to program great flexibility into the system. Incorporating great flexibility requires careful planning and often greater programming effort. Given that many SDSS developments take place under time and resource constraints, the possibility of developing these careful flexible applications is not common or always possible. In addition, one of the main constraints in SDSS reuse is the lack of continued software support and maintenance. Often, there may be support for development of the software but no resources allocated for future support and development. Thus the application comes as is, with future problems being the responsibility of the user. One major issue is when an SDSS uses various programs (e.g., GIS) that migrate to new versions. If the user upgrades to the new version, the code that was written for the original SDSS might not work with new versions of the GIS software.

Clearly, the decision of which software to include in the SDSS is crucial. If commercial software (such as GIS) is utilized, all potential users must have proper licenses. As was mentioned before, universities often have access to commercial software and site licenses because they receive academic discounts. Research at academic institutions is often the source of SDSS software. Potential users, such as private businesses or local, state, or federal agencies might not have either the access to or expertise in the necessary software. Agostini et al. (2009) pointed out several characteristics that make SDSS more likely to be reused, including ease of use, inclusion of commonly used software components, and a match between the system with regulatory and policy structures (2009). The challenges in developing useful SDSS software will be discussed further in the final chapter of this book.

5.2.3 Domain-Level SDSS

There are numerous SDSS that have been developed with functionality for addressing broader disciplines or domain areas. These systems are used to solve a set of closely related problems common for a specific discipline or domain area, such as ecosystem management, agriculture, or urban planning. Domain-oriented SDSS offer a greater degree of flexibility and a greater breadth of functionality compared to problem-specific SDSS software. As opposed to some of the application-specific SDSS, which have often been developed at least with some academic support, many of the domain-level SDSS have been developed commercially and are used by a wider variety of stakeholders. The use of these SDSS might be less likely to appear in the academic literature for this reason. The following section provides a few examples of widely used domain-level SDSS, including agricultural (SSToolbox), land use planning (INDEX and CommunityViz), and ecological and natural resources management (Ecosystem Management Decision Support [EMDS] and Marxan) software packages. An agricultural SDSS framework called SSToolbox was first introduced by Hey (1998). The SSToolbox was developed by, and is still sold by, the SST Development Group, Inc. for the purpose of supporting decision making by providing spatial and nonspatial data-handling tools to agricultural producers, farm input suppliers, agronomists, and crop consultants. The SSToolbox was designed to support decision making especially for precision farming operations. The system utilizes the desktop mapping functions of ArcView GIS and a specifically designed data management system called SST FarmCrawler. It contains analytical functions such as bivariate regression, correlation matrices, price scenario analysis, and subfield variety analysis. The SSToolbox software supports data formats from other agricultural, geospatial, and data management software. The SSToolbox has been used in precision agriculture studies in several publications (e.g., Baio and Balastreire 2002; Giles and Downey 2003).

There are a number of commercial SDSS that have been widely used in land use planning and have evolved over time with continual software upgrades. One of these is called INDEX, which is a suite of GIS planning tools that can be used for assessing community conditions, designing future scenarios, measuring scenarios success, ranking scenarios, and monitoring implementation of plans. The INDEX tools (Figure 5.4) were developed by Criterion Planners (www.crit.com/) and have been continuously developed since 1994 with the latest version running in ArcGIS 9.3 software. The INDEX software is being used by over 175 organizations in the United States as well as by Criterion in a consulting capacity. The main users are city and county planning departments. The INDEX software is used to benchmark existing conditions, design and visualize planning scenarios, analyze and score their performance, and compare alternatives. The software also allows evaluation of development proposals against plan goals. The developers cite the underuse of spatial analytical capabilities in GIS in planning processes and also the development of user friendly and portable GIS such as ArcView and Caliper's Maptitude, which has allowed the tools to be taken to stakeholders on laptop computers. Some characteristics or goals of the software include more objectivity in decision making; greater integration of land use, transportation, and environmental issues; support for the entire planning implementation process; real-time interactivity for process participants; and sensitivity to smart growth and sustainable development policies (Allen 2007). The system was built with ArcObjects and Visual Basic components and supports three levels of users, including software stewards (advanced GIS users), general users (practitioners), and citizens with basic computer literacy. The INDEX

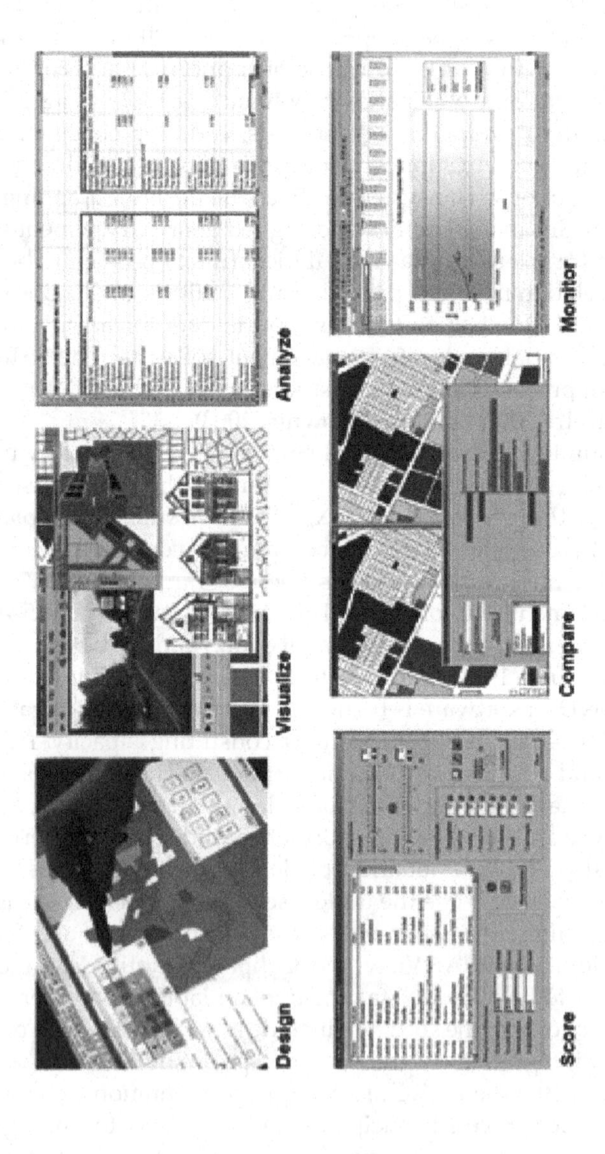

FIGURE 5.4
An overview of INDEX capabilities (Source: http://www.crit.com/documents/INDEX_InteractiveDesigner.pdf)

software comes with a set of ninety indicators for identifying an area's strengths and weaknesses, testing scenarios, and monitoring change in land use, urban design, transportation, and the environment. The indicators include those for demographics, land use, housing, employment, recreation, environment, and travel. The system allows users to set rankings and weightings in relation to goal achievement and also to interactively draw geographic features. The INDEX software is also intended to aid in participatory planning efforts. The software has been widely used for a variety of applications that are documented in many technical reports (see http://www.crit.com/) but not in many academic articles or conferences.

Another widely used planning SDSS tool is CommunityViz. Similar to INDEX, CommunityViz is built as an extension to ArcGIS software and includes a large suite of visualization, analytical, and communication tools. The CommunityViz software provides a wide range of specifically designed tools that operate from the ArcGIS environment. Given the name of the software, visualization capabilities are emphasized and include the ability to experiment with various scenarios and view results in a variety of formats, including charts, tables, reports, maps, and 3D models viewable in their own 3D rendering software (Figure 5.5) or Google Earth. Analysis capabilities include the ability to estimate socioeconomic and environmental impacts of user-defined development scenarios, calculation of development capacity of land, and multi-criteria evaluation of alternative scenarios. Tools available in CommunityViz include an allocator that estimates where growth will be most likely to occur; a build-out wizard that shows capacity of land based on land use regulations with numeric, spatial, and 3D analysis; and a common impacts wizard that can create estimates such as auto emissions, commercial and residential energy use, commercial and residential tax revenue, distance to places of interest, population, sensitive lands impact, and vehicle trips per day. There is also a land fragmentation analysis tool, a land use scenario and evaluation designer, an optimizing utility that helps users choose the best combination of features to satisfy certain goals, a site suitability analysis feature, and a future visualizer that projects development into the future based on assumptions set by the user. The CommunityViz system has been used in a wide range of applications, including the evaluation of development impacts on a naturalized floodplain (Nedovic-Budic et al. 2006), for a county comprehensive plan update (Lieske et al. 2009), and for investigating forest fragmentation in an area undergoing suburbanization (Parent et al. 2007).

The Ecosystem Management Decision Support (EMDS) system is another extension of ArcGIS software. It was originally developed in ArcView GIS software in the 1990s but has been migrated to the ArcGIS platform. The EMDS is an application framework for knowledge-based decision support of ecological assessments at any geographic

FIGURE 5.5
An example of a 3D model created in CommunityViz. (Source: http://www.placeways.com/communityviz/?p=3dex)

scale (Reynolds 2006). It integrates the logic engine of NetWeaver for performing landscape evaluations and the decision modeling engine Criterium DecisionPlus for evaluating management priorities as well as an Analytical Hierarchy Process (AHP) model for setting priorities, selecting alternatives, and allocation of organizational resources for integrated planning. The NetWeaver and DecisionPlus applications are separate software and are not integrated within ArcGIS software, but are linked to the ArcMap extension. ArcMap provides the spatial processing, analysis, data management operations, and output visualization functionality. The EMDS has a hot-link browser that can be used to trace the logic of evaluations in order that the system is not seen as a black-box evaluation. There is also a multi-criteria decision analysis tool called Priority Analyst (PA), which is used for priority setting, resource allocation, and trade-off analysis. The EMDS has been used in carrying out a knowledge-based assessment of watershed conditions and erosion processes (Reynolds et al. 2000), for integrated planning and restoration evaluation (Reynolds and Hessburg 2005), and for road system analysis in forests (Girvetz and Shilling 2003).

Marxan is a freely available spatial decision support system developed for conservation planning. It was initially developed around the year 2000 by researchers at the University of Adelaide and is supported and maintained by researchers at the University of Queensland. The software is meant to support planning problems such as designing new conservation reserve systems, reporting on the performance of existing reserve systems, and development of multiple-use zoning plans for management of natural resources. When planning for a reserve system, there are many ecological, social, and economic criteria and principles that need to be considered in the planning process. Marxan is designed to solve reserve design problems by achieving a minimum representation of biodiversity features for the smallest possible cost with the logic that these will more likely be adopted (Game and Grantham 2008). Marxan is meant to support decisions rather than provide definitive solutions. The Marxan software works on planning units which, if there are many, can lead to a very large number of possible configurations. The Marxan software uses a heuristic modeling technique called *simulated annealing* to find near-optimal solutions (Game and Grantham 2008). The overall objective function in the Marxan model is based on the total cost of the reserve network, the penalty for not representing desired conservation features, the total reserve boundary length, and the penalty for exceeding a given cost threshold. Preprocessing of data is very important in carrying out a Marxan analysis and includes setting polygon-based planning units (rectangular grid, hexagons, or natural divisions such as hydrological units) and assembling necessary data on conservation features within each planning unit. The Marxan software has basic graphical user interfaces for developing input parameter files with the actual modeling run from a DOS executable after the proper input files have been developed. The outputs are in the form of text files, which can be viewed in spreadsheet software as well as linked to the spatial data that was created in the preprocessing steps. The Marxan software is an example of a loosely coupled SDSS with several components. The GIS software is mainly used as a preprocessor, data manager, and a postprocessing visualization tool. Third-party developers have developed Marxan extensions that make the development of input parameters and visualizing of outputs easier. The Conservation Land Use Zoning (CLUZ) extension for ArcView links to Marxan for easier preprocessing of inputs and for viewing outputs. A program called Protected Areas Network Design Application for ArcGIS (PANDA) includes functionality for preprocessing spatial data, running Marxan, and viewing outputs. This software was developed for ArcGIS with Visual Basic and ArcObjects. The CLUZ extension and PANDA software provide a tighter coupling between the GIS spatial database management and processing capabilities as well as the graphical visualization capabilities and the Marxan modeling routines. The Marxan software has been used in a wide range of applications, including to investigate design of efficient fisheries management in order to support stocks and livelihoods of commercial fisheries (Ban and Vincent 2009) and for investigating conservation planning that accounts for ecosystem services (Chan et al. 2006). Watts et al. (2009) described an extension of Marxan called Marxan with Zones and illustrated its use with a number of case studies regarding the design of multiple-use marine parks in western Australia and California and the zoning of forest use in Indonesia.

Domain-level SDSS, which are used in several fields including agriculture (SSToolbox), land use planning (INDEX and CommunityViz), and ecological and natural resources management (EMDS and Marxan), have been examined in this section. The SSToolbox, INDEX, and CommunityViz software are commercial software. The INDEX software was developed and is sold by Criterion Planners, who also offer consulting services based on the system. CommunityViz is sold commercially by a company called Placeways, LLC. However, the software was originally developed and is still financially supported by the nonprofit Orton Family Foundation, allowing CommunityViz software to be offered at a more affordable price. The Placeways Company also offers consultation services based on the CommunityViz software. The SSToolbox software is sold as a part of a suite of data management, farm management, and other software by the company SST Software (http://www.sstsoftware.com/index.htm). The EMDS software was developed by the U.S. Department of Agriculture (USDA) Forest Service and is now developed in conjunction with the private companies InfoHarvest and Rules of Thumb, and is currently maintained by the Redlands Institute at the University of Redlands. Although the EMDS is free, the commercial software components NetWeaver Developer and Criterium Decision Plus are necessary to create and edit logic or decision models. The Marxan software is freely available and supported by the University of Queensland with good support in the form of documentation, online tutorials (under development), and courses. The SSToolbox, INDEX, and CommunityViz programs are mainly built within GIS software, while the EMDS is a system made up of tightly coupled GIS and other software. The Marxan software is stand-alone software that, without the use of third-party software, uses GIS as a preprocessor and for output visualization and mapping. These software programs can all be used for a variety of tasks within their specific domain and have proved very useful in practical applications often at regional or local levels. INDEX and CommunityViz have often been utilized in local planning activities that are documented in local reports, but are not usually submitted to peer-reviewed scientific publications. Similarly, the SSToolbox has likely often been used for farm management operations with no formalized reporting of its use appearing in publicly available literature. The EMDS and Marxan systems have a more academic origin and thus do appear in academic literature quite often.

Although domain-level SDSS are very useful for specific disciplines, there are still many research or management areas that do not have these types of SDSS. More generic SDSS, which could be utilized for a wide range of spatial problems, would be useful. In 1997, Rizzoli and Young addressed this when they said a fully general environmental decision support system was a very worthy goal but was still a long-term ambition. A truly generic and flexible SDSS or SDSS generator is still an ambition, although there are many software applications that provide utilities for building or developing SDSS.

5.2.4 Generic SDSS

There are a number of software systems that provide a wide range of tools that can be used in spatial decision-making situations and thus are qualified here as generic SDSS. Systems at this level allow users to address a wide range of problems through their flexible component-based design. Generic SDSS provide a rich set of decision-support components (i.e., spatial database management, analysis and processing, modeling, and user interfaces) as well as a framework allowing for rapid assembly of a specific SDSS. One of the strengths of these types of systems is that components can be combined without writing any code. There are powerful software components available in these systems that can be used in conjunction with each other to support decision making in unstructured spatial decision-making situations. These systems allow end users to directly participate in the process of SDSS design.

A powerful tool called *graphical modeling* is an example of technology that makes this level of SDSS software possible. Each of the software systems discussed below contains a graphical modeling user interface component that allows construction of useful and powerful models to be easily achieved without any programming or software development. Graphical modeling allows the representation of possible decision flows in a graphical form with modules and data stored as nodes of the graph and the connections between them as its vertices. The most obvious function of graphical modeling tools is to allow for the development and presentation of a clear layout of the decision process. Advanced modeling components provide additional control over the flow of the model execution. These graphical modeling tools allow for the creation of iterative processes in which batch processing of large amounts of data can be included in modeling processes. Another technique that provides power to the use of graphical modeling techniques is *logical branching*, which can be used to adjust the flow of the decision-making process based on intermediate results. These graphical modeling techniques allow the use and construction of a sequence of routines based on individual analysis techniques and specified data inputs as well as data outputs. This allows the building of simple to complicated powerful modeling routines that can be run from within these software systems.

Currently, there are only a few systems that can qualify as providing the necessary advanced functionality in spatial database management, user interface, and modeling components to allow construction of an SDSS. These systems are described in the following sections.

5.2.4.1 IDRISI Macro Modeler

Since 1987, researchers at Clark University in Clark Labs have been developing geospatial technologies for the support of effective decision making with an emphasis on environmental and natural resource issues. Clark Labs works with a variety of private, governmental, and nonprofit organization to advance decision support tool development. The best-known products from this work have been a long line of versions of the IDRISI software. Clark Labs defines IDRISI as "an integrated GIS and Image Processing software solution by Clark Labs." However, with a set of over 300 modules for analysis and display of spatial information, IDRISI goes beyond only the GIS and image processing functions to include generic and specialized modeling techniques. When taken as a whole, IDRISI software can be considered as a generic SDSS. It contains spatial and nonspatial database management components, modeling capabilities, and a common set of user interfaces for interaction with the user. All of these functions are provided in one suite of software that is sold by Clark Labs at a price that is generally below the main commercial GIS and image processing software. This has led to its use in academic, agency, and nonprofit organizations, especially in developing countries.

The functionality that truly separates IDRISI from other GIS and image processing is built-in analytical modeling capabilities. A large number of functions that allow multi-criteria and multi-objective analysis, time series modeling, trend analysis, and land change modeling are available in IDRISI to support spatial decision making. The multi-criteria functions include Boolean analysis, AHP, weighted linear combination (WLC), ordered weighted averaging (OWA), multi-objective land allocation, and a multiple ideal point procedure. A fuzzy set membership function is also available. In IDRISI, there is a menu item in the main interface called Modeling, which has the submenu items Model Deployment Tools, Empirical Model Development Tools, and Environmental/Simulation Models. Models available include logistic regression, multiple regression, multilayer perception classifier, Kohonen's Self-Organizing Map, classification tree analysis, land use change simulation, Revised Universal Soil Loss Equation (RUSLE), sedimentation, and runoff. Many of these types of models could be built with components in other GIS or geospatial processing software. However, the infrastructure already constructed within

IDRISI makes it easier for potential SDSS users and developers. Various custom models using these functions can be constructed through IDRISI's graphic modeling environment called Macro Modeler (Figure 5.6). The Macro Modeler tool allows a user to build algorithmic chains using simple graphical tools for accessing all of IDRISI's modules as objects. The Macro Modeler enables batch processing by using feedback capabilities that feed the output of a given operation back into the given model as new input. However, there are limitations in the modeling capabilities of the Macro Modeler. There are no options for logical branching, no preconditioned execution, and no error or sensitivity analyses functionality. Iteration capabilities are also limited by the user defining a number of iterations. There is also no ability to stop a loop with a Boolean condition.

IDRISI software has been used in many SDSS applications. Some of these applications used IDRISI mainly as the spatial database manager and processor with linkages to outside modeling functionality. With the expansion of modeling capabilities over the years as IDRISI has been developed, SDSS applications have been created using primarily the IDRISI software. Abu-Zeid (1998) used IDRISI in conjunction with an expert system and simulation model for evaluating the effects on water quality and quantity based on cropping patterns in Egypt. In this SDSS, IDRISI provided the spatial data management, processing, and display capabilities but was linked to an outside simulation model and expert system. Ianni et al. (2008) used IDRISI as the spatial data management and processing component of an SDSS for assessing management options for reducing agricultural nitrogen loads to a water body in Italy. The system utilized an outside multi-criteria analysis software. Dragan et al. (2003) used the IDRISI multi-criteria evaluation (MCE) and multi-objective land allocation (MOLA) modules for the modeling components of an SDSS that was constructed for assisting in the reduction of soil erosion in Ethiopia. This SDSS did not require any modeling capabilities outside of those provided by the IDRISI software. Mwasi (2001) used the MCE Weighted Linear Combination option (Figure 5.7) and the MOLA modules in an IDRISI-based SDSS for land use conflict resolution in Kenya.

5.2.4.2 ArcGIS ModelBuilder

The GIS software most commonly used in published SDSS applications, ArcGIS, also has a graphical modeling environment in which users can build SDSS for supporting spatial decision making. The ArcGIS ModelBuilder environment (Figure 5.8) graphical interface is similar to the IDRISI Macro Modeler interface. The user can insert spatial processing tools, data sets, variables, scripts, and scalars (numbers, strings, etc.) and then can move them around on the ModelBuilder canvas in order to create simple to complicated processing models. The user can directly

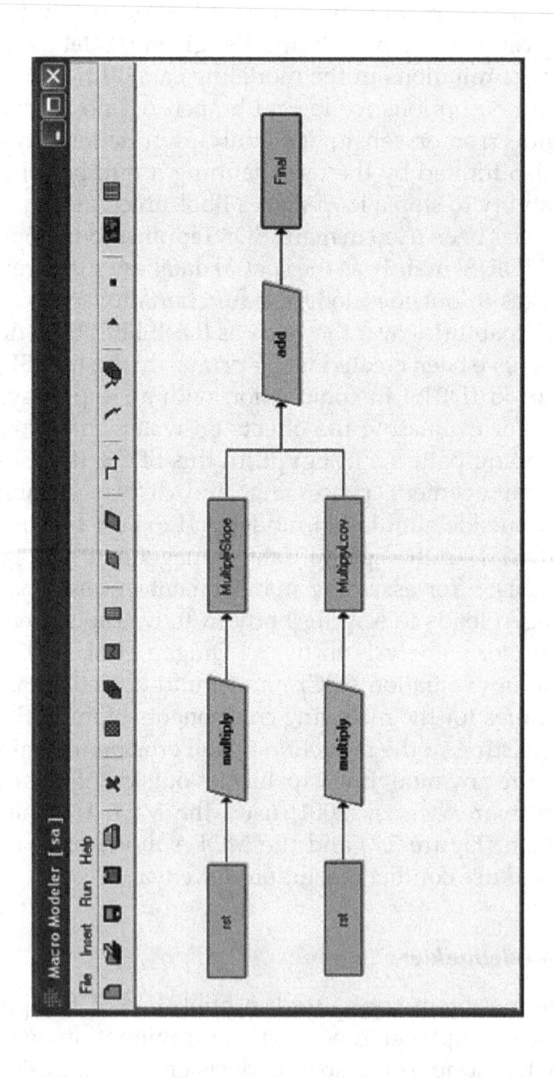

FIGURE 5.6
IDRISI Macro Modeler model interface. An example of a weighted linear combination model is shown.

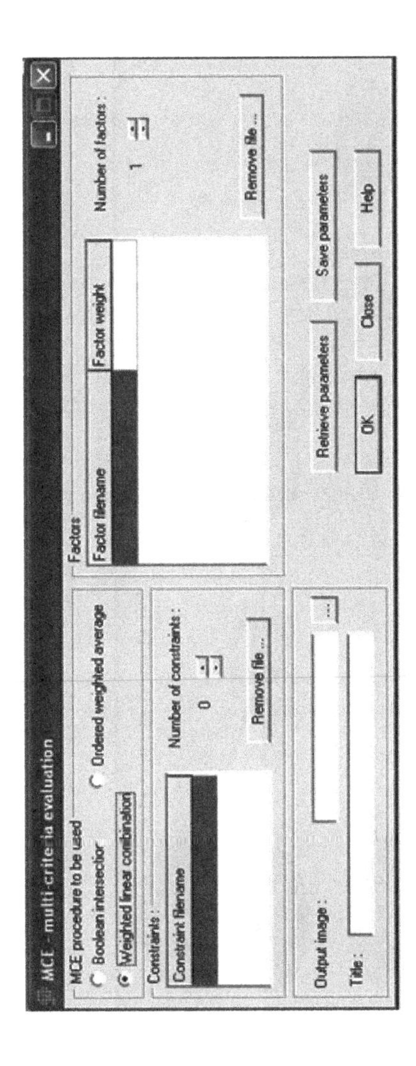

FIGURE 5.7
IDRISI Weighted Linear Combination model dialog.

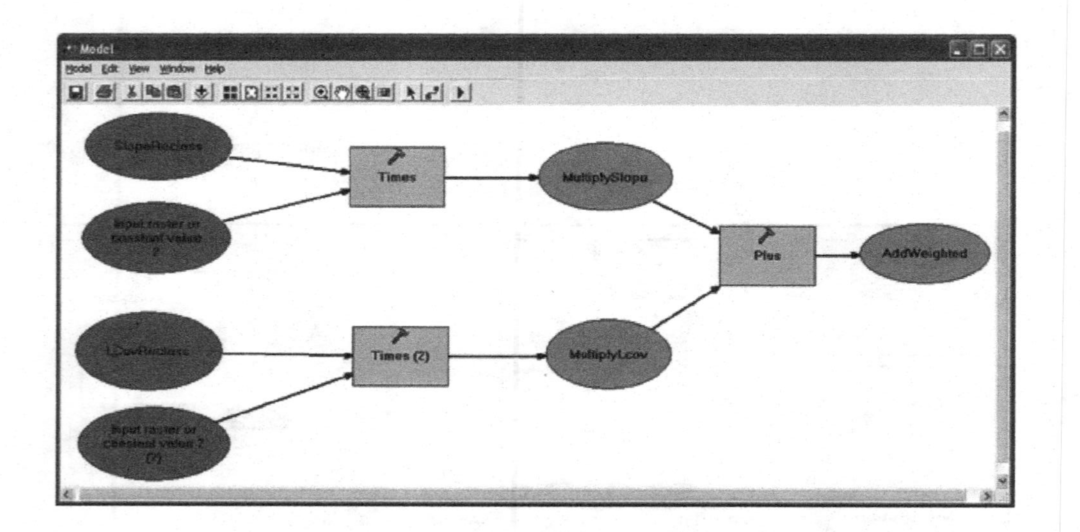

FIGURE 5.8
The ModelBuilder interface in the ArcGIS software. The same weighted linear combination model as shown for the IDRISI Macro Modeler in Figure 5.6 is shown as an example.

drag datasets from the ArcMap interface as well as any tool from the ArcToolbox interface directly onto the ModelBuilder canvas. Then they can use the connector tool to connect them, allowing the user to define the inputs (e.g., layer, field) to the tool. Alternatively, they can add a blank dataset or tool and fill in the datasets or scalars as necessary by double-clicking on the tool and setting the parameters in the tool dialog (Figure 5.9). Any of the hundreds of tools available in ArcToolbox can be incorporated into a model in the ModelBuilder environment. ModelBuilder has support for logical branching of model execution; it supports multiple modes of iteration, precondition variables, and different possibilities for batch processing. Any created model can be saved or exported to a script (Python, JScript, VBScript), which can then be adapted by adding loops or conditional statements. These scripts can then be added back into another model. Despite its extensive toolbox of spatial analysis modules, decision support functionality in ArcGIS is limited, out of the box, to multi-criteria evaluation tools such as the Weighted Overlay (Figure 5.10) and Weighted Sum tools. ESRI will probably continue to add more modeling functionality into the ArcGIS software. In addition, there are many extensions that have been created by third-party developers, which are freely available, that match some of the decision support tools in IDRISI. For example, there is a freely available extension for ArcGIS that provides AHP capabilities. The ModelBuilder tools have been used frequently in practical management and scientific research efforts including in SDSS. For example, Best et al. (2007) used ModelBuilder within a fairly complicated SDSS for predicting marine mammal habitats. This system provided information through Web browsers and Google Earth. Mbilinyi et al. (2007) used ModelBuilder as a platform for an SDSS that had the purpose of identifying potential sites for rainwater harvesting. Specifically, the weighted overlay process was used as the model in the SDSS. Schaller et al. (2009) detailed the development of ModelBuilder-based applications to support planning processes and to deliver results for political planning decisions in Bavaria, Germany. The authors stressed that the workflow diagrams presented from ModelBuilder were very useful in public participatory processes as they were easy to explain to the public, government agency personnel, and other stakeholders, which contributed to better acceptance of modeling results for planning decisions. The ModelBuilder environment is being used more frequently and will likely continue to be widely used and adopted for use in SDSS.

5.2.4.3 ERDAS IMAGINE

The ERDAS IMAGINE software, which is primarily an advanced image processing application with some GIS capabilities, also provides a graphical modeling environment. The Model Maker graphical interface, which is

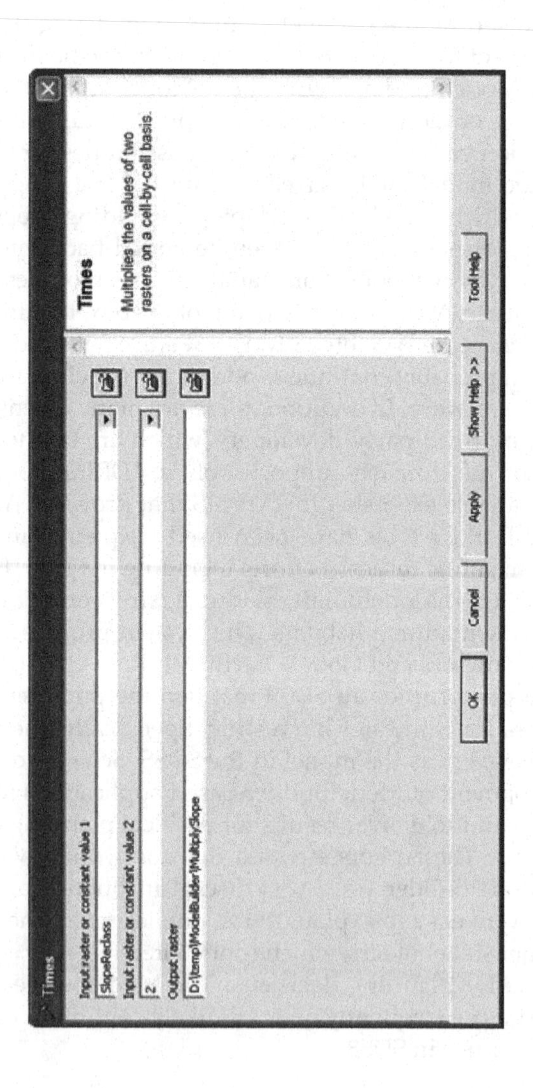

FIGURE 5.9
The tool Times dialog, which in this case was invoked by double-clicking on the Times tool graphic in the ModelBuilder canvas (Figure 5.8).

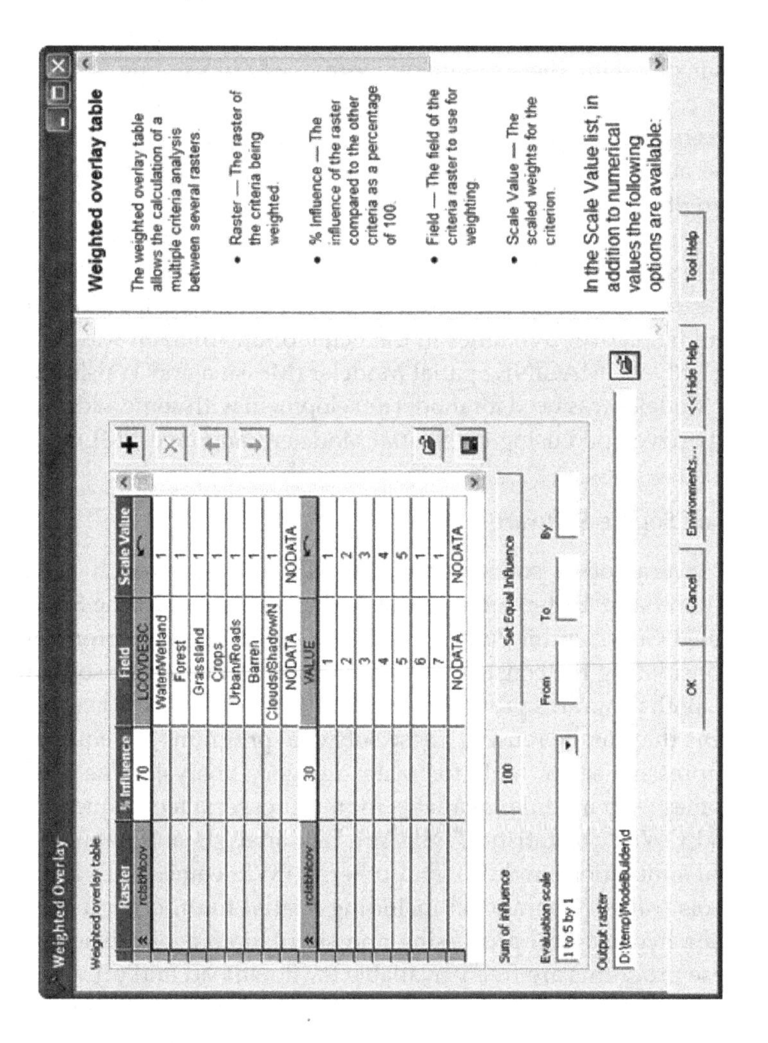

FIGURE 5.10
The Weighted Overlay tool from ArcGIS software.

part of ERDAS IMAGINE's Spatial Modeler module (Figure 5.11), is very similar to those in IDRISI and ArcGIS in that the user can add spatial and nonspatial datasets (i.e., tables, rasters, or vector layers), scalars, and processes or functions to the canvas. The user can then connect these graphically and dictate the way in which the model parameters interact. There is a wide range of spatial (and nonspatial) processing functions available for use in Model Maker. As the focus of ERDAS IMAGINE is image processing of remote sensing data, there are many more image processing functions available as compared to ArcGIS ModelBuilder and even IDRISI, which does have significant image processing functionality. ERDAS IMAGINE also has the ability to process a wide variety of spatial data formats, as compared to other software, which can eliminate many of the extra data formatting steps that are necessary in other software. ERDAS IMAGINE does not have some of the multi-criteria evaluation and other models that are available, especially compared to IDRISI. A framework for modeling land use and land cover dynamics in the Ecuadorian Amazon was developed using ERDAS IMAGINE Spatial Modeler (Messina and Walsh 2001). The Spatial Modeler was used for model development with some additional functionality developed using the Spatial Modeler Language (SML).

5.2.4.4 Open Source Software

There are several open source and free GIS programs, such as the Integrated Land and Water Information System (ILWIS) and the System for Automated Geoscientific Analysis (SAGA), which are very promising for those considering the construction of an SDSS from a single software framework. Both programs provide user-friendly mechanisms for customizing systems that can be reused. These software programs are equipped with an impressive set of tools for raster imagery analysis. The SAGA software comes with modules available for multi-criteria techniques such as AHP and OWA, geostatistics, wild fire risk analysis and simulation, hydrological simulation modules, and others. ILWIS comes with a wide range of tools (called Operations) including spatial multi-criteria evaluation (SMCE), hydrological processing and simulation, geostatistics, and others. These programs are freely available for download and provide an impressive amount of functionality, but will require expertise to use and to build an SDSS. The ILWIS SMCE tools (Figure 5.12) were used to facilitate the process of selecting a local park in Italy by Zucca (2008). It should be expected that these tools will gain greater acceptance as they grow. These applications were developed in academic environments, and they require a fairly steep learning curve. The most likely use of these systems will still be in academic environments.

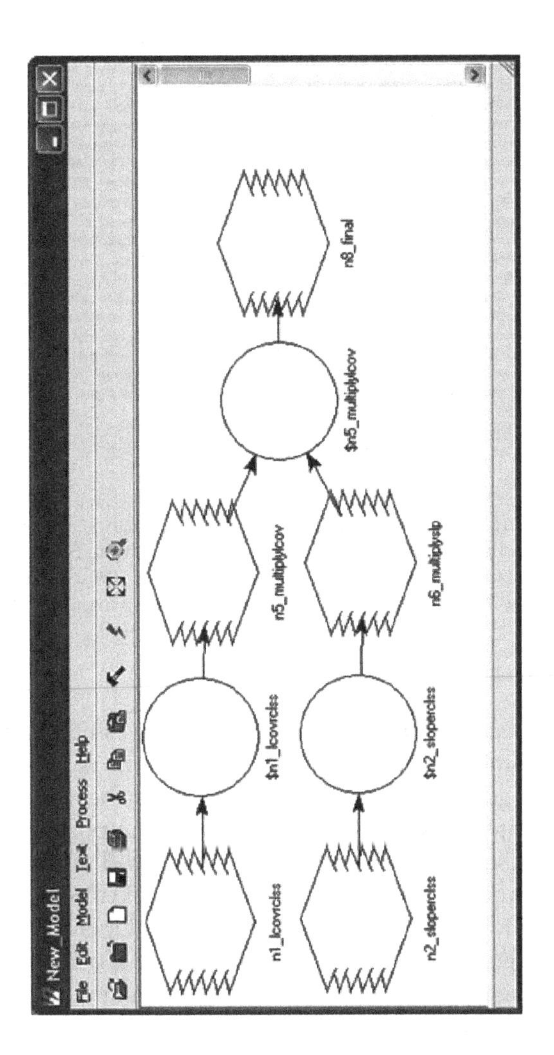

FIGURE 5.11
The ERDAS Imagine Model Maker interface. The same weighted linear combination model as shown for the IDRISI Macro Modeler in Figure 5.6 and ArcGIS ModelBuilder in Figure 5.8 is shown as an example.

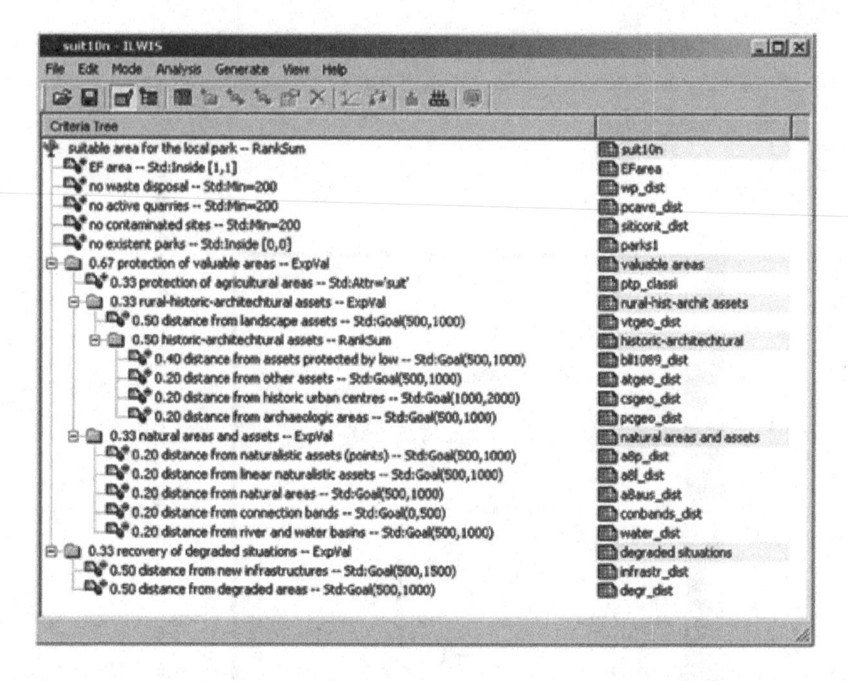

FIGURE 5.12
The criteria tree view in ILWIS software showing constraints, factors, and weights used in an application for deciding on a suitable area for a park. (Zucca, A., Sharifi, A. M., and A. G. Fabbri. 2008. Application of spatial multi-criteria analysis to site selection for a local park: A case study in the Bergamo Province, Italy. *Journal of Environmental Management* 88(4):752–769.)

5.2.4.5 Open-SDSS

A generic open source and free SDSS, called Open-SDSS, is presently undergoing development. The Open-SDSS framework is being created with the following goals in mind: the system must be generic (unrestricted), model oriented, flexible, and with extensive support for graphical modeling and end user customization. Open-SDSS will allow the inclusion of modules from open source software such as ILWIS and SAGA in user-developed applications. Some of the features of Open-SDSS include the generation, storage, and organization of scenarios, sensitivity analysis functions, and the aggregation and ranking of alternatives. The framework of Open-SDSS includes the following elements: user interface, data, models, scenarios, and graphical modeling environment. Open-SDSS is currently a prototype but upon completion will be quite useful to the inexperienced SDSS user. Some of the key technologies used in this project are GTK+ (a widget toolkit used to create the Open-SDSS graphical user interface), GooCanvas (a canvas widget that is the backbone of the Graphical Modeling System), GDAL (a raster geospatial data library used in some of the models), and the Python programming language which binds all of the components together.

5.3 Summary

When confronted by a spatial decision-making situation, those responsible for deciding how to address the problem will need to investigate possible software systems that could support the process. This can be a daunting task as there are many options available including building a new system from scratch, adopting an existing specific system that meets the user's needs, or adapting a generic system for a specific use. The task of deciding is made harder by having to fit the choice into an organization's existing software and hardware infrastructure.

The scientific literature is replete with examples of SDSS that have been developed for specific purposes. Many of these could be applied in different geographic locations with little or no adaptation, but in general these SDSS often seemed to have been applied only a single time or only by one set of users. Often for these problem-specific SDSS, users require access to the specific data sources. One of the main problems with these types of SDSS is that they are developed for a specific purpose and are, in general, not continuously supported. Thus, with the rapid evolution of dependent software components, the SDSS as originally developed is not updated to operate using new versions and can quickly become obsolete. An example of this would be that many SDSS in the late 1990s and early 2000s were developed using ArcView 3.x software. As many organizations will only be using or have access to ArcGIS software (the successor to ArcView 3.x), it is likely that those ArcView 3.x software-based SDSS will go unused unless an investment is made to port the old Avenue code to a more modern development environment.

There are a number of more general SDSS that are continually supported and thus gain success with repeated applications. These software programs are generally specific to a certain domain and have a broad range of functionality for addressing spatial problems within that domain. In this chapter, we have examined several of these types of domain-specific SDSS, including INDEX and CommunityViz, which are used for land use planning; Marxan and EMDS, which are ecologically and natural resource-related planning systems; and SSToolbox, which is an agricultural SDSS. Each of these systems has a variety of tools for addressing a broad range of problems with a spatial dimension. The INDEX and CommunityViz tools have been widely used, mainly in a management capacity by various government agencies, often at a local level. The Marxan and EMDS models have generally been used by land and water management agencies at a regional or federal level.

There are several existing software products that could be considered as providing a generic framework for developing SDSS. We have termed these generic SDSS in this chapter. These systems provide a wealth of tools or modules that can be constructed into an SDSS. In the case of IDRISI Macro Modeler, ArcGIS ModelBuilder, IMAGINE Spatial Modeler, and Open-SDSS, graphical modeling development environments are available. These environments allow the development of a range of SDSS based on multi-criteria evaluation, simulation, and statistical models. As these systems continue to evolve, greater amounts of modeling functionality that are useful in SDSS will be built in and exposed for SDSS development. These systems are all commercial, with IDRISI being the least expensive. Freely available and open source software such as SAGA and ILWIS actually expose greater modeling functionality to a user but require significant expertise in use. In the future, it should be expected that these generic SDSS frameworks will provide greater ease and flexibility and will become more widely used in the development of SDSS.

References

Abu-Zeid, K. M. 1998. A multicriteria decision support system for evaluating cropping pattern strategies in Egypt. In *Multiple objective decision making for land, water, and environmental management*, ed. S. El-Swaify, and D. S. Yakowitz, 105–120. Boca Raton: Lewis Publishers.

Agostini, P., S. Torresan, C. Micheletti, and A. Critto. 2009. Review of decision support systems devoted to the management of inland and coastal waters in the European Union. In *Decision support systems for risk-based management of contaminated sites*, ed. A. Marcomini, G. W. Suter II, and A. Critto, 1–19. New York: Springer.

Allen, E. 2007. Clicking toward better outcomes. In *Planning support systems for cities and regions*, ed. R. K. Brail, 139–166. Cambridge: Lincoln Institute of Land Policy.

Baio, F. H. R., and L. A. Balastreire. 2002. Evaluation of a site specific chemical application system based on spatial variability of weeds. In *Proceedings of the World Congress of Computers in Agriculture and Natural Resources*. Iguacu Falls, Brazil. American Society of Agricultural and Biological Engineers, St. Joseph, MI.

Ban, N. C., and A. C. J. Vincent. 2009. Beyond marine reserves: exploring the approach of selecting areas where fishing is permitted, rather than prohibited. *PLoS ONE* 4(7):e6258.

Barton, J., J. Plume, and B. Parolin. 2005. Public participation in a spatial decision support system for public housing. *Computers, Environment and Urban Systems* 29(6):630–652.

Best, B. D., P. N. Halpin, E. Fujioka, A. J. Read, S. S. Qian, L. J. Hazen, and R. S. Schick. 2007. Geospatial web services within a scientific workflow: Predicting marine mammal habitats in a dynamic environment. *Ecological Informatics* 2(3):210–223.

Bostian, C. W., L. W. Carstensen, and D. G. Sweeney. 2002. Using geographical information systems to predict coverage in broadband wireless systems, http://www.ursi.org/Proceedings/ProcGA02/papers/p2238.pdf (accessed December 2, 2009).

Chan, K. M. A., M. R. Shaw, D. R. Cameron, E. G. Underwood, and G. C. Daily. 2006. Conservation planning for ecosystem services. *PLoS Biol* 4(11):e379.

Dragan, M., E. Feoli, M. Fernetti, and W. Zerihun. 2003. Application of a spatial decision support system (SDSS) to reduce soil erosion in northern Ethiopia. *Environmental Modelling & Software* 18(10):861–868.

Engel, B. A., J-Y. Choi, J. Harbor, and S. Pandey. 2003. Web-based DSS for hydrologic impact evaluation of small watershed land use changes. *Computers and Electronics in Agriculture* 39(3):241–249.

Game, E. T., and H. S. Grantham. 2008. Marxan user manual for Marxan version 1.8.10. University of Queensland, St. Lucia, Queensland, Australia, and Pacific Marine Analysis and Research Association, Vancouver, British Columbia, Canada.

Giles, D. K., and D. Downey. 2003. Quality control verification and mapping for chemical application. *Precision Agriculture* 4(1):1573–1618.

Girvetz, E., and F. Shilling. 2003. Decision support for road system analysis and modification on the Tahoe National Forest. *Environmental Management* 32(2):218–233.

Gorr, W., M. Johnson, and S. Roehrig. 2001. Spatial decision support system for home-delivered services. *Journal of Geographical Systems* 3(2):181–197.

Haettenschwiler, P. 1999. *Neues anwenderfreundliches Konzept der Entscheidungsunterstützung. Gutes Entscheiden in Wirtschaft, Politik und Gesellschaft*, 189–208. Zurich: vdf Hochschulverlag AG.

Hoy, P. 1998. SSToolbox an agricultural spatial decision support system. Paper presented at Proceedings of the Eighteenth Annual ESRI User Conference, California.

Ianni, E., I. Ortolan, M. Scimone, and E. Feoli. 2008. Assessment of management options to reduce nitrogen load from agricultural source in the Grado. *Management of Environmental Quality: An International Journal* 19(3):318–334.

Johnson, M. P. 2001. A spatial decision support system prototype for housing mobility program planning. *Journal of Geographical Systems* 3(1):49–67.

Laudien, R., S. Brocks, S. Weyler, A. Christmann, N. Köhn, and G. Bareth. 2009. Tools for modeling water supplies. *ArcUser*, Fall 2009.

Lieske, S. N., S. Mullen, and J. D. Hamerlinck. 2009. Enhancing comprehensive planning with public engagement and planning support integration. In *Planning support systems best practice and new methods*, ed. S. Geertman and J. Stillwell, 295–315. New York: Springer.

Mbilinyi, B. P., S. D. Tumbo, H. F. Mahoo, and F. O. Mkiramwinyi. 2007. GIS-based decision support system for identifying potential sites for rainwater harvesting. *Physics and Chemistry of the Earth, Parts A/B/C* 32(15–18):1074–1081.

McKenzie, J. S., R. S. Morris, D. U. Pfeiffer, and J. R. Dymond. 2002. Application of remote sensing to enhance the control of wildlife-associated *Mycobacterium bovis* infection. *Photogrammetric Engineering & Remote Sensing* 68(2):153–159

McKenzie, J. S., R. S. Morris, C. J. Tutty, and D. U. Pfeiffer. 1997. EpiMAN-TB, a decision support system using spatial information for the management of tuberculosis in cattle and deer in New Zealand. Paper presented at the second annual conference of Geocomputation'97, Dunedin, New Zealand.

Messina, J. P., and S. J. Walsh. 2001. 2.5D morphogenesis: Modeling land use and land cover dynamics in the Ecuadorian Amazon. *Plant Ecology* 156:75–88.

Morris, R. S., R. L. Sanson, M. W. Stern, M. Stevenson, and J. W. Wilesmith. 2002. Decision-support tools for foot and mouth disease control. 21(3):557–567. *Review of Science and Technology Off. International Epiz.*

Mwasi, B. 2001. Land use conflicts resolution in a fragile ecosystem using multi-criteria evaluation (MCE) and a GIS-based decision support system (DSS). Paper presented at Proceedings of an International Conference on Spatial Information for Sustainable Development, Nairobi, Kenya.

Nedović-Budić, Z., R. G. Kan, D. M. Johnston, R. E. Sparks, and D. C. White. 2006. CommunityViz-based prototype model for assessing development impacts in a naturalized floodplain-EmiquonViz. *Journal of Urban Planning and Development* 132(4):201–210.

Parent, J., D. Civco, and J. Hurd. 2007. Simulating future forest fragmentation in a Connecticut region undergoing suburbanization. Paper presented at the annual conference of ASPRS, Tampa, Florida.

Power, D. J., and S. Kaparthi. 2002. Building web-based decision support systems. *Studies in Informatics and Control* 11(4):291–302.

Ray, J. J. 2007. A web-based spatial decision support system optimizes routes for oversize/overweight vehicles in Delaware. *Decision Support Systems* 43(4):1171–1185.

Reynolds, K. M. 2006. EMDS 3.0: A modeling framework for coping with complexity in environmental assessment and planning. *Science in China: Series E Technological Sciences* 49:63–75.

Reynolds, K. M. and P. F. Hessburg. 2005. Decision support for integrated landscape evaluation and restoration planning. *Forest Ecology and Management* 207:263–278.

Reynolds, K. M., M. Jensen, J. Andreasen, and I. Goodman. 2000. Knowledge-based assessment of watershed condition. *Computers and Electronics in Agiculture* 27:315–334.

Rizzoli A. E., and W. J. Young. 1997. Delivering environmental decision support systems: Software tools and techniques. *Environmental Modelling & Software* 12:237–249.

Salt, C. A., and M. C. Dunsmore. 2000. Development of a spatial decision support system for post-emergency management of radioactively contaminated land. *Journal of Environmental Management* 58(3):169–178.

Schaller, J., T. Gehrke, and N. Stout. 2009. ArcGIS processing models for regional environmental planning in Bavaria. In *Planning Support Systems Best Practice and New Methods,* ed. S. Geertman and J. Stillwell, 243–264. Heidelberg: Springer.

Scheibe, K. P. 2003. A spatial decision support system for planning broadband, fixed wireless telecommunication networks. PhD diss., Virginia Polytechnic Institute and State University.

Scheibe, K. P., L. W. Carstensen Jr., T. R. Rakes, and L. P. Rees. 2006. Going the last mile: A spatial decision support system for wireless broadband communications. *Decision Support Systems* 42(2):557–570.

Srinivasan, R., and B. A. Engel. 1994. A spatial decision support system for assessing agricultural nonpoint source pollution. *Water Resources Bulletin* 30(2):441–452.

Sugumaran, V., and R. Sugumaran. 2007. Web-based spatial decision support systems (WebSDSS): Evolution, architecture, and challenges. *Communications of the Association for Information Systems* 19:844–875.

Uran, O., and R. Janssen. 2003. Why are spatial decision support systems not used? Some experiences from the Netherlands. *Computers, Environment and Urban Systems* 27(5):511–526.

Wang, L., and Q. Cheng. 2006. Web-based collaborative decision support services: Concept, challenges and application. Paper presented at ISPRS Technical Commission II Symposium, Vienna.

Watts, M. E., I. R. Ball, R. S. Stewart, C. J. Klein, K. Wilson, C. Steinback, R. Lourival, L. Kircher, and H. P. Possingham. 2009. Marxan with zones: Software for optimal conservation based land- and sea-use zoning. *Environmental Modelling and Software* 24(12):1513–1521.

Zucca, A., A. M. Sharifi, and A. G. Fabbri. 2008. Application of spatial multi-criteria analysis to site selection for a local park: A case study in the Bergamo Province, Italy. *Journal of Environmental Management* 88(4):752–769.

6

Building SDSS Software

Learning Objectives

- Gain an understanding of advantages and limitations of and strategies for integrating different programs to build a spatial decision support system (SDSS).
- Learn about various technologies that are important in constructing new SDSS software, including programming languages, development environments, and spatial libraries.

In Chapter 5 we examined a wide range of existing SDSS software packages that have been used for various purposes. Problem-specific SDSS, more general domain-level SDSS software such as planning software (e.g., CommunityViz), and generic SDSS (e.g., IDRISI Macro Modeler, OpenSDSS) were examined. That chapter was designed to provide an overview of existing software that potential SDSS users might adopt for their spatial decision-making situations. In contrast, this chapter focuses on how to build new SDSS software using various technologies, techniques, and tools. This chapter focuses on both the technique of integrating existing software and how to build a new system from scratch. The chapter will attempt to give an overview of technologies to use and strategies to follow in the development of new SDSS.

6.1 Introduction

Spatial problems, discussed in previous chapters, are by their nature ill structured, complex, and have some aspects that are unique to a specific domain or geographical area. Thus, there are often inherent limitations in the reuse of existing tools that were developed for unique spatial problems in varying geographic locations. Specific issues in the use of existing

SDSS for new problems include difficulty in transfer and adoption to different geographic areas; availability of necessary software and data; lack of flexibility for defining user-specific applications; lack of proper software, documentation, and training opportunities; lack of realistic scenario development options; lack of applicability to specific aspects of a given problem situation; and the prescriptive and inflexible nature of existing SDSS. Most SDSS are not built with a tremendous amount of flexibility to allow new users to easily adapt them to their own unique situation. When a spatial problem of significant complexity requires the use of an SDSS and no available system meets the requirements necessary for solving the problem, then the development of a new SDSS is necessary. The development of a new SDSS can be achieved by varying methods ranging from (a) designing and developing a system completely from scratch, (b) building a system within a single software such as a geographical information science (GIS), or (c) using existing software in combination with each other in configurations based on the development of technologies to allow interaction between the different software (Djokic 1996; Denzer 2005; Herzig 2008).

The term *SDSS* can cover a very wide variety of systems that vary in their purpose, focus, intended audience, and technological implementation. In general, these systems have spatial data management and analysis functionality in conjunction with some analytical modeling functions for addressing complex and unstructured spatial decision-making situations. However, these systems are used to address spatial problems in many different domains, including land use planning, environment and natural resource management, transportation, business, and many other areas. Many unique database formats and structures, as well as modeling functions, have been developed for these different application domains. In addition, the stakeholders that might play a role in decision-making processes vary across domains and specific situations. Thus, it is difficult to comprehensively describe all of the possibilities in designing and building an SDSS. Other authors have discussed techniques and tools that can be used for building decision support systems (DSS) and SDSS (Keenan 1996, Malczewski 1999, Ungerer and Goodchild 2002). Increasingly, SDSS do not necessarily reside on a single machine and are not single software but are delivered over the Web or on networked machines.

This chapter will attempt to cover some of the relevant generic strategies for designing and building an SDSS but will also get into some of the specifics, such as technologies that can be used. The chapter will examine types of SDSS that can be built and the tools that are used to build them.

6.2 SDSS Software Components

Addressing complex spatial decision problems requires spatial data collection and processing, modeling, alternative evaluation, and solution presentation (Djokic 1996). In Chapters 3 and 4, the fundamental components of SDSS, such as the database management component (DBMC), the model management component (MMC), the dialog management component (DMC), and the knowledge management component (KMC), were explored. There are various computer programs that can be used to meet the requirements of these SDSS components. Computer programs such as GIS (DBMC, MMC, DMC), artificial intelligence software such as expert systems (KMC), relational database management systems (DBMC), output visualization software (DMC), and modeling utilities (MMC) have been used for constructing SDSS (Figure 6.1). Some of the common software programs used in SDSS are highlighted and reviewed in the paragraphs below. These paragraphs are not meant to be exhaustive but rather provide an overview of common software used in SDSS configurations as described in the literature.

6.2.1 Common Software for Utilization in SDSS Development

6.2.1.1 Spatial Data Collection, Management, Analysis, and Visualization Software

Geographic information systems and related software can often fulfill many of the roles needed in an SDSS. The roles that a GIS program can

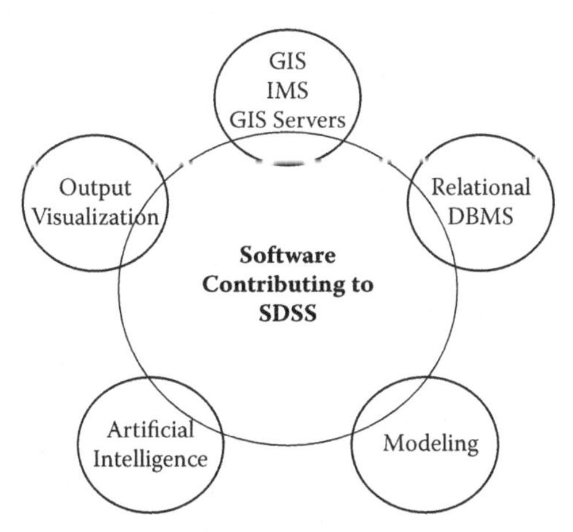

FIGURE 6.1
Common software types potentially contributing to SDSS development.

fulfill include database management, spatial data creation and processing, spatial analysis, and graphical visualization of results in the form of maps, charts, and 3D visualizations. As mentioned in Chapter 3, there are many open source and commercial GIS software available. Some open source desktop GIS software examples include Quantum GIS, Geographic Resource Analysis Support Systems (GRASS), Integrated Land and Water Information System (ILWIS), System for Automated Geoscientific Analyses (SAGA), OpenJUMP, gvSIG, and uDig. Steineger and Hay (2009) provide a good overview of these open source GIS with an emphasis on potential use for landscape ecology. An overarching index of open source and free GIS software can be seen at FreeGIS.org (http://freegis.org/) and at Open Source GIS (http://opensourcegis.org). The most popular commercial desktop GIS products include ESRI's ArcGIS, Manifold, Autodesk's line of products including AutoCAD Map 3D, Intergraph's GeoMedia, Pitney Bowes's MapInfo, Clark Lab's IDRISI, and General Electric's Smallworld. Historically, Environmental Systems Research Institute (ESRI) products from ArcInfo to ArcView to present day ArcGIS software have captured the greatest market share of desktop GIS with use commonly by business, government, academic, and nonprofit sectors. Other products have gained more specific market niches, such as Smallworld, which has been commonly used by public utilities. In addition to the base GIS desktop products, there are extensions or additional products that add significant geospatial functionality for specific data types or spatial analysis operations. For example, extensions available for use with ArcGIS software include Network Analyst (routing, transportation, etc.), 3D Analyst (visualization and analysis of 3D data), and Geostatistical Analyst (advanced statistical tools). Most GIS programs have output visualization capabilities in the form of cartographic production functionality, report generators, and chart creation utilities.

From the latter half of the 1990s, the delivery of geospatial information and services through the Internet has grown tremendously. This trend will likely continue with ever increasing sophistication in the delivery of both geospatial information and services. In the late 1990s, Internet map server technology became extremely popular and provided a mechanism for the widespread distribution of spatial information. These technologies began to appear in the SDSS literature in the late 1990s and early 2000s (Wan et al. 1999; Carver et al. 2001). The main purpose of these early Internet map servers was the presentation of spatial data online as interactive maps in which the user could navigate and query spatial data. The most common Internet map servers used in SDSS have been ESRI's ArcIMS and the University of Minnesota's free MapServer software. The technology is quickly evolving as more sophisticated functionality for providing geospatial services (e.g., spatial processing functionality) is being included in addition to simple online mapping. Presently, there is an ongoing software extension whose main functionality is to serve spatial data in navigable maps to also serving GIS or geospatial services. Examples of common open source map and geospatial server technologies include MapServer, GeoServer, and MapGuide, while ArcGIS Server, GeoMedia WebMap, and AltaMap Server are commercial products.

6.2.1.2 Relational Database Management Software

To support the often large amounts of spatial and nonspatial data and the size of datasets that are necessary in SDSS applications, there are many open source (e.g., PostGIS, MySQL, Spatial Lite) and commercial (Spatial Query Server, Oracle Spatial, IBM DB2, Microsoft Access, Microsoft SQL Server, ArcSDE) relational database management systems and related software (see Table 6.1). Technologies such as PostGIS (works with PostgreSQL) and ArcSDE (works with multiple relational database management systems such as Oracle and SQL Server) facilitate the handling of spatial data in relational databases. Karnatak et al. (2007) used ArcSDE and Oracle to manage spatial and nonspatial data in a multi-criteria SDSS. MySQL database software was used in a case study of a participatory SDSS in an urban planning application (Sidlar and Rinner 2007). Rao et al. (2007) used SQL Server in conjunction with ArcSDE software to manage multiple sets of data stored in different locations for managing and planning the Conservation Reserve Program administered by the United States Department of Agriculture. Similarly, Wong et al. (2007) used SQL Server in a Canadian SDSS called WILDSPACE for management of species at risk. An Oracle database was used in conjunction with ArcGIS and statistics software in an SDSS for prawn fishery management in Australia (Carrick and Ostendorf 2007).

6.2.1.3 Modeling Software

There is a large number of modeling programs or applications that can be used in conjunction with other programs in SDSS configurations. An overview of modeling techniques was presented in Chapter 4. However, there are several specialized software applications that are not GIS but carry out specific spatial modeling operations and have been or could be utilized in SDSS. The spatial pattern analysis software FRAGSTATS has been used in a number of ecological SDSS applications (Reynolds and Hessburg 2005; Munier et al. 2004). There are a number of statistical software applications that operate with spatial data, including CrimeStat, ClusterSeer, SaTScan, and others. Other software applications, such as AutoCad or Surfer (3D surface mapping), are often used with GIS and could be incorporated in an SDSS. There are also very niche-specific applications such as Site Recorder, which is used for maritime archaeology. The free Spatial

TABLE 6.1

Potential Programs for Integration in Desktop SDSS

Software Category	Distribution Model	Name	URL
Spatial data collection, management, analysis, and visualization	Commercial	ArcGIS	www.esri.com
		TransCAD	www.caliper.com/tcovu.htm
		MapInfo	www.pbinsight.com/
		IDRISI	www.clarklabs.org
		GeoMedia	www.intergraph.com/
		ERDAS Imagine	www.erdas.com
		Manifold	www.manifold.net/index.shtml
		Smallworld	www.gepower.com/home/index.htm
	Free	GRASS	http://grass.itc.it/
		UDig	http://undig.refractions.net
		SAMT	http://www.samt-lsa.org/
		ILWIS -Open	www.ilwis.org
		SPRING	www.dpi.inpe.br/spring/
		Quantum GIS	www.qgis.org/
		SAGA	saga-gis.org/en/index.html
Data Management Systems	Commercial	ArcSDE	www.esri.com/software/arcgis/arcsde/index.html
		Spatial Query Server	http://active.boeing.com/mission_systems/products/
		Oracle Spatial	http://www.oracle.com/technology/products/spatial/index.html
		IBM-DB2	www-01.ibm.com/software/data/db2/
		SQL Server 2008	www.microsoft.com/sqlserver/2008/
	Free	PostGIS	postgis.refractions.net/
		H2Spatial	http://geoserver.org/display/GEOS/H2+Spatial+Database
		SpatialLite	www.gaia-gis.it/spatialite/
		MySQL Spatial	dev.mysql.com/doc/refman/5.0/en/spatial-extensions.html

TABLE 6.1

Potential Programs for Integration in Desktop SDSS (Continued)

Software Category	Distribution Model	Name	URL
Modeling Related Software	Commercial	S-Plus	http://spotfire.tibco.com/Products/SPLUS-Client.aspx
		SPSS	http://www.spss.com/
		MATLAB	http://www.mathworks.com/products/matlab/
	Free	SME	http://www.uvm.edu/giee/SME3/
		openModeller	http://openmodeller.sourceforge.net/
		R	http://www.r-project.org/
		FRAGSTATS	www.umass.edu/landeco/research/fragstats/fragstats.html
Knowledge Component	Commercial	Jess	http://www.jessrules.com/
		NetWeaver	http://www.rules-of-thumb.com/
		Criterium DecisionPlus	http://www.infoharvest.com/

Analysis and Decision Assistance (SADA) software was developed at the University of Tennessee Institute for Environmental Modeling. SADA incorporates tools from environmental assessment fields and was used in an SDSS for a terrestrial ecological risk assessment of a former scrap metal yard (Purucker et al. 2008).

The use of mathematical and statistical software is common in SDSS as GIS and other software lack robust mathematical and statistical functionality. There have been many different statistical and mathematical programs such as Genstat, SAS, SPSS, R, MATLAB, Minitab, and S-Plus utilized in SDSS. Carrick and Ostendorf (2007) linked ArcGIS, GenStat, and S-Plus in their SDSS for managing a prawn fishery in Australia. The MATLAB program was used to provide fuzzy probability functions to an SDSS used for urban water system pipe replacement prioritization (Makropoulos and Butler 2005). Spreadsheet programs such as Microsoft Excel have been used in SDSS configurations. Examples include the use of Excel and GIS in an SDSS for afforestation (Gilliams et al. 2005), the coupling of Excel and MapObjects software for property development assistance (Li et al. 2004), and the development of an SDSS for environmental modeling and spatial data visualization within Excel (Berardi 2002).

6.2.1.4 Knowledge Management Software

Knowledge components are sometimes built into SDSS software in order to provide the applications with some automated intelligence. There are a variety of programs that can be used to build these types of capabilities into an SDSS, including expert systems and artificial intelligence shells and programming languages. Some examples of programs used for SDSS include Prolog, C Language Interface Production System (CLIPS), the NetWeaver logic engine, and Visual Rule Studio. The artificial intelligence programming language Turbo Prolog was used to build a prototype system called HydroLOGIC, which was used to guide decisions on locating a new water well (Crossland 1990). The expert system development tool CLIPS was used with ArcInfo and HARDY for the development of a generic SDSS that was used for strategic planning of land use (Zhu 1997) and also with ArcGIS in an SDSS for live hazard monitoring and detection (McCarthy et al. 2008). The Ecosystem Management Decision Support (EMDS) system was developed using the NetWeaver logic engine (Gartner et al. 2008). The Visual Rule Studio software was used to build an industrial site selection SDSS by Eldrandaly et al. (2003), and by Sugumaran et al. (2007) to construct an SDSS for snow-removal planning. These types of software assist in the building of rule-based expert knowledge into an SDSS.

6.2.2 SDSS Development by Software Integration

Spatial decision-making situations are complex, require a significant amount of disparate data, and involve a range of stakeholders. The development of SDSS to aid in addressing these types of issues requires a range of components that all serve specific, and sometimes multiple, purposes (Figure 6.1). As there are few software applications that are capable of providing all of the functions required in an SDSS, it is often necessary to link numerous software together in a comprehensive system. This can be done by creating interfaces or intermediary programs that allow data transfer between various software applications. This approach can reduce development time as it makes use of existing software capabilities, but often at the cost of reduction in control by the developers and potentially less flexibility in the final product. The majority of existing SDSS have been developed through the approach of linking separate pieces of software together. An alternative to this approach is to develop all of the necessary SDSS functionality through a single software application either through existing tools or by developing customized tools. The technique of developing all functionality within a single software package is less common because it is not very likely for a single piece of software to have tools out of the box to meet all the necessary functionality requirements of an SDSS, and it is often expensive in the short term to develop all functionality within a single piece of software. When this approach is utilized, GIS is the framework in which the SDSS is usually developed. GIS programs generally meet more of the functionality requirements of an SDSS than any other individual software packages. In GIS software, there is usually functionality for spatial data analysis, spatial modeling, spatial and non-spatial database management, report generation, visualization through maps and potentially 3D models, and often environments for customizing user interfaces. The following sections will examine the different technologies and approaches for developing SDSS by coupling different software or integrating all new functionality into an existing program (i.e., GIS).

6.2.2.1 Integration Technologies

There are various technologies that facilitate the communication between and integration of various computer programs into a single system. The basics of these technologies will be touched upon in order to provide insight into SDSS integration strategies. A few of the most important techniques and technologies include system calls, dynamic-link libraries (DLL), Dynamic Data Exchange (DDE), the Component Object Model (COM), and ActiveX. These technologies provide the ability for distinct programs to share data and processes, and to communicate efficiently with each other. System calls can be used to activate one program from another program through requests to the operating system. System calls are also used for basic interactions with the operating system, such as for controlling processes, file management, device management, and communication. System calls are a basic way to allow communication between programs. Wu et al. (2004) tested the use of system calls to GRASS GIS from a C program in comparison to a component-based GRASS GIS server architecture, finding the component methods were slightly slower but acceptable. Dynamic Data Exchange is a Microsoft technology used for communication and data sharing between multiple applications. There are numerous limitations to DDE, such as lack of support for networked computers and limits on data passed. These limitations have led to reduced use of DDE in relation to other technologies. However, DDE was used in many applications in the 1990s and 2000s. Jankowski et al. (1997) used DDE to link the GroupChoice and ArcView software in their Spatial Group Choice SDSS. DDE was used as the mechanism for controlling flow between programs (GIS, relational database, models) in a marine SDSS (West 1999).

A DLL is a library that contains code and data that can be used by multiple programs simultaneously. For example, in the Windows operating system, a common open dialog can be accessed from multiple programs at the same time. Each of the programs is accessing functionality contained in a DLL. Computer programs can be modularized by building separate modules and distributing them as DLLs. In addition, extensions to existing software can be packaged and delivered as DLLs. This is often the case with GIS software such as ESRI's ArcGIS, for which many third-party extensions are distributed as DLLs. Disadvantages of the DLL are that it can complicate the deployment process because many files may need to be copied to appropriate locations and DLLs must also be built with programming languages that are designed to work together. Bennett and Vitale (2001) used DLLs to couple functionality from ArcInfo, ArcView, and AGNPS. Eldrandaly et al. (2003) deployed an expert system as a DLL module in an SDSS for industrial site selection.

The Component Object Model (COM) is a platform-independent, distributed, object-oriented standard for creating binary software components that can interact with each other. It can be thought of as a higher-level version of the DLL. COM is language neutral, which means that any set of languages that understand COM can communicate. Although the COM platform has been superseded by the .NET framework, it is still supported and is a viable technology. Indeed, several GIS programs such as ArcGIS, MapInfo, and GeoMedia were built at least partially upon the base of the COM protocol (Li et al. 2005). The SDSS developed by Eldrandaly et al. (2003) integrated expert systems, GIS, and multi-criteria decision-making software using COM technology. Li et al. (2005) integrated six separate COM object libraries in a typhoon insurance pricing SDSS. The COM-compliant libraries were sourced from a combination of COM-compliant commercial software (e.g., ArcGIS), adaptation of open-source programs (e.g., CLIPS-based shell), and directly developed libraries (e.g., insurance pricing) using COM-compatible languages. COM-compliant components have been used in many SDSS, but this fact is not always explicitly mentioned in published applications.

Object Linking and Embedding (OLE) is a system built on top of COM. It enables programs to build documents that contain parts of documents created using other programs. An example would be the inclusion of a map viewer in a spreadsheet document as was done by Li et al. (2004) using MapObjects and Excel. Finally, the ActiveX framework is used to define reusable software controls that can be plugged into many different types of applications. GIS-type controls in ESRI MapObjects are examples of ActiveX controls and were used in many SDSS as mentioned previously.

6.2.2.2 Integration Strategies

The integration of multiple programs into a single SDSS is classified most commonly along a spectrum from no coupling, to loosely coupled, to tightly coupled, to full integration within a single software (Malczewski 1999; Chakhar and Mousseau 2008). In the instance of no coupling, the different programs are used in isolation with no automated interaction between them, while full coupling is when all functionality is built into a single software framework. We will focus on instances in which some effort is made to construct a new SDSS software configuration through automation of tasks and communication between different programs. Figure 6.2 presents the three coupling approaches as given by Chakhar and Mousseau (2008) in regard to GIS and multi- criteria program integration. In the following sections, these varying approaches are explained in detail with examples.

6.2.2.2.1 Loose Integration or Coupling

The loose coupling method usually involves the development of intermediate software for the conversion of data files for sharing between multiple applications, such as GIS and modeling software. In a multi-criteria SDSS, this would usually involve the development of intermediate software for restructuring the data obtained from GIS operations into a format that could be utilized by separate multi-criteria evaluation software. In this type of situation, other functionality might be developed to allow output files from the multi-criteria evaluation software to be reformatted for visualization of results in a spatial format supported by the GIS. Each part of the SDSS would have its own database and user interface in a loose coupling approach (Chakhar and Mousseau 2008). Giacomelli (2005) stated that the aim of loose integration is to facilitate the interaction of different tools that have unique data formats. The advantage of the loose coupling approach is that there is a limited investment in software development. The SDSS developer can develop small pieces of software that can convert data accordingly. However, there is not an intensive effort to develop common user interfaces and seamless integration of the different pieces of software. The disadvantages of this type of approach are that the interaction between the user and the software is not very efficient and the transferability of the system is limited due to the requirements of having multiple pieces of software.

There are many examples of SDSS that have been developed with a loose coupling approach. In his review of GIS-based multi-criteria decision analysis studies, Malczewski (2006) found that 33.2% of reviewed publications (N = 319) reported using a loosely coupled approach. These approaches have often used GIS software in conjunction with some sort of modeling software. The developers of the SDSS usually wrote some sort of computer program(s) that allowed for the transformation of data between two or more software systems. For example, Carrick and Ostendorf (2007) stated that they kept the linkages between software components of their SDSS as simple as possible. They used ArcGIS, Oracle, GenStat, and SPlus software. They used Open Database Connectivity drivers or formatted text files to facilitate the coupling of the different software components in the SDSS. They acknowledged that their SDSS relies on highly trained personnel for use because they did not invest in the development of sophisticated user interfaces. The authors pointed out that their modular and

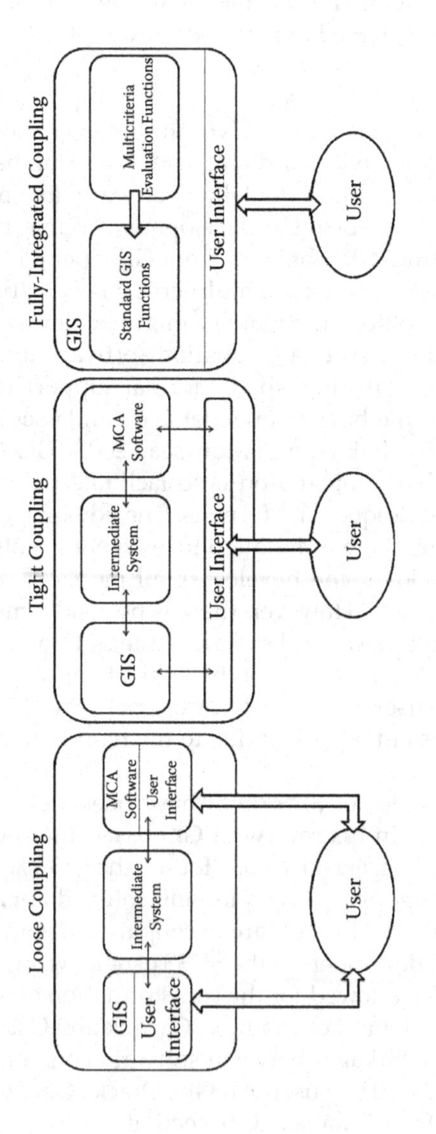

FIGURE 6.2
Diagrams representing loose, tight, and fully integrated coupling approaches (adapted from Chakhar and Mousseau 2008).

loosely coupled approach relies on powerful commercial GIS and statistical packages that are constantly being upgraded. Thus, according to the authors, with their modular approach, their system can take advantage of technological development without them having to carry out major software rebuilding.

Although the technology for developing more tightly coupled or integrated SDSS has improved, the limited initial development time offered by a loosely coupled approach is often the driving factor in their development. The development of applications that had user-friendly interfaces was more difficult before the introduction of development languages and environments that allowed relatively simple construction of programs with graphical user interfaces. With the introduction of technology such as Visual Basic in the 1990s, the development of programs with attractive graphical user interfaces became possible for a wider range of developers (i.e., with less expertise). With the development of some of the technologies mentioned above (i.e., COM, ActiveX, DLL), the reuse or sharing of software components and communication between software components became easier. Now, there are many development environments in which users can develop interfaces for interaction with multiple pieces of software using technologies such as COM. An example of difficulty in coupling software was detailed by Pidd et al. (1996). The authors were attempting to link ArcInfo GIS software with a custom-developed evacuation simulator programmed in C++ (Figure 6.3). They had difficulty accessing the INFO databases of ArcInfo software. In the end, they linked the different software by developing routines to periodically format data

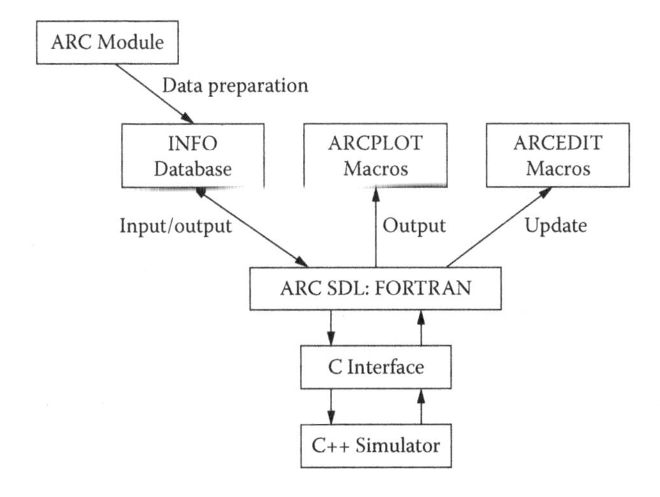

FIGURE 6.3
The general architecture of an SDSS using GIS and a simulation model for emergency evacuation (adapted from Pidd et al. 1996).

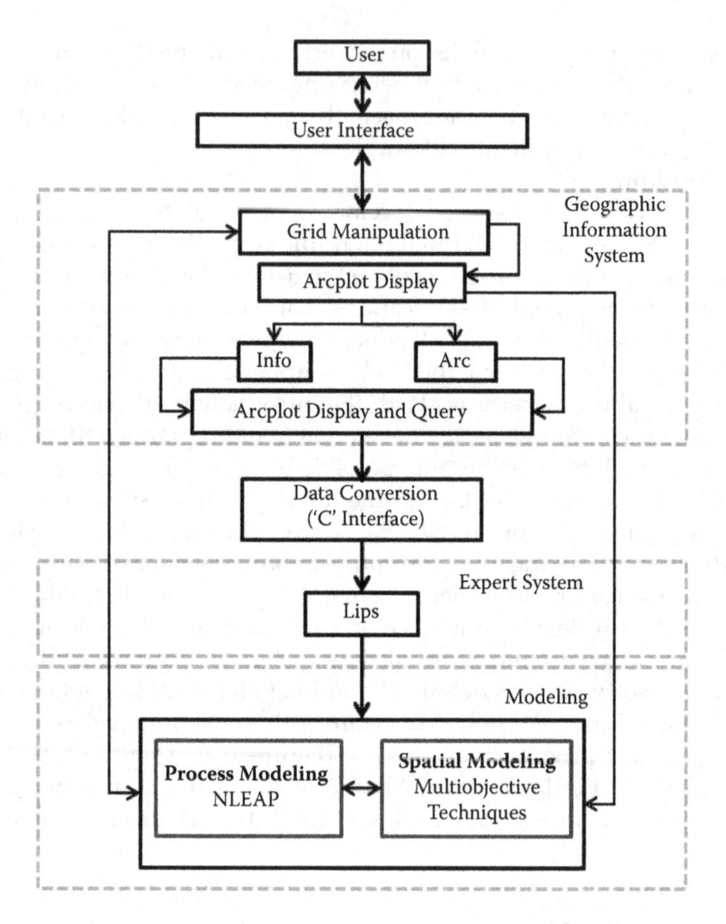

FIGURE 6.4
The architecture of the SDSS created for livestock production planning and environmental management with a data conversion utility to translate data between the GIS and modeling software (adapted from Jain et al. 1995).

for transfer between the two applications. The authors stressed the difficulties of developing efficient coupling routines using ArcInfo software. These types of issues have been alleviated to a great extent with more modern software, including succeeding generations of ESRI's GIS software.

Another example of loose coupling from the 1990s was given by Engel et al. (1993), who coupled the Agricultural Non-Point Source Pollution Model (AGNPS) environmental model with GRASS GIS. They developed input and output tools in the C language. The input tools assisted with the preparation and extraction of data from the GIS database for use in the AGNPS model. The output tools extracted the distributed parameter output from the model ASCII output files and built these into GIS layers for use in GRASS. Figure 6.4 demonstrates a classic loose coupling structure used in a livestock management SDSS (Jain et al. 1995). The system couples ArcInfo GIS, Nitrate Leaching and Economic Analysis Package (NLEAP, a nitrogen leaching model), and an expert system with a data conversion utility developed with the C language (middle Figure 6.4). Falcão et al. (2006) loosely integrated numerous software components including GIS, a relational database management system, decision models, and a 3D forest landscape visualization program in a forest ecosystem management decision support system. A loose coupling approach was used by Gilliams et al. (2005) with GIS and multiple criteria decision methods in a system meant to provide support for policy and planning decisions pertaining to afforestation of agricultural land in Europe.

Changes in the portability of data formats in general and specifically in GIS has led to easier development of component interaction in SDSS. Also, the packaging of easier-to-use and higher-level programming languages and development environments with GIS software has allowed the development of SDSS with loose coupling based around the GIS without the need for lower-level programming languages such as C or C++. An example was the development of an SDSS for evaluating urban road networks in which the links between the MapInfo GIS and the multi-criteria models were developed with MapBasic programming that came with the MapInfo GIS software (Klungboonkrong and Taylor 1998). ArcView GIS was used extensively in the late 1990s and early 2000s in loosely coupled SDSS because of the greater ease in working with the shapefile vector format and the ease in designing interactions with outside applications through the Avenue programming language and the development of custom data exchange mechanisms such as DDE and DLL. Leao et al. (2004) developed a loosely coupled SDSS that used ArcView as its core, but retrieved data from cellular automata and spreadsheet submodules. Sugumaran (2002) developed a desktop SDSS called the Integrated Range Management Decision Support System (IRMDSS) by linking ArcView GIS software to other software through a DLL developed in C++ to allow communication between the different programs. The other software included were Microsoft Access as the nonspatial database management system, Crystal Report as the report generator, and Sictus Prolog as the knowledge management system (Figure 6.5). An SDSS for urban water management developed by Makropoulos et al. (2003) consisted of a loosely coupled framework of ArcView GIS and MATLAB with the exchange of spatial data from ArcView to MATLAB being in the form of ASCII files. Thorp et al. (2008) used customized ArcGIS tools to process spatial soils and agricultural data and to format this data in ASCII files for implementing crop growth models that were run outside the GIS environment.

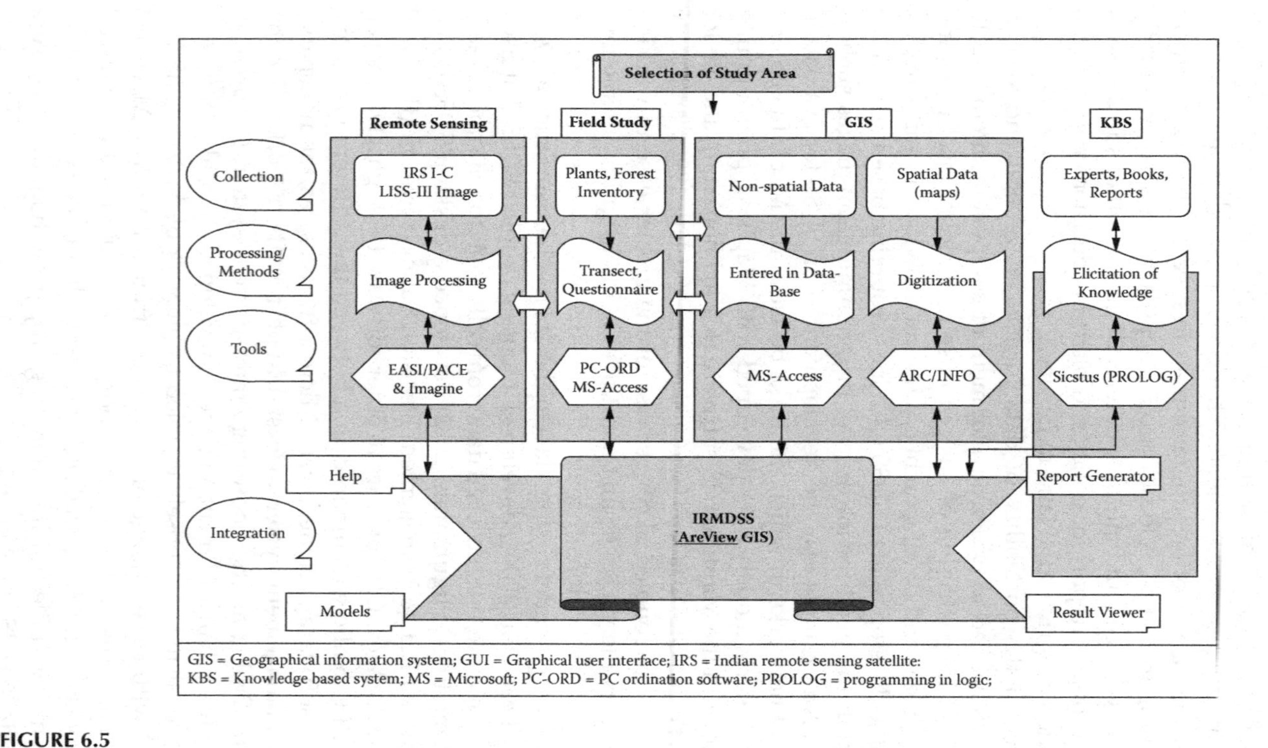

FIGURE 6.5
Architecture of the IRMDSS (Sugumaran, R. 2002. Development of a range management decision support system using remote sensing, GIS and knowledge based system. *Computers and Electronics in Agriculture* 37:199–205.)

6.2.2.2.2 Tight Integration or Coupling

Tight coupling is a strategy in which two or more separate software programs are utilized but with a common user interface developed for users to interact with both. According to Malczewski (2006), a tight coupling approach means that there is a single data and model manager that controls how communication files are shared between different programs. Tight coupling has the advantages of more efficient user interaction based on common user interfaces and efficient data handling. Stevens et al. (2007) indicated that the difference between a tightly coupled system and a loosely coupled system is that the tightly coupled system would allow, for example, a model to directly access and manipulate spatial or attribute data that GIS software has open in memory. The loosely coupled system would only allow access to data stored on the hard disk. They pointed out that this type of arrangement provides greater performance and efficiency. Crossman et al. (2007) identified that tight coupling with GIS enables fast spatial queries and analysis of outputs through maps. The disadvantages of tight coupling approaches can include greater time and resources for system development. With tight coupling, there is likely to be more interface development and considerably more programming. In his review of GIS-based multi-criteria decision analysis studies, Malczewski (2006) found that 29.8% of reviewed papers (N = 319) reported using a tightly coupled approach.

There are many examples of SDSS that utilized a tightly coupled approach. In a wildland fire prevention SDSS, Guarnieri and Wybo (1995) developed a system made up of GIS, a relational database management system, numerical modeling routines, and qualitative modeling. This system used common interfaces and, behind the scenes, an information manager utility to manipulate spatial and nonspatial data between the GIS and numerical modeling program (Figure 6.6).

A tightly coupled SDSS was developed by Bennett and Vitale (2001) for evaluating nonpoint source pollution policy. User interfaces were developed in ArcView GIS for running outside special-purpose software through DLL connections. The special-purpose software was developed to convert files from GIS format to AGNPS input format and also to convert from AGNPS output formats to GIS format for display in ArcView. Similarly, Sengupta and Bennett (2003) tightly coupled a farm-based economic model (GEOLP) and the AGNPS model with ArcView GIS. Their Distributed Intelligent Geographic Modelling Environment (DIGME) SDSS transformed data in GIS spatial formats to formats necessary to run the models. Zhu et al. (1996) developed a tightly coupled strategic land use SDSS. The system, called the Islay Land Use Decision Support System (ILUDSS), contained a knowledge base, expert system, GIS, and analytical procedures. The ILUDSS was built with three software tools: the CLIPS

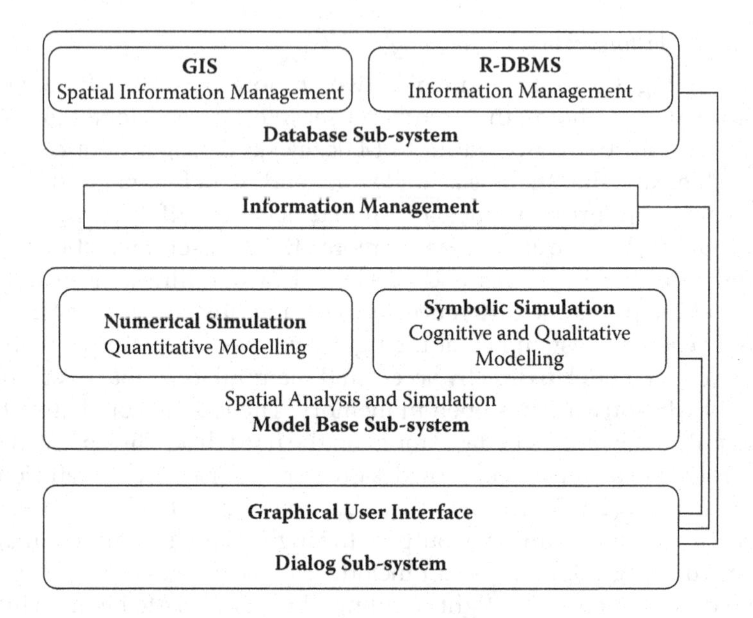

FIGURE 6.6
The architecture of a wildland fire SDSS with the Information Manager serving as the bridge in the tight coupling approach in which data is transferred and reformatted between the components of the SDSS (adapted from Guarnieri and Wybo 1995).

expert system development tool, the HARDY diagramming tool, and ArcInfo GIS. The developers programmed links for interaction between these software systems that were invisible to the user, giving the appearance of a single system.

The tight coupling of the Conservation Reserve Evaluation and Design Optimisation System (CREDOS) with ArcGIS software and integer programming analytical software was described by Crossman et al. (2007). With this SDSS, the user interacts with interfaces accessed from ArcGIS, allowing the execution of spatial operations in ArcGIS and also providing the ability to specify the optimization routines in the integer programming analytical software. The Configurable Emergency Management and Planning Simulator (CEMPS) was a tightly coupled SDSS developed with ArcInfo GIS and an evacuation simulation model (de Silva and Eglese 2000). An *integration link interface* was developed to facilitate the interchange of communication signals for co-coordinating simultaneous functions of both the GIS and evacuation simulation model. This interface also carries out conversion of data between applications. A single user interface is used to access all of the components. Several technologies were used to develop this system, including the ArcInfo Arc Macro Language (AML), custom C programs, and ARC software development language. A system tightly coupling MapObjects and Excel was developed in order to support property professionals (Li et al. 2004). In this system, a MapObjects component was added to Excel for spatial visualization and processing. The built-in development environment of Visual Basic for Applications (VBA) was used to develop functionality to access both MapObject ActiveX controls and to run statistical analyses available in Excel.

A tightly coupled system for rural land use planning called Land Allocation Decision Support System (LADSS) was developed by Matthews et al. (1999). Their system kept the GIS (Smallworld) and knowledge-based system (KBS, developed with G2 software) and the modeling environment as separate applications but with a common user interface that was developed using the G2 environment. The bridge allowing Smallworld to pass required information to the KBS was developed using C programming (see Figure 6.7). A GIS-based planning support system for rural land use allocation called the Rural Land-use Exploration System (RULES) was developed by Santé-Riveira et al. (2008). They used the GeoMedia Professional GIS from Intergraph. A customized optimization application was integrated within the GIS to solve linear programming models with a heuristic algorithm for spatial allocation of land uses. Other modules are accessed from the GIS interface, although the actual algorithms reside in other software. For example, the area optimization module was developed from LINDO (optimization software) libraries but is accessed from the GIS-based SDSS interface (Figure 6.8). A GIS-based DSS was developed by Zeng et al. (2007) by integrating a forest growth and yield model (SIMA) and a mechanistic wind damage model (HWIND) with ArcGIS 8.2. The customized SDSS was developed using ArcObjects and Microsoft

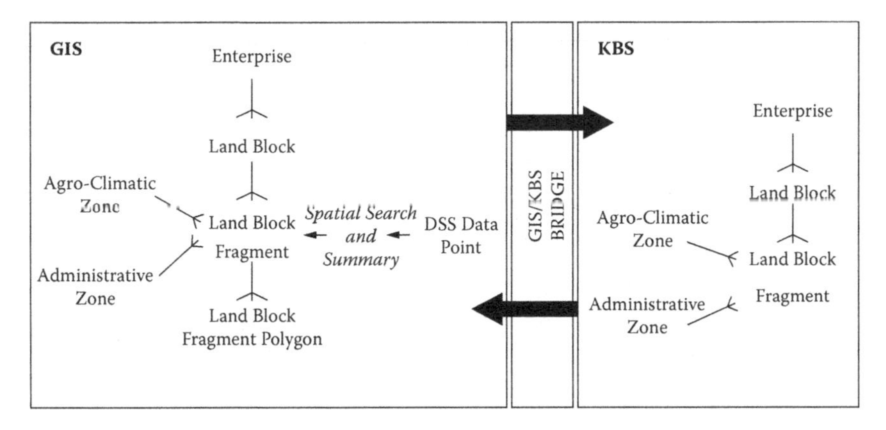

FIGURE 6.7
The coupling of Smallworld GIS and G2 knowledge based systems in the LADSS (Matthews, K. B., A. R. Sibbald, and S. Craw. 1999. Implementation of a spatial decision support system for rural land use planning: Integrating geographic information system and environmental models with search and optimisation algorithms. *Computers and Electronics in Agriculture* 23(1):9–26.).

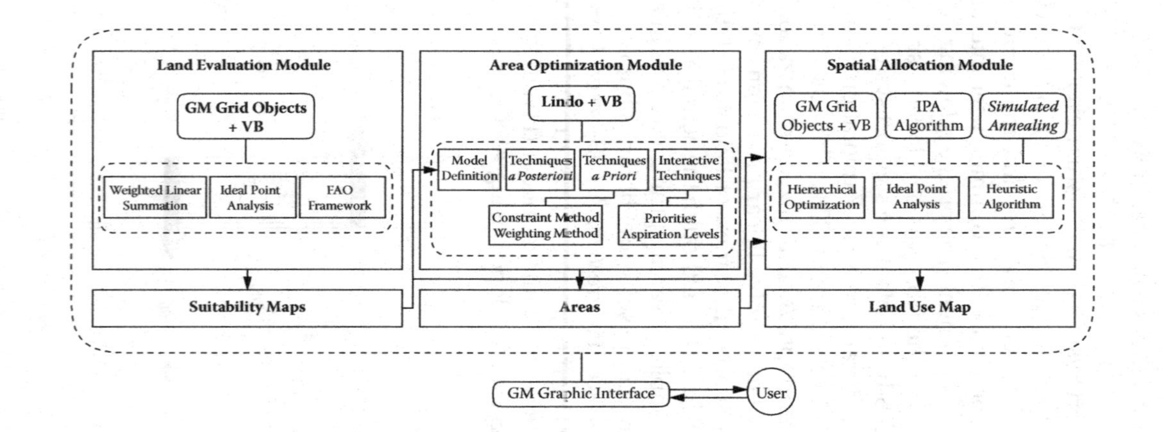

FIGURE 6.8
The achitecture of the RULES SDSS. (*Source*: Santé-Riveira, I., R. Crecente-Maseda, and D. Miranda-Barrós. 2008. GIS-based planning support system for rural land-use allocation. *Computers and Electronics in Agriculture* 63(2):257–273).

Visual Basic 6. The tools were packaged as a DLL that was embedded as an extension into ArcMap. What makes this a tightly coupled application, as opposed to a completely integrated application, is that the HWIND and SIMA models still exist as executables that are called from the software components developed for the ArcGIS extension. Tightly coupled approaches have been widely used and are facilitated by technological developments such as COM technology in which software can more easily be integrated than could be done previously.

6.2.2.2.3 Full Coupling or Embedded

Embedded or full coupling is characterized by all components of the SDSS being built into a single software system. The most common method is to develop an embedded system within the GIS software. Fully integrated SDSS perform like a single program with data interfacing being seamless with a single database management system (Ozan et al. 2003). In a multi-criteria SDSS context, the multi-criteria method is invoked from the GIS interface and the GIS database is extended to support spatial and necessary nonspatial data (Chakhar and Mousseau 2008). This approach usually requires the development of user-specified routines using programming languages that can access functionality in the GIS package (Malczewski 2006). The main advantage of the fully integrated approach is the efficiency gained by having a single database and interface with which the users can interact. The main disadvantage is the cost of development. Significant effort is required to develop the modeling routines, database management functionality, and user interfaces that provide for an efficient and useful system. Malczewski (2006) found only 11% (N = 319) of multi-criteria SDSS utilized a fully integrated option.

Fully integrated SDSS have become more common since the late 1990s. One of the earliest SDSS based entirely in GIS software was for chemical emergency preparedness and response in an urban environment (Chang et al. 1997). The SDSS was developed for ArcInfo software by developing chemical dispersion modeling algorithms within the GIS using AML along with customized user interfaces. Another early, fully integrated SDSS was a sustainable tourism planning SDSS developed by Beedasy and Whyatt (1999). They originally developed their SDSS in ArcInfo on a UNIX-based workstation, but later migrated the system to ArcView software because of the more user-friendly interface development possible. The SDSS was developed using ArcView's Avenue programming language to build multi-criteria analysis techniques into ArcView GIS software. The ArcView interface was modified in order to make the user interfaces as intuitive and transparent as possible. A multi-criteria evaluation planning SDSS was developed as an extension of ArcView GIS by Pettit and Pullar (1999). The graphical user interfaces were developed with the ArcView Dialog Designer extension, while specific functionality was developed using the Avenue programming language. The integration of hydrological modeling into ArcView GIS was carried out by Huang and Jiang (2002). They integrated the TOPography-based hydrological MODEL (TOPMODEL) within ArcView and named the final product AVTOP. The implementation used customized Avenue scripts, the dialog designer for developing user interfaces, the Spatial Analyst extension for raster hydrological processing, and made available 3D visualization capabilities through customized ArcView 3D Analyst tools. Outputs of AVTOP included hydrographs showing predicted stream flow, 3D animations, and 2D maps of modeled hydrological conditions. A non-GIS based SDSS was developed completely within Microsoft Excel by Berardi (2002). The system, called ASTROMOD, includes a user-friendly interface for visualizing spatial data, a parameter database, and programming modules within Excel. The system was developed using Visual Basic and allows for the importation of raster data into Excel and the visualization of these values according to customized symbolization routines. The system included vegetation dynamics modeling capabilities developed to operate on cells within Excel spreadsheets.

A GIS-based DSS called the Drainage Runoff Input of Pesticides in Surface Water (DRIPS) was developed in Germany by Ropke et al. (2004). The system was built by integrating a variety of model components into an ArcView 3.2 extension. Model algorithms built into DRIPS included those to estimate surface runoff, tile drainage, and spray drift. User-friendly interfaces were developed for easy calculation of spatially distributed scenarios for risk assessment. The output formats include maps covering the territory of Germany. Rinner and Heppleston (2006) developed a home buyer's SDSS using ArcGIS 9.0 and VBA. The SDSS utilizes information set by the user through customized interfaces on twelve non-spatial decision criteria such as price, lot size, year built, etc., and eleven spatial decision criteria such as distance to public transit, distance to parks, distance to schools, distance to hospitals, etc. In this SDSS, the user sets preference weights for the different parameters with sliders, and these weights are used in a spatially adjusted simple additive weighting methodology.

An SDSS for assistance in the control of the medfly, a significant pest to citrus crops, was developed in Israel (Cohen et al. 2008). The researchers integrated a rule-based decision tree based on binary, linear, logarithmic, and biological-based models into the ArcGIS environment. The SDSS produces a certainty factor for spraying that ranges from zero to 100%, which can be used by decision makers to define spraying schedules. An ArcGIS-based SDSS called the Poultry Litter Decision Support System (PLDSS) was developed by Kang et al. (2008) for nutrient management planning in Alabama. The system provides nutrient management planning functionality for confined animal feeding operations, assistance with record keeping of poultry litter applications, and provides transportation analysis functions. The system was developed using VBA and ArcObjects. The system interfaces with a MySQL poultry litter database.

The development of fully integrated SDSS has increased since the latter half of the 1990s and has most commonly taken place with GIS software serving as the integration platform. The development of GIS software with more user-friendly graphical user interfaces facilitated the role of GIS as an SDSS platform in the second half of the 1990s. In addition, the inclusion of customization capabilities through proprietary programming languages and user interface development tools in GIS software provided much greater capabilities for building modeling and dialog components directly into the GIS software. This was evidenced especially in the use of ArcView GIS software as a platform for GIS-based SDSS software in the late 1990s and early 2000s (e.g., Beedasy and Whyatt 1999; Pettit and Pullar 1999; Ropke et al. 2004). In recent years, GIS software has been reengineered to become more open to user development based on standards such as the Component Object Model (COM), which is a set of programming standards that allows code written in one language to work with code written in another language (Burke 2003). The development of COM-compliant GIS software has further opened up the development of customized applications. For example, ArcGIS software is built upon ArcObjects, which were created with C++ (Burke 2003). However, ArcObjects can be accessed through any COM-compliant programming or development environment such as Visual Basic, Visual Basic for Applications, C++, C#, or others. ArcGIS software has VBA packaged with it, making it the most convenient way for developers to design custom SDSS based in the ArcGIS software. This technique was utilized by several researchers (e.g., Rinner and Heppleston 2006 and Kang et al. 2008). Specific programming languages and development environments will be discussed in more detail later in the chapter.

6.2.3 Integration Issues

A wide range of SDSS applications exhibiting some sort of integration of multiple software components have been presented. The level of integration ranges from no software integration to complete integration. The efficiency in data interchange, user interaction, and use of computing resources generally increases with greater integration levels. On the other hand, development costs generally also increase with greater levels of integration. The amount of user expertise necessary might also be greatest for the lower levels of integration as the user would require expertise in using several pieces of software without common, user-friendly interfaces developed to aid the use of a given system. If an SDSS is likely to be used only in a one-time application or in situations where the developers of the system will likely be the only ones operating the software, then a loose coupling approach might be the most practical. However, if a general-purpose SDSS is being developed for use by a range of stakeholders in varying geographic locations, then a more tightly coupled or embedded approach would likely be more efficient and useful. A disadvantage to full integration is that it might be necessary to recreate some of the modeling routines in the GIS software if they are not in a format capable of being embedded.

Traditionally, it has been difficult to match data models and formats used in different software to each other. For instance, as pointed out by Rizzoli and Young (1997), GIS data models do not match the time and space models used in various simulation models. Increasingly, GIS programs are able to handle continuous temporal data in order to carry out some simulations at fine time steps (e.g., daily instead of a single annual simulation). For example, the Tracking Analyst extension for ArcGIS software allows the creation of time-series visualization, and with the Tracking Server, real-time data can be integrated to facilitate applications such as fleet management, sensor network monitoring, emergency response, and resource management. The introduction of these types of technologies into SDSS has been limited to this point in time. Although there have been improvements in standards and data format conversion utilities, SDSS developers usually still have to concern themselves with data conversion utility development in coupled software systems.

There are inherent difficulties in developing and supporting SDSS. One of the main problems is the inevitable evolution of software components on which the SDSS depends. When new versions of the GIS, modeling, or expert systems software become available, the SDSS might no longer function properly. Planning SDSS such as CommunityViz and INDEX were discussed in Chapter 5 and represent systems that are fully integrated into a GIS environment. These two systems are examples of SDSS that are successful commercially and are continually supported. This type of model requires a commitment to updating code for new versions of the ArcGIS software in which they are based. This can regularly require considerable programming effort. In academic situations, it is difficult to continually support new development and upgrades of SDSS software, which limits their shelf life. An example of a major software change was the move by ESRI from the ArcView 3.x platform to the ArcGIS platform. Any customized tools and systems developed in the Avenue programming language for ArcView 3.x software were not easily upgraded for use in the ArcGIS platform. This meant a significant expense in reengineering these customizations for the new ArcGIS platform. Although the ArcView 3.x software and customization might potentially still work, many organizations have abandoned that GIS technology. Many organizations have moved from using the shapefile format that was used in ArcView 3.x to the geodatabase, which is the standard for ArcGIS software. Even though an SDSS that was created in ArcView 3.x might still work on shapefiles, many organizations will not be utilizing data in that format anymore, thus making the use of the legacy SDSS impractical.

6.3 Design and Development of SDSS from Scratch

If the development of an integrated system does not seem to match the necessities of a given spatial decision-making situation, then it is possible to design and develop an SDSS from scratch. To do this, however, requires a substantial amount of time, financial and human resources, and knowledge of various tools and techniques. Developing a complete SDSS also requires a wide range of expertise, including skills in advanced computer programming, database design, spatial modeling, visualization, map display, and report generation. The advantages of this type of approach are flexibility and the amount of control available to the developers in building a high-quality system with the possibility for future extension. Before undertaking the development of a new stand-alone SDSS, many questions need to be considered. The first, of course, is whether it is necessary. The potential developer(s) should investigate if existing tools could be used to construct a system that would meet the particular needs of the situation. If this fundamental question leads to a developer wanting to develop a new SDSS, then other questions must be investigated. For example, what computer platform will most effectively meet the requirements of the system (i.e., desktop or Web based)? The SDSS planners must decide if the system will be developed for a single user or several individuals or if it might be used in group settings. They must also decide what the abilities and expertise level of the users are for all components of the SDSS. So the developer might need to consider if users will have experience in spatial database management, spatial processing of data, simulation modeling, scenario development, and understanding outputs such as maps, charts, or 3D visualizations. The user interface design is critical to the effectiveness of the SDSS. The user interface provides the communication between the user and the computer system and is one of the most important aspects in the successful development and use of any decision support system. Careful user interface design considerations are important as the user interface will play a large role in the user's satisfaction with the system. All of these factors need to be considered in the context of what kind of development resources and software development expertise are available for SDSS development. Accordingly, the development of an SDSS must be considered carefully before any actual design work begins.

6.4 Enabling Technologies for the Development of Desktop SDSS

A desktop application, in general, allows for more intensive use of local disk space, memory, and processing power in comparison to a Web platform, which is limited by the transfer of data according to bandwidth and other hardware concerns. Thus, the desktop platform allows the development of computationally intensive applications operating on large datasets with complex user interaction. There is continuing and accelerated evolution in software and hardware technologies as well as standards that very much influence the technologies that can be used to effectively develop SDSS. It is not possible to discuss all of these technologies here. However, there are various programming languages and development environments available to develop desktop SDSS. The following sections highlight some of the important technologies used in the development of desktop SDSS.

6.4.1 Programming Languages

There are many programming languages that can be used to develop SDSS or SDSS components, including Visual Basic, Visual Basic for Applications, C, C++, C#, Java, Delphi, and others. In the 1990s, C was common while the object-oriented C++ has gradually overtaken C in popularity over the course of the last decade. Visual Basic became popular in the late 1990s and continues to be used widely. Visual Basic for Applications (VBA) was integrated with a number of software applications in recent years, including Microsoft Office products and ArcGIS software. This integration allows for easy customization of these programs (i.e., ArcGIS, Microsoft Excel) as well as the development of components that can be used between software. Java and Microsoft's C# also became popular development languages over the last decade or so. Each programming language has advantages and disadvantages. The C and C++ languages are very powerful, and many commercially successful software applications have been built with these languages. Visual Basic became very popular because of its relative simplicity and the ability to easily develop appealing graphical user interfaces. Delphi is a language that rivals Visual Basic, which is also especially useful for developing applications that interact with databases. Java is an open source development language (developed by Sun Microsystems) that is powerful and allows the development of applications that can run on different operating systems (i.e., Windows, Linux) and hardware such as mobile devices. The C# language is similar to Java but was developed by Microsoft. The C# language has gained popularity over the last few years.

These languages have been used at various levels in SDSS development. Although there were no examples of an SDSS being built completely from scratch with Delphi, it was widely used in SDSS component development, especially outside of the United States. Delphi was used with MapObjects in an SDSS for modeling crop yields by Lagacherie et al. (2000) in France. A landscape model implemented in Delphi was linked with ArcView GIS in an SDSS for supporting landscape management decisions (Rudner et al. 2007). O'Brien et al. (2004) used Borland Delphi 6 along with MapObjects LT to develop an agricultural SDSS for use in Central America. There were numerous SDSS applications that used the C language in some aspect of their development (e.g., Srinivasan and Engel 1994, Engel et al. 2003). The C++ language has been used in many applications, especially in the last ten years or so, for purposes such as the development of mathematical models (Arampatzis et al. 2004), development of agents or objects in an agent-based modeling SDSS (Sengupta et al. 2005), and for transforming exported spatial data (Downs and Horner 2008). Java was used by Ballas et al. (2007) to develop the Micro-MaPPAS software, which is a spatial microsimulation modeling and predictive policy analysis system. Kaster et al. (2005) used Java to develop case-based reasoning facilities that were coupled with IDRISI. Most SDSS developments that utilized Visual Basic involved customizations of existing GIS software such as MapInfo (Abdullah et al. 2004), ArcGIS (e.g. Dye and Shaw 2007), and GeoMedia (Santé-Riveira et al. 2008).

Different programming languages can be used along with GIS and mapping components which are exposed through COM or ActiveX. For example, Symeonidis et al. (2004) used Visual Basic with MapX to develop a customized emission inventory application with mapping and some spatial analysis capabilities but without full GIS capabilities. In their system, they used Visual Basic to develop all user interfaces, emission calculation modules, and also for interaction with MapX to define the mapping and spatial analysis capabilities. Similarly, Vlachopoulou et al. (2001) used Visual Basic with MapObjects to develop a stand-alone system for warehouse site selection. MapObjects was developed by ESRI as a set of embeddable mapping and spatial analysis components which, along with development frameworks such as Visual Basic, could be used to develop lightweight applications with spatial data management and analysis as well as mapping capabilities. MapObjects has since been supplanted by software called ArcGIS Engine. ArcGIS Engine is a collection of GIS components and developer resources that allow the embedding of GIS capabilities in existing software or in new custom applications. The software exposes ArcObjects (objects with which the ArcGIS software is built) to any application programming interface for COM, including .NET languages such as Visual Basic, C++, or C#. The advantage of ArcGIS Engine and similar software is that they allow the construction of lightweight applications that don't require the users to purchase or have a full copy or license of the GIS software. Rather, the developers buy a developer's license and the user purchases a runtime license, which is a fraction of the cost of full GIS software. Software such as ArcGIS Engine provides potential SDSS developers with a way to access spatial analysis, spatial data management, and mapping capabilities without requiring the potential SDSS users to invest in a full (and costly) GIS license. Yang et al. (2007) used ArcGIS Engine to develop the client side interface and mapping component in an SDSS for epidemic disease prevention in China.

There are many freely available programming languages, including Perl, Smalltalk, Eiffel, Ada, Ruby, Python, Java, and many others. These languages have had somewhat limited application in SDSS developments. However, there will likely be a wider range of SDSS developments in the future that will at least have components developed with many of these languages, especially in Web-based applications. To date, the open source languages that have been most often used include Python, Java, and Perl. Ballas et al. (2007) developed the Micro-MaPPAS using the Java programming language. Java was used to develop the Workflow-based Spatial Decision Support System (WOODSS), which works in conjunction with IDRISI software by capturing user interactions with IDRISI in real time and documenting them in scientific workflows (Seffino et al. 1999). Python was used by Best et al. (2007) to access processes in ArcGIS and the R statistical package in an SDSS for predicting marine mammal habitats.

6.4.2 Application Development Environments

For desktop application development, including for SDSS software, there are several commercial and open source development platforms available that incorporate one or more development languages. The Microsoft Visual Studio development platform integrates a number of programming languages, including Visual C#, Visual C++, C, and Visual Basic. With the Visual Studio development environment, users have a choice of which language they want to use, but there are common tools for developing sophisticated graphical user interfaces, writing code (e.g., IntelliSense) and debugging utilities, and common interfaces for developing applications. Microsoft Visual Studio can be used to develop desktop applications that run on Windows computers. Ebarcadero Technologies describes their All-Access software as an "on-demand multi-platform tool chest." Developers can use Delphi, C++, Java, PHP, Ruby, and numerous other languages through this package.

6.4.3 Spatial Libraries

There are a number of spatial analytic capabilities that are freely available in the form of desktop GIS extensions as well as digital libraries. These extensions, tools, or libraries can be integrated into SDSS configurations. For example, ESRI provides a clearinghouse for third-party developers to deposit their custom tools and extensions for free downloading by potential users (http://arcscripts.esri.com/). The user can search by keyword and see which tools are available. The Geospatial Data Abstraction Library (GDAL) is an open source data translator that can read many raster data sources and also provides a variety of useful geospatial analysis utilities. Example utilities include those for development of contours from a digital elevation model (DEM), mosaicing multiple rasters, vector rasterization, raster proximity calculations, and many other conversion utilities (http://www.gdal.org/index.html). GeoTools is an open source Java GIS toolkit (http://www.geotools.org/) made up of a series of libraries, plug-ins, and extensions. Sidlar and Rinner (2007) used the GeoTools toolkit in a participatory SDSS configuration. The GeoTools libraries provide capabilities for reading a variety of spatial data formats, drawing maps, and display control. The plug-ins also provide database support capabilities. The extensions in GeoTools provide network capabilities such as finding shortest routes, spatial data renderers, and other functionality. The OpenMap package is an open source JavaBeans-based programmer's toolkit (http://openmap.bbn.com/). OpenMap provides the ability to access and visualize spatial data and has been used for a variety of applications, such as marine navigation software for recreational users, network modeling for transportation, and various mapping applications. The Java Topology Suite (JTS) is an application programming interface (API) for modeling and manipulating 2D linear geometry. It contains implementations of fundamental 2D spatial algorithms and spatial analysis methods (http://www.vividsolutions.com/jts/jtshome.htm). The JTS software has been used to build other software programs such as the Unified Mapping Platform (JUMP – now OpenJUMP), which is used for viewing and processing spatial data. The Geometry Engine, Open Source (GEOS) is a C++ version of the JTS (http://trac.osgeo.org/geos/). All of these applications require strong computer science as well as geospatial skill sets for their potential use in an SDSS framework.

6.4.4 SDSS Generator—Geonamica

The Research Institute for Knowledge Systems (RIKS) in The Netherlands has developed an object-oriented application framework that can be used to develop new SDSS. The Geonamica software is used as an in-house development platform. Thus, RIKS does not sell the Geonamica framework, but uses it on a consultancy basis. The Geonamica platform is used to build decision support systems based on spatial modeling and visualization of geographic data (Hurkens et al. 2008). The Geonamica framework provides the tools for building the three essential components of an SDSS: a database management system, a model management system, and user interfaces. Geonamica has components for storing spatial and nonspatial data, including time series data. It also has a modeling framework and a model controller. The modeling framework is based on model blocks, which are encapsulated parts of models that can communicate with each other. Finally, Geonamica provides a base set of user interfaces and a library of user interface components (Hurkens et al. 2008). The Geonamica framework has been used to build numerous SDSS. Rutledge et al. (2008) built a multiple-scale (local, district, and region) planning SDSS that incorporates aspects related to the economy, environment, and society. Engelen et al. (2003) described the use of Geonamica to develop the Environment Explorer, an SDSS for assessment of socioeconomic and environmental policies in the Netherlands.

6.5 Web-Based SDSS Development and Architecture

The Web has revolutionized application development. The ubiquitous nature of the Web, along with characteristics of Web applications such as centrally managed applications, platform independency, and reductions in distribution costs and maintenance problems, has facilitated the deployment of complex applications such as SDSS over the Web (Sugumaran et al. 2004; Peng and Tsou 2003). A Web-based SDSS (WBSDSS) usually includes a Web-based GIS as a problem solver, which facilitates geographic data retrieval, display, and analysis. It combines several different components, including HTML user interfaces, Internet interface programs, computational models, and geographic databases. There are two ways to set up a WBSDSS: (1) server-side processing and (2) client-side processing. The server-side approach uses a thin client, meaning most of the processing, including spatial data access and manipulation, is performed on the server side. The resulting information and image objects representing outputs are then sent to the client to be rendered. The typical components of a server-side WBSDSS are shown in Figure 6.9. The client-side processing approach uses a thick client in which GIS functionality is preloaded on the client machine and only the geographic data is accessed from one or more servers.

In a server-side configuration, the server-side environment typically includes a Web server (e.g., Apache, IIS) and a map server (ArcIMS, ArcGIS

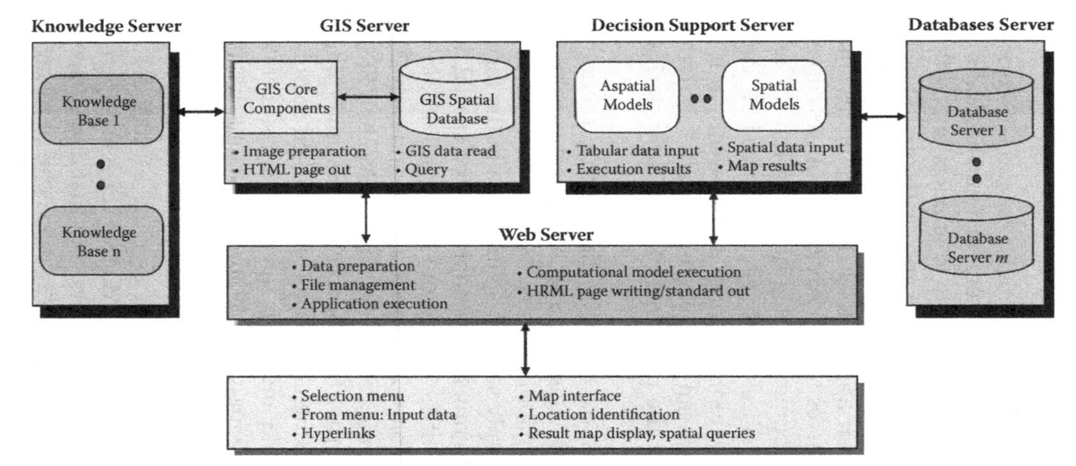

FIGURE 6.9
(See color insert following page 74.) Schmatic representation of Web-based SDSS components. (Source: Sugumaran, V. and R. Sugumaran. 2007. Web-based spatial decision support systems (WebSDSS): evolution, architecture, and challenges. *Communications of the Association for Information Systems* 19:844-875.)

Server, MapGuide, Mapserver) that provides GIS services. The map server software establishes a common platform for the exchange of Web-enabled GIS data and services. The Web server transfers spatial and nonspatial data between the client side (Web browser) and the map server. There are several server-side languages, such as PHP, JSP, Perl, Python, Ruby, and Ruby on Rails, which can be used to connect with map servers and a variety of databases.

GIS Web services provide hosted spatial data and GIS functionality via the Internet to Web applications and users. In a nutshell, GIS Web services provide GIS content and functionalities to applications without the user having to invest in costly GIS software and platforms. The clients do not have to host the GIS data or develop sophisticated tools to incorporate GIS capabilities within their applications. This facilitates smaller organizations with limited resources to take advantage of GIS capabilities without having to incur development costs (including time). Gonzales (2003) predicted that GIS Web services would revolutionize how companies use and interact with geospatial information, and presently this transition is ongoing. User communities can gain extensive spatial analytic value from GIS Web services without the problems of physically storing and maintaining spatial databases.

Web-based GIS services are used for distributing spatial information as well as spatial data management and analytical capabilities through the Internet. The exact capabilities vary by software, but in many cases are advancing beyond simply serving spatial data in the form of Internet map services, to providing much more sophisticated spatial data management and analysis capabilities. Numerous examples of GIS Web service technology are listed in Table 6.2. In the late 1990s and early 2000s, serving maps on the Internet became commonplace. ArcIMS by ESRI superseded the original ArcView Internet Map Server and became quite commonly used, especially by government agencies who wanted to publicly display their data holdings. ArcIMS and other Internet map servers became well established and were used for a variety of applications, including as part of SDSS configurations. For example, Halls et al. (2003) used ArcIMS to provide an interactive mapping capability in an SDSS for water quality management. Java GeoTools was used as early as 2001 for serving GIS data over the Web (Carver et al. 2001). MapServer is an Open Source project developed at the University of Minnesota that allows development of dynamic spatial maps over the Internet using vector or raster spatial data. The MapServer application has been used in multiple SDSS mainly for the display of data in dynamic maps (Engel et al. 2003; Sidlar and Rinner 2007). For example, Best et al. (2007) used MapServer for both displaying maps and for accepting user-defined, spatially explicit areas of interest, which is done by drawing a polygon in the browser-based map display. ArcGIS Server from ESRI extends beyond Internet mapping to providing

TABLE 6.2

Potential Software for Specific Integration in Web SDSS

Software Category	Distribution Model	Name	URL
Internet Map Servers	Commercial	ArcIMS	http://www.esri.com/software/arcgis/arcims/index.html
		Manifold Internet Map	www.manifold.net/info/ims.shtml
		ArcGIS Server	www.esri.com/software/arcgis/arcgisserver/
		GeoMedia WebMap	www.intergraph.com/sgi/products/default.aspx
	Free and open source	MapServer	http://mapserver.org/
		Mapnik	http://www.mapnik.org/
		GeoServer	http://geoserver.org/display/GEOS/Welcome
		MapGuide	http://mapguide.osgeo.org/

GIS services across the Internet. Depending on the license level bought for ArcGIS Server, Web services can be developed for mapping, data management, spatial analysis for vector and raster data, and mobile applications. There are various API that can be used to develop ArcGIS Server applications, including API for Flex, API for JavaScript, and API for Microsoft Silverlight. These APIs allow for the easy development of Web GIS and mapping services. Developers can also access ArcObjects components through various programming interfaces such as Visual Basic or C# to develop more sophisticated GIS services over the Web. The development of SDSS using ArcGIS Server has begun in earnest only very recently. One example is Park et al. (2008), who used ArcGIS Server in a public-participation SDSS. ArcGIS Server allowed the user to spatially select features and carry out some geoprocessing based on these selections. It should be expected that the use of these types of Web GIS server applications for SDSS development will be adopted more frequently in the future.

6.5.1 Cloud Computing

Many spatial decision-making situations require the inclusion of sophisticated analytical tools as well as the processing of large amounts of spatial and nonspatial data, which can be spread across various organizations. In general, there are ever greater amounts of data being collected and analyzed in more sophisticated analytical processing approaches. One approach to address these issues is that of *cloud computing*. Cloud computing is defined by Williams (2009, p. 2) as "computing in which dynamically scalable computing hardware and software resources are provided as services over the Internet." The concept is fairly broad with software that can be accessed over the Web, such as Google Docs, qualifying as cloud computing. Spatially related applications in which developers can utilize functionality from other computers over the Internet, such as Google Earth or Google Maps, are also a form of the cloud. Applications using Web GIS services on servers that are distant from the user also qualify. Given the great amount of spatial data inherent in many business operations, there is a great potential for the development of cloud-based computing efforts. Indeed, for example, ESRI and Microsoft cooperate to make the use of Bing Maps available through ArcGIS Server applications. Williams (2009) describes their GeoPortal components application as a spatial cloud computing solution that provides the ability to "map-enable" their business systems and data. This GeoPortal platform uses Google Maps data and imagery to serve as the base mapping system. Their software introduces spatial and nonspatial related functionality through a browser and allows the user to add his or her own data sources for consumption over the Internet. The use of cloud computing is experiencing a strong push from business, will advance in academics and government, and will lead to the application of SDSS using cloud computing resources in the future.

6.6 Summary

While there are some ready-to-use SDSS available, it is more common for potential SDSS developers to design and construct a new system. This chapter has provided an overview of some of the technologies and strategies that have been and can be utilized in the development of SDSS. The most common method of developing SDSS is by coupling two or more separate pieces of software together into one system. There are a range of strategies for doing this, from no coupling (i.e., use the discrete software separately with no digital communication between them) to full integration in a single software. Loose coupling approaches, in which custom software is developed to allow data file sharing between two or more software components, has been the most common method for developing SDSS. This approach requires a minimum of development time while still utilizing functionality of the individual coupled programs. There is inherent limited flexibility in this approach, and it is unlikely to be an efficient method if the SDSS is meant to be used by a range of stakeholders with varying levels of expertise. Generally, with increased coupling comes greater investment in time and resources by the SDSS developer but also greater efficiency in the use of the developed SDSS. The assumed increased efficiency in the use of the SDSS is based on the assumption that the developers carefully plan and build the application with contributions from potential users during the development process.

There are a variety of tools used for building SDSS applications, and the technology continues to evolve. Traditionally, many SDSS were GIS-centric as GIS provided the necessary spatial data management and analysis functionality. Disadvantages of GIS-centric SDSS have been that GIS software is expensive and generally requires expertise to use. The development of open source and free GIS software is to some extent overcoming the first problem, while the development of GIS services over the Web and the development of digital spatial analytical and mapping libraries or modules are helping to address both issues. The use of these technologies allows the development of SDSS that only utilize a portion of the functionality available in a full-fledged GIS, and thus limits cost and reduces the level of expertise necessary to use the applications. This type of modular structure is not limited to GIS software but other potential SDSS components as well. This chapter has presented a range of potentially useful software for inclusion in an SDSS, including database, modeling, statistical, and expert systems. In addition, a range of programming languages, application programming interfaces, and other development tools were discussed. The future of SDSS will entail the continued development of desktop SDSS, but will also be more commonly Web based and use distributed services over the Internet using innovations such as cloud computing. Nevertheless, the one thing that will hold true of any SDSS development is that the inclusion of potential SDSS users during the entire development cycle is crucial.

References

Abdullah, A., M. F. Abdullah, and M. N. A. Shahbudin. 2004. Collaborative decision support for spatial planning and asset management: IIUM total spatial information system. Paper presented at 11th International Symposium on Spatial Data Handling, Leicester, Britain.

Arampatzis, G., C. T. Kiranoudis, P. Scaloubacas, and D. Assimacopoulos. 2004. A GIS-based decision support system for planning urban transportation policies. *European Journal of Operational Research* 152(2):465–475.

Ballas, D., R. Kingston, J. Stillwell, and J. Jin. 2007. Building a spatial microsimulation decision support system. *Environment and Planning A* 39(10):2482–2499.

Beedasy, J., and D. Whyatt. 1999. Diverting the tourists: A spatial decision-support system for tourism planning on a developing island. *International Journal of Applied Earth Observation and Geoinformation* 1(3–4):163–174.

Bennett, D., and A. J. Vitale. 2001. Evaluating nonpoint pollution policy using a tightly coupled spatial decision support system. *Environmental Management* 27(6):825–836.

Berardi, A. 2002. ASTROMOD: A computer program integrating vegetation dynamics modelling, environmental modelling and spatial data visualisation in Microsoft Excel. *Environmental Modelling & Software* 17(4):403–412.

Best, B. D., P. N. Halpin, E. Fujioka, A. J. Read, S. S. Qian, L. J. Hazen, and R. S. Schick. 2007. Geospatial web services within a scientific workflow: Predicting marine mammal habitats in a dynamic environment. *Ecological Informatics* 2(3):210–223.

Burke, R. 2003. *Getting to know ArcObjects: programming ArcGIS with VBA*. Redlands, CA: ESRI Press.

Carrick, N. A., and B. Ostendorf. 2007. Development of a spatial decision support system (DSS) for the Spencer Gulf penaeid prawn fishery, South Australia. *Environmental Modelling & Software* 22(2):137–148.

Carver, S., A. Evans, R. Kingston, and I. Turton. 2001. Public participation, GIS, and cyberdemocracy: Evaluating on-line spatial decision support systems. *Environment and Planning B: Planning and Design* 28(6):907–921.

Chakhar, S., and V. Mousseau. 2008. Multicriteria spatial decision support systems. In *Encyclopedia of GIS*, ed. S. Shekhar, and H. Xiong, 753–758. New York: Springer.

Chang, N-B., Y. L. Wei, C. C. Tseng, and C. Y. J. Kao. 1997. The design of a GIS-based decision support system for chemical emergency preparedness and response in an urban environment. *Computers, Environment and Urban Systems* 21(1):67–94.

Cohen, Y., A. Cohen, A. Hetzroni, V. Alchanatis, D. Broday, Y. Gazit, and D. Timar. 2008. Spatial decision support system for Medfly control in citrus. *Computers and Electronics in Agriculture* 62(2):107–117.

Crossland, M. D. 1990. HydroLOGIC—A prototype geographic information expert system for examining an artificial intelligence application in a GIS environment. Paper presented at the Annual GIS/LIS Conference, Anaheim, California.

Crossman, N. D., L. M. Perry, B. A. Bryan, and B. Ostendorf. 2007. CREDOS: A conservation reserve evaluation and design optimisation system. *Environmental Modelling and Software* 22(4):449–463.

de Silva, F. N., and R. W. Eglese. 2000. Integrating simulation modelling and GIS: Spatial decision support systems for evacuation planning. *Journal of the Operational Research Society* 51(4):423–430.

Denzer, R. 2005. Generic integration of environmental decision support systems—State-of-the-art. *Environmental Modelling & Software* 20:1217–1223.

Djokic, D. 1996. Toward a general-purpose decision support system using existing technologies. In *GIS and environmental modeling: Progress and research issues*, ed. M. F. Goodchild, L. T. Steyaert, B. O. Parks, C. Johnston, D. Maidment, M. Crane, and S. E. Glendinning, 353–356. Ft. Collins, CO: GIS World, Inc.

Downs, J. A., and M. W. Horner. 2008. Spatially modelling pathways of migratory birds for nature reserve site selection. *International Journal of Geographical Information Science* 22(6):687–702.

Dye, A. S., and S-L. Shaw. 2007. A GIS-based spatial decision support system for tourists of Great Smoky Mountains National Park. *Journal of Retailing and Consumer Services* 14(4):269–278.

Eldrandaly, K., N. Eldin, and D. Sui. 2003. A COM-based spatial decision support system for industrial site selection. *Journal of Geographic Information and Decision Analysis* 7(2):72–92.

Engel, B. A., J.-Y. Choi, J. Harbor, and S. Pandey. 2003. Web-based DSS for hydrologic impact evaluation of small watershed land use changes. *Computers and Electronics in Agriculture* 39(3):241–249.

Engel, B. A., R. Srinivasan, and C. Rewerts. 1993. A spatial decision support system for modeling and managing agricultural non-point source pollution. In *Environmental Modeling with GIS*, ed. M. F. Goodchild, B. O. Parks, and L. T. Steyaert, 231–237. New York: Oxford University Press.

Engelen, G., R. White, and T. de Nijs. 2003. The environment explorer: Spatial support system for integrated assessment of socio-economic and environmental policies in the Netherlands. *Integrated Assessment* 4(2):97–105.

Falcão, A. O., M. Próspero dos Santos, and J. G. Borges. A real-time visualization tool for forest ecosystem management decision support. *Computers and Electronics in Agriculture* 53:3–12.

Gärtner, S., K. M. Reynolds, P. F. Hessburg, S. Hummel, and M. Twery. 2008. Decision support for evaluating landscape departure and prioritizing forest management activities in a changing environment. *Forest Ecology and Management* 256:1666–1676.

Giacomelli, A, 2005. Integration of GIS and simulation models. In *GIS for sustainable development*, ed. M. Campagna, 181–190. Boca Raton, FL: CRC Press.

Gilliams, S., D. Raymaekers, B. Muys, and J. V. Orshoven. 2005. Comparing multiple criteria decision methods to extend a geographical information system on afforestation. *Computers and Electronics in Agriculture* 49(1):142–158.

Gonzales, M. 2003. The new GIS landscape. *Intelligent Enterprise*, February: 21–24.

Guarnieri, F., and J. L. Wybo. 1995. Spatial decision support and information management application to wildland fire prevention The WILFRIED System. *Safety Science* 20(1):3–12.

Halls, J. N. 2003. River run: An interactive GIS and dynamic graphing website for decision support and exploratory data analysis of water quality parameters of the lower Cape Fear river. *Environmental Modelling & Software* 18:513–520.

Herzig, A. 2008. A GIS-based module for the multiobjective optimization of areal resource allocation. Paper presented at the 11th AGILE International Conference on Geographic Information Science, University of Girona, Spain.

Huang, B., and B. Jiang. 2002. AVTOP: A full integration of TOPMODEL into GIS. *Environmental Modelling and Software* 17(3):261–268.

Hurkens, J., B. Hahn, and H. van Delden. 2008. Using the GEONAMICA software environment for integrated dynamic spatial modelling. Paper presented at the International Congress on Environmental Modelling and Software, Barcelona, Spain.

Jain, D. K., U. S. Tim, and R. W. Jolly. 1995. A spatial decision support system for livestock production planning and environmental management. *Applied Engineering in Agriculture* 11:711–719.

Jankowski, P., T. L. Nyerges, A. Smith, T. J. Moore, and E. Horvath. 1997. Spatial group choice: A SDSS tool for collaborative spatial decision-making. *International Journal of Geographical Information Science* 11(6):577–602.

Kang, M. S., P. Srivastava, T. Tyson, J. P. Fulton, W. F. Owsley, and K. H. Yoo. 2008. A comprehensive GIS-based poultry litter management system for nutrient management planning and litter transportation. *Computers and Electronics in Agriculture* 64:212–224.

Karnatak, H. C., S. Saran, K. Bhatia, and P. S. Roy. 2007. Multicriteria spatial decision analysis in web GIS environment. *Geoinformatica* 11:407–429.

Kaster, D. S., C. B. Medeiros, and H. V. Rocha. 2005. Supporting modeling and problem solving from precedent experiences: The role of workflows and case-based reasoning. *Environmental Modelling & Software* 20(6):689–704.

Keenan, P. B. 1996. Using a GIS as a DSS Generator. In *Perspectives on DSS*, ed. J. Darzentas, J. S. Darzentas, and T. Spyrou, 33–40. Samos, Greece: University of the Aegean Press.

Klungboonkrong, P., and M. A. P. Taylor. 1998. A microcomputer-based-system for multicriteria environmental impacts evaluation of urban road networks. *Computers, Environment and Urban Systems* 22(5):425–446.

Lagacherie, P., D. R. Cazemier, R. Martin-Clouaire, and T. Wassenaar. 2000. A spatial approach using imprecise soil data for modelling crop yields over vast areas. *Agriculture, Ecosystems & Environment* 81(1):5–16.

Leao, S., I. Bishop, and D. Evans. 2004. Spatial-temporal model for demand and allocation of waste landfills in growing urban regions. *Computers, Environment and Urban Systems* 28(4):353–385.

Li, L., J. Wang, and C. Wang. 2005. Typhoon insurance pricing with spatial decision support tools. *International Journal of Geographical Information Science* 19(3):363–384.

Li, Y., Q. Shen, and K. Li. 2004. Design of spatial decision support systems for property professionals using MapObjects and Excel. *Automation in Construction* 13(5):565–573.

Makropoulos, C. K., and D. Butler. 2005. A neurofuzzy spatial decision support system for pipe replacement prioritisation. *Urban Water Journal* 2(3):141–150.

Makropoulos, C. K., D. Butler, and C. Maksimovic. 2003. Fuzzy logic spatial decision support system for urban water management. *Journal of Water Resources Planning and Management* 129(1):69–77.

Malczewski, J. 1999. *GIS and multicriteria decision analysis.* New York: John Wiley & Sons, Inc..

Malczewski, J. 2006. GIS-based multicriteria decision analysis: A survey of the literature. *International Journal of Geographical Information Science* 20(7):703–726.

Matthews, K. B., A. R. Sibbald, and S. Craw. 1999. Implementation of a spatial decision support system for rural land use planning: integrating geographic information system and environmental models with search and optimisation algorithms. *Computers and Electronics in Agriculture* 23(1):9–26.

McCarthy, J. D., P. A. Graniero, and S. M. Rozic. 2008. An integrated GIS-expert system framework for live hazard monitoring and detection. *Sensors* 8:830–846.

Munier, B., K. Birr-Pedersen, and J. S. Schou. 2004. Combined ecological and economic modelling in agricultural land use scenarios. *Ecological Modelling* 174(1–2):5–18.

O'Brien, R., M. Peters, A.Schmidt, S. Cook, and R. Corner. 2004. Helping farmers select forage species in Central America: The case for a decision support system. Paper presented at CIAT (Centro Internacional de Agricultura Tropical), Cali, Colombia.

Ozan, E., P. Kauffmann, and Y. Sireli. 2003. How to design multicriteria spatial decision support systems. Paper presented at Proceedings of 2003 National Conference, American Society for Engineering Management, St. Louis, Missouri.

Park, S., Y. Choi, J. Y. Nam, and J. Lee. 2008. To improve a public participation decision support system based on Web 2.0 technology. Paper presented at the Twenty-Eighth Annual ESRI User Conference, San Diego, California.

Peng, Z-R, and M-H. Tsou. 2003. *Internet GIS: Distributed geographic information services for the Internet and wireless networks*. New York: John Wiley & Sons, Inc.

Pettit, C., and D. Pullar. 1999. An integrated planning tool based upon multiple criteria evaluation of spatial information. *Computers, Environment and Urban Systems* 23(5):339–357.

Pidd, M., F. N. de Silva, and R. W. Eglese. 1996. A simulation model for emergency evacuation. *European Journal of Operational Research* 90(3):413–419.

Purucker, S. T., R. N. Stewart, and C. J. E. Welsh. 2008. SADA: Ecological risk based decision support system for selective remediation. In *Decision support systems for risk-based management of contaminated sites*, ed. A. Marcomini, G. W. Suter II, and A. Critto, 1–18. New York: Springer.

Rao, M., G. Fan, J. Thomas, G. Cherian, V. Chudiwale, and M. Awawdeh. 2007. A web-based GIS decision support system for managing and planning USDA's conservation reserve program (CRP). *Environmental Modelling & Software* 22(9):1270–1280.

Reynolds, K. M., and P. F. Hessburg. 2005. Decision support for integrated landscape evaluation and restoration planning. *Forest Ecology and Management* 207(1–2):263–278.

Rinner, C., and A. Heppleston. 2006. The spatial dimensions of multi-criteria evaluation: Case study of a home buyer's spatial decision support system. Paper presented at the 4th International Conference on Geographic Information Science, Munster, Germany.

Rizzoli, A. E., and W. J. Young. 1997. Delivering environmental decision support systems: Software tools and techniques. *Environmental Modelling & Software* 12(2–3):237–249.

Ropke, B., M. Bach, and H-G. Frede. 2004. DRIPS—a DSS for estimating the input quantity of pesticides for German river basins. *Environmental Modelling & Software* 19(11):1021–1028.

Rudner, M., R. Biedermann, B. Schrader, and M. Kleyer. 2007. Integrated grid based ecological and economic (INGRID) landscape model—a tool to support landscape management decisions. *Environmental Modelling & Software* 22(2):177–187.

Rutledge, D., M. Cameron, S. Elliott, T. Fenton, B. Huser, G. McBride, G., et al. 2008. Choosing regional futures: Challenges and choices in building integrated models to support long-term regional planning in New Zealand. *Regional Science Policy & Practice* 1(1):85–108.

Santé-Riveira, I., R. Crecente-Maseda, and D. Miranda-Barrós. 2008. GIS-based planning support system for rural land-use allocation. *Computers and Electronics in Agriculture* 63(2):257–273.

Seffino, L. A., C. B. Medeiros, J. V. Rocha, and B. Yi. 1999. WOODSS—a spatial decision support system based on workflows. *Decision Support Systems* 27(1–2):105–123.

Sengupta, R., C. Lant, S. Kraft, J. Beaulieu, W. Peterson, and T. Loftus. 2005. Modeling enrollment in the Conservation Reserve Program by using agents within spatial decision support systems: An example from southern Illinois. *Environment and Planning B: Planning and Design* 32:821–834.

Sengupta, R. R., and D. A. Bennett. 2003. Agent-based modelling environment for spatial decision support. *International Journal of Geographical Information Science* 17(2):157–180.

Sidlar, C. L., and C. Rinner. 2007. Analyzing the usability of an argumentation map as a participatory spatial decision support tool. *URISA Journal* 19(1):47–55.

Srinivasan, R., and B. A. Engel. 1994. A spatial decision support system for assessing agricultural nonpoint source pollution. *Journal of the American Water Resources Association* 30(3):441–452.

Stevens, D., S. Dragicevic, and K. Rothley. 2007. iCity: A GIS-CA modelling tool for urban planning and decision making. *Environmental Modelling & Software* 22(6):761–773.

Sugumaran, R. 2002. Development of a range management decision support system using remote sensing, GIS and knowledge based systems. *Computers and Electronics in Agriculture* 37:199–205.

Sugumaran, V. and R. Sugumaran. 2007. Web-based spatial decision support systems (WebSDSS): evolution, architecture, and challenges. *Communications of the Association for Information Systems* 19:844-875.

Sugumaran, R., S. Ilavajhala, and V. Sugumaran. 2007. Development of a web-based intelligent spatial decision support system WEBSDSS: A case study with snow removal operations. In *Emerging spatial information systems and applications*, ed. B. N. Hilton, 184–202. Hershey, PA: Idea Group.

Sugumaran, R., J. Meyer, and J. Davis. 2004. A web-based environmental decision support system (WEDSS) for environmental planning and watershed management. *Journal of Geographical Systems* 6:1–16.

Symeonidis, P., I. Ziomas, and A. Proyou. 2004. Development of an emission inventory system from transport in Greece. *Environmental Modelling & Software* 19:413–421.

Taweepworadej, W., W. Kanarkard, R. G. Adams, N. Davey, and D. Hormdee. 2006. Development of a spatial decision support system (DSS) for the point-source pollution. Paper presented at TENCON 2006 IEEE Region 10 Conference, Hong Kong.

Thorp, K. R., K. C. DeJonge, A. L. Kaleita, W. D. Batchelor, and J. O. Paz. 2008. Methodology for the use of DSSAT models for precision agriculture decision support. *Computers and Electronics in Agriculture* 64:276–285.

Ungerer, M. J., and M. F. Goodchild. 2002. Integrating spatial data analysis and GIS: A new implementation using the component object model (COM). *International Journal of Geographical Information Science* 16(1):41–53.

Vlachopoulou, M., G. Silleos, and V. Manthou. 2001. Geographic information systems in warehouse site selection decisions. *International Journal of Production Economics* 71(1–3):205–212.

Wan, Q., J. Zhang, H. Lin, and C. Beijing. 1999. On-line group spatial decision support system for investment environment analysis. Paper presented at the International Conference of GeoInformatics, Ann Arbor, Michigan.

West, L. A., Jr. 1999. Florida's marine resource information system: A geographic decision support system. *Government Information Quarterly* 16(1):47–62.

Williams, H. 2009. Spatial cloud computing (SC2): A new paradigm for geographic information services. White Paper, SKE, Inc., http://www.skeinc.com/pages/SC2/SKE_SC2_White_Paper.pdf (accessed January 21, 2010).

Wong, I. W., R. Bloom, D. K. McNicol, P. Fong, R. Russell, and X. Chen. 2007. Species at risk: Data and knowledge management within the WILDSPACE® decision support system. *Environmental Modelling & Software* 22(4):423–430.

Wu, X., S. Zhang, and S. Goddard. 2004. Development of a component-based GIS using GRASS. Paper presented at the FOSS/GRASS Users Conference, Bangkok, Thailand.

Yang, K., S. Peng, Q. Xu, and Y. Cao. 2007. A study on spatial decision support systems for epidemic disease prevention based on ArcGIS. In *GIS for health and the environment*, ed. W. Cartwright, G. Gartner, L. Meng, and M. P. Peterson, 30–43. New York: Springer.

Zeng, H., A. Talkkari, H. Peltola, and S. KellomÃki. 2007. A GIS-based decision support system for risk assessment of wind damage in forest management. *Environmental Modelling & Software* 22(9):1240–1249.

Zhu, X. 1997. An integrated environment for developing knowledge-based spatial decision support systems. *Transactions in GIS* 1(4):285–300.

Zhu, X., R. J. Aspinall, and R. G. Healey. 1996. ILUDSS: A knowledge-based spatial decision support system for strategic land-use planning. *Computers and Electronics in Agriculture* 15(4):279–301.

7

Building Desktop SDSS

Learning Objectives

- Understand important considerations when developing SDSS software.
- Be introduced to the iterative process of SDSS development, which is inclusive of a variety of stakeholders.
- Carry out all necessary steps to develop and run simple SDSS extensions in both Microsoft Excel and ArcGIS software.
- Be exposed to two new SDSS tools (a Microsoft Excel extension called SpreadsheetSDSS and a new generic SDSS-generating utility called OpenSDSS).

7.1 Introduction

In Chapter 5, we provided an overview of existing spatial decision support systems (SDSS) types (problem specific, domain oriented, and generic) and also provided examples of each with the purpose of giving potential users an idea of the software that is already available for potential use. In Chapter 6, we examined various software applications that could be integrated into a new SDSS and also specific software tools that could be used for developing new SDSS. In this chapter, we will first investigate some of the considerations that need to be taken into account when building SDSS software within a decision-making situation, and also look at the general process that should be followed. The second part of the chapter will provide examples of SDSS software construction. The purpose of these examples is to demonstrate software development techniques with practical illustrations. These examples could potentially be used as templates for future SDSS tool construction.

7.2 SDSS Development Considerations

There are many considerations and important questions to be answered before development of any SDSS. The stakeholders involved should attempt to answer the following questions as clearly as possible: (1) what problem(s) will the SDSS be used to solve, (2) what are the roles of various stakeholders, including potential decision makers, (3) what technologies are currently available to the developers, (4) are adequate time and resources available, (5) who will maintain the system through updates and upgrades, and (6) what technologies can be used to meet system requirements (e.g., programming languages, spatial data analysis, modeling capabilities, development platform—desktop or Web based)?

As mentioned in Chapter 4, the human stakeholder component is crucial to the successful development and utilization of any SDSS. Most SDSS development requires the involvement of individuals from multiple disciplines with different individuals and groups fulfilling different roles. For example, urban decision support systems need to encompass information regarding population and land use/cover change patterns, economic conditions, policy drivers, and the availability of infrastructure. These aspects call for expertise from a wide range of disciplines. For example, urban planners, scientists (modelers), economists, software architects (programmers), geographical information science (GIS) analysts, and policy or decision makers (end users) all must participate in the decision-making process for which the SDSS should facilitate efficient solutions.

In discussing DSS in relation to integrated water resource management, Hahn (2005) presented a useful discussion on the stakeholder component and outlined some ideas as to what will and will not lead to successful use of SDSS in a spatial decision-making situation. He identified four major actors necessary for the process, including end users, scientists, IT specialists, and the overall architect of the project. According to Hahn, end users define problems and policy context, dictate the potential usage of the SDSS, and help formulate functional requirements. These individuals might be planners, managers, or policy makers. Other stakeholders, according to Hahn, include scientists who develop modeling approaches, IT specialists who assist in designing state-of-the-art software architectures and technologies, and finally, the project architect, who is the main player and is responsible for integrating other stakeholders and managing the project or system. One of the requirements for the architect is a very good understanding of the application domain and skill in bridging the methodological and knowledge gaps among the other three stakeholder groups. Hahn stressed that the

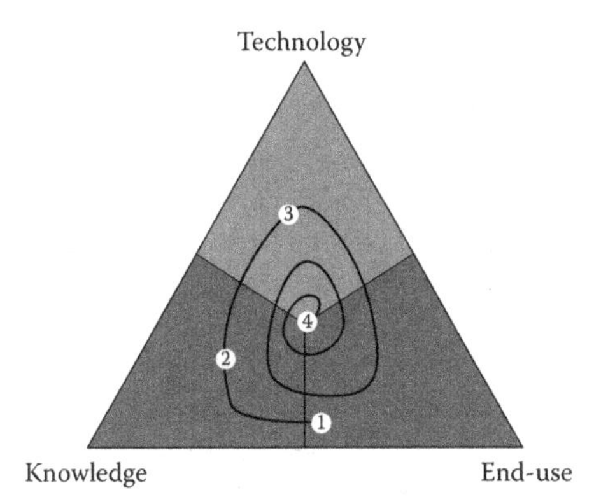

FIGURE 7.1
The representation of an iterative SDSS development process. (Hahn, B. 2005. The need for a science and knowledge integration. Paper presented at the Success and Failure of DSS for Integrated Water Resource Management Conference, Venice, Italy.)

SDSS development process should be iterative and must pass through the end users, the knowledge specialists (scientists), and the technology specialists (Figure 7.1). As Figure 7.1 indicates, the process does not follow a straight-line pattern but a cyclical one that must include all stakeholders throughout. He pointed out that the biggest failure in the use of SDSS is not beginning the process with the end users and also not involving each of the groups throughout the iterative process. Successful SDSS development requires interdisciplinary team efforts with input from all these actors. While the ideas of inclusivity in stakeholder involvement and iterative development processes should be followed, each situation will be unique. Sometimes the architect might work within the same organization as the scientists, modelers, programmers, and GIS specialists, while other times, the GIS and programming work might be outsourced to a contractor. The architect must take responsibility for making sure that the inclusivity and iteration principles are followed. As discussed earlier, many SDSS have not been successfully implemented because of a lack of collaboration among the multiple actors, including decision makers, throughout the SDSS development process.

Proper planning among all stakeholders, in order to develop a systematic process for SDSS development and use, can help to eliminate or reduce problems and technical issues (e.g., which software to use, design of efficient user interfaces). This planning process must take into account realistic time and resource constraints. The challenges facing, and limitations affecting successful SDSS applications will be further examined in Chapter 10. The next section of this chapter provides an overview of a useful general process to be considered when developing an SDSS.

7.3 SDSS Development Process

There has been considerable discussion of methodologies or processes for developing decision support software, but not many specifically for SDSS software development. There are too many classifications of different methodologies to cover in depth here, but a couple will be discussed to give an overview. Veronica (2007) talked about several different methodologies, including phased development, the evolutionary method, prototyping, and end user development. The *phased approach* follows a linear path of problem definition, analysis of requirements, design of software, programming and development, and finally, implementation. This type of methodology provides limited flexibility in iterative design and lacks continual involvement of potential end users.

The *evolutionary method* combines analysis, design, construction, and implementation steps into a single step that is iterated repeatedly. In this approach, small subproblems are addressed by analysis, design, construction, and implementation steps until that subproblem is satisfactorily addressed. Then the developers and end users move on to another subproblem. In this way, functionality is added to the system over time. This approach requires a high degree of stakeholder involvement, but provides greater ability to more completely capture stakeholders' needs.

The *prototyping method* is similar to the evolutionary method, but instead of developing finished pieces of the system one at a time, a rough idea of the functionality of the SDSS is approximated with various tools but without complete programming of the system. The overall functionality of the system is constructed in a rough version in order for end users to evaluate it. Then in an iterative process, the developers and end users can get closer to the final requirements through prototype development and evaluation cycles.

In the *end user approach*, the user also serves as the developer. In this case, some sort of decision support systems (DSS) generator software, such as a spreadsheet or for SDSS, a GIS might be used. This type of approach is likely only viable for spatial problems or issues of a smaller magnitude. Veronica (2007) suggested that, for the development of complex DSS, the evolutionary or prototyping approaches are the best.

Power (2002) came to similar conclusions as Veronica based on his classification of three methodologies or processes: systems development life cycle (SDLC), rapid prototyping, and end user development. The SDLC is akin to the phased development defined by Veronica (2007) and is considered to not provide enough flexibility in the development process, while the end user approach is not applicable when somewhat complex DSS are to be developed. Power (2002) considered *rapid prototyping* to be the best methodology. This methodology is characterized by the quick, rough-draft type development of systems based on preliminary requirements definition, subsequent testing and review by users, repetition of these steps, and finally, when end users are satisfied, final development and implementation.

The development of SDSS software is usually carried out to address complex problems. As emphasized by Power and Veronica, in these situations, it is clear that software development requires a carefully planned and iterative process in order to attain a successful product. This type of iterative process is represented in Figure 7.2. We suggest that, after the overall problem is defined and stakeholders identified, an iterative process of requirements definition, design and development of prototypes, and testing should be followed. After the iteration cycle results in a system that meets the satisfaction of stakeholders, the final implementation can take place. This approach requires the inclusion of a variety of stakeholders (decision makers or end users, domain experts like modelers or scientists, IT/GIS specialists, programmers) throughout the process. Although this can lead to greater costs in development stages, it can lead to a product that is much more effective and acceptable.

Within this type of process, logistical questions and constraints must be addressed. The nature of the problem, number and variety of stakeholders, whether the software development is outsourced, and other issues can influence the feasibility of this type of process in SDSS development and effective application. Specific aspects and issues related to the steps identified in Figure 7.2 are discussed below. It must be kept in mind that the

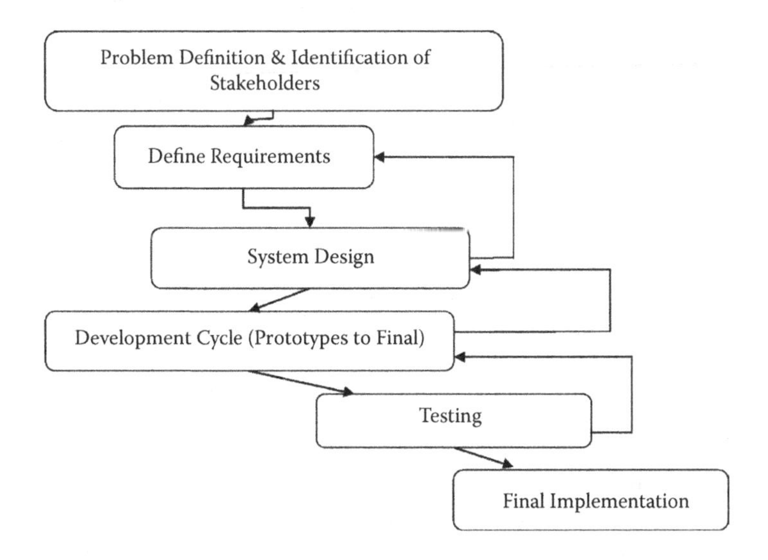

FIGURE 7.2
Overall SDSS development process.

stages or phases discussed below should not proceed literally, but rather form parts of an iterative process in which they are revisited throughout the process. It is also noted that there is no restriction to revisit only the step directly above, but it is possible to go from, for example, the testing step back to the define-requirements step when an end user recognizes an aspect of the issue that is not being addressed.

The first step leading to the development of SDSS (as outlined above) is the recognition of a problem and the motivation of an individual or organization to address that problem with a computerized decision support system. Regardless of who is the originator of the idea of using an SDSS, it is important to establish a champion within the organization. The champion should have the status necessary to make decisions, including those of resource allocation, and must also have contacts and influences for dealing with potential stakeholders both within and outside a given organization. The early proponent(s) or champion(s) should begin the process of identifying important stakeholders who need to be included early in the process. Also at this early stage, the strategic value of the potential system needs to be defined in relation to the policy and management context of potential end users (Van Delden 2009). Questions that should be addressed at this stage include whether the culture of a given end user organization will be amenable to the use of a given SDSS, and how the system will be used and incorporated into an organization (Van Delden 2009). During initial discussions among potential stakeholders, goals and objectives can begin to be defined, tentative measures identified, and also spatial, temporal, economic, and other boundaries of the system defined (Hahn et al. 2005). During this phase, it is possible that further identification of important stakeholders, such as domain experts, will happen. At this initial stage, communication protocols and mechanisms between stakeholders should be established.

During the second step, as highlighted in Figure 7.2, more formal requirement definition must take place. This is not a single step in the process, but is a step that likely will be revisited when new ideas or issues arise in the system design and development cycle steps. The requirements definition stage is crucial in the process because this is the time when stakeholders come together and identify what they would like to derive from the potential system. Depending on the scope of the SDSS and number of stakeholders, this can be a significant step. The stakeholders should identify their own goals and needs before any meeting with other stakeholders. Then all stakeholders should meet to identify a hierarchy of requirements. In a large project utilizing SDSS in New Zealand, Rutledge et al. (2008) noted that there were five high-level outcomes, 38 more detailed outcomes, and an associated set of 75 core indicators that were used to help guide the development of SDSS. At this stage, it will likely be necessary to translate broad ideas or goals into concrete or quantitative measures that can be modeled or analyzed in the SDSS. Indeed, Rutledge et al. (2008) noted that the 75 core indicators were more applicable for informing the development of the SDSS as they were quantifiable. During this phase, the technical stakeholders (i.e., scientist, modeler, programmer, GIS specialist) can begin to scope potential spatial and nonspatial data sources, selection of modeling techniques, software requirements and availability, and output needs.

More concrete design steps, the third stage in Figure 7.2, can be initiated when there is sufficient agreement on the general and specific requirements of the system. The system design process should occur in an iterative process with lessons learned in a prototyping process in the development cycle feeding back into new system design. In the preliminary iteration, technical questions should be addressed such as which platform will be most effective (desktop vs. Web based), which models or modeling techniques are useful and feasible, which software component(s) would meet requirements and also are available to necessary end users, and which development environments or programming languages will be used. The question of whether to build a new software system from scratch or to develop one that integrates existing programs must be decided at this stage. In addition, data sources should be assembled at this point and database structures defined. The initial system design will feed into initial prototyping, which will lead to new conclusions about datasets, assumptions, modeling techniques, user interface design, and other aspects that can be included in the next iteration of system design.

An iterative process of development and testing, stages four and five in Figure 7.2, can move the project from initial prototypes through to the final system distribution. Several authors suggest the use of a prototyping process for the development of DSS and SDSS (Power 2002; Veronica 2007; Hahn et al. 2005; Van Helden 2009). The responsibility for progress in the iterations of these stages will be that of the modelers, scientists, IT/GIS specialists, and programmers. These stakeholders will have to use information and input gained from the design stage and testing phases in the iterative process to guide the development of prototypes and eventually the final system. The testing stage should include end users who can formally determine if the system is meeting the necessary requirements. The testing phase should also include systematic software testing, especially in later prototypes, to discover problems with the software. Before final distribution of the software, proper resources such as user documentation and online help facilities should have been developed and tested. Sufficient training sessions should also be arranged before final implementation of the SDSS. There are important post-implementation activities such as the update and maintenance of the system (Veronica 2007). These activities are necessary to ensure continued successful use of a developed SDSS.

7.4 SDSS Development Examples

The intention of the remaining part of this chapter and also Chapter 8 is to provide several examples of actual SDSS software tool development. Specifically, the goal is to demonstrate the use of some of the development tools and environments that are available and methods that can be used to develop SDSS tools. These examples are not full-fledged SDSS used for real decision-making processes, but are simple SDSS-type tools. We will provide step-by-step instructions on how to develop several example SDSS software tools, thus allowing readers (if they have the proper software) to reproduce them. The hope is that these examples will offer developers and others interested in developing an SDSS a starting point and potentially reusable ideas and code samples. These examples can serve as a template for future SDSS development. The components presented should be easy to adopt and quickly modified for other applications. The authors are not aware of any other books that provide detailed examples for the development of SDSS. Figure 7.3 highlights the examples of SDSS that will be presented in this book. These examples were chosen to represent the variety of technologies that are used in SDSS based on the comprehensive literature review carried out for the writing of this book (please refer to Chapter 2 and Chapter 9). In particular, both desktop and Web-based SDSS tools are presented as well as those developed using both commercial and open source software. We also present SDSS developed by customizing existing

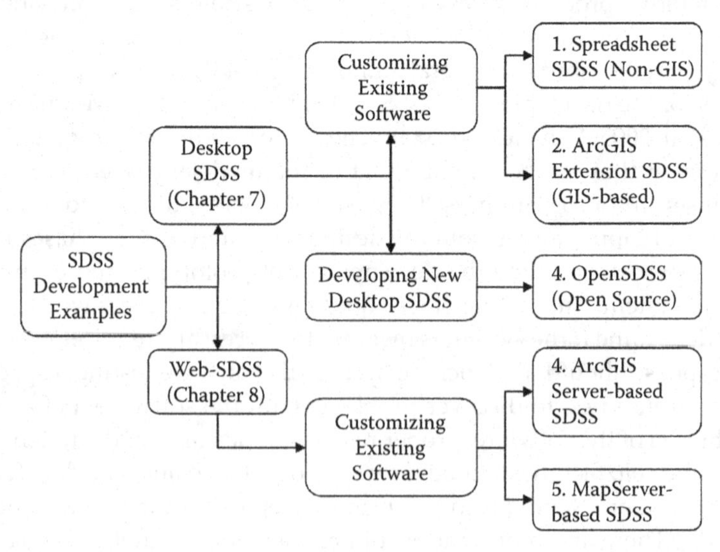

FIGURE 7.3
SDSS development examples.

software and provide examples of two new SDSS that were developed from scratch. In addition, GIS-based SDSS and non-GIS-based SDSS will be presented. This chapter will focus on examples of desktop SDSS development, whereas Chapter 8 will offer examples of Web-based SDSS development.

This chapter demonstrates the development of SDSS using three different approaches that use different software programs and different computer languages. These three systems demonstrate a variety of important technologies that can be used in SDSS creation, including GIS and non-GIS platforms, commercial and open source software, and development with different programming languages. The first SDSS implemented here was developed in Microsoft Excel using the Visual Basic for Applications (VBA) programming language. This SDSS incorporates the Analytic Hierarchy Process (AHP) model into Microsoft Excel. The advantage of a Microsoft Excel–based system is that it does not require GIS expertise or software that provides accessibility to a greater number of users. The second SDSS implementation discussed is an ArcGIS-based desktop SDSS that implements a weighted linear combination (WLC) model. Although not provided in extremely detailed form, the necessary steps to reproduce these first two (Excel and ArcGIS-based) SDSS tools are provided so the readers can attempt to build their own versions. These examples do assume some familiarity with Excel and ArcGIS. A short explanation will also be given of a more fully developed SDSS extension to Microsoft Excel. Finally, a new generic system called OpenSDSS will be described. The OpenSDSS software is currently under development as an open source, generic SDSS.

7.4.1 Spreadsheet-Based AHP SDSS (Microsoft Excel)

An example in which a commonly used, existing spreadsheet software—Microsoft Excel—is customized to provide a simple SDSS is provided here. In a real-world situation, the development presented here might serve as a prototype developed after initial brainstorming among stakeholders (e.g., city planners, business managers, software developer) about requirements (e.g., easy to use and demonstrate to others, not GIS-based, allows for consideration of multiple criteria) and initial system design and development. Some advantages of customizing existing software include the fact that users are already familiar with using the software and that the cost of development can be kept down as existing functionality of the customized software can be utilized. The goal of this specific example is to demonstrate the integration of AHP modeling within Microsoft Excel, a program that is used widely by a variety of users in academics, business, and government agencies. Many people in business and other disciplines are much more familiar with Excel than GIS software and are the target audience for this type of SDSS program. Although Microsoft Excel is not normally used for handling spatial data, it does use a cellular-based format that is analogous to the raster data model in GIS, which allows spatial data to be incorporated. Excel is a fairly flexible program that can handle a large number of data values (over a million rows by over 16,000 columns), display data visually, and allow for custom developments to be carried out using Visual Basic for Applications or other programming languages. There have been some previously developed Excel-based SDSS. Berardi (2002) constructed the ASTROMOD program using Microsoft Excel for integrating vegetation dynamics modeling, environmental modeling, and spatial data visualization for analyzing the development of forest ecosystems inside the Astroni Crater State Nature Reserve in Naples, Italy. Li et al. (2004) developed an SDSS for property professionals using MapObjects and Microsoft Excel. Spreadsheet-based DSS are actually quite common and have been the subject of one recent book, *Developing Spreadsheet-Based Decision Support Systems Using Excel and VBA for Excel* (Şeref et al. 2007). However, using spreadsheets (Excel in particular) for SDSS has not been nearly as widespread. Here we present the steps, code, and details necessary for creating this Excel-based SDSS with AHP and WLC. The purpose of this example is to demonstrate how a simple Excel-based SDSS can be developed and adapted, thus providing potential SDSS developers with a basic template for using it or creating a similar system.

In this section we will present the steps to build the user interface component of the spreadsheet-based SDSS. This example requires Microsoft Excel 2007 and that Visual Basic for Applications (VBA) is enabled within Excel. Microsoft Excel 2007 includes VBA, but it is possible that during installation the VBA option was not enabled. If VBA is not accessible, the user will not be able to complete the steps necessary to create the spreadsheet-based SDSS. If VBA is not available in Excel, the reader should install VBA with Excel. Before beginning the example, it is a good idea to save the workbook you will be working with as an Excel Macro-Enabled Workbook (*.xlsm) to a directory of your choice.

To open the **VBA Developer Ribbon** (a Ribbon is a feature in Microsoft Office that is basically a set of toolbars that organize groups of tools logically, e.g., Home, Insert, Data) go to the **Excel Options** (found at bottom of the main menu in Excel) dialog and check the checkbox **Show Developer** tab in the Ribbon under the **Popular** category. Select the **Developer** tab to activate it and then click the **Visual Basic** button. After opening the VB editor, the first step is to insert a new User Form by choosing the **Insert** menu and then **UserForm** from the drop-down menu. After adding the form, the user can resize the form as he or she wants (Figure 7.4). Users can change properties of the form in the Properties window when the form is the active control (controls are the form and any tools or display items that are added to it). In our example, the form's properties will be set automatically when the code given below is executed.

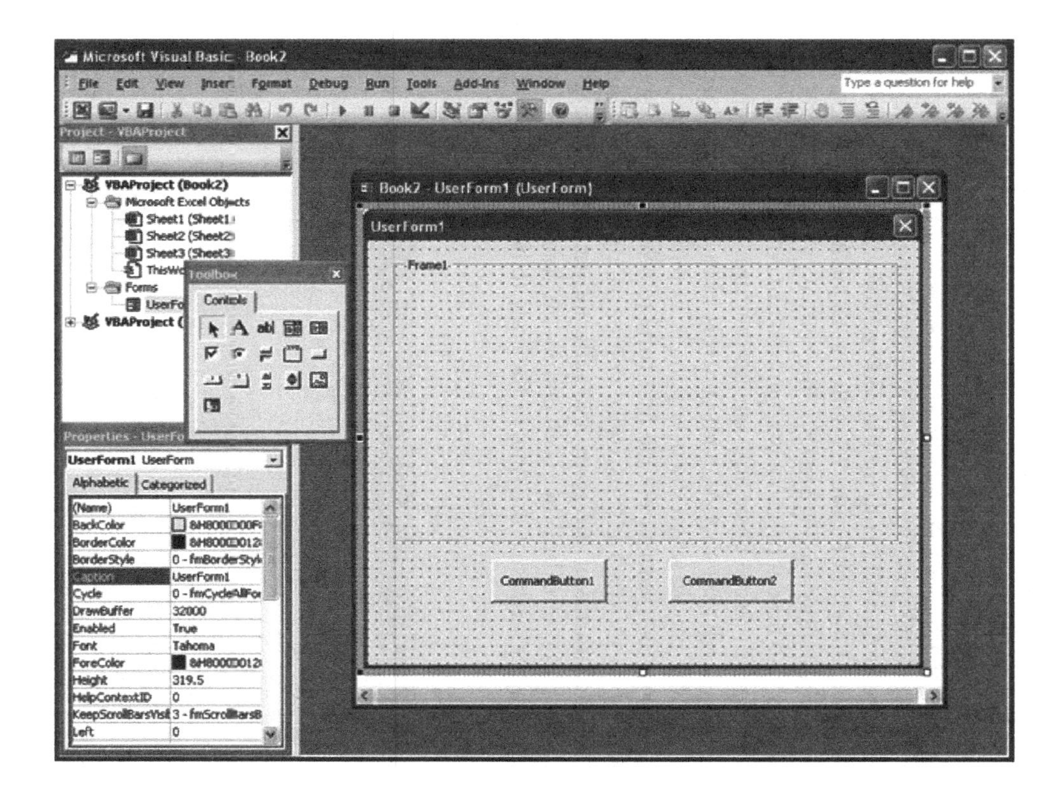

FIGURE 7.4
The Visual Basic for Applications interface in Microsoft Excel. In this figure, a new UserForm has been added and two new buttons and an empty frame added to it.

To add controls to the form, you will use the **Controls Toolbox** (small floating window in Figure 7.4). First, add a command button to the form by selecting the **CommandButton** control (you can see the control type name by hovering your cursor over it) and then drawing a rectangle onto the form canvas. Repeat this step to add another command button. Next, add a frame to the canvas by choosing the **Frame** command in the **Controls Toolbox** and drawing a rectangle on the canvas. You could adjust the properties of these controls manually by using the Properties window. In this case, we will set the properties through the following code. You should see a form in your VB interface similar to that shown in Figure 7.4 at this point. Double-clicking (or right-clicking and choosing **View Code**) on a blank part of the form will bring up the code editor window (Figure 7.5). Replace the code shown in Figure 7.5 with the following code (copy and paste over the code shown in Figure 7.5). This code is available at the Code > Chapter 7 > Excel link at http://www.geotree.uni.edu/SDSSbook. This code was written in Visual Basic and sets up the form for use when activated, controls all user interaction with controls in the form, and carries out the AHP/WLC model when the user runs it. In reality, the same form created previously is used twice—the first time to select the worksheets and the second to set the AHP pairwise comparisons. The code below controls the appearance of the form. We will not get into the detail of examining the code here, but there are comments (indicated by a single quotation at the beginning of the line) in the code to help give an idea of

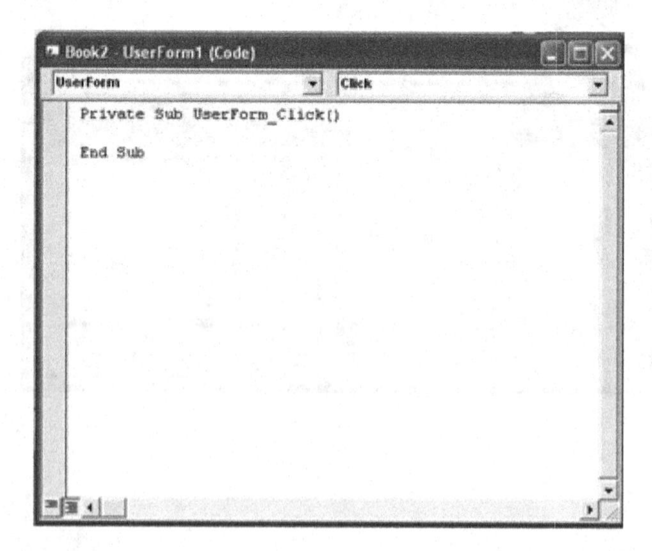

FIGURE 7.5
Code window associated with user form (UserForm1) created in Visual Basic editor.

what is happening. Again, this code is available at the Code > Chapter 7 > Excel link at http://www.geotree.uni.edu/SDSSbook.

```
'We want index to start from 1
Option Base 1

'Use names collection to store the selected sheet

Dim names As New Collection

'Define the Eigen vector which is used in AHP
Dim EigenVector

'Our Listbox
Dim lBox As MSForms.ListBox

'Do this before showing the UserForm:
'Fill the list box with available worksheet names and setup
the form
Private Sub UserForm_Initialize()
  With Me
    .Caption = "Select Sheets"
    .Height = 219.75
    .Width = 571.5
  End With

  With Me.Frame1
    .Caption = "Select Sheets"
    .Height = 144
    .Width = 546
    .Left = 12
    .Top = 12
  End With

Dim c As Control
  Set c = Me.Frame1.Controls.Add(bstrProgId:="Forms.
  ListBox.1", _
  Name:="lstBox", Visible:=True)
  With c
    .Height = 117.15
    .Width = 528.75
    .Left = 6
    .Top = 12
  End With
  Set lBox = c
  lBox.MultiSelect = fmMultiSelectMulti
```

```
  With Me.CommandButton1
    .Caption = "Select"
    .Height = 24
    .Width = 72
    .Left = 402
    .Top = 162
  End With

  With Me.CommandButton2
    .Caption = "Exit"
    .Height = 24
    .Width = 72
    .Left = 482
    .Top = 162
  End With

  For i = 1 To ActiveWorkbook.Sheets.count
    lBox.AddItem ActiveWorkbook.Sheets.Item(i).Name
  Next i
End Sub

'Do this if Button "Select" is pressed.

Private Sub CommandButton1_Click()
  'If list box is visible it means we are selecting
  'Fill the collection with the selected names of sheets
  If lBox.Visible = True Then
    Set names = New Collection
    For i = 0 To lBox.ListCount - 1
      If lBox.Selected(i) = True Then
        names.Add lBox.List(i)
      End If
    Next i

    If names.count <= 1 Then
      MsgBox "You should select at least two sheets..."
    Else
      lBox.Visible = False
      Call Construct_AHP_GUI
    End If
  'If list box is not visible it means
  'the AHP form is showing and thus the AHP calculation will
  be carried out
  ElseIf lBox.Visible = False Then
    Call Compute
  End If
End Sub
```

```
'AHP calculation ==========================================

Private Sub Compute()
  Dim values
  values = Array(9, 7, 5, 3, 1, 1 / 3, 1 / 5, 1 / 7, 1 / 9)

  'Initialize the matrix
  ReDim matrix(names.count, names.count) As Double

 'The diagonal of matrix is set to 1
  For i = 1 To UBound(matrix, 1)
    matrix(i, i) = 1
  Next i
  Dim ctrl As Control
  Dim optButton As MSForms.optionButton
  Dim x As Integer
  x = 0
  For i = 1 To UBound(matrix, 1)
    For j = i + 1 To UBound(matrix, 1)
      For counter = 1 To 9
        For Each ctrl In Me.Frame1.Controls
          If ctrl.Name = ("OptionButton." + CStr(counter + x
          * 9)) Then
            Set optButton = ctrl
            If optButton.Value = True Then
              If ((counter + x * 9) Mod 9) = 0 Then
                matrix(i, j) = values(9)
              Else
                matrix(i, j) = values((counter + x * 9) Mod 9)
              End If
              matrix(j, i) = 1 / matrix(i, j)
            End If
          End If
        Next
      Next counter
      x = x + 1
    Next j
  Next i

  'Initialize the sum array to store the sum
  'of each column of reciprocal matrix
  ReDim col_total(UBound(matrix, 1)) As Double
  For i = 1 To UBound(matrix, 1)
    For j = 1 To UBound(matrix, 1)
      col_total(i) = col_total(i) + matrix(j, i)
    Next j
  Next i
```

```
'Initialize the Eigen vector and compute it
ReDim EigenVector(UBound(matrix, 1)) As Double

For i = 1 To UBound(matrix, 1)
  For j = 1 To UBound(matrix, 1)
    EigenVector(i) = EigenVector(i) + matrix(i, j) / col_
    total(j)
  Next j
  EigenVector(i) = EigenVector(i) / UBound(matrix, 1)
Next i

'Principal Eigen value - L_Max
Dim L_Max As Double
Dim temp As Double
For i = 1 To UBound(matrix, 1)
  temp = 0
  For j = 1 To UBound(matrix, 1)
    temp = temp + matrix(i, j) * EigenVector(j)
  Next j
  temp = temp / EigenVector(i)
  L_Max = L_Max + temp
Next i
L_Max = L_Max / UBound(matrix, 1)

'Check for correct parameterization - Consistency Index
- CI
Dim CI As Double
CI = (L_Max - UBound(matrix, 1)) / (UBound(matrix, 1) - 1)
'Random Consistency Index - RI
Dim RI
RI = Array(0, 0, 0.58, 0.9, 1.12, 1.24, 1.32, 1.41, 1.45,
1.49, 1.51, 1.48, 1.56, 1.57, 1.59)

'Consistency Ratio - CR
Dim CR
If UBound(matrix, 1) > 15 Then
  MsgBox "RI for sheets more than 15 is unknown to me..."
Else
  CR = CI / RI(UBound(matrix, 1))
  CR = Round(CR * 100, 1)

  If (CR * 10) <= 100 Then
    CreateNewSheet
  Else
    MsgBox "Sorry, but your consistency ratio is " & CR & _
      vbLf & "which is unacceptable..."
  End If
End If
```

```
End Sub

'To create a our Result sheet ================================
Private Sub CreateNewSheet()
  'Creating the formula to compute
  'the result using WLC with weights obtained from AHP
  Dim my_formula As String
  my_formula = ""

  With ActiveWorkbook
    .Sheets.Add After:=.Sheets(.Sheets.count)
    .ActiveSheet.Name = "ourResult"
  End With

  Dim range As Excel.range
  Set range = ActiveWorkbook.Sheets(names.Item(1)).UsedRange

  ActiveWorkbook.Sheets("ourResult").range(range.Address).
  Select

  For i = 1 To names.count
    If i = names.count Then
      my_formula = my_formula & names.Item(i) & "!" & range.
      Address & "*" & EigenVector(i)
    Else
    my_formula = my_formula & names.Item(i) & "!" & range.
    Address & "*" & EigenVector(i) & "+"
    End If
  Next i

  Selection.FormulaArray = "=" & my_formula

  'Selecting the range and changing:
  '-its RowHeight and ColumnWidth
  '-coloring it using the Conditional Formatting ColorScale
  '-the zoom of the page

  ActiveWorkbook.Sheets("ourResult").range("A1").Select
  ActiveWorkbook.Sheets("ourResult").range(range.Address).
  Select

  Selection.RowHeight = 5
  Selection.ColumnWidth = 0.5
  Selection.FormatConditions.AddColorScale ColorScaleType:=3
  Selection.FormatConditions(Selection.FormatConditions.
  count).SetFirstPriority
```

```
  Selection.FormatConditions(1).ColorScaleCriteria(1).Type =
  xlConditionValueLowestValue
  With Selection.FormatConditions(1).ColorScaleCriteria(1).
  FormatColor
    .Color = 8109667
    .TintAndShade = 0
  End With
  Selection.FormatConditions(1).ColorScaleCriteria(2).Type =
  xlConditionValuePercentile
  Selection.FormatConditions(1).ColorScaleCriteria(2).Value
  = 50
  With Selection.FormatConditions(1).ColorScaleCriteria(2).
  FormatColor
    .Color = 8711167
    .TintAndShade = 0
  End With
  Selection.FormatConditions(1).ColorScaleCriteria(3).Type =
  xlConditionValueHighestValue
  With Selection.FormatConditions(1).ColorScaleCriteria(3).
  FormatColor
    .Color = 7039480
    .TintAndShade = 0
  End With
  ActiveWorkbook.Sheets("ourResult").range("A2").Select
  ActiveWindow.Zoom = 43
  Unload Me
End Sub

'To construct the AHP GUI
Private Sub Construct_AHP_GUI()
  Me.Frame1.Caption = "Pairwise Comparison, AHP"
  Dim MSForm_Frame As MSForms.Frame
  Dim MSForm_Label As MSForms.Label
  Dim ctrl As Control
  Dim x As Integer
  x = 0

  Dim captions
  captions = Array("9", "7", "5", "3", "1", "3", "5", "7",
  "9")

  Dim Def_Label As MSForms.Label
  Set ctrl = Me.Controls.Add(bstrProgId:="Forms.Label.1", _
    Name:="Label.Definition", Visible:=True)
  ctrl.Top = 172
  ctrl.Left = 12
  ctrl.Width = 250
  ctrl.Height = 15
```

```
Set Def_Label = ctrl
Def_Label.Caption = "9-Extreme; 7-Very Strong; 5-Strong;
3-Slightly; 1-Equal."
Def_Label.Font.Bold = True

For i = 1 To names.count
  For j = i + 1 To names.count
    '<Add left side label to Frame1>
    Set ctrl = Me.Frame1.Controls.Add( _
      bstrProgId:="Forms.Label.1", _
      Name:="Label.Row:" + CStr(i) + "Col:1", _
      Visible:=True)
    ctrl.Top = x * 34 + 10
    ctrl.Left = 20
    ctrl.Width = 100
    ctrl.Height = 24

Set MSForm_Label = ctrl
MSForm_Label.Caption = names.Item(i)
MSForm_Label.BorderColor = RGB(0, 0, 0)
MSForm_Label.BorderStyle = fmBorderStyleSingle
MSForm_Label.Font.Bold = True
'</Add left side label to Frame1>

'<Add frame to Fram1 which consists of options>
Set ctrl = Me.Frame1.Controls.Add( _
  bstrProgId:="Forms.Frame.1", _
  Name:="Frame." + CStr(i), _
  Visible:=True)
ctrl.Top = x * 34 + 10
ctrl.Left = 140
ctrl.Width = 230
ctrl.Height = 24

Set MSForm_Frame = ctrl
MSForm_Frame.BackColor = RGB(Int(Rnd * 255), Int(Rnd *
255), Int(Rnd * 255))
For k = 1 To 9
  Dim optButton As MSForms.optionButton
  Set ctrl = MSForm_Frame.Controls.Add( _
    bstrProgId:="Forms.OptionButton.1", _
    Name:="OptionButton." + CStr(x * 9 + k), _
    Visible:=True)
  ctrl.Top = 3
  ctrl.Left = (k - 1) * 20 + 20
  Set optButton = ctrl
  optButton.Caption = captions(k)
```

```
    If k = 5 Then
      optButton.Value = True
    Else
      optButton.Value = False
    End If
  Next k
  '</Add frame to Fram1 which consists of options>

  '<Add right side label to Frame1>
  Set ctrl = Me.Frame1.Controls.Add( _
    bstrProgId:="Forms.Label.1", _
    Name:="Label.Row:" + CStr(i) + "Col:2", _
    Visible:=True)
  ctrl.Top = x * 34 + 10
  ctrl.Left = 390
  ctrl.Width = 100
  ctrl.Height = 24

  Set MSForm_Label = ctrl
    MSForm_Label.Caption = names.Item(j)
    MSForm_Label.BorderColor = RGB(0, 0, 0)
    MSForm_Label.BorderStyle = fmBorderStyleSingle
    MSForm_Label.Font.Bold = True
    '</Add right side label to Frame1>
    x = x + 1
  Next j
Next i
  Me.Frame1.ScrollHeight = x * 34 + 20
  Me.Frame1.ScrollBars = fmScrollBarsVertical
  Me.Repaint
End Sub
'Do this if Button "Exit" is pressed ======================
Private Sub CommandButton2_Click()
  Unload Me
End Sub
```

At this point, the form can be opened and the command buttons in the form can be used. In order to open and run the form from Excel, there are several methods available. For example, you could create a new tab and button in the Excel interface, or add a button on the interface and run a macro. These methods are not intuitive in Microsoft Excel 2007. The simplest way is to create a new macro (a small set of steps that can be stored as code in a module in Visual Basic) and add a small amount of code to open the form. In the Microsoft **Visual Basic Interface**, choose the **Insert** menu and select **Module**. Then type the code exactly as seen below into the module window (Figure 7.6).

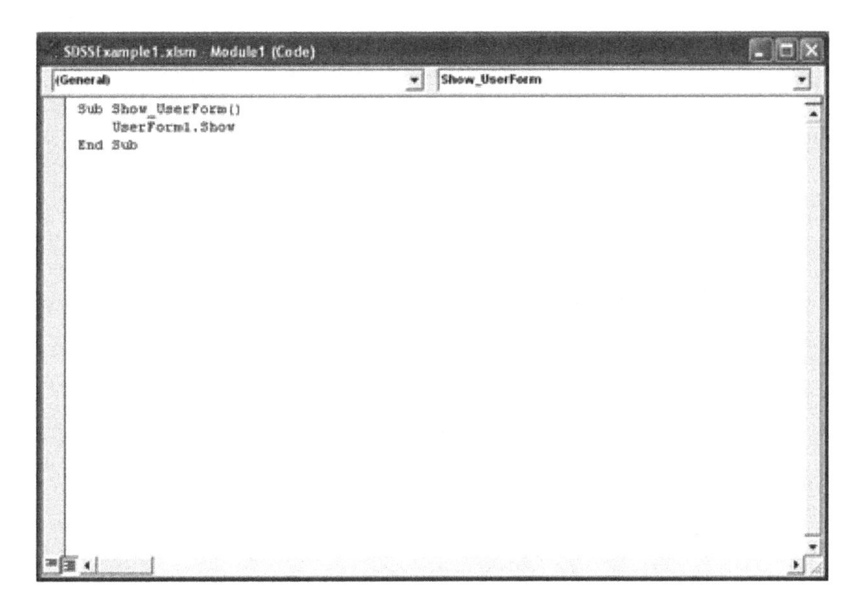

FIGURE 7.6
The macro module in the Visual Basic editor with code added to open the form created in Figure 7.4.

In the Microsoft Excel main interface, choose the **Developer** tab and click on the **Macros** button (Figure 7.7). This will activate the Macro interface as shown in Figure 7.8. Select the macro named **Show_UserForm** and click **Run**. Now you should see a Windows dialog or form open. This is the form that was created in the previous steps. The properties of the form were adapted on the fly through the code that was copied and pasted in the form's code window. At this point, the form will list the sheets available in the workbook. The idea in running the application would be to select the sheets holding the relevant data for running the AHP model. In the following paragraphs we will explain how to import that data into Excel in order that the form and program can actually be used. For now, click **Exit**.

The next paragraphs will present a hypothetical site suitability study applying the AHP model in Excel. We will examine the problem of locating a grocery store in a small urban area. In order to use the Excel tool, we first have to import spatial data into Excel. If this were going to be a full SDSS package, it is likely that a custom tool would be developed in order to make the importing of spatial data easier (indeed, in the SpreadsheetSDSS described later, this is true). In this example we will demonstrate the steps for manually importing spatial data, in particular, ASCII files created from Environmental Systems Research Institute (ESRI) grids, into Excel worksheets, and how to view the spatial data inside Excel.

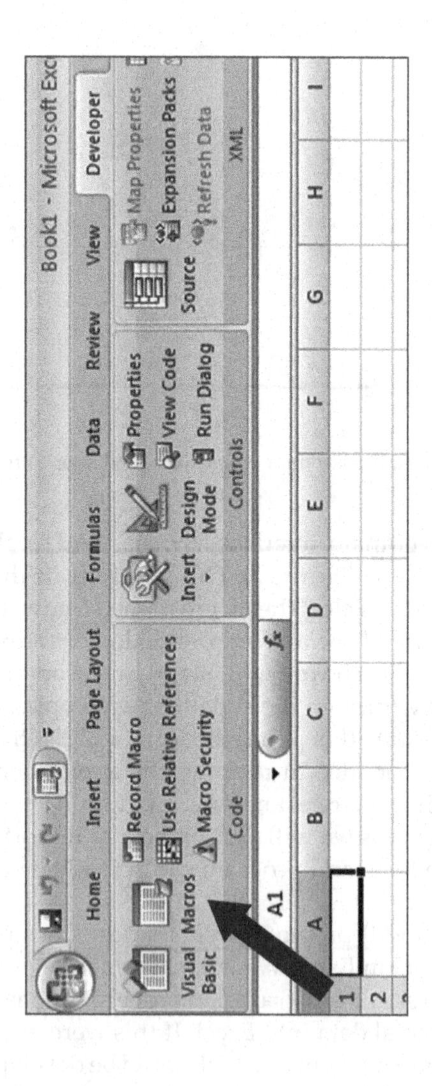

FIGURE 7.7
The Microsoft Excel interface with Developer tab chosen. To open the Macro dialog, click the Macros button.

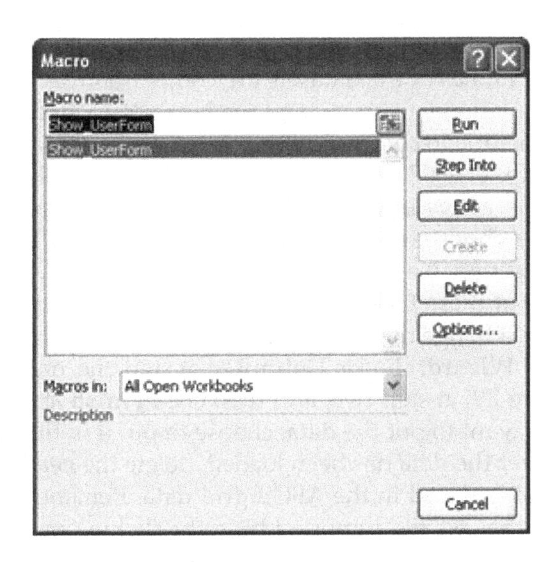

FIGURE 7.8
The Macro dialog in Microsoft Excel.

In this case study we will utilize multiple layers in order to determine which area(s) would be most suitable for locating a grocery store. It is estimated that a competitive grocery store would require approximately 10 acres of land (including the building and the parking lot and the areas to unload goods into the store). We will use local parcel data to determine where the area of any parcels is equal to or greater than 10 acres. We will also use the location of existing grocery stores as a basis for site selection. This will be accomplished by creating a buffer of 1.5 km from the center of the existing store buildings and "blacking out" these areas to exclude them as suitable locations for building a new store. Next, we will use reclassified land cover data. This data was reclassified from 17 classes to 2 classes (suitable or not suitable). In the suitable category, we included the following land use/cover classes: ungrazed grassland, grazed grassland, planted grassland, alfalfa/hay, corn, soybeans, other row crops, commercial industrial, and barren. Also included in the evaluation is a population density dataset. This is divided into three classes: lightly populated, moderately populated, and densely populated. The final layer included in our analysis is a buffered roads layer. We buffered the roads with a 100-m buffer with the idea that a new business would be built in an area with easy access for customers and deliveries. In a real SDSS development process, the discussion as to what spatial data to use would be begun at the requirements definition stage, scoped in the system design phase, included in the development cycle, and refined in the iterative process. These datasets are available at the Data > Chapter 7 > Excel SDSS link at the Web site that accompanies this book (http://www.geotree.uni.edu/SDSSbook). The names of the datasets are competition.asc, landcover.asc, parcels.asc, populationdensity.asc, and roads.asc. You can download these to your local machine to continue with this example.

In Microsoft Excel 2007, we can import data from ASCII files. With an empty worksheet active in the .xlsm file in which you created the form and added code, choose the **Data** tab, then choose **From Text** in the **Get External Data** section. Navigate to the directory with the ASCII grid files and choose one of these files to import (when the files have a .asc extension, you need to change the **Files of Type** drop-down list to **All Files**). In the **Text Import Wizard**, choose **Delimited** at step one, only **Space** as the delimiter (Figure 7.9) at step two, and then click **Finish** at step three. If it asks where you want to put the data, choose to put it in the A1 cell in the active sheet. After the data has been loaded, delete the header (Figure 7.9) information that is stored in the ASCII grid data. Rename the worksheet to the name of the .asc file imported by right-clicking on the **Sheet** tab, choosing **Rename**, and typing in the new name. Move on to the next worksheet and import another layer by repeating the steps discussed. Repeat until all of the input data has been loaded and formatted.

In the next step, the worksheets will be formatted for viewing in a map-like format. With all cells highlighted (click in the upper-left corner of the worksheet), under the **Cells** section in the **Home** tab, choose the **Format** drop down menu and choose **Row Height**, and set the row height to **5**. From the same drop-down menu, choose **Column Width** and set the column width to **0.67**. Then from the **Styles** section in the **Home** tab, select **Conditional Formatting > Color Scales** and choose one style (suggested style: **Red, Yellow, Green**). Repeat these steps for each worksheet. The user can explore different scale effects using the zooming tools available in Excel, including the zoom sliding toolbar at the bottom right of the Excel interface.

Once the ASCII grid data has been imported, you can run the AHP model that was created in the previous steps using the Visual Basic for Applications framework. The Analytic Hierarchy Process (AHP) model is useful for determining the importance of various factors in solving complex decisions. It provides a rational, structured framework for solving a decision problem. It can help in quantifying different elements of the decision-making process. Or, like the example here, it can be used to give weights to various layers, which may have been difficult to assign on an individual basis. It works by allowing the user(s) to decide which elements are more important in comparison to others through pairwise comparisons. As soon as the user specifies how different layers relate to each other, the AHP calculates the weighting of the various layers and the final analysis is carried out (in our example, weighted linear combination [WLC] is used). It basically converts the evaluation of the comparisons of

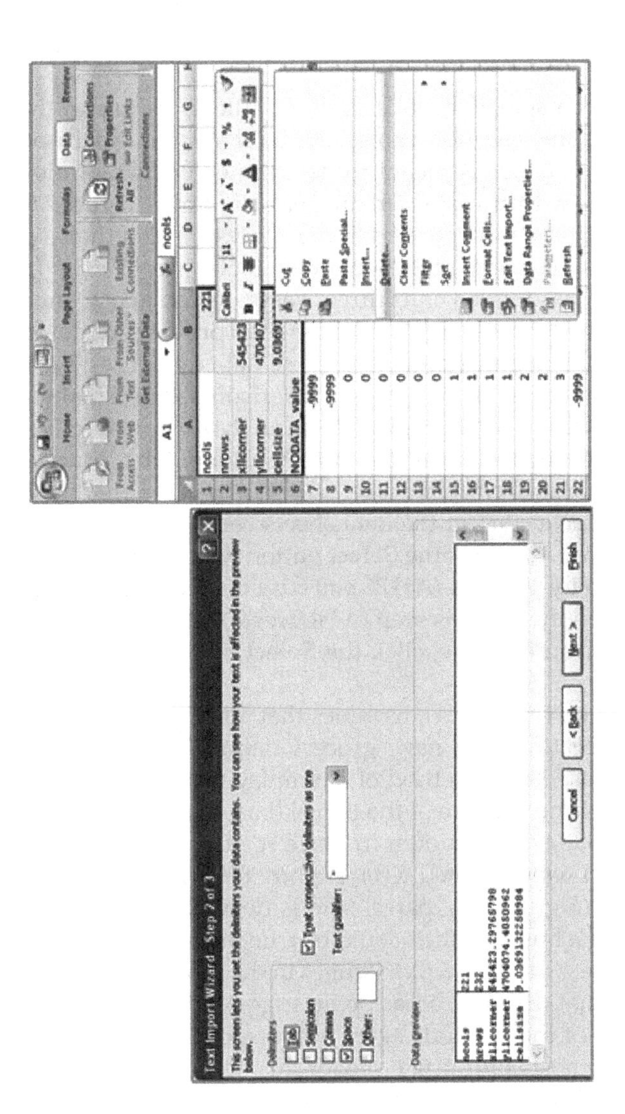

FIGURE 7.9
Importing spatial data into Excel.

different factors into numerical values, which can then be used to carry out a weighted analysis. However, using AHP to determine the weighting for variables relies on the user providing consistent comparisons. In the AHP model, the consistency ratio determines if the user's selections of relative importance in the different layers are consistent. In some cases, it is possible for the user's selections to result in an error because the consistency ratio was violated.

The creation of a spreadsheet-based SDSS allows us to run the hypothetical business site selection using AHP and WLC in Microsoft Excel. Specifically, this exercise is meant to determine suitable locations for a new grocery store within a small urban area's city limits. Figure 7.10 shows the five layers that are incorporated in the analysis. All of the layers were divided into two classes (suitable or not suitable) except for population density, which was divided into three classes (least dense, moderately dense, and highly densely populated areas). We provide an illustration of three example AHP simulations using these data in Figures 7.11 through Figures 7.13 with the AHP pairwise comparison settings shown on the left and results shown on the right. These analyses can be re-created by the reader by running the macro (choose **Developer tab > Macros > Run ShowUserForm** macro) created earlier and shown in Figure 7.8. When the original form appears, select the data sheets containing the ASCII grids that were imported and click the **Select** button when chosen. Then, in the **Analytic Hierarchy Process (AHP)** pairwise comparison dialog, set the rankings between the pairs as seen in Figures 7.11 through 7.13. When set properly for a given example, click the **Select** button for the analysis to take place.

The first analysis (Figure 7.11) assumes that all of the layers are equally important in locating the new grocery store. The second analysis (Figure 7.12) is based on a method of downplaying the importance of the distance to roads and the size of the parcel (based on the logic that a new store could likely develop its own frontage road and parking lot and the size of the parcel was assumed to be slightly flexible). Finally, in the last analysis shown (Figure 7.13), parcel size is deemed the most important factor in locating a business (this could be true in areas of the town where space is limited, e.g., downtown). Changes in the inputs in the AHP model result in somewhat different areas being suggested as suitable locations. Interestingly, all of the methods agree on a region (darkest red area) that was deemed the most appropriate site for this new business to be located. In fact, this is the only area in town that meets the criteria for all of the initial layers (not near other existing grocery stores, in an appropriate land cover class, large enough parcel size, high population density, and near enough to the existing road network). The areas on the map that are the closest to red or dark orange are the most suitable locations. However, there are other areas in the city that could also be adequate for locating a

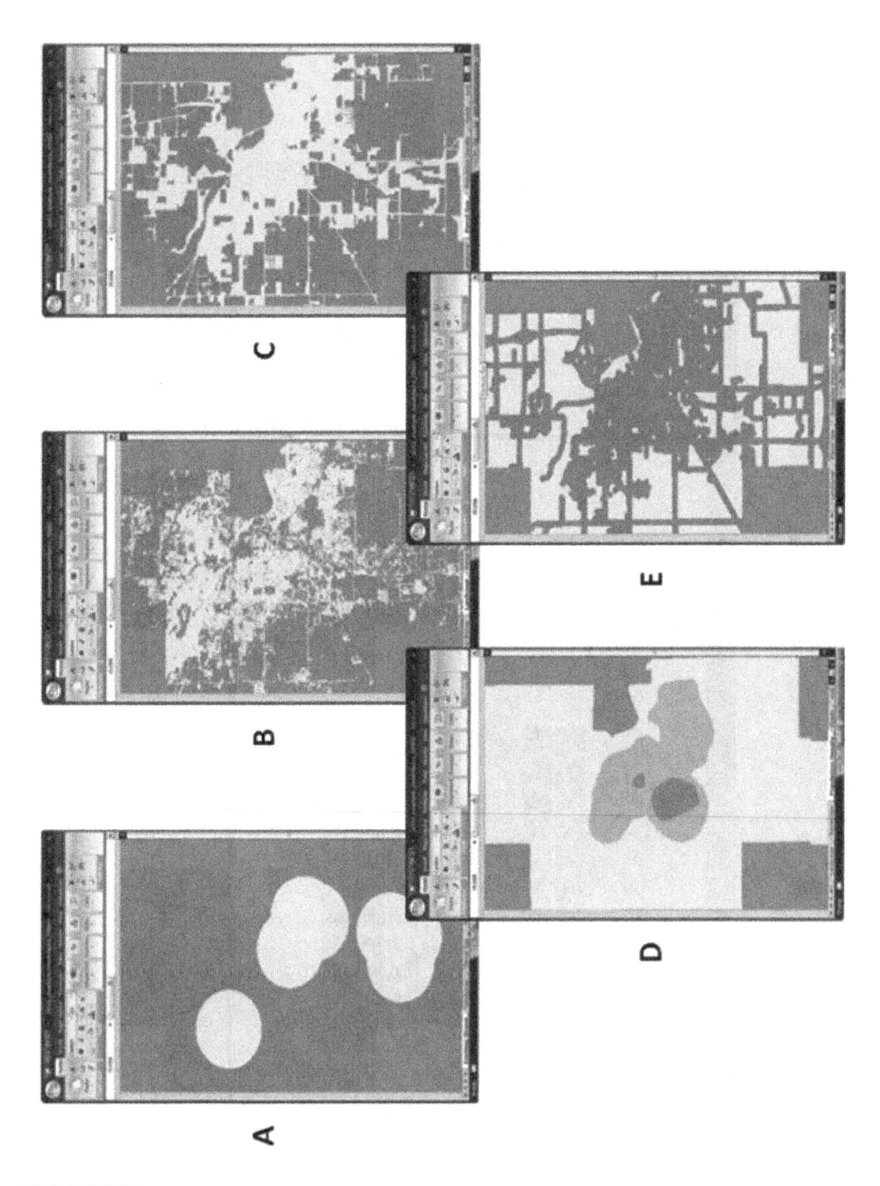

FIGURE 7.10
(See color insert following page 74.) Representations of (A) existing grocery stores with a 1.5-km buffer, (B) suitable landcover classes, (C) areas with suitable parcel size, (D) population density classes, and (E) existing roads with a 100-m buffer.

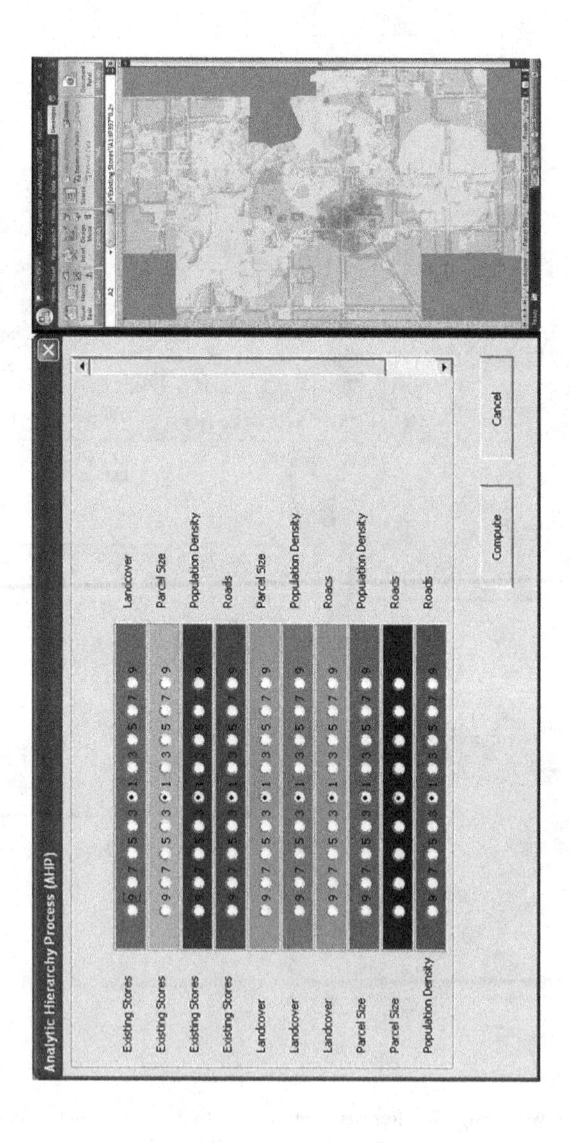

FIGURE 7.11
(See color insert following page 74.) AHP set at default with all layers set as equally important to each other.

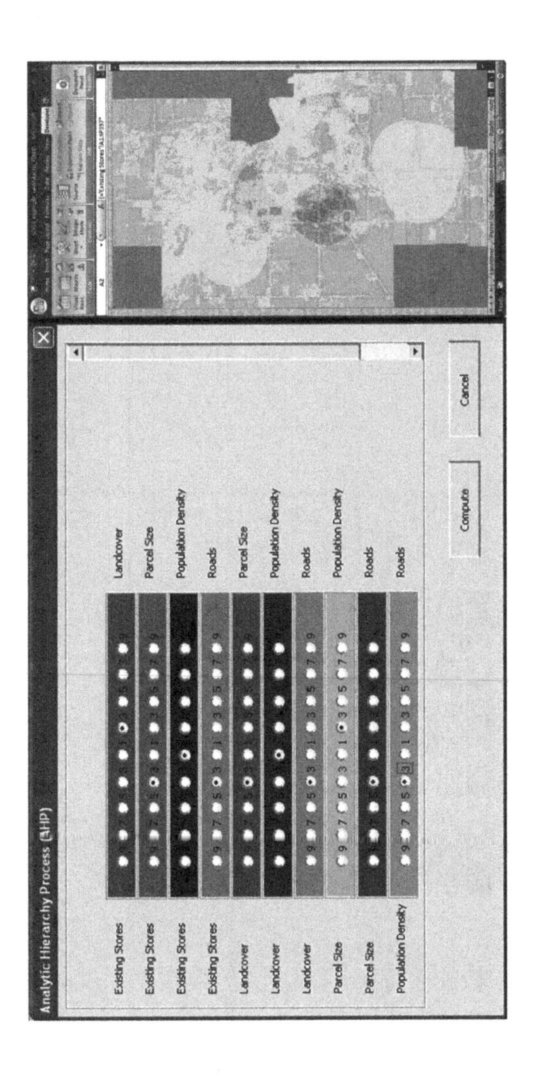

FIGURE 7.12
AHP setup with alternative selection of layer importance favoring land cover, population density, and existing stores.

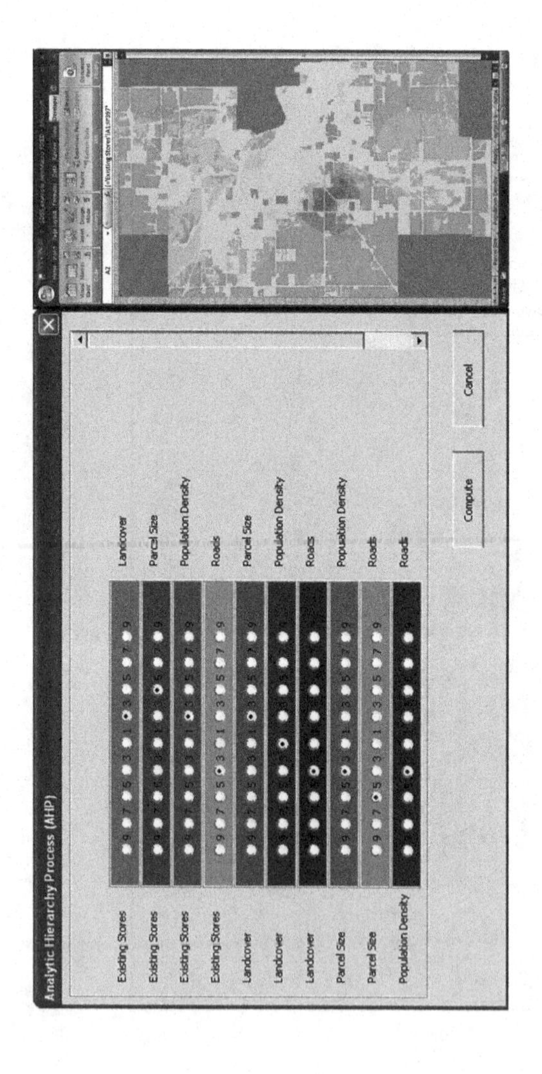

FIGURE 7.13
AHP settings for food store site suitability with a bias toward parcel size.

business. It is, of course, up to the business owners, land managers, city planners, council members, and so on, to reach a final absolute location on which to build. The steps discussed and results presented here could be part of the testing and final implementation phase in a real SDSS process. The parameterization of the AHP model directly affects the results and the potential decision, so it is important to determine the pairwise comparisons closely and consider different stakeholders' views in the selection process.

In this Excel-based AHP model, the user interface consists of a couple of forms. First, there is the Select Sheets form, which allows the user to select which layers (stored in Excel worksheets) will be included in the AHP analysis. The second form is the Analytic Hierarchy Process (AHP) pairwise comparison form. This form allows for the pairwise comparison of every layer included from the Select Sheets form. On one side is the name of the first layer and on the other side is the name of a comparison layer. In comparing two layers in this form, the users must select from the numbers 9, 7, 5, 3, 1, 3, 5, 7, 9. If the user believes that one of the layers being compared is more influential or important in the selection of the criteria, then the user selects a higher value on the side of the more important layer. There is a fundamental scale used for pairwise comparisons that is necessary in order to complete the AHP analysis. Selecting 1 would indicate that the user believes that the two elements contribute equally to the objective. Selecting 3 would indicate that one element is slightly favored over the other. Selecting 5 would suggest that one element is strongly favored over the other. Selecting 7 would say that the experience and judgment of the user shows that one element is very strongly favored over the other and implies that the first element's dominance has been demonstrated in practice. Selecting 9 would define one element as being extremely important and indicates that evidence favors one element over the other to the highest possible order of affirmation. The AHP model uses these pairwise comparisons to arrive at weights for the layers. In our example, these weights are then used in a weighted linear combination method. In this manner, weights are not assigned by the user, but the relationship between the layers is discovered and appropriate weighting is given to them based on the AHP analysis.

The result from running the AHP analysis with WLC is an Excel worksheet that demonstrates the results in the form of a color-coded map. The new Excel sheet is titled "ourResult" and is displayed in the red, yellow, green color scale and with row height and column width adjusted for easy viewing. There is no direct output into a GIS-acceptable data format, but the resultant map could be saved as a text file, and after copying, the appropriate header could be added into most GIS software as an ASCII file.

The spreadsheet-based SDSS incorporating AHP demonstrated that it is quite possible to use Excel to develop an SDSS. The results of a hypothetical site selection for a grocery store showed the applicability of the program. However, many types of analysis are possible, not just site selection. We showed here how to import spatial data into Excel (ASCII grids), how to incorporate AHP with WLC for spatial analysis, and how the results can be interpreted to help solve a problem (e.g., grocery store site selection). This spreadsheet-based SDSS is unique and allows the potential researcher an opportunity to perform GIS-like analyses and provides SDSS capabilities for decision making in the business-familiar environment of Microsoft Excel.

There are some limitations to this spreadsheet-based SDSS. The main limitation in using Excel as an SDSS is that it does not provide cartographic capabilities. There is neither the ability to overlay multiple datasets in one view nor functionality for creating sophisticated cartographic outputs. One limiting factor concerns the number of columns available in Excel 2007. The number of rows (according to the Excel Help function) is 1,048,576, but the number of columns is 16,384. This column limit could restrict analyses of large raster datasets. However, in the previous example analyses, the cell size was approximately 30 m and there were 250 columns and 397 rows. The column limitation will only preclude analysis of large raster spatial data (e.g., large study area with high-resolution data). Another limitation involves the formula bar used to create the output for the WLC model. In order to overcome some of these limitations, a plug-in for Excel using external programming languages has been developed. The next section will highlight the main characteristics of this Excel extension.

7.4.2 SpreadsheetSDSS Plug-in

The previous section demonstrated the capability of a simple spreadsheet-based SDSS module developed in Microsoft Excel using Visual Basic for Applications. Here, we will briefly discuss the development of a more complete set of tools in an Excel-based plug-in called SpreadsheetSDSS, which incorporates many new features and models compared to the AHP-only spreadsheet-based SDSS discussed previously. These new features include a spatial data management module with a direct importing utility, symbolization options, a model management module with multiple models (WLC, ordered weighted averaging [OWA], WLC using AHP, Boolean Overlay), cartographic capabilities (legend parameters, north arrow, text), a print preview tool, a report generator, and zoom functionality. The menu bar for the SpreadsheetSDSS is shown in Figure 7.14. The SpreadsheetSDSS plug-in is not described in detail in this book. However, the code and a detailed description of the functionality available with the SpreadsheetSDSS can be found at our Web site (http://www.geotree.uni.edu/SDSSbook).

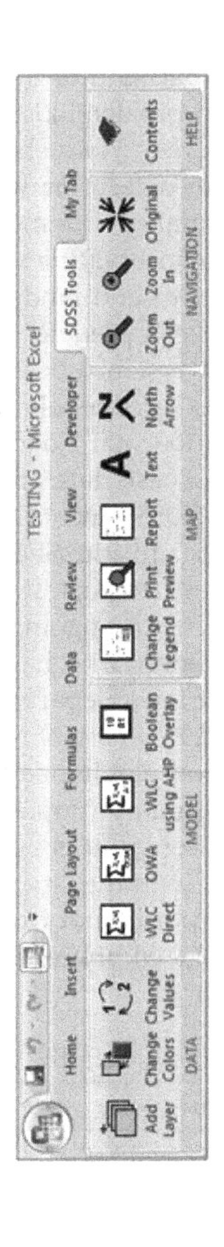

FIGURE 7.14
(See color insert following page 74.) The main menu of the Spreadsheet SDSS in the Microsoft Excel interface.

7.4.3 Customizing Existing Desktop GIS (ArcGIS)

We have already discussed how ESRI GIS software (ArcInfo, ArcView, and ArcGIS) have been used in many desktop SDSS configurations. The ability to develop new customized tools using Visual Basic for Applications (VBA) facilitates the development of ArcGIS-based SDSS (Table 7.1). Here we show how a simple WLC model can be incorporated into ArcMap using VBA code. An example application of the WLC model in ArcGIS is demonstrated with a real-world problem: identifying environmentally sensitive areas in the Dry Run Creek Watershed in northeastern Iowa. Utilizing the WLC model, we can determine which areas of the watershed could be focused on in potential environmental restoration activities or those that are most valuable in conservation activities. The WLC model is executed using various land cover and landscape data and assigning different weights based on the user's predetermined expectation of the importance of the different data layers. Included in this example of a desktop-based SDSS incorporating WLC are the following layers: slope, hydric soil group, green space, FEMA's 100-year floodplain boundaries, and a buffered impervious areas layer. These data are available as ESRI grids at the Data > Chapter 7 > ArcGIS SDSS link at the Web site that accompanies this book (http://www.geotree.uni.edu/SDSSbook). The names of the grids are test_slope, test_hysoil, test_green, test_fema, and test_bufimp.

This SDSS example is just part of an ongoing, small, real-world SDSS application example. The stakeholders who have been involved include city officials, the general public, the Iowa Department of Natural Resources (DNR), and researchers from the University of Northern Iowa (UNI). The city's officials are responsible for developing new storm-water policies in the watershed to improve water quality (impaired waters in the watershed). They are considered end users in this example. The Iowa DNR is the regulatory authority working with the city to improve water quality in the watershed. The watershed coordinator from the DNR served as the architect of the project and was crucial in keeping the other stakeholders

TABLE 7.1

SDSS Components of the ArcGIS-Based SDSS

Component	Value	Comments
Spatial Database	ESRI Grid	
Model	WLC	Weighted linear combination
User Interface	VBA	Visual Basic for Applications
Report Generator	Crystal Report	
Software	ArcGIS - ArcView	
Programming	VBA	
Types/Cost	Commercial	

engaged with each other. The researchers from UNI were contracted to develop the GIS-based SDSS. All of the stakeholders engaged in an iterative process as illustrated in Figure 7.2. Numerous meetings between the main stakeholders were held to define requirements and present and test initial prototypes. A public meeting was also held to present the system to interested individuals or groups from the general public. The development of this SDSS is ongoing but has been an example of a successful small SDSS.

The material presented below shows how to first develop and then use this ArcGIS-based SDSS utilizing WLC in an example of locating environmentally sensitive areas. The user must begin a new ArcMap session (it is a good idea to save to a new .mxd at this time). A new blank tool must be created first. To do this, first select the **Tools** menu, choose **Customize**, and then click **New** with the **Toolbars** tab active. Give your new toolbar a name (e.g., WLCTool) and save it in the present map document (i.e., *not the Normal.mxt*). This means that the new toolbar will not be available to all map documents, just the present one. When you click **OK**, a new empty toolbar will be visible in ArcMap.

The next step requires the selection of UIControls (in the Categories frame) after switching to the **Commands** tab in the **Customize** dialog. After selecting and highlighting **UIControls**, click the **New UIControl** button. This brings up the **New UIControl** dialog, which has four choices for types of controls to choose. Select the **UIButtonControl** option and click **Create**. In the **Commands:** frame (in the **Commands** tab of the **Customize** dialog), a new control appears that is called **Project.UIButtonControl1**. Change the second half of the name of this tool to "SDSS" (i.e., highlight the tool and change **UIButtonControl1** to **SDSS**). This new control can be dragged to the new toolbar by single-clicking it and dragging and dropping it onto the new toolbar just created. You may customize the appearance of the tool if you wish. You can also set a new image for the button: right-click on it (in the new toolbar), choose **Change Button Image**, and selecting a new icon. You can also decide if the tool is represented in the toolbar with just an image, just text (i.e., its name), or with both.

In order to add some functionality to the new button, we will use VBA. Right-click on the new button in the toolbar and select the **View Source** button to open the Microsoft Visual Basic environment (Figure 7.15). In the Visual Basic environment, we will set up a user form or dialog that will be used to choose input data and set weights for the WLC modeling activity in the ArcGIS-based SDSS. The first step is to create a new form. To do this, right-click **ThisDocument** in the **Project** window (see Figure 7.16) and choose **Insert > UserForm**. Resize the new form similar to Figure 7.17 so it will accommodate all of the controls that can be seen in Figure 7.18. Also rename the form in the **Properties** window by setting the **(Name)** property to **frmSDSS**. Also set the **Caption** property to **MySDSS**. The

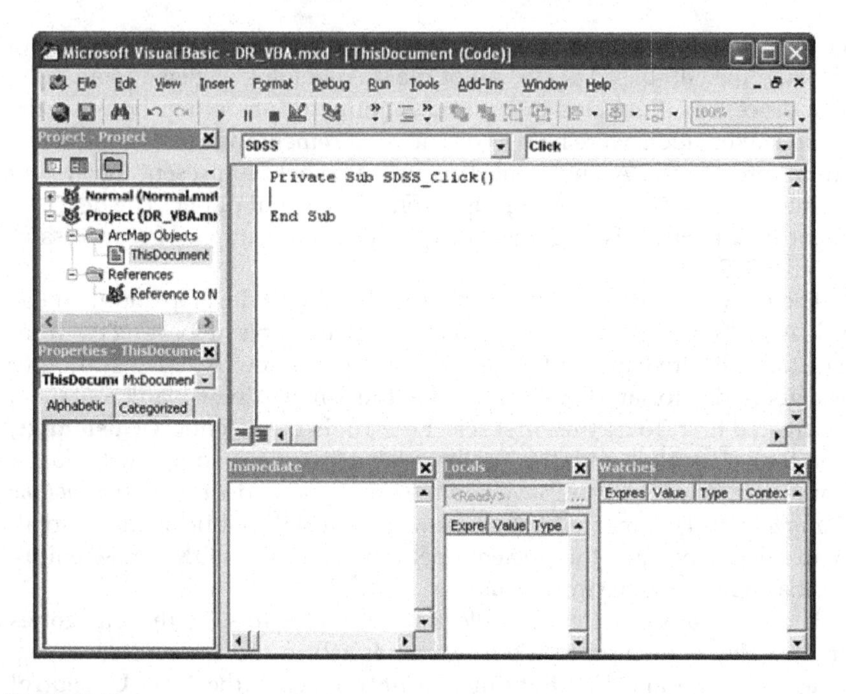

FIGURE 7.15
The Visual Basic interface seen after choosing View Source when right-clicking on the new button in the new toolbar created.

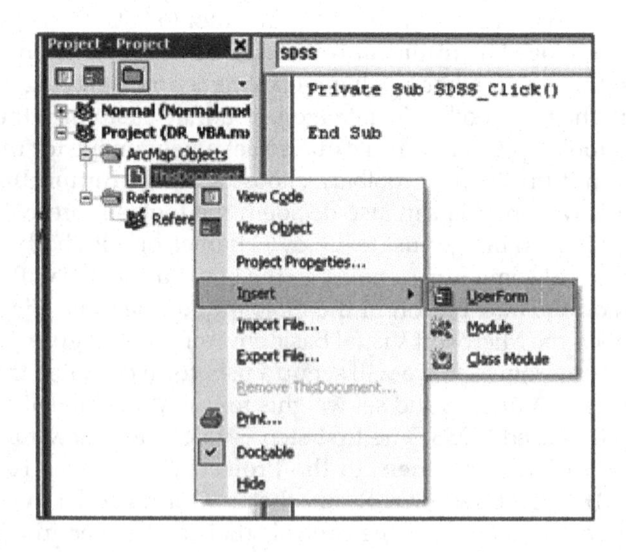

FIGURE 7.16
The step to add a new user form in the Visual Basic environment.

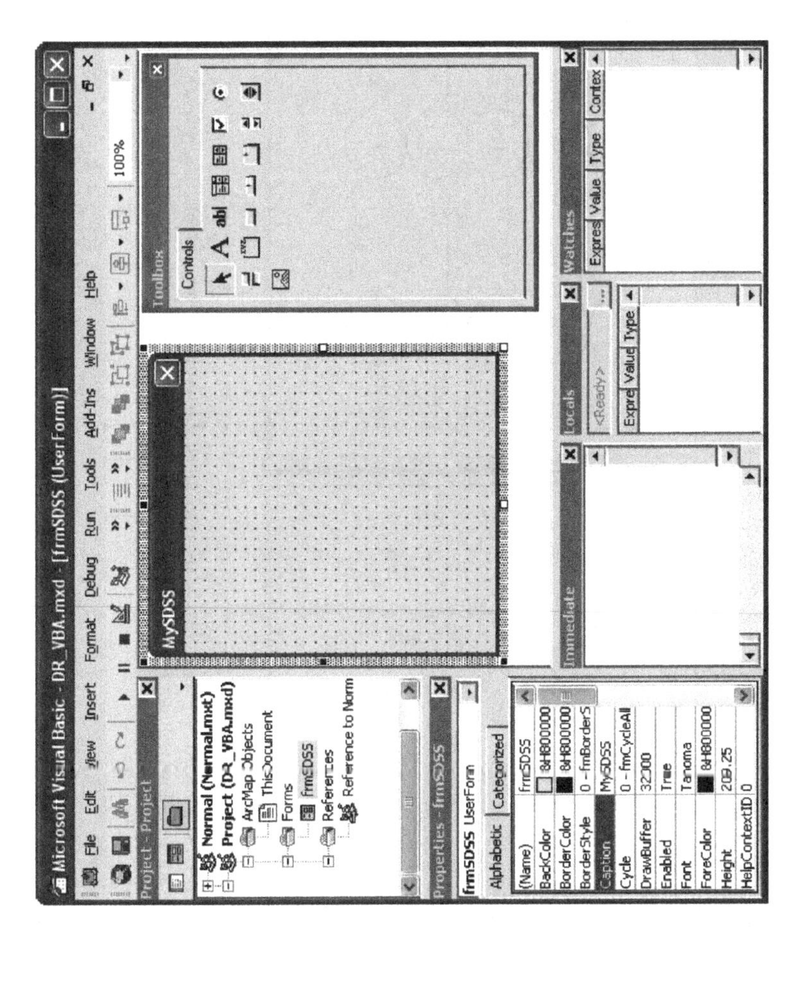

FIGURE 7.17
The Visual Basic environment and specifically the Toolbox dialog that will be used to create new controls on the form.

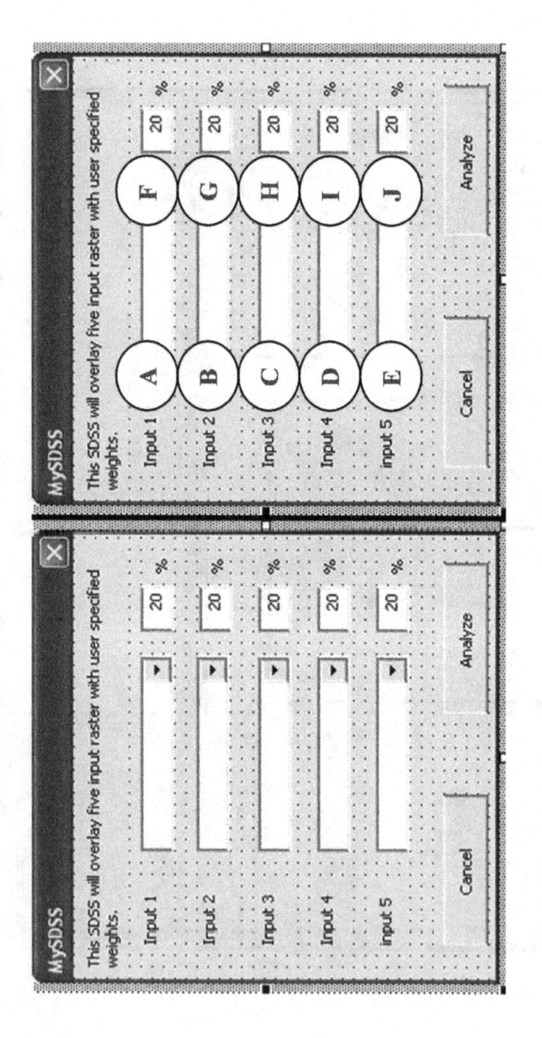

FIGURE 7.18
The form with all controls added. The added A–J labels can be used to check the names of the controls. The relevant controls must have the correct name or the code that will be added below will not work.

next step involves adding combo boxes, text boxes, and buttons, and then specifying the code (following) that makes the whole tool functional.

We will add a set of controls to the form. It is important that you name these controls in the Properties window exactly as indicated in the text below or the program will not work. Using the **Toolbox** dialog with the MySDSS form active, select the label control (big A on toolbox) and draw a new label in the upper-left-hand corner of the form. Change the **Caption** property to **Input 1**. Right-click on that control and select **Copy**. Click on a blank part of the form and choose **Paste** after right-clicking. Change the **Caption** property of the new label to **Input 2**. Repeat these steps until you have five label controls and have arranged them so they are similar to Figure 7.18. Now add a new ComboBox control to the form: in the **Toolbox**, choose **ComboBox**, drag a rectangle on the form, and place it right next to the Input 1 label. Change the **(Name)** property to **cmb1**. Create four more of these ComboBox controls by copying and pasting like you did with the labels. Make sure to name the controls cmb2, cmb3, cmb4, and cmb5, matching the number in the label next to it. These combo boxes will be used to choose the layers that will be included in a WLC analysis. Now add a text box to the right of the cmb1 control (Figure 7.18). Change this text box's **(Name)** property to **txt1**. Also set the **Text** property to **20**. Create four more of these TextBox controls by copying and pasting like you did with the labels. Make sure to name the controls txt2, txt3, txt4, and txt5, matching the number in the label next to it. These text boxes will allow the user to set the weights in the WLC procedure. Add five labels next to the text boxes that have the same Caption property, % (see Figure 7.18). Now add a new CommandButton control by clicking on that control type in the **Toolbox** dialog and dragging a rectangle on the form. In the **Properties** window, set the **(Name)** to **cmdCancel** and the **Caption** property to **Cancel**. Repeat these steps to add a new button with a **(Name)** of **cmdAnalyze** and **Caption** property of **Analyze**. The form should now have all of the controls seen in Figure 7.18. Check to make sure the combo and text boxes have the correct names as seen in Table 7.2. If they are named improperly, the code will not work.

Now we will add the code that will allow the form to be opened from the ArcMap interface by clicking on the new tool button created earlier. In the Visual Basic interface, open the ThisDocument code window by double-clicking **ThisDocument** (or right-clicking on it and choosing **View Code**) in the **Project** window. Now, in that code window, *replace all of the existing code* with the following code. This code is run when the user clicks on the new tool button you created in ArcMap. All of the code for this example are available from the Code > Chapter 7 > ArcGIS link at http://www.geotree.uni.edu/SDSSbook.

TABLE 7.2
Control Names for the MySDSS Form

Label Figure 7.18	Control Name
A	cmb1
B	cmb2
C	cmb3
D	cmb4
E	cmb5
F	txt1
G	txt2
H	txt3
I	txt4
J	txt5

```
Option Explicit
Private pMap As IMap
Private Sub SDSS_Click()
    'this is code for when user clicks on SDSS tool button
    in ArcMap
  Dim pMxdoc As IMxDocument
  Set pMxdoc = ThisDocument

  Set pMap = pMxdoc.FocusMap

  Set frmSDSS.pMap = pMap
  Set frmSDSS.pMxdoc = pMxdoc

  frmSDSS.frmSDSS_Initialize
  'open the form
  frmSDSS.Show
End Sub
```

The code that will allow the user to carry out a WLC analysis by providing functionality to the form and controls created in Visual Basic must be copied in. To insert the following code, right-click in an empty part of the user form (MySDSS) in the Visual Basic interface and choose **View Code**. Now select all of the following code and paste it into the **frmSDSS** (Code) window *over any existing code in the window*. The code was written in Microsoft Visual Basic using ArcObjects in order to control interaction between the user and the controls on the form while also accessing objects from ArcMap. ArcObjects are the software components upon which ArcGIS software is built. ArcObjects allow the user to control interaction with all aspects of the ArcGIS software (e.g., maps, layers, layouts, and symbology) through programming. ArcObjects can be accessed using VBA and many other development environments and programming languages.

```
Option Explicit

Public pMap As IMap
Public pMxdoc As IMxDocument

Public Sub frmSDSS_Initialize()
  'add layer names from map to comboboxes by calling
  function
  AddRasterLayerFromMapToComboBox cmb1, pMap
  AddRasterLayerFromMapToComboBox cmb2, pMap
  AddRasterLayerFromMapToComboBox cmb3, pMap
  AddRasterLayerFromMapToComboBox cmb4, pMap
  AddRasterLayerFromMapToComboBox cmb5, pMap

End Sub
'code that runs when user clicks Analyze button on the WLC
form
Public Sub cmdAnalyze_Click()
  If dataCheck() Then
    'get input
    Dim RawInputRaster As New Collection
    GetInputRaster RawInputRaster, pMap

    Dim intWeight As New Collection
    getWeight intWeight

    Dim weightedInput As New Collection
    WeightInputs RawInputRaster, intWeight, weightedInput

    Dim pEnv As IRasterAnalysisEnvironment
    Set pEnv = New RasterSettings 'RasterAnalysis

    Dim pmathop As IMathOp
    Set pmathop = New RasterMathOps
    Set pEnv = pmathop
    'carry out the WLC calculation
    Dim pInter As Raster
    Set pInter = weightedInput(1)
    Dim a As Integer
    For a = 2 To weightedInput.Count
      Set pInter = pmathop.Plus(pInter, weightedInput(a))
    Next
```

```
    Dim pcolor1 As IColor
    Set pcolor1 = New RgbColor
    pcolor1.RGB = RGB(255, 204, 204)
    Dim pcolor2 As IColor
    Set pcolor2 = New RgbColor
    pcolor2.RGB = RGB(219, 0, 0)
    'set up symbology for the output data
    AddRasterWithStretchColorRampRender pInter, "output",
    pcolor1, pcolor2, pMxdoc

    Unload Me
  End If
End Sub
'close form
Private Sub cmdCancel_Click()
  Unload Me
End Sub
'set up weight rasters
Private Sub WeightInputs(input1 As Collection, input2 As
Collection, output As Collection)
  Dim a As Integer
  Dim inRaster As Raster

  For a = 1 To input1.Count
    Set inRaster = TimesRaster(input1(a), CInt(input2(a)))
    output.Add inRaster
  Next
End Sub
'do the raster algebra multiplication
Private Function TimesRaster(pRaster As IRaster, iConstant
As Integer) As IRaster
  On Error GoTo erh
  'Create a RasterMapAlgebraOp operator
  Dim pMapAlgebraOp As IMapAlgebraOp
  Set pMapAlgebraOp = New RasterMapAlgebraOp

  'Bind a raster
  pMapAlgebraOp.BindRaster pRaster, "R1"
  Set TimesRaster = pMapAlgebraOp.Execute("([R1]) * " &
  CStr(iConstant))

  Set pMapAlgebraOp = Nothing
  Exit Function
erh:
  MsgBox "TimesRaster has problem: " & Err.Description
End Function
'get raster layers from form
```

```
Private Sub GetInputRaster(input1 As Collection, pMap As
IMap)
  Dim pInput1 As IRaster
  Set pInput1 = GetRasterLayer(cmb1.Text, pMap).Raster
  input1.Add pInput1

  Dim pInput2 As IRaster
  Set pInput2 = GetRasterLayer(cmb2.Text, pMap).Raster
  input1.Add pInput2

  Dim pInput3 As IRaster
  Set pInput3 = GetRasterLayer(cmb3.Text, pMap).Raster
  input1.Add pInput3

  Dim pInput4 As IRaster
  Set pInput4 = GetRasterLayer(cmb4.Text, pMap).Raster
  input1.Add pInput4

  Dim pInput5 As IRaster
  Set pInput5 = GetRasterLayer(cmb5.Text, pMap).Raster
  input1.Add pInput5
End Sub
'get weights from form
Private Sub getWeight(input2 As Collection)
  input2.Add CInt(txt1.Text)
  input2.Add CInt(txt2.Text)
  input2.Add CInt(txt3.Text)
  input2.Add CInt(txt4.Text)
  input2.Add CInt(txt5.Text)
End Sub
'function called to get raster layer names
Private Sub AddRasterLayerFromMapToComboBox(cboBox As
ComboBox, m_map As IMap)
  On Error GoTo ErrorHandler
  Dim iLyrCount As Integer
  Dim iLyrIndex As Double
  Dim pLyr As ILayer
  Dim iGroupLyrCount As Integer
  Dim pCompositeLayer As ICompositeLayer
  Dim pLyrInG As ILayer

  iLyrCount = m_map.LayerCount

  If iLyrCount > 0 Then
    For iLyrIndex = 0 To iLyrCount - 1
      Set pLyr = m_map.Layer(iLyrIndex)
      If (TypeOf pLyr Is IRasterLayer) Then
        cboBox.AddItem pLyr.Name
```

```
      'this part consider group layer
      ElseIf (TypeOf pLyr Is ICompositeLayer) Then
        Set pCompositeLayer = pLyr
        For iGroupLyrCount = 0 To pCompositeLayer.Count - 1
          Set pLyrInG = pCompositeLayer.Layer(iGroupLyrCount)
          If (TypeOf pLyrInG Is IRasterLayer) Then
            cboBox.AddItem pLyrInG.Name
          End If
          Set pLyrInG = Nothing
        Next iGroupLyrCount
        Set pCompositeLayer = Nothing
      End If
      Set pLyr = Nothing
      Next iLyrIndex
    End If

    Set pLyr = Nothing
    Set pLyrInG = Nothing
    Set pCompositeLayer = Nothing
    Exit Sub
ErrorHandler:
  MsgBox "AddRasterLayerFromMapToComboBox has error " & Err.
  Description
End Sub
'get the raster names
Private Function GetRasterLayer(sName As String, m_map As
IMap) As IRasterLayer
  On Error GoTo erh
  Dim LyrCount As Integer
  Dim pLayer As ILayer
  Dim pCompositeLayer As ICompositeLayer
  Dim pLyrInG As ILayer
  Dim sLayerName As String
  Dim i As Integer
  Dim j As Integer

  LyrCount = m_map.LayerCount
  If LyrCount <> 0 Then
    For i = 0 To LyrCount - 1
      Set pLayer = m_map.Layer(i)
      If TypeOf pLayer Is IRasterLayer Then
        sLayerName = pLayer.Name
        If (StrComp(sName, sLayerName, vbTextCompare) = 0)
        Then
          Set GetRasterLayer = pLayer
          Exit Function
        End If
      ElseIf (TypeOf pLayer Is ICompositeLayer) Then
```

```
        Set pCompositeLayer = pLayer
        For j = 0 To pCompositeLayer.Count - 1
          Set pLyrInG = pCompositeLayer.Layer(j)
          If TypeOf pLyrInG Is IRasterLayer Then
            sLayerName = pLyrInG.Name
            If (StrComp(sName, sLayerName, vbTextCompare) = 0)
            Then
              Set GetRasterLayer = pLyrInG
              Exit Function
            End If
          End If
          Set pLyrInG = Nothing
        Next j
        Set pCompositeLayer = Nothing
      End If
      Set pLayer = Nothing
      Next i
    Else
    Set GetRasterLayer = Nothing
  End If

  Set pLayer = Nothing
  Set pCompositeLayer = Nothing
  Set pLyrInG = Nothing
  Exit Function
erh:
  MsgBox "Failed in getting raster layer by name:" & Err.
  Description
End Function
'function for setting up symbology for output of WLC
Private Sub AddRasterWithStretchColorRampRender(pRaster As
IRaster, pName As String, pfromcolor As IColor, pToColor As
IColor, pMxdoc As IMxDocument)

  'Create renderer and QI RasterRenderer
  Dim pStretchRen As IRasterStretchColorRampRenderer
  Set pStretchRen = New RasterStretchColorRampRenderer
  Dim pRasRen As IRasterRenderer
  Set pRasRen = pStretchRen

  'Set raster for the renderer and update
  Set pRasRen.Raster = pRaster
  pRasRen.Update

  'Create color ramp
  Dim pRamp As IAlgorithmicColorRamp
  Set pRamp = New AlgorithmicColorRamp
  pRamp.size = 255
```

```
  pRamp.FromColor = pfromcolor
  pRamp.ToColor = pToColor
  pRamp.CreateRamp True

  'Plug this colorramp into renderer and select a band
  pStretchRen.BandIndex = 0
  pStretchRen.ColorRamp = pRamp

  'Update the renderer with new settings and plug into layer
  pRasRen.Update

  Dim pLayer As IRasterLayer
  Set pLayer = New RasterLayer
  pLayer.CreateFromRaster pRaster
  pLayer.Name = pName
  Set pLayer.Renderer = pStretchRen
  pMxdoc.FocusMap.AddLayer pLayer

  pMxdoc.ActiveView.Refresh
  pMxdoc.UpdateContents

  'Release memeory
  Set pLayer = Nothing
  Set pStretchRen = Nothing
  Set pRasRen = Nothing
  Set pRamp = Nothing
End Sub
'check that parameters are ok after user clicks Analyze
button
Private Function dataCheck() As Boolean
  On Error GoTo erh
  If txt1.Text = "" Or txt2.Text = "" Or txt3.Text = "" Or
  txt4.Text = "" Or txt5.Text = "" Then
    MsgBox "A weight is missing"
    dataCheck = False
    Exit Function
  ElseIf cmb1.Text = "" Or cmb2.Text = "" Or cmb3.Text = ""
  Or cmb4.Text = "" Or cmb5.Text = "" Then
    MsgBox "A input layer is missing"
    dataCheck = False
    Exit Function
  ElseIf (CInt(txt1.Text) + CInt(txt2.Text) + CInt(txt3.
  Text) + CInt(txt4.Text) + CInt(txt5.Text)) <> 100 Then
    MsgBox "your total weight is not equal to 100"
    dataCheck = False
    Exit Function
  End If
  dataCheck = True
```

```
  Exit Function
erh:
    MsgBox "data checking has problem: " & Err.Description
End Function
'check user is entering numbers in text boxes
Private Sub txt1_KeyPress(ByVal KeyAscii As MSForms.
ReturnInteger)
  If IsNumeric(Chr(KeyAscii)) = False And Not KeyAscii = 8
  And Not KeyAscii = 9 Then
    MsgBox "Only numeric numbers is accepted"
    KeyAscii = 0
    txt1.SetFocus
  End If
End Sub

Private Sub txt2_KeyPress(ByVal KeyAscii As MSForms.
ReturnInteger)
  If IsNumeric(Chr(KeyAscii)) = False And Not KeyAscii = 8
  And Not KeyAscii = 9 Then
    MsgBox "Only numeric numbers is accepted"
    KeyAscii = 0
    txt2.SetFocus
  End If
End Sub

Private Sub txt3_KeyPress(ByVal KeyAscii As MSForms.
ReturnInteger)
  If IsNumeric(Chr(KeyAscii)) = False And Not KeyAscii = 8
  And Not KeyAscii = 9 Then
    MsgBox "Only numeric numbers is accepted"
    KeyAscii = 0
    txt3.SetFocus
  End If
End Sub

Private Sub txt4_KeyPress(ByVal KeyAscii As
MSForms.ReturnInteger)
  If IsNumeric(Chr(KeyAscii)) = False And Not KeyAscii = 8
  And Not KeyAscii = 9 Then
    MsgBox "Only numeric numbers is accepted"
    KeyAscii = 0
    txt4.SetFocus
  End If
End Sub

Private Sub txt5_KeyPress(ByVal KeyAscii As MSForms.
ReturnInteger)
```

```
  If IsNumeric(Chr(KeyAscii)) = False And Not KeyAscii = 8
  And Not KeyAscii = 9 Then
    MsgBox "Only numeric numbers is accepted"
    KeyAscii = 0
    txt5.SetFocus
  End If
End Sub
```

At this point, the tool is generic enough to be used for any multi-criteria evaluation that involves simple additive weighting. For demonstration, however, the tool will be used to examine environmentally sensitive areas based on user-specified weights. The study area is the Dry Run Creek watershed in Black Hawk County, Iowa. The following layers (Figure 7.19) are used to demonstrate the functionality of the tool: slope, hydrologic soil group, green space, floodplain areas, and a buffer of the impervious areas derived from land cover data. To run the tool, the user should add each of these raster layers into the ArcMap interface (downloadable from the Data > Chapter 7 > ArcGIS SDSS link at the Web site that accompanies this book: http://www.geotree.uni.edu/SDSSbook). The user should then click the **SDSS** button that was added to the new toolbar earlier. When this button is clicked, the MySDSS form should open. The user can then choose the layers to include in the analysis by selecting one from each of the combo boxes (Figure 7.20). The user must specify the importance or weighting of each of the environmental layers. The weighting is expressed in this case as a percentage. The five percentages must add up to 100. When all the layers are chosen and weights set, the user should click the **Analyze** button and the analysis will be carried out. The environmentally sensitive areas are then defined and output into a layer for viewing in ArcMap. Figures 7.20 through 7.22 demonstrate different weightings and the effect on output. Figure 7.20 shows equal weighting, Figure 7.21 displays the results of excluding a variable (by setting % equal to zero) from the analysis, and Figure 7.22 shows a varied weighting scheme involving all of the variables.

The goal of this exercise has been to demonstrate the development and application of a simple ArcGIS-based SDSS utilizing WLC. This type of simple ArcGIS-based SDSS provides a simple example on which the reader could build. This example SDSS was presented as very basic for demonstration purposes. The real system being developed for more than demonstration purposes had more sophisticated user interfaces, help utilities, and also significantly more code to control for potential errors.

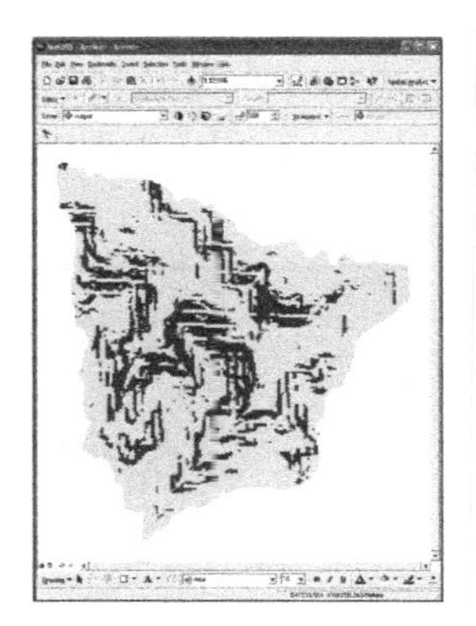

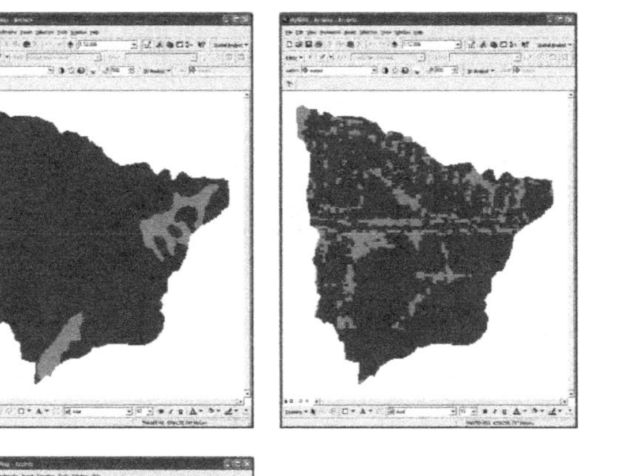

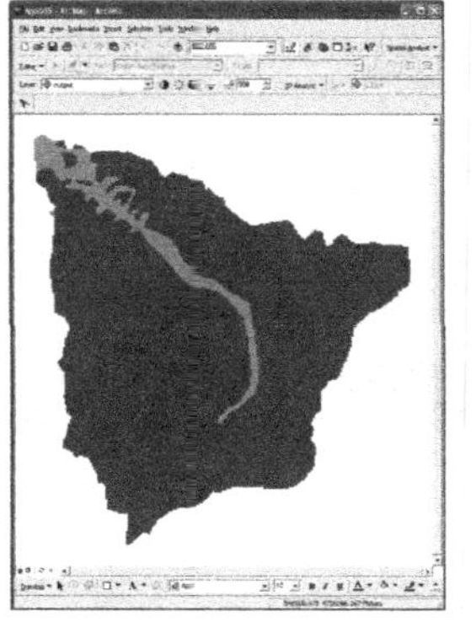

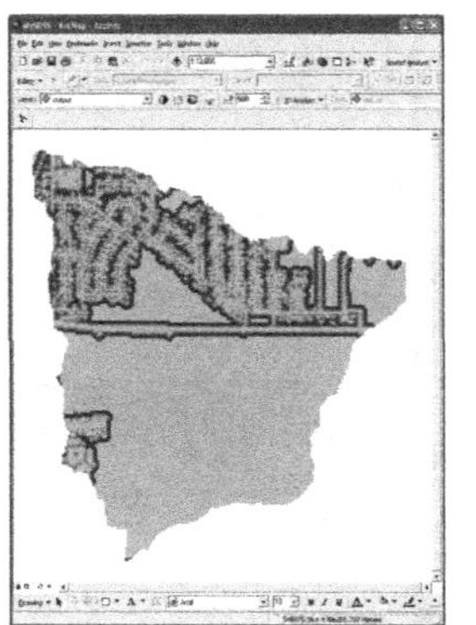

FIGURE 7.19
(See color insert following page 74.) Original layers included in the WLC analysis of environmentally sensitive areas. From top left to bottom right: slope, hydrological soil group, green space, FEMA 100-year floodplain, and buffered impervious areas.

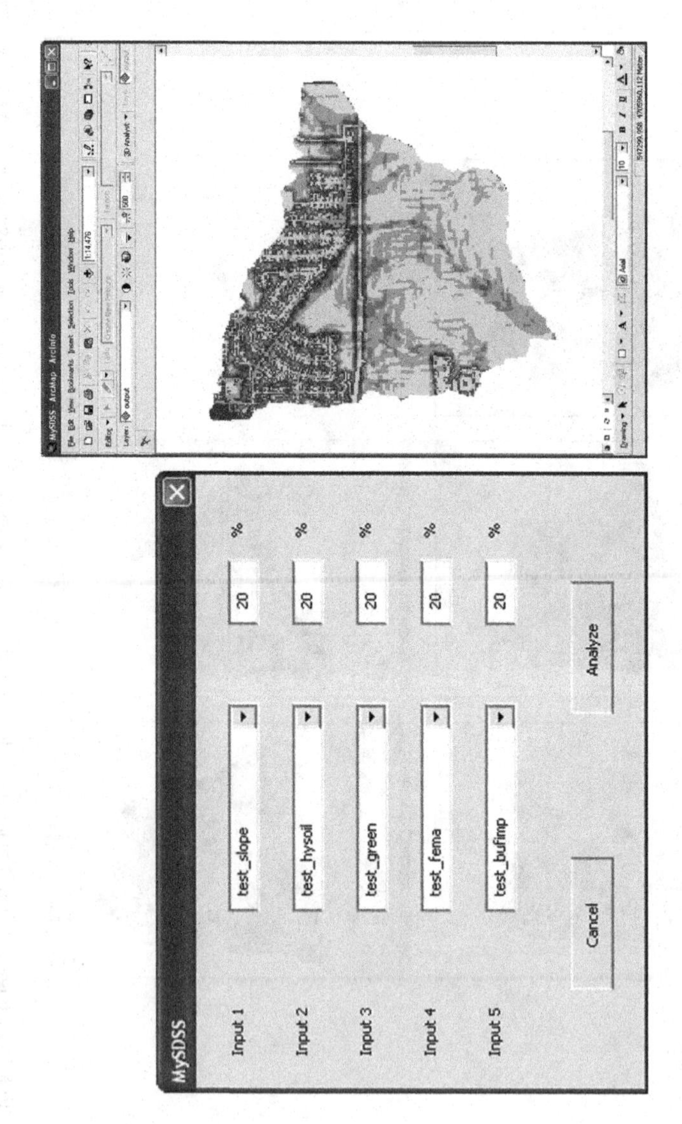

FIGURE 7.20
A WLC analysis with equal weighting for all layers.

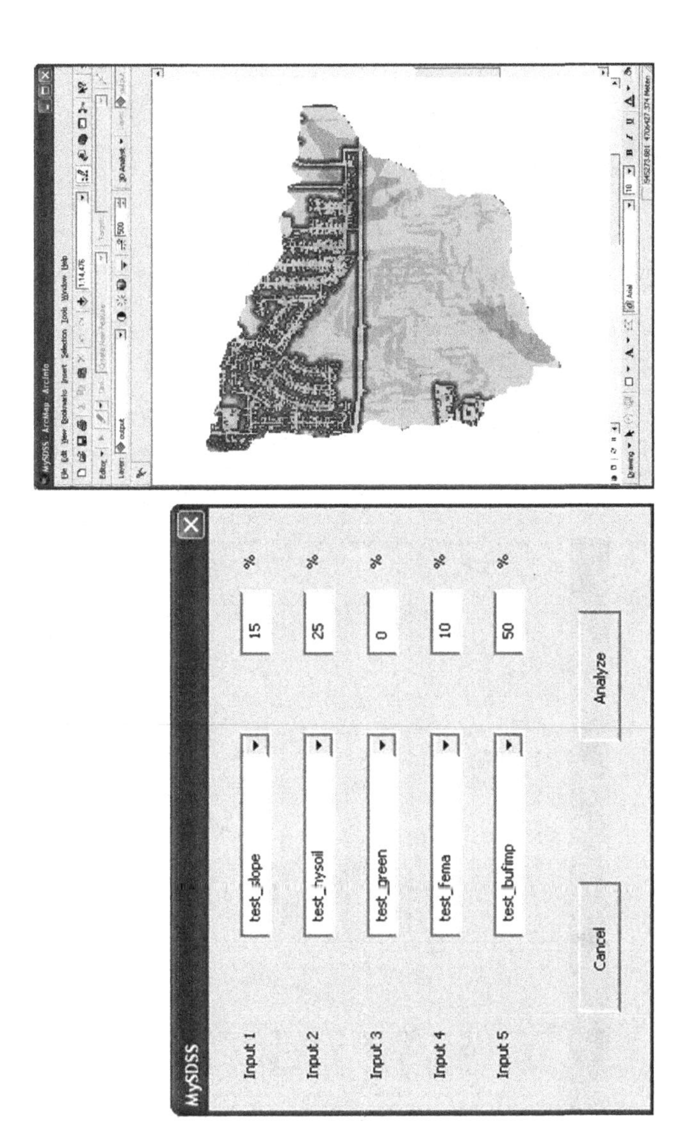

FIGURE 7.21
A WLC analysis. The green space layer is excluded by giving it a weight of zero.

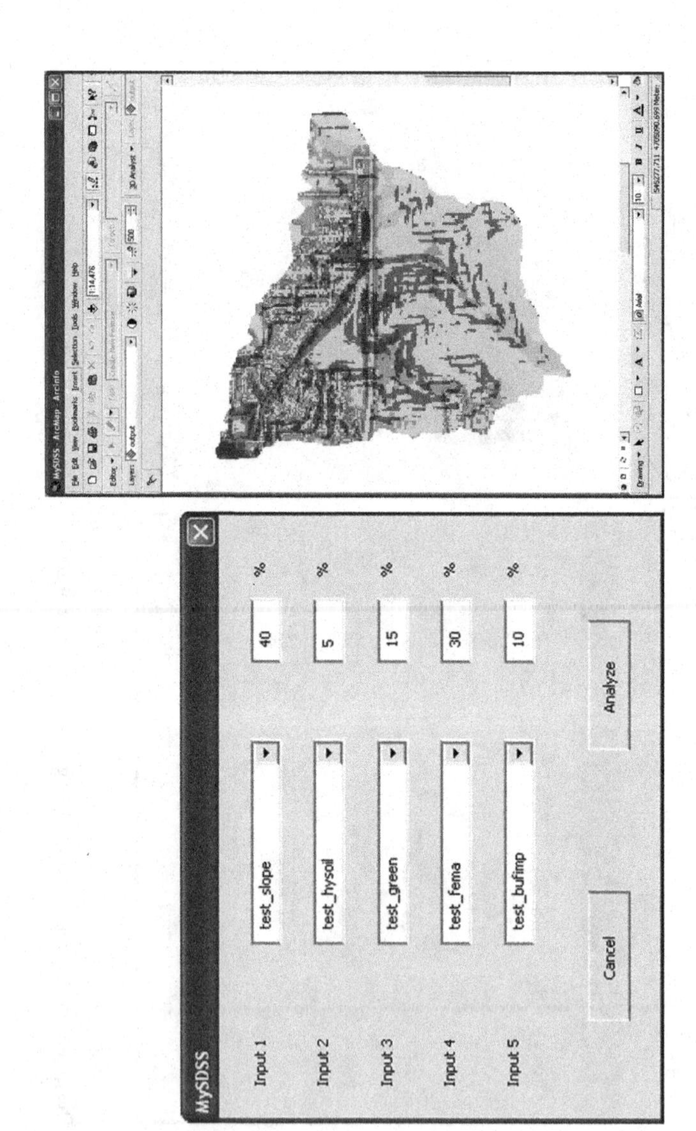

FIGURE 7.22
A WLC analysis. The slope layer is emphasized and the soils layer is given the lowest weight.

7.4.4 Creation of a New Generic SDSS Program

This section will describe the characteristics, development process, and functionality of an SDSS application called OpenSDSS. This software is undergoing development presently by the authors of the book and colleagues. This software was created with the following goals in mind: open source and free, generic (unrestricted), model-oriented, flexible, and with extensive support for graphical modeling and end user customization. It also allows for the inclusion of modules from open source software such as the Integrated Land and Water Information System (ILWIS) and the System for Automated Geoscientific Analyses (SAGA). Some of the features of OpenSDSS include the generation, storage, and organization of scenarios, sensitivity analyses, as well as the aggregation and ranking of alternatives. The extensive capabilities for graphical modeling supported by OpenSDSS allow for a high degree of end user participation. A large set of decision support models are included. The framework of OpenSDSS includes the following elements: the user interface, data, models, scenarios, and graphical modeling. Some of these elements have wide definitions (e.g., a data object can represent a spatial dataset or a number, a model can be a stand-alone executable file or a shared function from a software library, and scenarios are obtained as a flexible combination of data, models, and the interactions between them).

Figure 7.23 demonstrates the flow of components in the OpenSDSS system. The model environment stores frequently used file system paths to data, models, user interface settings, and for the remaining customizable software options. The component manager is necessary in order to organize

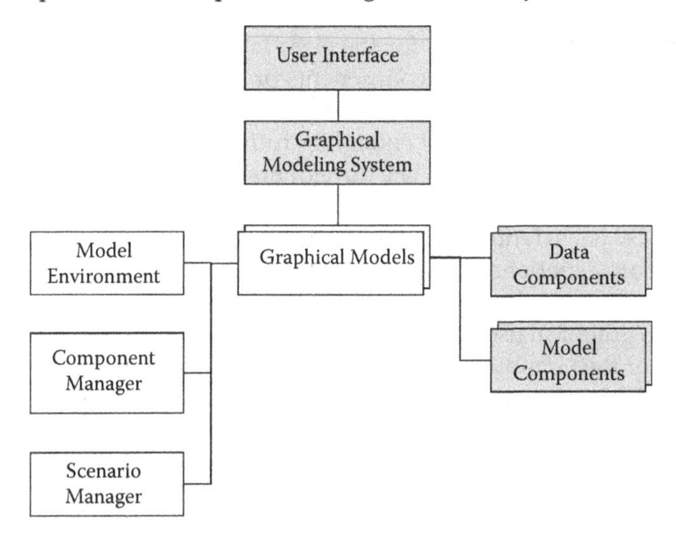

FIGURE 7.23
The basic design of OpenSDSS.

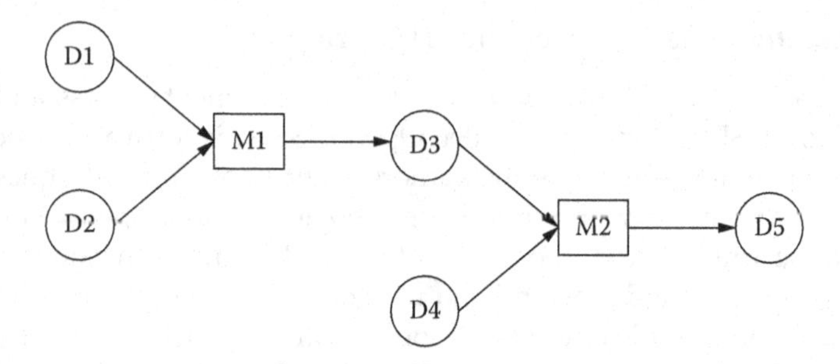

FIGURE 7.24
Hypothetical abstract model related to the scenario manager.

the insertion and deletion of components from the graphical model. It keeps track of the various connections between the model elements, their movements, and user interactions. Finally, the scenario manager oversees different decision-making scenarios. These scenarios are not separate from the graphical models themselves. An example of an abstract model is shown in Figure 7.24. This model has five datasets (shown as circles) and two models (shown as squares). The first model (M1) takes datasets 1 and 2 (D1 and D2) as inputs and stores its output to the third dataset (D3). As the model M1 is executed, it appends a special record to the dataset D3 that outlines its origins. In this case, the record would look like this: $M1_{D1, D3}$. The same record for the fifth dataset, D5, would look like this: $M2_{M1, D4}$. If we think of the scenario as an account of a course of action taken by the model, we will see that such records can be used to describe, store, and replay the decision-making scenarios in a clear and easy-to-understand fashion.

Graphical modeling allows for the clear layout of the decision-making process. Advanced modeling components provide additional control over the flow of the model execution. To date, only a few systems could possibly qualify as being advanced enough to fulfill the requirements of being generic SDSS (see Chapter 5). The developers of OpenSDSS hope that this software will qualify as a truly generic SDSS.

OpenSDSS is currently a prototype, but upon completion will be quite useful to both experienced and inexperienced SDSS users. Some of the key technologies used in this project are GTK+ (a widget toolkit used to create the OpenSDSS graphical user interface), GooCanvas (a canvas widget that is the backbone of the Graphical Modeling System), Geospatial Data Abstraction Library (GDAL; a raster geospatial data library used in some of the models), and the Python programming language, which binds all of the components together. The OpenSDSS interface is demonstrated in Figure 7.25.

The main application window consists of a menu bar, a toolbar, and a canvas area. The menu bar provides access to the import and export functions of the graphical models, allows the addition and removal of datasets and models, and provides tools for output analysis and evaluation.

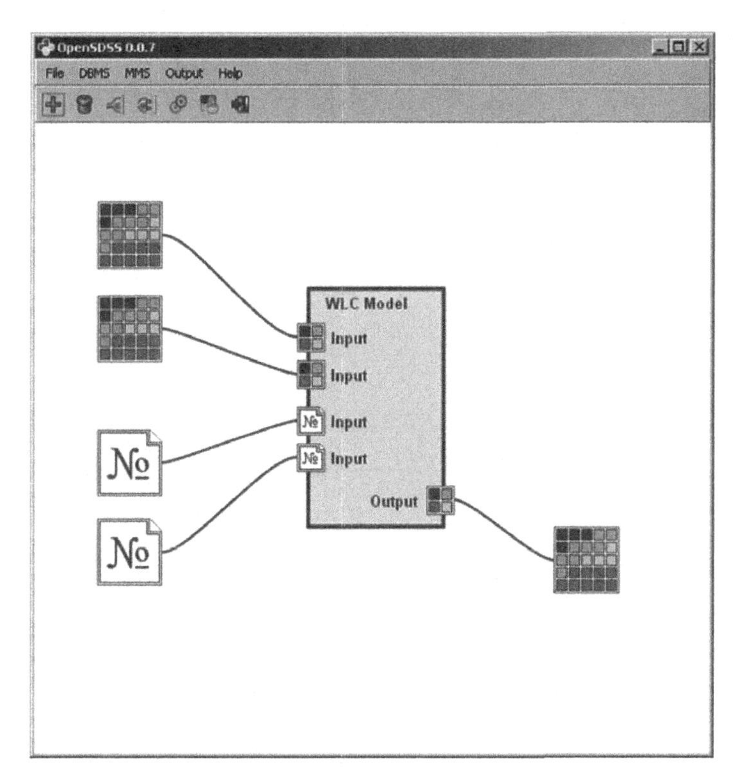

FIGURE 7.25
Prototype OpenSDSS interface.

The toolbar contains graphical shortcuts to some of the functions available through the menu bar and those most often used in graphical models design, such as connection and disconnection of different items, execution of the model, and so forth. The canvas area is the part of the interface where the graphical models are created. Different components of the graphical model can be freely positioned on the canvas, connected in the desired way, and later executed. In the top of Figure 7.26, a Weighted Linear Combination model is set up that accepts two raster layers with corresponding weights.

Some of the advanced modeling capabilities are shown in Figure 7.26. In this case, the output of two separate models is used as input for a second level of WLC. One of the weights of the final WLC model is held constant, while the other one (inside dashed box) is iterated over with three different values, providing for simple sensitivity analysis. The resulting scenarios are analyzed with a histogram component that shows the distribution of raster values across different categories.

In an example, real-world data is incorporated into OpenSDSS to exemplify the use of this system for spatial decision support analyses. Four

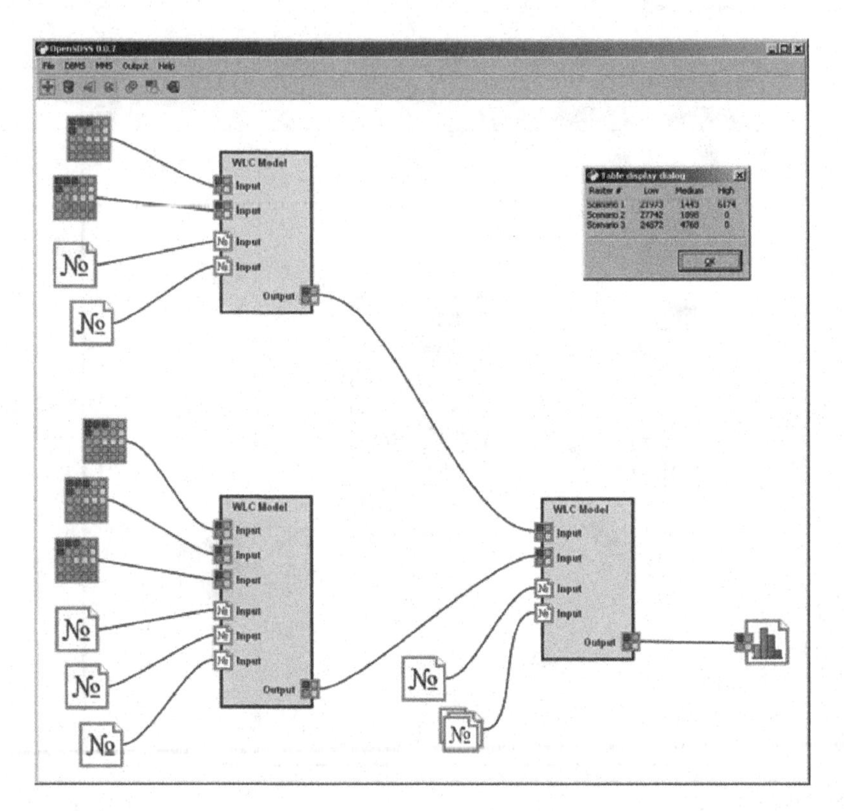

FIGURE 7.26
(See color insert following page 74.) An example of a more complex model in OpenSDSS.

spatial data layers are investigated that describe land use, parcel distribution, population change, and school locations for the city of Cedar Falls, Iowa. All data is transformed to address the issue of placing a new school. Land use is ranked based on the suitability for new building construction, parcels are classified according to their size, and spatial buffers are created around the existing schools to enforce even distribution of the facilities (Figure 7.27).

A simple WLC model was built to perform weighted overlay using the spatial data mentioned previously. The resulting model is presented in Figure 7.28.

All four layers used in the analysis for the school site suitability analysis were combined with numerical weight factors associated with each dataset. One of the scenarios obtained is shown in Figure 7.29. As is evidenced by the analysis, there are only a handful of areas that would make a suitable location for a new school in the study area. The best location is the large white area in the southwest of the city. This occurred because the combination

FIGURE 7.27
The spatial layers used in OpenSDSS example: land use, parcel distribution, population change, and existing school locations (clockwise, starting from the upper left).

of spatial layers and weights agreed that this area is experiencing the largest population growth, contains many land use classes that are suitable for development (mainly agricultural land), is not located within the specified buffer distance of existing schools, and contains parcel sizes adequate for the construction of a school with its associated sporting fields (i.e., a track, football field,) and parking lots. This example and the description were meant to provide readers with an introduction to the development of a new generic SDSS-producing software called OpenSDSS. OpenSDSS is under development presently and will be made freely available in the near future. Check the Web site associated with this book for updates on OpenSDSS (http://www.geotree.uni.edu/SDSSbook).

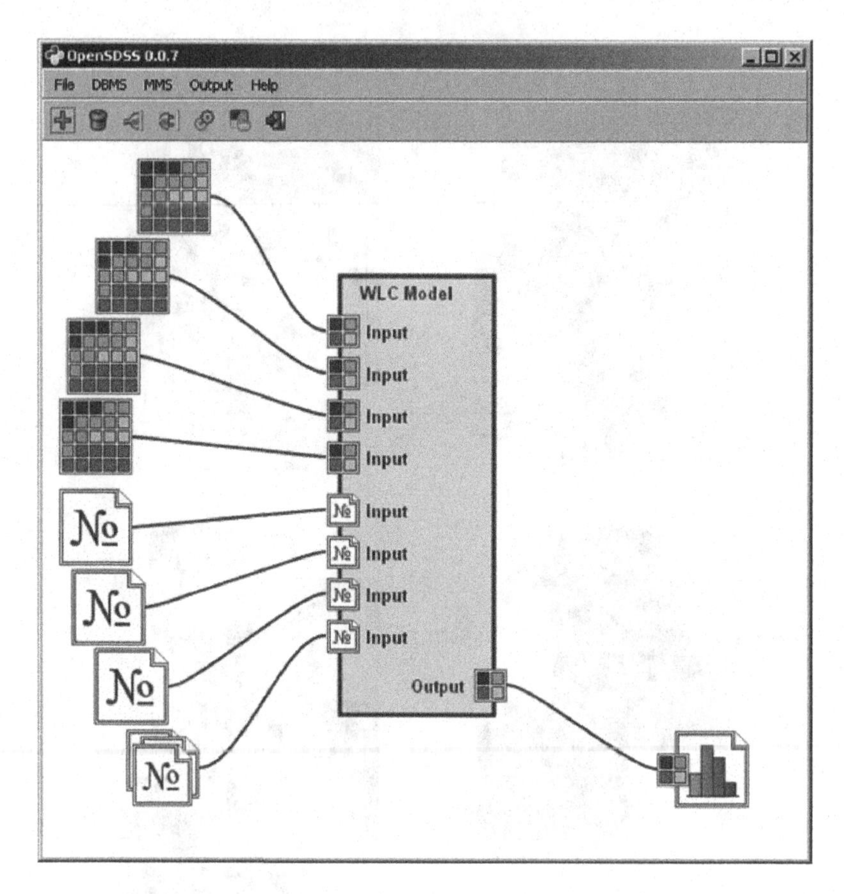

FIGURE 7.28
Example site suitability analysis using OpenSDSS.

7.5 Summary

In this chapter, we have attempted to accomplish three main goals:

- To provide an overview of some important considerations when developing a new SDSS
- To give an overview of the SDSS development process
- To provide several examples of actual SDSS software development

There are many considerations that must be taken into account when developing an SDSS. A multidisciplinary team is usually necessary. Thus, it is important to assemble the necessary project team at an early stage of the process. The development of a stakeholder group and their continued

FIGURE 7.29
The OpenSDSS WLC model output for school site suitability analysis. Suitability ranges from appropriate (white) to inappropriate (black).

involvement in the process is very important. Careful planning must take place. This process must take into account legitimate time and resource constraints and must also provide plans for how developed software will be updated and maintained in the future. We identified six major steps in the SDSS development process

- Problem definition and identification of stakeholders
- Requirements definition

- System design
- Development cycle
- Testing
- Final implementation

This process must be an iterative one in which a step is revisited after lessons are learned in subsequent steps. For example, the system design step should be revisited after the prototype development and testing stages. It is crucial that end users are included in this iterative development process along with the technical specialists (programmers, GIS, etc.) and domain specialists (scientists, planners).

We provided several examples of SDSS software applications. We provided all of the necessary instruction, code, and data to develop and run both a Microsoft Excel-based AHP/WLC SDSS and an ArcGIS-based WLC tool. The Microsoft Excel application provides not only a working AHP/WLC SDSS but also a useful template for developing other SDSS capabilities in a spreadsheet environment. The advantage of the Microsoft Excel platform is that it is less expensive and much more widely used than GIS software. This implies a greatly expanded potential audience for SDSS applications. That is the logic behind the development of SpreadsheetSDSS, which has more extensive generic SDSS functionality. This Microsoft Excel plug-in (currently under development by the authors) was introduced in this chapter, and more information concerning it can be found on the Web site that accompanies this book (http://www.geotree.uni.edu/SDSSbook). Steps for developing an ArcGIS-based SDSS with a WLC model were given in this chapter as well as data for carrying out an example application. Again, this example provides a working module but also a useful template for developing further SDSS functionality within ArcGIS. Finally, an overview of a new generic SDSS program (OpenSDSS) was provided. This software is under development presently and will eventually provide a freely available generic framework for developing specific SDSS.

References

Berardi, A. 2002. ASTROMOD: A computer program integrating vegetation dynamics modeling, environmental modeling and spatial data visualization in Microsoft Excel. *Environmental Modelling & Software* 17:403–412.

Hahn, B. 2005. The need for a science and knowledge integration. Paper presented at the Success and Failure of DSS for Integrated Water Resource Management Conference, Venice, Italy.

Hahn, B., G. Engelen, J. Berlekamp, and M. Matthies. 2005. Towards a generic tool for river basin management: IT framework report Elbe River Basin feasibility study—phase 4. Working paper, Institute of Environmental Systems Research. http://elise.bafg.de/servlet/is/3473/DSS_MBS_Report4.pdf (accessed January 29, 2010).

Li, Y., Q. Shen, and H. Li. 2004. Design of spatial decision support systems for property professionals using MapObjects and Excel. *Automation in Construction* 13:565–573.

Power, D. J. 2002. *Decision support systems: Concepts and resources for managers.* Westport, CT: Quorum Books.

Rutledge, D., M. Cameron, S. Elliott, T. Fenton, B Huser, G. McBride, et al. 2008. Choosing regional futures: Challenges and choices in building integrated models to support long-term regional planning in New Zealand. *Regional Science Policy & Practice* 1(1):85–108.

Şeref M. M. H., R. K. Ahuja, and W. L. Winston. 2007. *Developing spreadsheet-based decision support systems using Excel and VBA for Excel.* Belmont, MA: Dynamic Ideas.

Van Delden, H. 2009. Lessons learnt in the development, implementation, and use of integrated spatial decision support systems. Paper presented at the 18th World IMACS/MODSIM Congress, Cairns, Australia.

Veronica, R. R. 2007. Decision support systems development. *Journal of the Faculty of Economics* 2:882–885.

8

Building Web-Based SDSS

Learning Objectives

- Be introduced to Web-based SDSS including implementation considerations.
- Learn how to implement a Web-based SDSS using ArcGIS Server.
- Be exposed to an example of Web-based SDSS implementation based upon open source software.

8.1 Introduction

In Chapter 7, the implementation of desktop spatial decision support systems (SDSS) was discussed by presenting two examples of available SDSS programs and two examples of developing new custom SDSS tools in Microsoft Excel and ArcGIS. This chapter will focus on Web-based SDSS. Traditional desktop SDSS implementations often require expensive software and also powerful desktop computers. In addition, they also often rely upon existing GIS, modeling, and other software, thus calling upon users to have access to and a certain level of experience with these programs. These characteristics make it difficult for organizations to effectively place SDSS within their institutional context. With the tremendous growth in the use of the Web, there has been a move to develop both Web-based decision support systems DSS and SDSS. The Web is playing a huge role in SDSS application development mainly because of advantages such as platform independency, reductions in distribution costs and maintenance problems, ease of use, and widespread access (Peng and Tsou 2003). These advantages help explain the increase in Web-based SDSS applications over the last decade or so. In the database of SDSS publications collated and described in earlier chapters, there were approximately fifty

Web-based SDSS documented, with the earliest appearing in 1999. For more detail, Rinner and Jankowski (2002), Rinner (2003), and Sugumaran and Sugumaran (2007) have described technical foundations and applications of Web-based SDSS.

Although there are many advantages presented by using Web-based SDSS, there are also many unique technological challenges. Issues that can be of particular importance to Web-based SDSS include performance, integration of various technologies, security, and interoperability. A Web-based SDSS generally requires numerous components, including HTML user interfaces, Internet interface programs, computational models, and geographic databases (Sugumaran and Sugumaran 2007). At the most fundamental level, the distribution of functions between server (i.e., a distant computer that provides services) and client (i.e., the computer the user is operating, which consumes services from a server) needs to be decided. A server-side approach means that most of the processing takes place on the server while the client is used for gathering input from the user and presentation of an HTML-based interface including static maps. The client-side processing approach has functionality, including spatial processing, preloaded on the client machine while the geographic data is accessed from the server. A server-side Web-based SDSS only requires a browser as the spatial processing, modeling, database, and other SDSS components will be located on the server. However, frequent communication between client and server are necessary (Sugumaran and Sugumaran 2007). The server-side model is the more common approach. Web-based SDSS rely on a range of technologies such as JavaScript, Java-based applets and viewers, Web servers, map servers, geographical information system (GIS) servers, and others. The establishment of standards in relation to Web mapping and GIS services has facilitated the development of Web-based SDSS. Examples include the Open Geospatial Consortium (OGC) Geographic Markup Language (GML), which is a vendor-neutral format for storing geographic information, and the OGC Web services standard, which is meant to enable seamless integration of online geoprocessing and location services (Sugumaran and Sugumaran 2007). The full extent of technological issues in relation to Web-based SDSS cannot be given here. However, the two examples detailed here will provide some insight into the importance of some of these standards and technologies. Chapter 10 provides a discussion of some of the issues to be considered in relation specifically to Web-based SDSS.

The focus of this chapter is to provide two illustrative examples of Web-based SDSS implementation. The first provides step-by-step instructions for implementing a Web-based SDSS using the commercial ArcGIS Server software. The second example demonstrates how to implement a Web-based SDSS using open source Web mapping software called MapServer. MapServer is free and has been widely used in SDSS applications. ArcGIS

Server is capable of providing powerful GIS services through the Internet and will likely be used in a variety of SDSS applications in the future.

8.2 Web-Based SDSS Developed with ArcGIS Server

8.2.1 Web-Based SDSS for Environmentally Sensitive Areas

The goal of this example is to implement an environmentally sensitive areas analysis over the Web using ArcGIS Server. This example demonstrates the development of the same weighted linear combination (WLC) model that was shown for the ArcGIS-based example from Chapter 7. Both examples involve the same spatial processing on the same datasets and both use the same ArcObjects components. The difference lies in the software with which the user interacts and which software he or she is required to have. For the Web-based SDSS, the user only needs access to a Web browser (and Internet connectivity with sufficient bandwidth capabilities), whereas the desktop version requires both ArcGIS and the Spatial Analyst extension. Repeating the same example gives the reader a perspective of the issues involved in developing desktop versus Web-based SDSS tools. This Web-based example highlights the great potential of opening SDSS use to a larger and wider audience through Web-based architectures, but also some of the difficulties in doing so. Figure 8.1 presents the general architecture of the ArcGIS Server SDSS example.

In the text that follows, we present the steps necessary to construct a simple Web-based SDSS for identifying environmentally sensitive areas and to run that application. When finished, this application will run from your local machine. This is a Windows-based example that was created using ArcGIS Server 9.2 and with programming carried out in Visual Basic through the Visual Studio 2005 application development framework. It requires the developer to have ArcGIS Server 9.2, ArcGIS 9.2 desktop, and the Spatial Analyst extension. There would be minor but essential changes to run the example for ArcGIS Server 9.3 or later versions. This example also requires access to Visual Studio 2005 or a later version. This example will also require you to have administrative privileges on your machine. The example assumes significant experience in using ArcMap, the Spatial Analyst extension, and Visual Studio. In order to understand the code properly, experience in Visual Basic, ArcObjects, JavaScript, HTML, and ASP would all be beneficial. The steps needed to construct this Web-based SDSS are provided below. These steps attempt to provide enough information that a user with the proper access to and experience

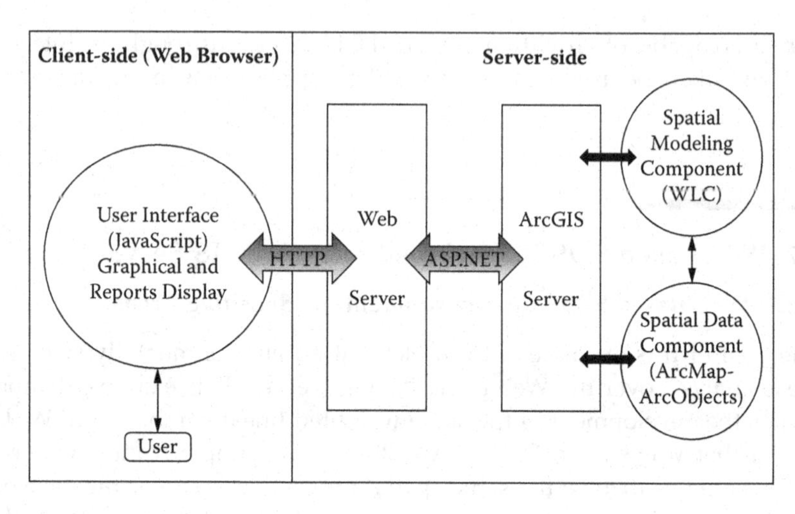

FIGURE 8.1
Description of the client-side and server-side arrangement for the Web-based SDSS with ArcGIS Server example provided in this chapter.

using the given software could reproduce the application. Specifically, the steps assume familiarity with ArcMap, ArcCatalog, and Visual Studio.

Set up the ArcMap document: The initial stage of this process is to create a map composition in ArcMap, which will be used in the ArcGIS Server application. Open **ArcMap** and add the six input datasets to the map. These datasets include a shapefile representing the watershed boundary and five raster datasets stored as Environmental Systems Research Institute (ESRI) grids: FEMA (Federal Emergency Management Agency) floodplain, slope, hydric soils, impervious areas, and green space derived from land cover data. These data are available from the http://www.geotree.uni.edu/SDSSbook Web site. The names of the datasets are ws_boundary.shp, test_fema, test_slope, test_hysoil, test_bufimp, and test_green. Make sure you have downloaded these raster datasets and the shapefile to your local machine and add them to the ArcMap document. Rearrange the layers in the table of contents so the order matches the way you would like them to appear in your ArcGIS Server application. Also, save the ArcMap document to your local machine to a directory of your choice. Save it as **WebSDSS.mxd**.

We need to ensure that the Spatial Analyst extension is enabled (choose **Tools > Extensions >** and check **Spatial Analyst**). If not already available in the ArcMap interface, activate the Spatial Analyst extension (choose **View > Toolbars > Spatial Analyst**). Open the **Spatial Analyst Options** dialog (choose **Spatial Analyst > Options**). Under the **General** tab, click the drop-down arrow for **Analysis Mask** and choose the **ws_boundary** layer. Next, activate the **Extent** tab and in the **Analysis extent:** drop-down box, choose **Same as Layer ws_boundary**. Finally, select the **Cell Size** tab and choose **Same as Layer test_slope** in the drop-down box for **Analysis cell size:**. Click **OK** to close the Options dialog.

Create the Map Service: Open the **ArcCatalog** application. Within ArcCatalog, in the **Catalog Tree** panel, expand the **GIS Servers** category. This will open a list that should include your local machine. If you don't see the name of your local machine in the list, then double-click on **Add ArcGIS Server**, select **Manage GIS Services** in the dialog, and then select **Next**. Then type **http://localhost** in the **Server URL:** field and type **localhost** in the **Host Name:** text box. After you click **Finish**, you should see that your local machine is available (should say localhost(admin)). Now, right-click on your machine name under **GIS Servers** and choose **Add New Service**. In the **Add GIS Service** dialog, select **Map Service** as the **Type** and give the SDSS a name (e.g., WebSDSS1) and a description (e.g., "This is a WLC SDSS for environmentally sensitive areas delineation."). Click **Next**.

In the next step of the **Add GIS Service** dialog, click **Browse** next to the **Map Document:** text box. Navigate to and choose the recently created .mxd file (WebSDSS.mxd). Leave the other settings as they are and click **Next**. Leave all parameters as they are in next dialog and click **Next**. At the dialog that has the **Pooling** frame, select **Not pooled…**; leave the minimum instance at 1 and the maximum number of instances at 2. Click **Next**. In the last **Add GIS Service** dialog, review the parameters and then select **Yes, start the service right now**. Click **Finish**.

Creating the Web site: Open **Visual Studio** and use the following steps to create your Web site. In Visual Studio (Figure 8.2a), under the **Recent Projects** frame (Figure 8.2b), click on **Web Site** next to where it says **Create:**. In the **New Web Site** dialog, choose **Web Mapping Application**. Choose **HTTP** in the **Location:** drop-down list. Next, in the drop-down box next to Location:, make sure it says **http://localhost/WebSDSS** (see Figure 8.3). Type it in if necessary. Also make sure that **Visual Basic** is the language chosen in the **Language:** combo box. Click **OK** to close the dialog.

Visual Studio will carry out some processing (setting up the Web site based on a template, etc.). When this processing is done, open the **Default.aspx** file by double-clicking on it in the **Solution Explore** window. If the file opens as code, switch it to Design view by choosing the **View** menu and then **Designer**. The layout may not look exactly like that shown in Figure 8.4 but will be similar.

In the design view of the Default.aspx file, single-click on the **MapResourceManager** control to make it active. Next, in the **Properties** window, navigate down to the **Resources > ResourceItems** property and click on the browse button (the button with ellipses […] in Figure 8.5a). In the **MapResourceItem Collection Editor** dialog (Figure 8.5b), click the

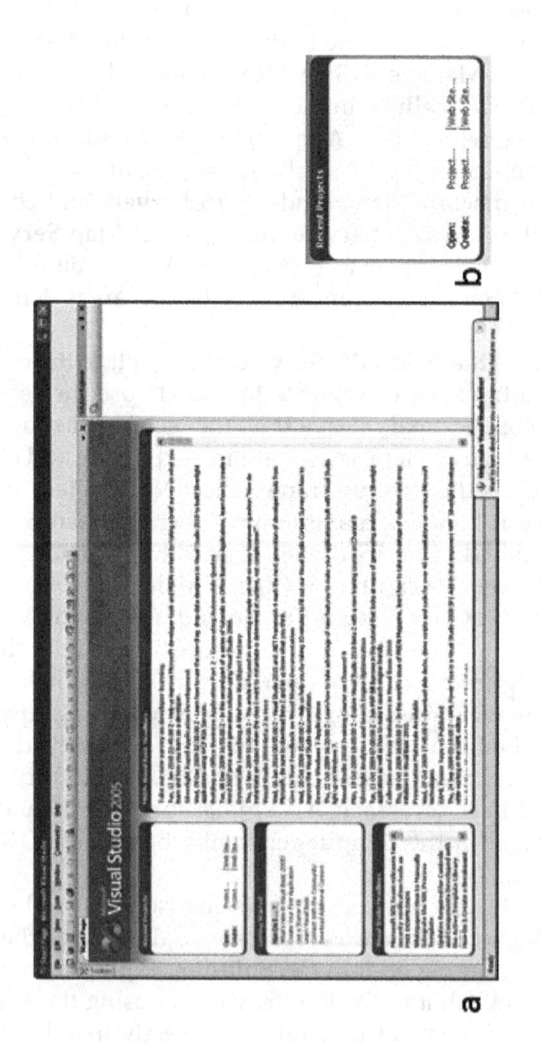

FIGURE 8.2
(a) The Visual Studio 2005 interface and (b) the Recent Projects frame.

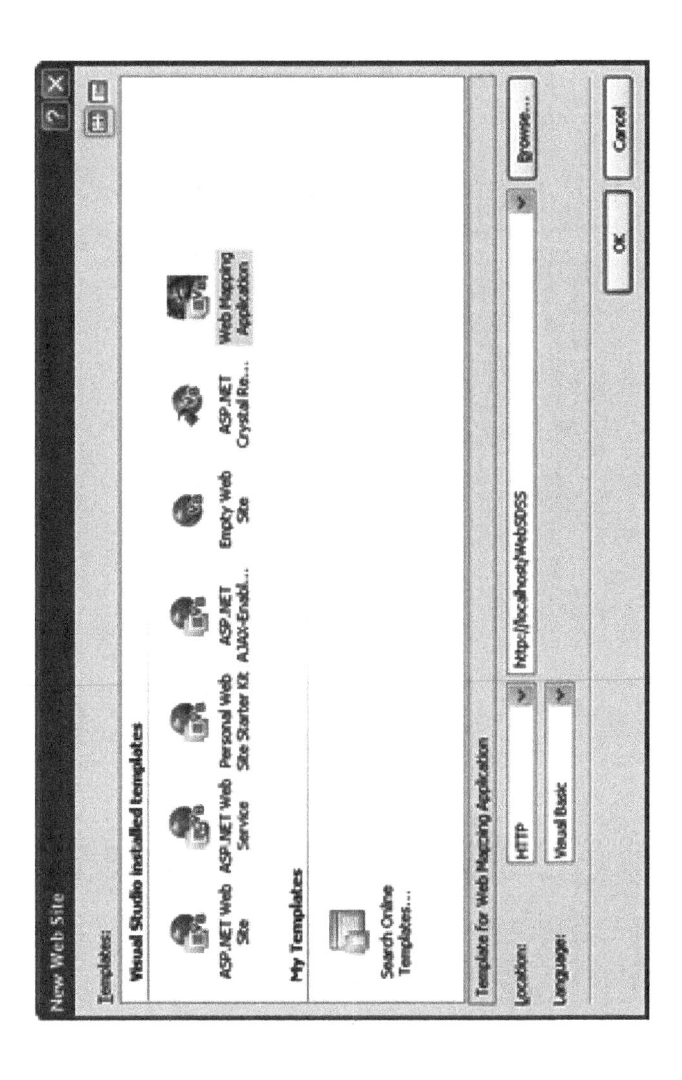

FIGURE 8.3
The dialog for creating a new Web mapping application Web site in Visual Studio.

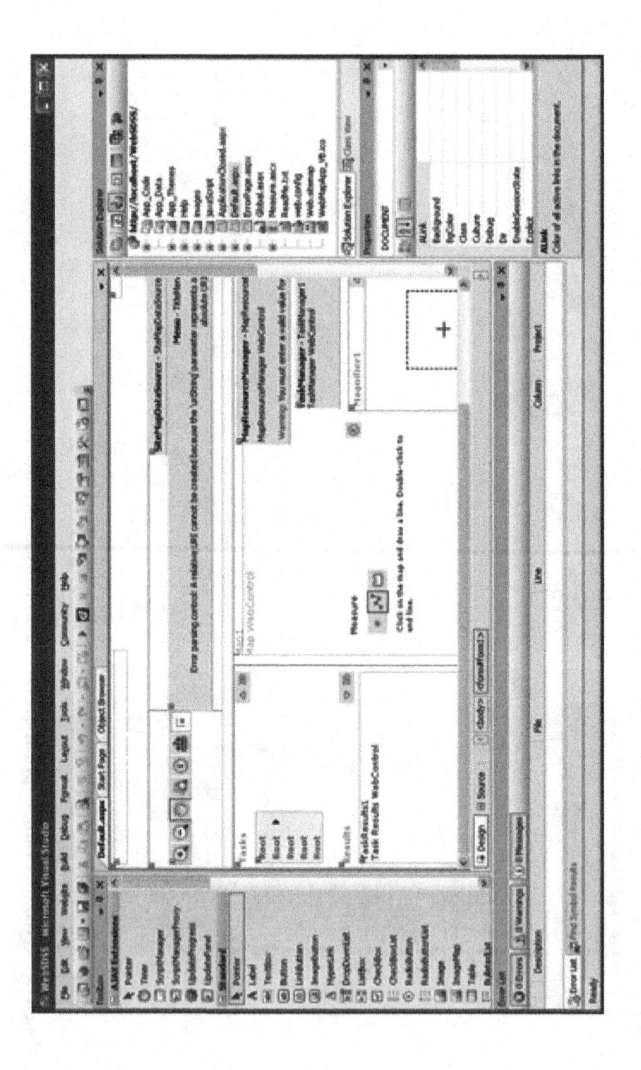

FIGURE 8.4
The Default.aspx file in the Design view in Visual Studio.

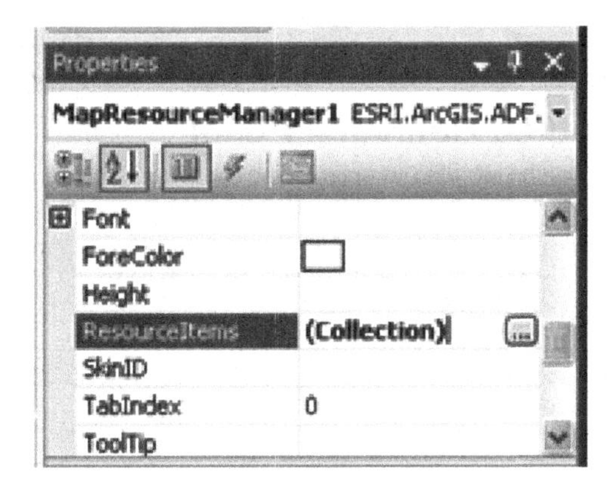

FIGURE 8.5a
The Properties window with the ResourceItems property selected.

Add button to add a new item under the **Members:** frame. Select the new item and then change the **Name** property in the right frame to **WebSDSS**. Then click on the **Definition** property and you will see a new browse button (button with [...]) appear. Click on that button to bring up the **Map Resource Definition Editor** (Figure 8.5c). Define the parameters as follows: **Type** = **ArcGIS Server Local, Data Source** = the name of your computer (in the case of the computer on which this example was prepared = geotree240-8); and **Resource** = **WebSDSS1** (this is set by clicking the browse button to the right of the textbox to open the **ArcGIS Resources Definition Editor** form (Figure 8.5d) and choosing the name of the GIS Service that

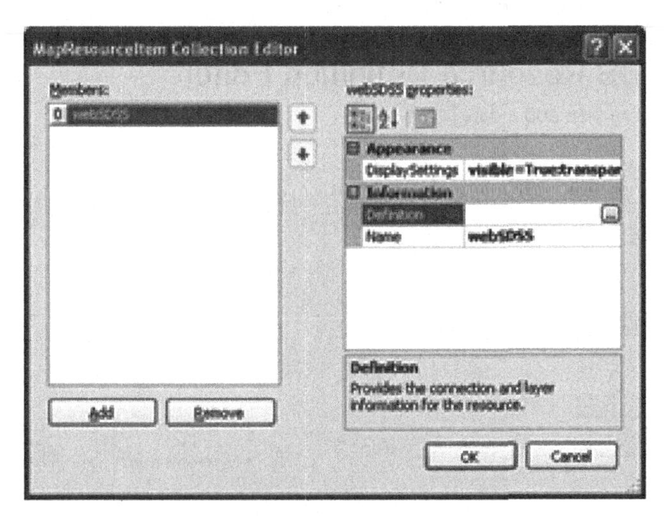

FIGURE 8.5b
The MapResourceItem Collection Editor.

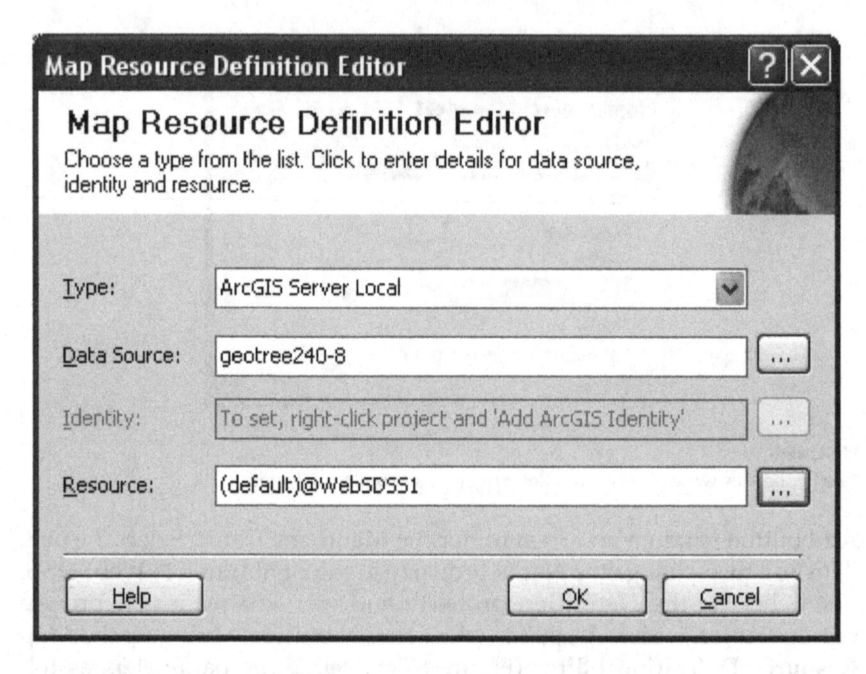

FIGURE 8.5c
The Map Resource Definition Editor.

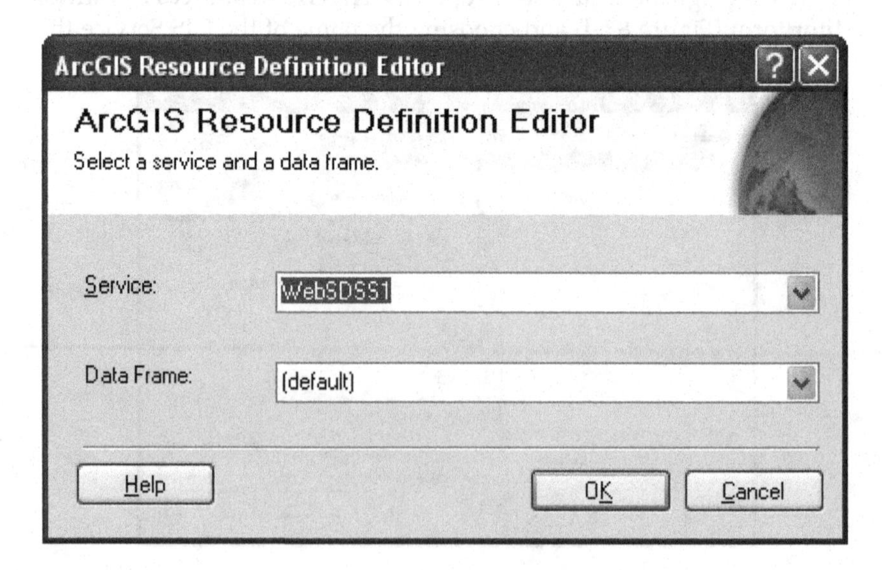

FIGURE 8.5d
The ArcGIS Resource Definition Editor.

was created earlier—in our example, WebSDSS1—in the **Service:** drop-down box). You can leave the Data Frame: drop-down box setting as it is. (It might say Default or it might say the name of the Data Frame from your ArcMap.mxd document. Click **OK** on the **ArcGIS Resource Definition Editor.**) Now click **OK** on the **Map Resource Definition Editor**. Finally, click **OK** on the **MapResourceItem Collection Editor** form.

Open the **web.config** file from the **Solution Explorer** window in Visual Studio by double-clicking the file name. Find the following code in the web.config file:

```
<compilation debug="true" strict="false" explicit="true">.
```

It is possible that this line of code will say something different, such as:

```
<compilation debug="false" >.
```

Add the following line of code directly above that code:

```
<identity impersonate="true" userName="userName"
password="**********"/>
```

In this new line of code, you must change the **userName** and **password** to the proper ones for your machine (i.e., your userName and password for the machine you are working on).

Now you can actually build and run your application, which will open the ArcGIS Server Web site. To do this, choose the **Debug** menu and choose **Start Debugging**. If there are no errors in your Visual Studio project, then your default browser should open and look similar to Figure 8.6. The actual appearance of the data may vary depending on the order of the layers as you added them to your ArcMap .mxd earlier and also how they were symbolized. You can investigate the map contents by clicking on the **Map Contents** tab and exploring the layers present in the map. When finished, close the browser window, and click **Stop Debugging** under the **Debug** menu in Visual Studio.

If there is an error in the browser window when you run the ArcGIS Server application, close the browser window, stop debugging, click the **Save All** button, and then choose **Rebuild Web Site** from the **Build** menu. After the rebuild is finished (look in lower left corner of Visual Studio for a status message), try to run the application again by choosing the **Debug** menu and choosing **Start Debugging**.

At this point this is just a standard ArcGIS Server Web map. We are now going to add a new tool to the interface that will allow us to run the WLC operation from our ArcGIS Server application. To add a new tool, navigate to Design mode by choosing **View > Designer**, and opening

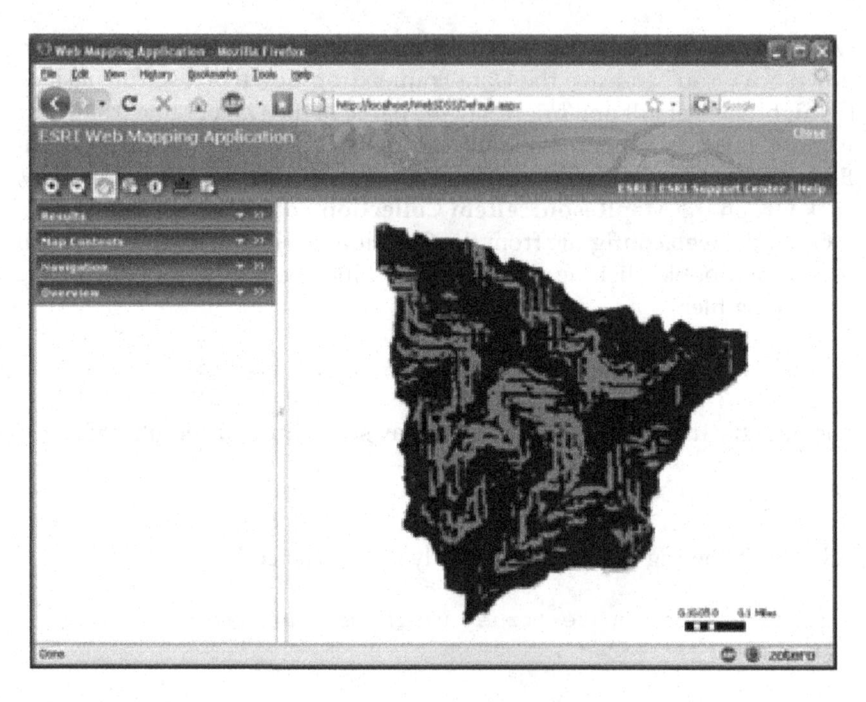

FIGURE 8.6
The ArcGIS Server Web application before any new functionality is added.

the **Default.aspx** page (double-click on it in the **Solution Explorer** window). Then select the toolbar (small area with all of the tools on it) and in the **Properties** window, select the **Toolbar Items** row and then click the browse button (button with **[…]**) from that row (Figure 8.7a). In the **ToolbarCollectionEditorForm** dialog (Figure 8.7b), click **Tool** in the **Toolbar Items:** frame and then click the **Add** button. With the new tool highlighted in the **Current Toolbar Contents:** frame, set the **Name, Text,** and **ToolTip** properties to **WLC**. Set the **DefaultImage** property to **find.png** by clicking the browse button next to the DefaultImage property and selecting **find.png** under the **Images** category (if find.png is not available, choose another .png file such as crosshair.png). Set the **ClientAction** property by choosing **Custom** from the drop-down list. This opens the **Custom Client ToolAction Editor** dialog. In this dialog, type **WLC()** in the **Enter JavaScript to execute as custom ClientAction:** frame (Figure 8.7c). Click **OK** on the **Custom Client ToolAction Editor** dialog. Click **OK** in the **ToolbarCollectionEditorForm** dialog. The toolbar in Default.aspx should now include the new tool (Figure 8.7d).

Right-click on the **JavaScript** folder in the **Solution Explorer** window and choose **Add New Item**. In the **Add New Item** dialog, choose **Jscript File** and name it **wlc.js** in the **Name:** text box. Then click the

FIGURE 8.7a
The toolbar properties window.

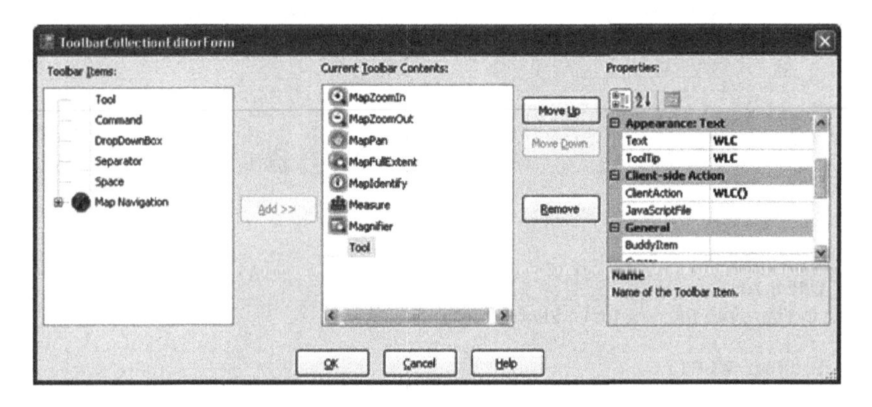

FIGURE 8.7b
The ToolbarCollectionEditor Form..

Add button. Now enter the following JavaScript code in the **wlc.js** window. This code is available at the http://www.geotree.uni.edu/SDSSbook Web site. When you run this code, it will open a pop-up window named WLC_aspx.

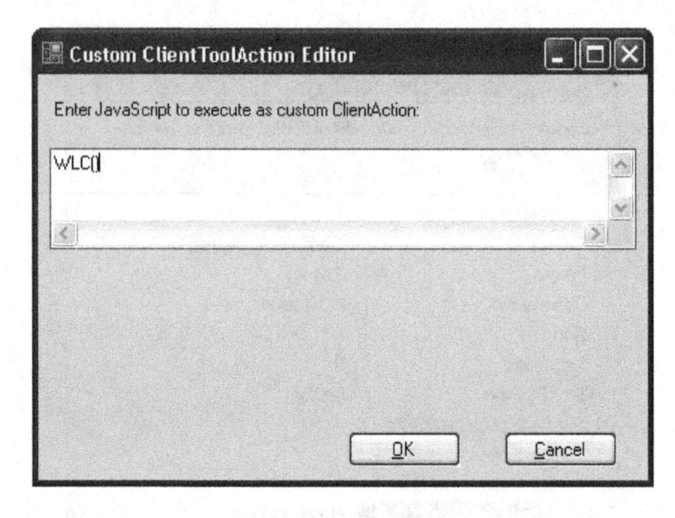

FIGURE 8.7c
The Custom ClientToolAction Editor dialog in which the name of the JavaScript code to run when clicking on the new tool created is entered.

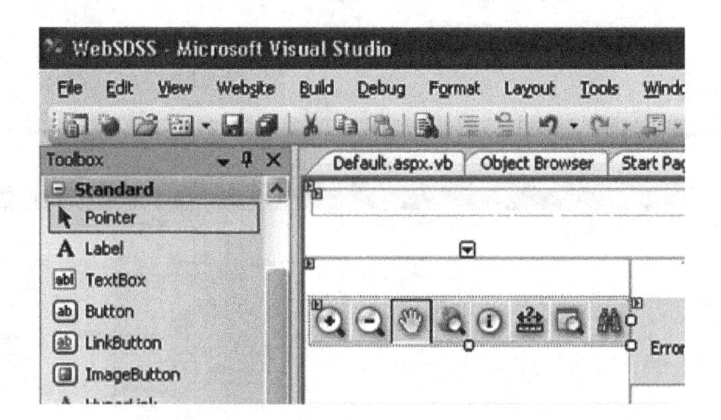

FIGURE 8.7d
The toolbar with the new tool added (binoculars).

```
function WLC()
{
  var WLCdiv = document.getElementById("MapDiv_Map1");
  if (WLCdiv!=null)
    var f = document.forms[0];
    f.minx.value=zleft;
    f.miny.value=ztop;
    var today = new Date();
    var rand = today.getTime();
    var winId = "WLCWindow_" + sessionId;
```

```
    var url = "WLC.aspx?WLCx=" + zleft + "&WLCy=" + ztop +
    "&WLCtype=new&random=" + rand;
    idWin = window.open(url, winId, "width=700,height=600,sc
    rollbars,status, resizable=yes, left=150,top=100");
    return false;
}
```

Open the source code for Default.aspx (double-click on **Default.aspx** in **Solution Explorer** and choose **Markup** from the **View** menu). Find the following line of code. To find it, use the **Quick Find** tool, which can be accessed by pressing **Ctrl-F** or from the **Edit** menu by choosing **Quick Find**.

```
<script language="javascript" type="text/javascript"
src="javascript/WebMapApp.js"></script>
```

Then add this line of code on the line directly below it:

```
<script language="javascript" type="text/javascript"
src="javascript/WLC.js"></script>
```

Now we must create the WLC.aspx page or pop-up Web form. Right-click on the project (the top of the Solution Explorer where it will say something like http://localhost/WebSDSS) in the **Solution Explorer** window and choose **Add New Item**. In the **Add New Item** window, choose **Web Form** and name it **WLC.aspx** (Figure 8.8). Make sure the **Place code**

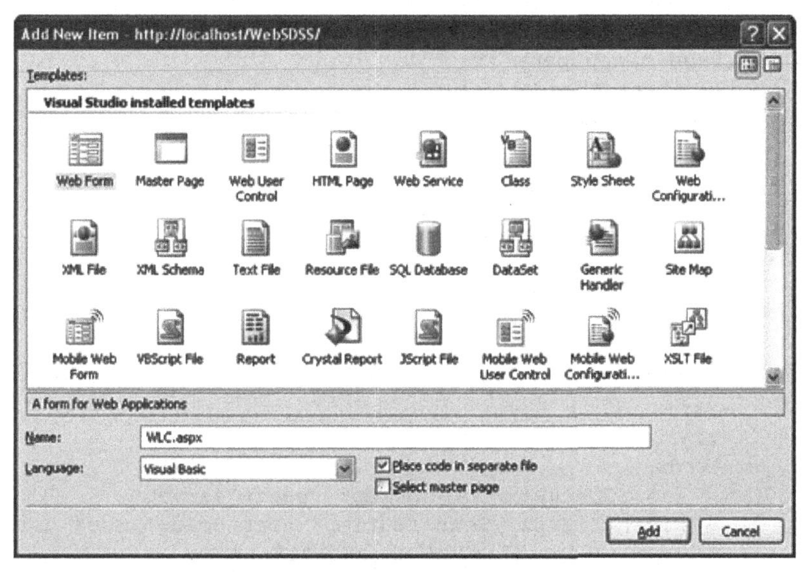

FIGURE 8.8
The Add New Item form for adding the WLC Web form.

in separate file checkbox is checked and that **Visual Basic** is chosen in the **Language:** combo box. Click the **Add** button.

The WLC.aspx file should now be open. If it is in Design mode it should be blank. Activate the Code page if it is not already active by choosing **Code** from the **View** menu. Now replace all of the code in this page with the following code (code can be found at the http://www.geotrcc.uni.cdu/SDSSbook Web site). This code sets properties of the WLC form that will be activated when the new tool is clicked in our ArcGIS Server application. Some of the characteristics of this form are hard-coded for the specific datasets that are in this ArcGIS Server application. If this application was adapted for other data or for a more generic application, some of this code would need to be changed.

```
<%@ Page Language="VB" AutoEventWireup="false"
CodeFile="WLC.aspx.vb" Inherits="WLC" %>

<!DOCTYPE html PUBLIC "-//W3C//DTD XHTML 1.0 Transitional//
EN" "http://www.w3.org/TR/xhtml1/DTD/xhtml1-transitional.
dtd">

<html xmlns="http://www.w3.org/1999/xhtml" >
<head runat="server">
  <title>WebSDSS WLC</title>
</head>
<body>
  <div>Web Weight Linear Combination</div>
  <form id="form1" runat="server">
    <asp:Panel ID="Panel1" runat="server"
    CssClass="inputpanel">
      <div class="header">
        <div class="help">
           </div>
      </div>
      <div class="input">
        <table class="inputtable">
          <tr class="tableheader">
          <td style="height: 21px">
            Input layers</td>
          <td style="height: 21px">
            Weight</td>
        </tr>
        <tr>
          <td>
            <asp:CheckBox ID="chkImp" runat="server"
            Width="184px" Text="Bufimp" Font-Bold="True"
              Font-Size="Small" Height="18px"
              Checked="True"></asp:CheckBox></td>
          <td class="inputvalue">
```

```
            <asp:TextBox ID="txtImpW" runat="server"
            CssClass="txt" Width="49px" Height="18px"
              Columns="5">1</asp:TextBox></td>
        </tr>
        <tr>
          <td>
            <asp:CheckBox ID="chkFEMA" runat="server"
            Width="185px" Text="FEMA" Font-Bold="True"
              Font-Size="Small" Height="18px"
              Checked="True"></asp:CheckBox></td>
          <td class="inputvalue">
            <asp:TextBox ID="txtFEMAW" runat="server"
            CssClass="txt" Width="49" Height="18px"
              Columns="5">1</asp:TextBox></td>
        </tr>
        <tr>
          <td>
            <asp:CheckBox ID="chkGL" runat="server"
            Width="185px" Text="Green land" Font-Bold="True"
              Font-Size="Small" Height="18px"
              Checked="True"></asp:CheckBox></td>
          <td class="inputvalue">
            <asp:TextBox ID="txtGreenW" runat="server"
            CssClass="txt" Width="49px" Height="18px"
              Columns="5">1</asp:TextBox></td>
        </tr>
        <tr>
          <td>
            <asp:CheckBox ID="chkSoil" runat="server"
            Width="288px" Text="Hydric Soil"
              Font-Bold="True" Font-Size="Small" Height="18px"
              Checked="True"></asp:CheckBox></td>
          <td class="inputvalue">
            <asp:TextBox ID="txtSoilW" runat="server"
            CssClass="txt" Width="49px" Height="18px"
              Columns="5">1</asp:TextBox></td>
        </tr>
        <tr>
          <td>
            <asp:CheckBox ID="chkSlope" runat="server"
            Width="272px" Text="Slope"
              Font-Bold="True" Font-Size="Small" Height="18px"
              Checked="True"></asp:CheckBox></td>
          <td class="inputvalue">
            <asp:TextBox ID="txtSlopeW" runat="server"
            CssClass="txt" Width="49px" Height="18px"
              Columns="5">1</asp:TextBox></td>
        </tr>
```

```
      </table>
      </div>
           <asp:Button ID="btnAna" runat="server"
           Text="Analyze" /></asp:Panel>
    </form>
</body>
</html>
```

If you now choose **Designer** from the **View** menu, you will see that the code just added created controls on the WLC Web form (Figure 8.9). These controls include a title, a panel, a table, and a button. There are also checkboxes for each of the potential layers for the WLC analysis and text boxes for entering weights for those layers. It would have been possible to design this form manually in the Design view using the controls from the Toolbox, but here it was done using the code.

We will now add ArcGIS references to the Visual Studio project. The purpose of this step is to make available the ArcObjects (those objects on which ArcGIS is built) that will be necessary to carry out the WLC analysis in this ArcGIS Server application. By adding these references, the code that we will use later can access the necessary ArcObjects. We have to add several references because ESRI stores the ArcObjects in separate files (e.g., DLLs) based on categories (e.g., Display, Geodatabase, Geometry, etc.). To add the references, right-click on **http://localhost/WebSDSS/** in the **Solution Explorer** window and choose **Add ArcGIS Reference**. You

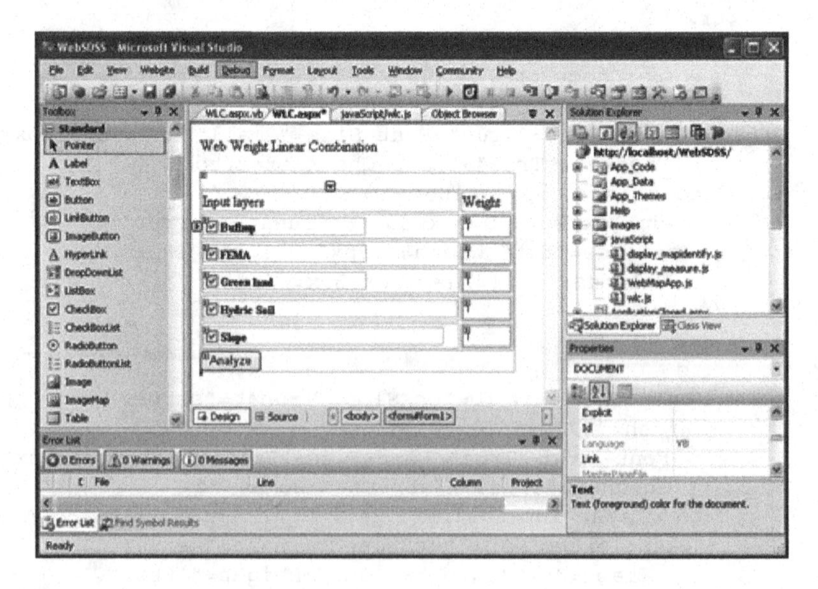

FIGURE 8.9
The Web form (WLC.aspx) in Designer view.

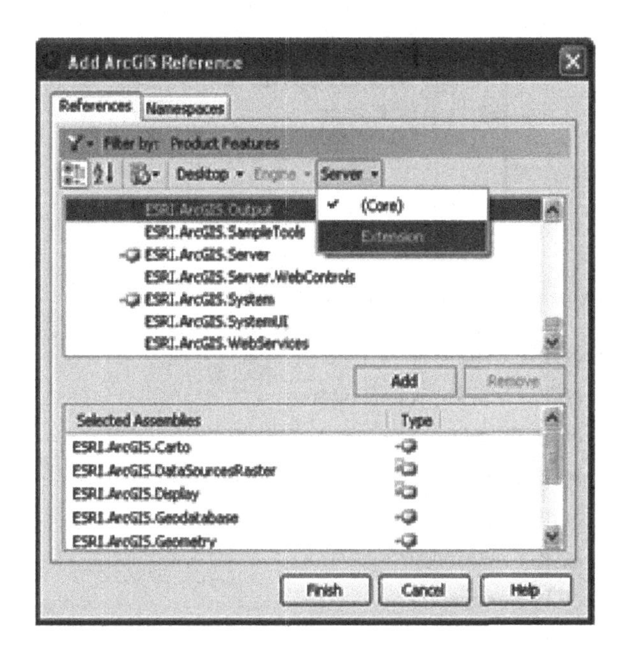

FIGURE 8.10
The Add ArcGIS Reference form.

will see the Add ArcGIS Reference form (Figure 8.10). First, expand the **Desktop ArcMap** category so you see the full list under this category. Select the following references one-by-one and click the **Add** button after each selection (**ESRI.ArcGIS.Carto, ESRI.ArcGIS.DataSourcesRaster, ESRI.ArcGIS.Display, ESRI.ArcGIS.Geodatabase, and ESRI.ArcGIS. Geometry**). After you have added these references, collapse the **Desktop ArcMap** category and expand the **Server (Core)** category. Select the following references one at a time and click the **Add** button after each selection (**ESRI.ArcGIS.Server and ESRI.ArcGIS.Server.WebControls**). Finally, select the **Server** drop-down menu (Figure 8.10) and make sure the **Extension** checkbox is checked. Collapse the **Server (Core)** category and expand the **Server Extension** category. Add the last two references under the **Server Extension** category (**ESRI.ArcGIS.GeoAnalyst** and **ESRI.ArcGIS.SpatialAnalyst**). Alternatively, you can click the in the toolbar that has an A and a Z and a downward pointing arrow to display all references in alphabetical order. Then they can be added one at a time from this complete list. You should have the following references chosen. Click **Finish** after you have verified that the proper references have been added.

FIGURE 8.11
Open the WLC.aspx.vb by double-clicking on it.

- ESRI.ArcGIS.Carto
- ESRI.ArcGIS.DataSourcesRaster
- ESRI.ArcGIS.Display
- ESRI.ArcGIS.GeoAnalyst
- ESRI.ArcGIS.Geodatabase
- ESRI.ArcGIS.Geometry
- ESRI.ArcGIS.Server
- ESRI.ArcGIS.Server.WebControls
- ESRI.ArcGIS.SpatialAnalyst

Open the **WLC.aspx.vb** file (Figure 8.11) from the **Solution Explorer** window by double-clicking on the file (you might first have to expand WLC.aspx). We will now add code to this form that will carry out the spatial analysis in the WLC method.

Replace the code in the WLC.aspx.vb code module window with the following code (available at the http://www.geotree.uni.edu/SDSSbook Web site). There are comments in the code that start with a single quotation mark to give the reader an idea of what is happening.

```
'these lines import references making it easier to write
code without having to write 'out the full 'path to object.
Imports ESRI.ArcGIS.Server.WebControls
Imports ESRI.ArcGIS.Geometry
Imports ESRI.ArcGIS.Carto
Imports ESRI.ArcGIS.esriSystem
Imports ESRI.ArcGIS.Server
Imports ESRI.ArcGIS.display
Imports ESRI.ArcGIS.Geodatabase
Imports ESRI.ArcGIS.SpatialAnalyst
Imports ESRI.ArcGIS.DataSourcesRaster
Imports ESRI.ArcGIS.GeoAnalyst

Partial Public Class WLC
```

```
  Inherits System.Web.UI.Page

#Region " Web Form Designer Generated Code "

  'This call is required by the Web Form Designer.
  <System.Diagnostics.DebuggerStepThrough()> Private Sub
  InitializeComponent()

  End Sub

  Private m_displaylist As New ArrayList
  'NOTE: The following placeholder declaration is required
  by the Web Form Designer.
  'Do not delete or move it.
  Protected Toc1 As ESRI.ArcGIS.ADF.Web.UI.WebControls.Toc '
ESRI.ArcGIS.Server.WebControls.Toc
  Protected Map1 As ESRI.ArcGIS.ADF.Web.UI.WebControls.Map '
ESRI.ArcGIS.Server.WebControls.Map
  'Protected webmap1 As ESRI.ArcGIS.Server.WebControls.
  WebMap
  Protected sc As ESRI.ArcGIS.Server.ServerContext
  Protected TableNameLabel As System.Web.UI.WebControls.
  Label
  Protected ShowTable As System.Web.UI.WebControls.Table
  Protected TocTitleTable As System.Web.UI.WebControls.Table
  Protected TimeoutPanel As System.Web.UI.WebControls.Panel
  Protected CloseButton As System.Web.UI.HtmlControls.
  HtmlInputButton
  Protected mapSI As ESRI.ArcGIS.Carto.IMapServerInit
  Protected pMap As ESRI.ArcGIS.Carto.IMap

  Private designerPlaceholderDeclaration As System.Object
  Private idOption As esriIdentifyOption =
  esriIdentifyOption.esriIdentifyTopmost
  Private idOptionValue As String = "_TopMost_"
  Private m_dataset As System.Data.DataSet = Nothing
  Private m_extent As ESRI.ArcGIS.Server.WebControls.Extent
  Private table_index As Integer = 0
  Private row_index As Integer = 0
  Private id_layer As Integer = -1
  Private idx As Integer = 0
  Private idy As Integer = 0
  Private m_width As Integer
  Private m_height As Integer
  Private m_serverobject As String = [String].Empty
  Private m_dataframe As String = [String].Empty
  Private m_lastdataframe As String = [String].Empty
  Private m_host As String = [String].Empty
```

```
  Private m_contextname As String = [String].Empty
  Private m_featurelayeridlist As New ArrayList
  Private m_featurelayernamelist As New ArrayList

  'output path can be set with session and serverobject's
  property
  'Private sOutput As String = "C:\inetpub\wwwroot\wfmis_
  try\work_try\"
  'all the variables related with land use

  Private WebSDSS_output As New Collection

  Private Sub Page_Init(ByVal sender As System.Object, ByVal
e As System.EventArgs) Handles MyBase.Init
    'CODEGEN: This method call is required by the Web Form
    Designer
    'Do not modify it using the code editor.
    InitializeComponent()
  End Sub

#End Region
  Private Sub Page_Load(ByVal sender As Object, ByVal e As
System.EventArgs) Handles MyBase.Load

  Try
    'initialize PI and corresponding value control
    If Not (m_lastdataframe = [String].Empty) And Not (m_
dataframe = m_lastdataframe) Then
      'this request is using a different dataframe from the
previous request
      'so reset the selected option to the default... top-
most... and clear out any selected layer
      idOptionValue = "_TopMost_"
      Session.Add("idOption", esriIdentifyOption.
      esriIdentifyTopmost)
      Session.Add("idLayer", "-1")
    End If
    Session.Add("IdOptionValue", idOptionValue)

    Map1 = CType(Session("map1_ss"), ESRI.ArcGIS.ADF.Web.
    UI.WebControls.Map)
    If Not (Map1.GetFunctionality(0) Is Nothing) Then
      Dim mapFunc As ESRI.ArcGIS.ADF.Web.DataSources.
ArcGISServer.MapFunctionality = CType(Map1.
GetFunctionality(0), ESRI.ArcGIS.ADF.Web.DataSources.
ArcGISServer.MapFunctionality)
      Dim mapD As ESRI.ArcGIS.ADF.ArcGISServer.MapDescription
= mapFunc.MapDescription
```

```
     Dim mrl As ESRI.ArcGIS.ADF.Web.DataSources.
ArcGISServer.MapResourceLocal = mapFunc.Resource
     sc = mrl.ServerContextInfo.ServerContext

     Dim mi As ESRI.ArcGIS.ADF.Web.DataSources.ArcGISServer.
MapInformation = mrl.MapInformation
     Dim mapS As ESRI.ArcGIS.Carto.IMapServer = mrl.
     MapServer
     Dim mapSObj As IMapServerObjects = mapS
     pMap = mapSObj.Map(mapS.DefaultMapName)
     Session.Add("map_last_dataframe", m_dataframe)
     mapSI = CType(mapS, ESRI.ArcGIS.Carto.IMapServerInit)

     'start to set basic varialbles
     'Dim webMap1 As ESRI.ArcGIS.Server.WebControls.WebMap =
     Nothing
     Dim m_ctx As IServerContext = Nothing
     Dim o As Object = Session(m_contextname)
     Dim imgd As New ImageDescriptor(WebImageFormat.BMP,
     m_width, m_height)
     If Not (o Is Nothing) Then
       m_ctx = o
       'webmap1 = New WebMap(m_ctx, m_host, m_dataframe)
     Else
       'webmap1 = New WebMap(New ESRI.ArcGIS.Server.
WebControls.ServerConnection(m_host, True), m_serverobject,
m_dataframe, imgd)
     End If
     End If
   Catch ex As Exception
     Response.Write("page load has problem: " & ex.ToString
     & Err.Description)
   End Try
  End Sub 'Page_Load

  Private Sub btnAna_Click(ByVal sender As System.Object,
ByVal e As System.EventArgs) Handles btnAna.Click
   'deals with output layers
  Session.Add("WLC_output", WebSDSS_output)
   'all the input layers

 Try
  Dim pRasters As New Collection ' this one collects all
rasters that will be overlayed
  Dim pEnvironment As ESRI.ArcGIS.GeoAnalyst.
IRasterAnalysisEnvironment =
sc.CreateObject("esriGeoAnalyst.RasterAnalysis")
  pEnvironment.RestoreToPreviousDefaultEnvironment()
```

```
  'get all the imput layers and also times it with its
  weight
  If chkImp.Checked Then
    Dim rlBufimp As ESRI.ArcGIS.Carto.IRasterLayer =
GetRasterLayer("test_bufimp", pMap)
    pRasters.Add(TimesRaster(sc, rlBufimp.Raster,
    CDbl(txtImpW.Text)))
  End If
  If chkFEMA.Checked Then
    Dim rlfema As ESRI.ArcGIS.Carto.IRasterLayer =
GetRasterLayer("test_fema", pMap)
    pRasters.Add(TimesRaster(sc, rlfema.Raster,
    CDbl(txtFEMAW.Text)))
  End If
  If chkGL.Checked Then
    Dim rlGreenland As ESRI.ArcGIS.Carto.IRasterLayer =
GetRasterLayer("test_green", pMap)
    pRasters.Add(TimesRaster(sc, rlGreenland.Raster,
    CDbl(txtGreenW.Text)))
  End If
  If chkSoil.Checked Then
    Dim rlSoil As ESRI.ArcGIS.Carto.IRasterLayer =
GetRasterLayer("test_hysoil", pMap)
    pRasters.Add(TimesRaster(sc, rlsoil.Raster,
    CDbl(txtSoilW.Text)))
  End If
  If chkSlope.Checked Then
    Dim rlSlope As ESRI.ArcGIS.Carto.IRasterLayer =
GetRasterLayer("test_slope", pMap)
    pRasters.Add(TimesRaster(sc, rlslope.Raster,
    CDbl(txtSlopeW.Text)))
  End If
  ''--------------------sum up all layers------------------
  Dim pROutput As IRaster

  If pRasters.Count > 0 Then
    pROutput = pRasters(1)
    Dim i As Integer
    For i = 2 To pRasters.Count
      pROutput = AddTogether(sc, pROutput, pRasters(i))
    Next
  End If

  Dim pLayer As IRasterLayer = sc.CreateObject("esriCarto.
  RasterLayer")
  pLayer.CreateFromRaster(proutput)
  pLayer.Name = "output"
```

```
  WebSDSS_output.Add(pLayer)

  Catch ex As Exception
    Response.Write("main procedure has problem: " &
ex.ToString & Err.Description)
    End Try
  End Sub
  '******************** support codes ********************
  Private Sub SessionEnd(ByVal sender As Object, ByVal e As
EventArgs)
    Dim context As IServerContext
    Dim i As Integer
    For i = 0 To Session.Count - 1
      context = Session(i)
      If Not (context Is Nothing) Then
        context.ReleaseContext()
      End If
    Next i
    Session.RemoveAll()
  End Sub

  'this function find the feature layer by its name in a map
  (even in layer group)
  Function GetRasterLayer(ByVal sName As String, ByVal m_map
As ESRI.ArcGIS.Carto.IMap) As ESRI.ArcGIS.Carto.IRasterLayer
    Dim LyrCount As Integer
    Dim pLayer As ESRI.ArcGIS.Carto.ILayer
    Dim pCompositeLayer As ESRI.ArcGIS.Carto.ICompositeLayer
    Dim pLyrInG As ESRI.ArcGIS.Carto.ILayer
    Dim sLayerName As String
    Dim i As Integer
    Dim j As Integer

    LyrCount = m_map.LayerCount
    If LyrCount <> 0 Then
      For i = 0 To LyrCount - 1
        pLayer = m_map.Layer(i)
        If TypeOf pLayer Is ESRI.ArcGIS.Carto.IRasterLayer
        Then
          sLayerName = pLayer.Name
          If (StrComp(sName, sLayerName, vbTextCompare) = 0)
          Then
            GetRasterLayer = pLayer
            Exit Function
          End If
        ElseIf (TypeOf pLayer Is ESRI.ArcGIS.Carto.
        ICompositeLayer) Then
          pCompositeLayer = pLayer
```

```
        For j = 0 To pCompositeLayer.Count - 1
          pLyrInG = pCompositeLayer.Layer(j)
          If TypeOf pLyrInG Is ESRI.ArcGIS.Carto.
          IRasterLayer Then
            sLayerName = pLyrInG.Name
            If (StrComp(sName, sLayerName, vbTextCompare) =
            0) Then
              GetRasterLayer = pLyrInG
              Exit Function
            End If
          End If
          pLyrInG = Nothing
        Next j
          pCompositeLayer = Nothing
        End If
        pLayer = Nothing
      Next i
    Else
      GetRasterLayer = Nothing
    End If
  End Function

  Public Function TimesRaster(ByVal sc As ESRI.ArcGIS.
  Server.ServerContext, ByVal pRaster As IRaster, ByVal
  iConstant As Double) As IRaster
    'Create a RasterMapAlgebraOp operator
    Dim pMapAlgebraOp As IMapAlgebraOp = sc.CreateObject("e
sriSpatialAnalyst.RasterMapAlgebraOp")

    'Bind a raster
    pMapAlgebraOp.BindRaster(pRaster, "R1")
    TimesRaster = pMapAlgebraOp.Execute("([R1]) * " &
    CStr(iConstant))
  End Function

  'this function is an extened "plus" which can use a mask
  Public Function AddTogether(ByVal sc As ESRI.ArcGIS.
Server.ServerContext, ByVal raster1 As IRaster, ByVal raster2
As IRaster) As IRaster
    Dim pMathOp As IMathOp = sc.CreateObject("esriSpatialAn
    alyst.RasterMathOps")
    'Calls the method
    AddTogether = pMathOp.Plus(raster1, raster2)
  End Function
End Class
```

Next, open the **Default.aspx.vb** file by double-clicking on the file in the **Solutions Explorer** window (expand **Default.aspx** to see **Default.aspx.**

vb). Find these exact lines in the code. You can search using the Quick Find tool (choose **Edit > Quick Find** or **Ctrl-F**).

```
Partial Class WebMapApplication
  Inherits System.Web.UI.Page
  Implements ICallbackEventHandler
  Dim identify As MapIdentify
  Public m_newLoad As String = "false"
  Public m_closeOutCallback As String = ""
  Public m_copyrightCallback As String = ""
```

Add the following lines directly below them.

```
  Public ExtentList As New ArrayList
  Private m_extenthistory As ArrayList
  Private isPooled As Boolean = False
  Private sessionId As String = ""
  Private m_MapDescriptSessName As String = ""

  'this variable is used to create popup window, suggestion
from ESRI
    Public sCallBackFunctionInvocation As String
```

Within the Default.aspx.vb file, replace the page load sub with the code below. Select all of the code beginning with the line **Protected Sub Page_Load** completely through the line **End Sub Page_Load**. Replace the selected code with the following code.

```
  Protected Sub Page_Load(ByVal sender As Object, ByVal e As
EventArgs) Handles MyBase.Load
    If Not Page.IsCallback And Not Page.IsPostBack Then
      If Map1.MapResourceManager Is Nothing Or Map1.
      MapResourceManager.Length = 0 Then
        callErrorPage("No MapResourceManager defined for the
        map.", Nothing)
    End If
    If MapResourceManager1.ResourceItems.Count = 0 Then
      callErrorPage("The MapResourceManager does not have a
valid ResouceItem Definition.", Nothing)
    ElseIf MapResourceManager1.ResourceItems(0) Is Nothing
    Then
      callErrorPage("The MapResourceManager does not have a
valid ResouceItem Definition.", Nothing)
    End If
    m_newLoad = "true"
  End If
```

```
  m_closeOutCallback = Page.ClientScript.
GetCallbackEventReference(Page, "argument",
"CloseOutResponse", "context", True)
  m_copyrightCallback = Page.ClientScript.
GetCallbackEventReference(Page, "argument",
"processCallbackResult", "context", True)
  'initiate identify class and set link to TaskResults1 for
response
  identify = New MapIdentify(Map1)
  identify.ResultsDisplay = TaskResults1
  identify.NumberDecimals = 4
'***********************************************************
  '''''My codes added to the template

  If Not Page.IsPostBack Then

    'Is this a new session?
  If Session.IsNewSession Then
    ' Save extent history to Session
    m_extenthistory = New ArrayList
    m_extenthistory.Add(Map1.Extent)
    Session.Add("extenthistory", m_extenthistory)
    Session.Add("index", 0)

  Else
    Try
      'this part is planned to move all the new layers back
here and add to map
      If Not Session("WLC_output") Is Nothing Then
 'addlayers()
        Dim imgd As New ImageDescriptor(WebImageFormat.BMP,
Map1.Width.Value - Map1.BorderWidth.Value * 2, Map1.Height.
Value - Map1.BorderWidth.Value * 2)
        Dim pLayer As ESRI.ArcGIS.Carto.ILayer
        Try
          Dim Outputlayers As Collection =
CType(Session("WLC_output"), Collection)
          For i As Integer = 1 To Outputlayers.Count
            pLayer = CType(Outputlayers(i), ESRI.ArcGIS.Carto.
ILayer)
            AddLayer2Map(pLayer, Map1, Toc1)
          Next
          Session.Remove("WLC_output")
        Catch ex As Exception
          Response.Write(ex.ToString())
        Finally

        End Try
```

```
      End If
    Finally
     End If
  Else

    'make sure that the session is still going
    If Session("extenthistory") Is Nothing Then
      'Send it to the appropriate error page
      callErrorPage("Your session has timed out.", Nothing)
    End If
  End If
  Session.Add("map_width", Map1.Width.Value - Map1.
  BorderWidth.Value * 2)
  Session.Add("map_height", Map1.Height.Value - Map1.
  BorderWidth.Value * 2)
  Session.Add("Map1_CurrentExtent", Map1.Extent)
  Session.Add("Table_of_Contents", Toc1)

  '********used in WLC
  Session.Add("map1_ss", Map1)

  sessionId = Session.SessionID
  Dim scriptString As String = ControlChars.Lf + "<script
language=javascript>sessionId = '" + sessionId + "';</
script>" + ControlChars.Lf
  'Page.RegisterStartupScript("SessionIdScript",
scriptString)
  'get name of session object holding map description
  Dim pagePath As String = Page.Request.FilePath
  Dim pageName As String = ""
  Dim lastSlash As Integer = pagePath.LastIndexOf("/")
  If lastSlash > -1 Then
  pageName = pagePath.Substring((lastSlash + 1))
  End If
  m_MapDescriptSessName = pageName + Map1.ID + "_md"
  ''''My codes added to the template
'************************************************************
End Sub 'Page_Load
```

Finally, almost at the end of the code in Default.aspx.vb, but directly above the line **End Class**, add the following code method:

```
Public Sub AddLayer2Map(ByVal pLayer As ESRI.ArcGIS.Carto.
ILayer, ByVal map1 As
  ESRI.ArcGIS.ADF.Web.UI.WebControls.Map, ByVal Toc1 As
  ESRI.ArcGIS.ADF.Web.UI.WebControls.Toc)
  Try
```

```
    Dim mapFunc As ESRI.ArcGIS.ADF.Web.DataSources.
    IMapFunctionality = map1.GetFunctionality(0)
    Dim mrl As ESRI.ArcGIS.ADF.Web.DataSources.ArcGISServer.
    MapResourceLocal = mapFunc.Resource
    Dim sc As ESRI.ArcGIS.Server.IServerContext
    sc = mrl.ServerContextInfo.ServerContext

    map1.CallbackResults.AddRange(Toc1.CallbackResults)

    Dim mapS As ESRI.ArcGIS.Carto.IMapServer = mrl.MapServer
    Dim mapSObj As ESRI.ArcGIS.Carto.IMapServerObjects =
    mapS
    Dim pMap As ESRI.ArcGIS.Carto.IMap
    pMap = mapSObj.Map(mapS.DefaultMapName) 'mi.DataFrame

    pMap.AddLayer(pLayer)
    mrl.RefreshServerObjects()
  Finally
    map1.Refresh()
    Toc1.Refresh()
  End Try
End Sub
```

At this point it should be possible to test your application. With the Default.aspx.vb file displayed, go to the **Build** menu and choose **Build Web Site**. If there were no errors, you can now run the application. Choose **Start Debugging** from the **Debug** menu. The new ArcGIS Server application should open in your default Web browser (Figure 8.12a). The new WLC tool that was added should appear as the rightmost tool in the toolbar (Figure 8.12a). Again, depending on the order of the layers (and their symbology) as you originally added them to the ArcMap .mxd you saved earlier, your map might look different than the one in Figure 8.12a. Click on the new tool (the binoculars icon) to bring up the **WLC Web** pop-up form (Figure 8.12b). You may need to tell your browser to allow pop-ups in order for it to work. Set the weights for the layers as you would like and then click the Analyze button. The processing will take a moment; when you see Done at the bottom of the pop-up Web WLC form, you can close that form. Now refresh the browser (F5) in order for the map to be redrawn. If you choose the **Map Contents Tab** you should see that there is a new raster layer called **output** in your map (Figure 8.12c). This represents the output from the WLC analysis carried out using the weights set in Figure 8.12b.

This example has attempted to provide an illustration of the applicability of building a Web-based SDSS tool using ESRI's ArcGIS Server. This specific example was used for an analysis of environmentally sensitive

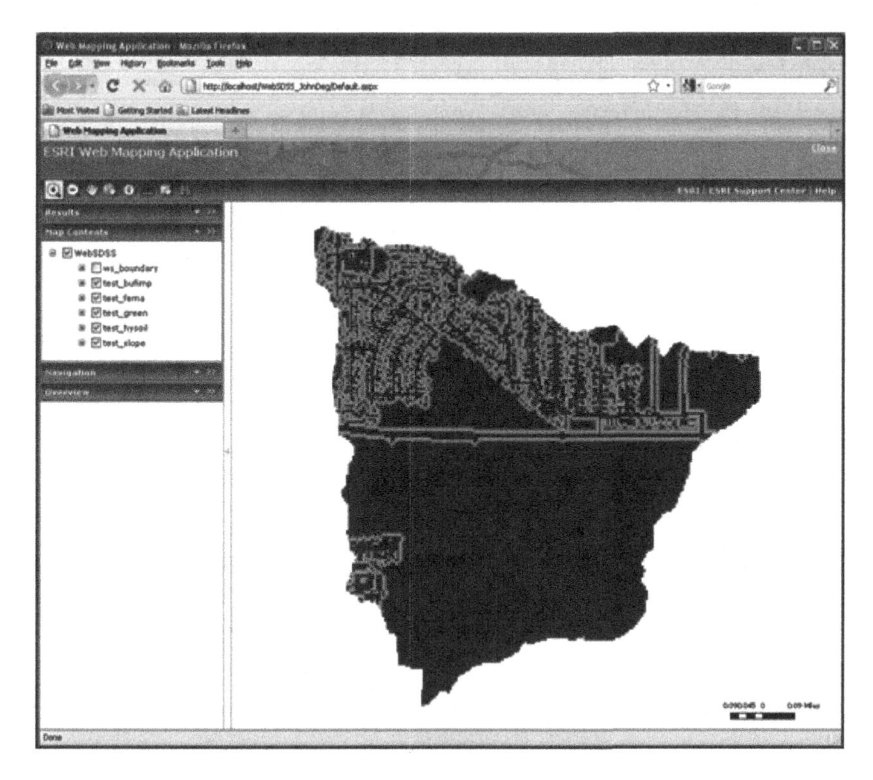

FIGURE 8.12a
The new ArcGIS Server application open in a Firefox browser. The new tool that was created is on the right end of the toolbar.

areas. However, this example could be used as a template in order to build a Web-based WLC SDSS tool using any data for any purpose for which the technique is relevant. In order to do this, the user would have to use an ArcMap .mxd file at the beginning that had other raster layers. The user would also have to adjust some things in the code (or through design mode) in Visual Studio to reflect the changes in the data used. For example, the characteristics (specifically the names and text) of the controls in the Web form would have to be changed. These could be changed in the code or design mode for WLC.aspx. In this ArcGIS Server example there were also instances in which the names of the layers (i.e., test_bufimp, test_fema, test_green, etc.) were hard-coded (explicitly set in the code). Thus, if a new application with different layers was to be used, some of these things would need to be changed in the code (in WLC.aspx.vb). If this was going to be a more polished Web-based SDSS, greater effort would be made to build a more generic system.

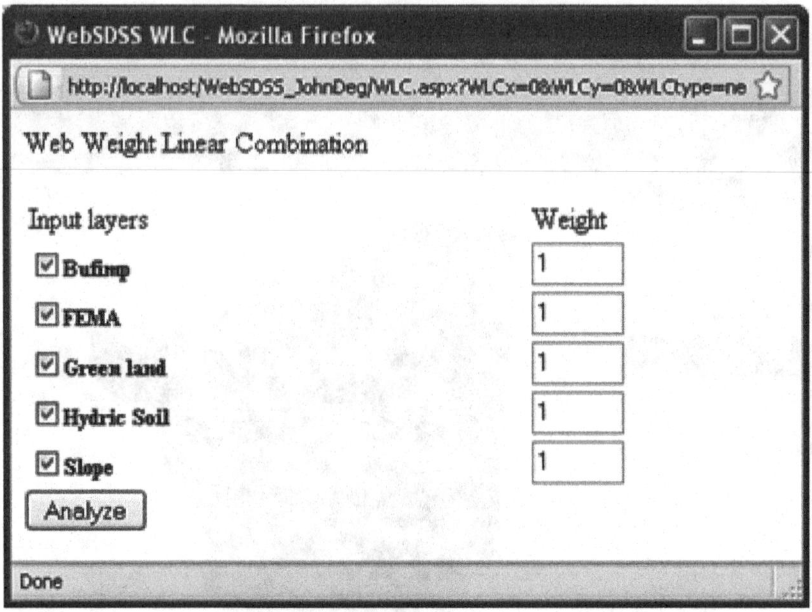

FIGURE 8.12b
The WLC Web form for setting the weights to be used in the WLC analysis.

8.3 Web-Based SDSS Development with Open Source Software

The next example uses all free and open-source software applications for developing a Web-based SDSS that demonstrates a WLC method for identifying locations of wind farms in Iowa. This example is different from the others (i.e., Excel and ArcGIS-based in Chapter 7 and the ArcGIS Server example in this chapter) in that all of the necessary files (code, data, etc.) for running the application are provided as a zipped file via the book's Web site. The reader will be able to download the zipped file, unzip it, and carry out a simple command to run the application. The files and code used to produce the application will then be discussed, but no step-by-step instructions will be given. This application was built for the Linux operating system and used a variety of open source programs to achieve the results. To develop this type of application, a person would need experience using the Linux operating system as well as a good understanding of how software packages are distributed in Linux. Those with little or no experience with Linux, or the other software used, would have trouble following the details of the code in this example but can gain an

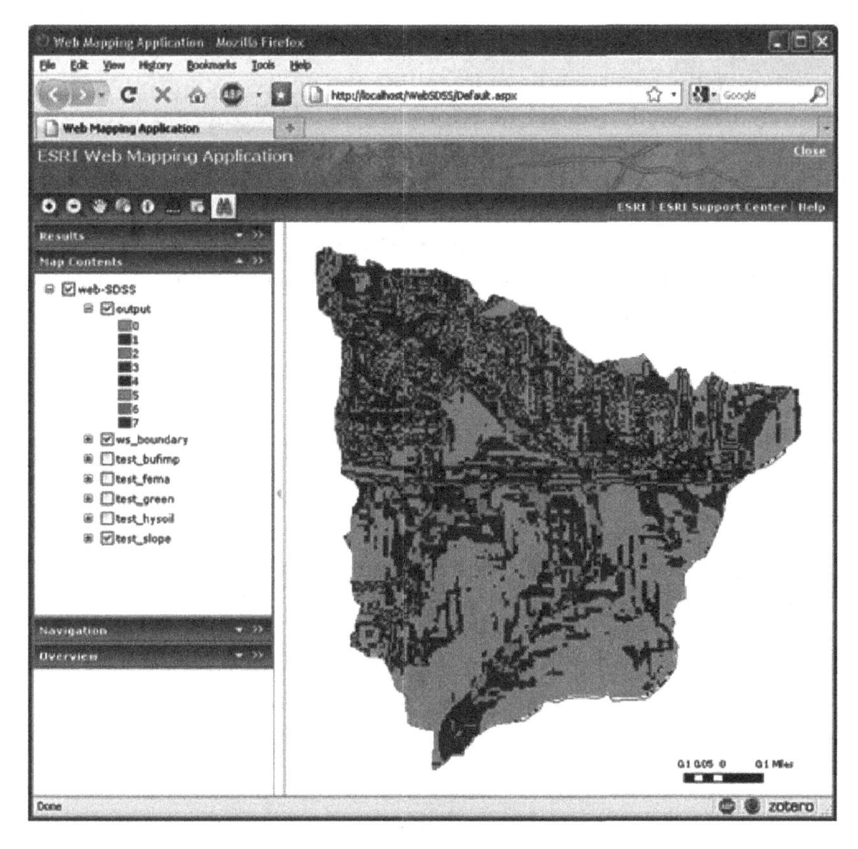

FIGURE 8.12c
The results from carrying out the WLC using weights from above.

understanding of the nature of this type of Web-based SDSS. Experience with a package manager (e.g., Synaptic) in Linux can make the installation of necessary software much easier. Also necessary is some experience in working with the Linux shell.

This section will first give instructions for installing the necessary software components. Then it will provide an overview of these software components and give an overview of the architecture of the system. Next, instructions for downloading all of the necessary files and for running the application are provided. Finally, the code with which the application was created is presented along with an explanation of the functionality achieved with individual chunks of code. Those with the necessary experience and desire could attempt to reproduce the example. Those without that experience can browse through the rest of the example to get an overview of how the example is constructed.

8.3.1 Software Installation

We will give the basic details of installing the necessary software. Again however, this section is written for those with significant experience using the Linux operating system. This example was carried out using an installation of Ubuntu 9.10 (Karmic Koala version). To install the necessary software, first open the **Software Sources** administration application and make sure that the **universe** and **multiverse** repositories are enabled. Next, to install the necessary software, choose from and carry out ***one*** of the following steps (*not both*):

1. Run the following command to install the software needed:

```
sudo apt-get install gdal-bin libgdal1-1.5.0 python-
gdal python-mapscript cgi-mapserver mapserver-bin
mapserver-doc python-numpy apache2 python-django
```

   You will be asked for confirmation before downloading and installing the packages. After giving confirmation, wait for the installations to complete.

Or

2. Open **Synaptic** and double-click on the following packages:
   - gdal-bin
   - libgdal1-1.5.0
   - python-gdal
   - python-mapscript
   - cgi-mapserver
   - mapserver-bin
   - mapserver-doc
   - python-numpy
   - apache2
   - python-django

   Click **Apply** and wait for the packages to be installed.

8.3.2 Software Used

The goal of this project is to develop a Web-based SDSS using several open source software packages including the Web application framework called Django, the open mapping software MapServer originally developed at the University of Minnesota, and OpenLayers, which is used to

TABLE 8.1

Different Components Used for the Development of the Open Source Web-Based SDSS

Components	Value
Spatial Database	Raster files (ERDAS Imagine and Geo TIFF)
Spatial Processing	GDAL
Model	Weighted Linear Combination (WLC)
User Interface	JavaScript
Software	Django, MapServer, and OpenLayers
Programming Language	JavaScript and Python
WLC Calculation	NumPy

display maps on the Web. Table 8.1 describes the specific components used in this example. A short description of each application is given below. We demonstrate an example implementation that provides functionality for finding suitable locations for placing potential wind turbines based on user-defined criteria.

MapServer is an open source platform for publishing spatial data and interactive mapping applications on the Web. The MapServer software is available for users at http://mapserver.org/. Some of the advantages of MapServer include the fact that it runs on all major platforms, supports many raster and vector data formats, supports important Web geographic standards such as the OGS Web Map Service Interface Standard (WMS), OGC Web Feature Service (WFS), and OGC Web Coverage Service, is accessible through many ready-to-use open source application environments, and it also features MapScript, a powerful scripting interface for popular languages such as PHP, C#, Java, Perl, Python, and Ruby. MapScript allows developers to add geospatial functions to any application. In our example, MapServer is used to host a WMS service. Django is an open source, high-level Python Web framework that encourages rapid development and clean, pragmatic design of Web applications. For a detailed description, please visit http://www.djangoproject.com/. Django is used in this project to generate an HTML user interface. OpenLayers makes it easy to put dynamic maps in any Web page (http://openlayers.org/). It is a pure JavaScript library for displaying map data in most modern Web browsers, with no server-side dependencies. In this project, OpenLayer is used to interact with the WMS service provided by MapServer. GDAL stands for Geospatial Data Abstraction Library. It is a translator library for raster geospatial data formats (http://www.gdal.org/) and it is used to read raster data and write WLC output files. Finally, NumPy is a Python library for working with large arrays of numerical data (http://numpy.scipy.org/). It is used to carry out the WLC model calculations.

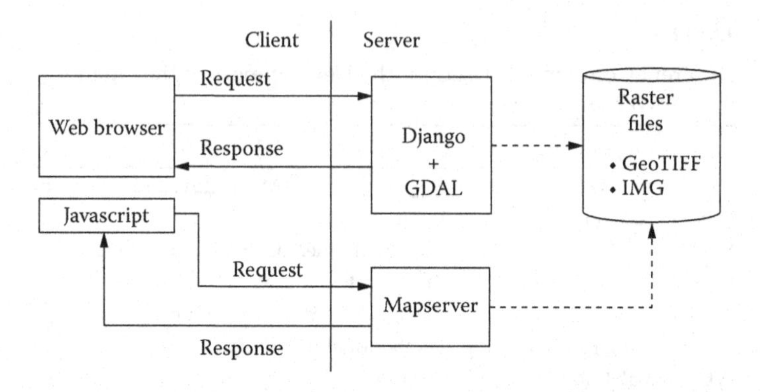

FIGURE 8.13
Architecture of the SDSS client and server communications in our second Web-based SDSS example.

8.3.3 Architecture Used and Implementation

In this SDSS application, JavaScript is utilized on the client side while Python is implemented on the server side. Figure 8.13 shows the overall architecture used in the project. The Web browser starts by sending a request to the Django Web application. Then, the Django Web application parses the map file used by MapServer and responds to the request with information used to interact with MapServer. The Web browser then uses a JavaScript library to communicate with MapServer. In this case, the Web browser initiates the process by sending the request. Django receives the request, parses the current MapServer map file, and then generates one or more new raster map files. Finally, a new map file is generated and written to the disk.

8.3.4 Open Source SDSS Download and Execution

All of the files necessary to run the open source SDSS example have been zipped into a single file, which is available from the Web site that accompanies this book (http://www.geotree.uni.edu/SDSSbook). Download that file and extract all of the contents. Next run **python manage.py syncdb** from the project directory. Next, run **python manage.py runserver 0.0.0.0:8000**. Finally, open a browser and copy this link into it: **http://localhost:8000/**.

`MapServer`

There are six steps necessary to carry out an analysis for placing wind turbines using the Web-based SDSS:

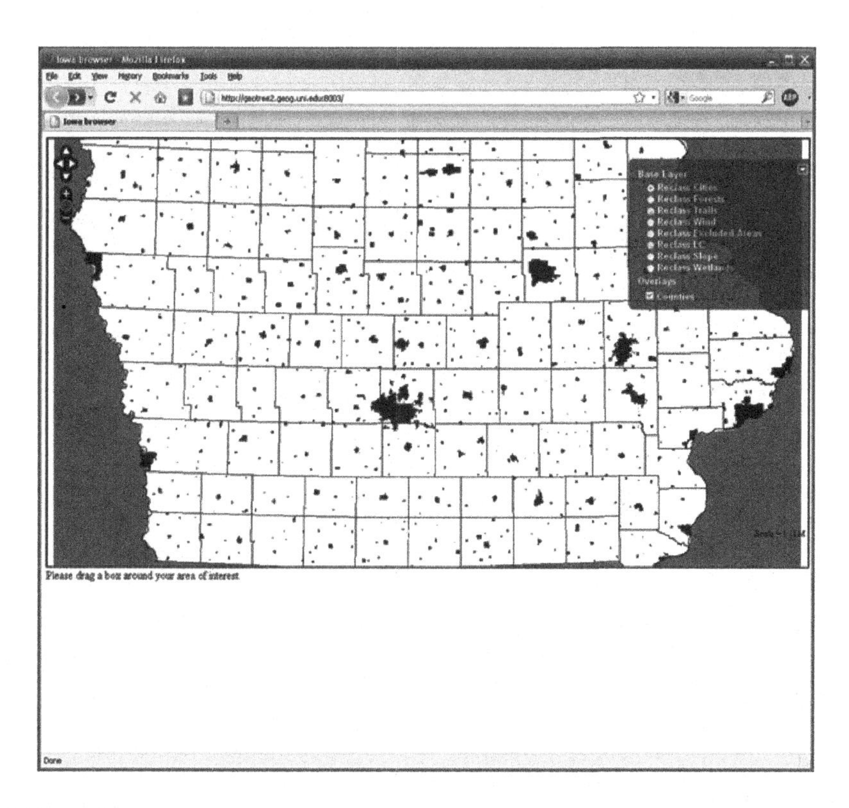

FIGURE 8.14
User interface of the Web-based SDSS running on the Mozilla Firefox Web browser.

1. User defines the area of interest by drawing a polygon on the map (client side).
2. The relevant spatial data layers are clipped using that polygon. Spatial data layers include city boundaries, wooded areas, trails, an estimate of wind potential, slope, wetlands, and land cover (server side).
3. Spatial data from the clipped area are displayed in the Web browser (client side).
4. The user defines the weights for the spatial layers that will be used in the WLC analysis for the study area (client side).
5. The WLC model is run and the final map is created (server side).
6. The results are displayed in the Web browser (client side).

When the Web application (Figure 8.14) is started (after opening http://localhost:8000/ in a browser), the browser is essentially given a list of layers. The client can then use this information to construct the map interface

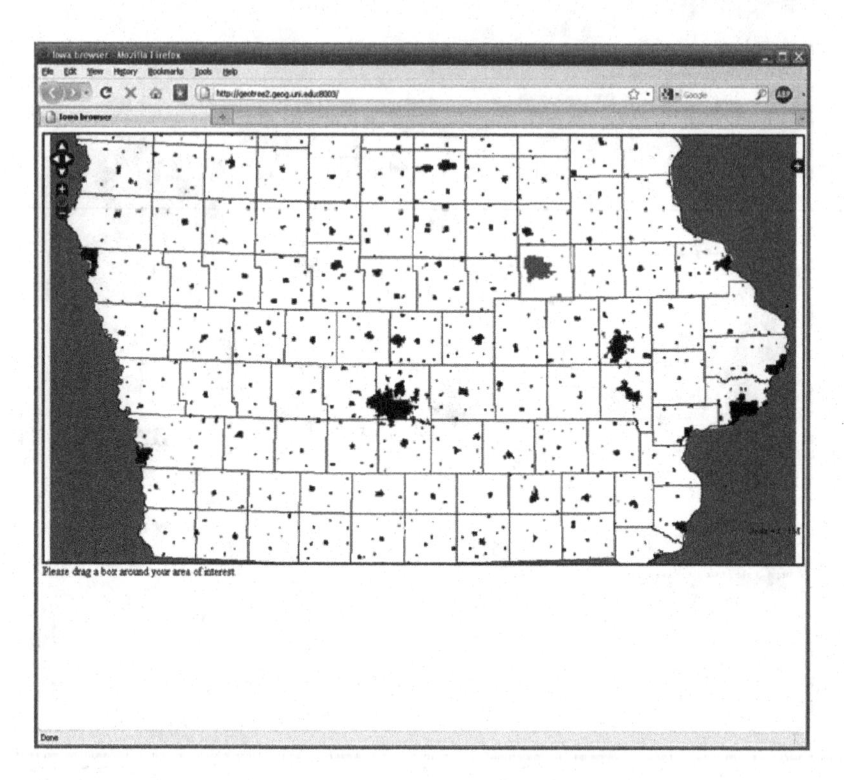

FIGURE 8.15
The selection of the user's area of interest is highlighted.

and converse with MapServer. The user can view the input data using the Base Layer selector (i.e., turn layers on and off).

The user selects the area of interest by drawing a rectangle that should encompass their area of interest. The application sends a clip request to the server. The server sets up a temporary location for the clipped files. If the clipping operation is successful, the browser is redirected to the standard map page. Figure 8.15 shows the user selecting an area around Black Hawk County, Iowa, by drawing a box (Figure 8.15) around the county's outline.

Once clipped, the user must specify the weighting of each layer included in the analysis. In this example, the following layers are included: cities, forested areas, recreational trails, wind speed, excluded areas (inappropriate land cover classes), reclassified (appropriate) land cover classes (barren, grasslands, and agricultural areas), slope, and wetlands. The user must decide what weights to apply to the various layers based on expert opinion. Figure 8.16 shows an example weighting of the layers for Black Hawk County, Iowa. After entering the weights and clicking the Submit button, the Web application performs the WLC calculation and places it with the other clipped data files (if used).

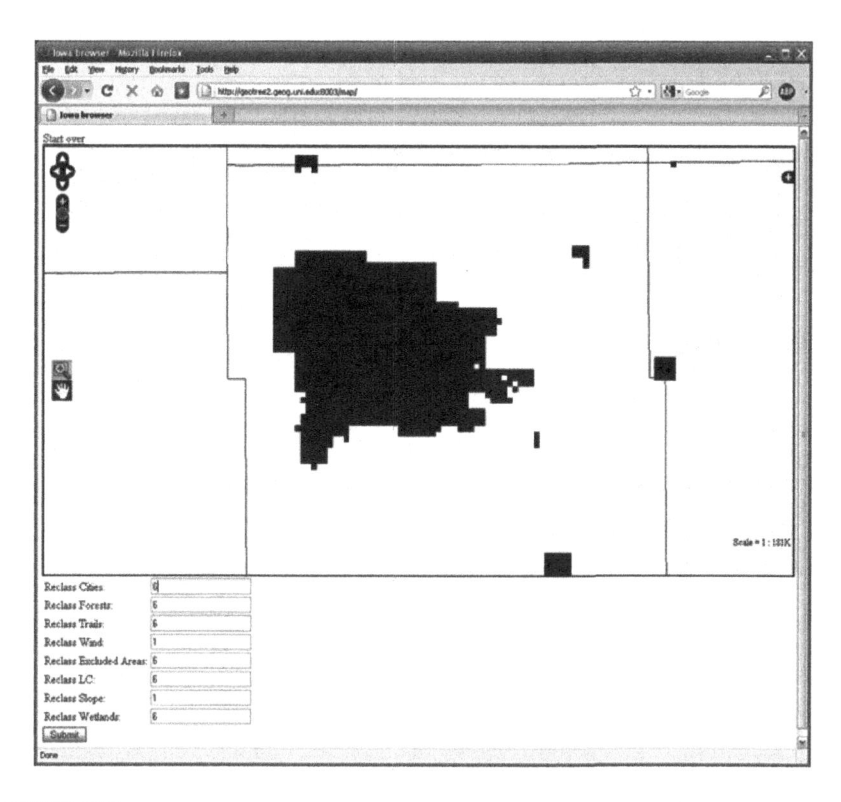

FIGURE 8.16
Area of interest with user-specified weights for the various layers.

A modified mapfile (MapServer file) that contains a reference to the WLC result layer replaces the old mapfile. The client's browser is again redirected to the standard map viewing page. The resultant map displays areas determined to be suitable or unsuitable (dark colors represent better suitability for a wind turbine and lighter areas represent less appropriate areas to locate wind turbines). These results may be viewed, or they may be used in additional calculations. Figure 8.17 shows the results of the weighting of the layers and the WLC computation for the weights set up in Figure 8.16.

8.3.5 Detailed Explanation and Code

This section will provide the code that was used to create this application and provide some explanation of the process of setting up this SDSS. It provides the reader with usable code to carry out this or a similar SDSS operation. First, the various Python modules are briefly discussed and the code is given below the descriptions. Second, the view modules will be

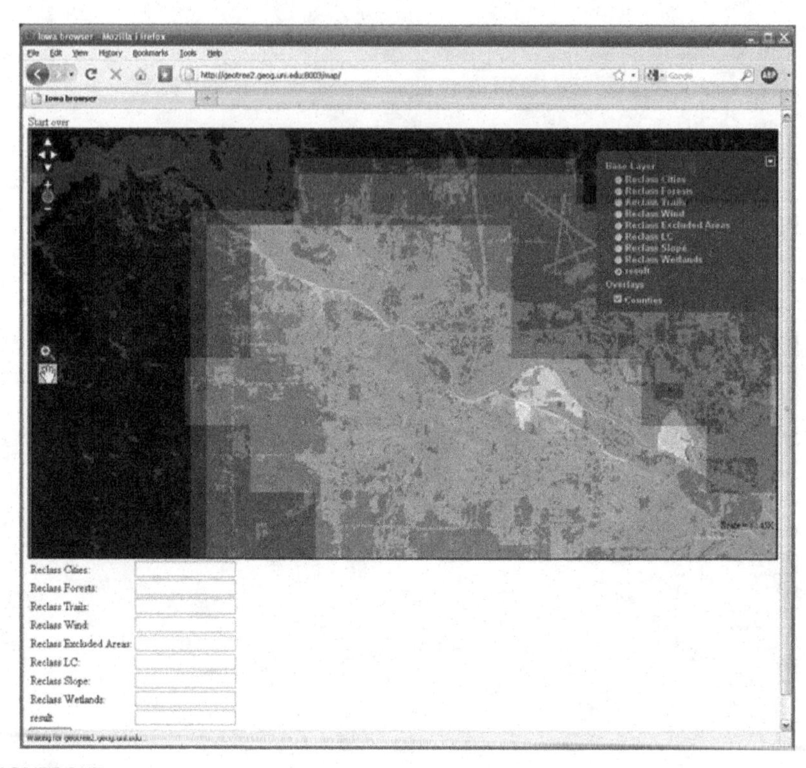

FIGURE 8.17
(See color insert following page 74.)Results of the WLC calculation on the study area. Dark areas suggest high suitability and light areas highlight areas of low suitability for the placement of wind turbines.

introduced along with their associated code. Figure 8.18 shows the hierarchy of files that have been provided and explained below.

8.3.5.1 Python Modules

8.3.5.1.1 The settings.py Module

The **settings.py** module is used to configure the Django application and is provided here:

```
from os import path

DEBUG = True
TEMPLATE_DEBUG = DEBUG

ADMINS = ()

MANAGERS = ADMINS
```

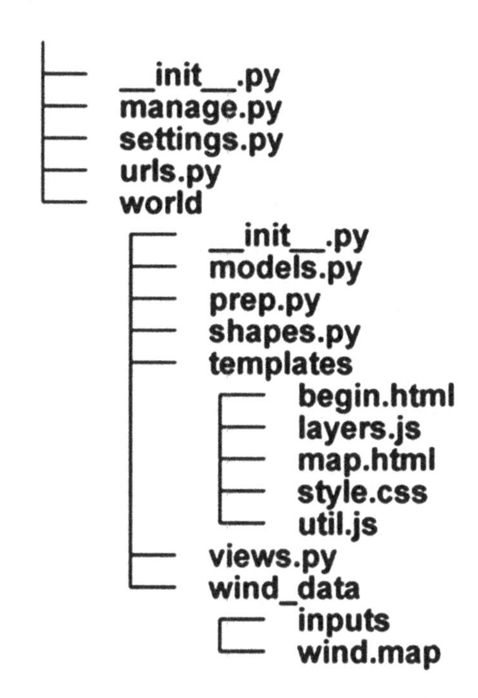

FIGURE 8.18
Directory structure of the necessary files for this Web-based SDSS.

```
DATABASE_ENGINE = 'sqlite3'
DATABASE_NAME = 'windproject'
DATABASE_USER = ''
DATABASE_PASSWORD = ''
DATABASE_HOST = ''
DATABASE_PORT = ''

TIME_ZONE = 'America/Chicago'

LANGUAGE_CODE = 'en-us'

SITE_ID = 1

USE_I18N = True

MEDIA_ROOT = ''

ADMIN_MEDIA_PREFIX = '/media/'

SECRET_KEY = '4w5$o4zm(35^p!968d#*xbu=9w9=6t@=eiyhqtf4&n^_e!
3kgq'
```

```
TEMPLATE_LOADERS = (
  'django.template.loaders.filesystem.load_template_source',
  'django.template.loaders.app_directories.load_template_
  source',
)

MIDDLEWARE_CLASSES = (
  'django.middleware.common.CommonMiddleware',
  'django.contrib.sessions.middleware.SessionMiddleware',
  'django.contrib.auth.middleware.AuthenticationMiddleware',
)

ROOT_URLCONF = 'urls'

TEMPLATE_DIRS = (
  path.join(path.dirname(__file__), 'world', 'templates'),
)

INSTALLED_APPS = (
  'django.contrib.auth',
  'django.contrib.contenttypes',
  'django.contrib.sessions',
  'django.contrib.sites',
)
```

8.3.5.1.2 The urls.py Module

The urls.py module contains the information needed for Django to dispatch Web requests to the Web application. It does this by matching the address of the Web request to a pattern and using this match to determine which view function to use. The left value in the urlpatterns list is the string to which the request is matched, and the right value is the view function. Each view function is given in the views module.

```
from django.conf.urls.defaults import *
from world import views

urlpatterns = patterns('',
  (r'^$', views.begin),
  (r'^clip/(\d+\.?\d*)/(\d+\.?\d*)/(\d+\.?\d*)/(\
  d+\.?\d*)/', views.clip),
  (r'^wlc.*', views.wlc),
  (r'^map/$', views.show_map),
  (r'^util.js$', views.util),
  (r'^layers\.js$', views.layers),
 (r'^style\.css$', views.css),
)
```

8.3.5.1.3 The views.py Module

The views.py module contains functions that deal with Web requests. These functions handle requests by pulling information out of request objects and passing it on to more generic functions.

```
from django.shortcuts import render_to_response
from django.http import HttpResponseRedirect
from os import path, tmpnam
from mapscript import mapObj
import models

missingValue = 'Please enter numerical values for each
raster.'

def begin(request):
  """This is the entry point for the application. This
displays the default data set."""
  # clear the mapfile if there is one.
  request.session['mapfile'] = None
  # render
  return render_to_response('begin.html')

def css(request):
  """Render our CSS file."""
  return render_to_response('style.css', None,
  mimetype='text/css')

def clip(request, bottom, top, left, right):
  """Prepare a temporary location and clip the raster
  files."""
  # get the default mapfile
  mapfile = path.join(path.dirname(__file__), 'wind_data',
  'wind.map')
  # read the mapfile
  map = mapObj(mapfile)
  rasters, shapes = models.read(map)
  # prepare a new place for this data
  tmppath = tmpnam()
  newmapfile = models.prepare_temp_data(mapfile, tmppath)
  # clip the rasters
  models.clip(rasters,
    path.join(path.dirname(mapfile), map.shapepath),
    bottom, top, left, right, path.join(tmppath, map.
    shapepath))
  # set the session variable
  request.session['mapfile'] = newmapfile
  # redirect to /map/
```

```
  return HttpResponseRedirect('/map/')

def end(request):
  pass

def util(request):
  """Render our javascript utility file."""
  return render_to_response('util.js', None, mimetype='text/
  javascript')

def layers(request):
  """Render our LayerData objects as JSON."""
  # find the current mapfile
  mapfile = current_mapfile(request.session)
  if mapfile is None:
    mapfile = path.join(path.dirname(__file__), 'wind_data',
    'wind.map')
  map = mapObj(mapfile)
  rasters, shapes = models.read(map)
  models.add_mapfile_to_layers(rasters, mapfile)
  models.add_mapfile_to_layers(shapes, path.join(path.
  dirname(__file__), 'wind_data', 'wind.map'))
  return render_to_response('layers.js', {'layers': shapes +
  rasters}, mimetype='text/javascript')

def show_map(request, error=None):
  """Render the layers and present the form for setting
  weights."""
  # find the current mapfile
  mapfile = current_mapfile(request.session)
  if mapfile is None:
    return begin(request)
  # read the mapfile
  map = mapObj(mapfile)
  rasters, shapes = models.read(map)
  # add mapfile data
  models.add_mapfile_to_layers(rasters, mapfile)
  models.add_mapfile_to_layers(shapes, path.join(path.
  dirname(__file__), 'wind_data', 'wind.map'))
  # render
  return render_to_response('map.html', {'layers': shapes +
  rasters, 'error': error})

def current_mapfile(session):
  """Retrieve the current mapfile from the session."""
  ret = None
  if 'mapfile' in session:
    ret = session['mapfile']
```

```
  return ret

def nothingiszero(s):
  """float(f) does not like empty strings so we have to
  check for those and convert them to zeros."""
  if s == "":
    return 0
  return s

def get_multipliers(request, layers):
  """Retrieve the weights from the post data."""
  if request.method == 'POST':
    return [float(nothingiszero(request.POST[d.file])) for d
    in layers]
  raise Exception("not a post request")

def wlc(request):
  """Perform the WLC operations."""
  # find the current mapfile
  mapfile = current_mapfile(request.session)
  if mapfile is None:
    return begin(request)
  # read the mapfile
  map = mapObj(mapfile)
  rasters, shapes = models.read(map)

  try:
    # get the post values for each raster layer
    weights = get_multipliers(request, rasters)

    # prepare the input data with prepare data
    wlc = models.WlcFactory()
    preparer = wlc.create_layer_preparer(rasters)
    calc = preparer.prepare_rasters(path.join(path.
dirname(mapfile), map.shapepath))

    # do numpy ops
    saver = calc.perform_calculations(weights)

    # output resulting data
    output_file = 'result.img'
    saver.save(path.join(path.dirname(mapfile), map.
shapepath, output_file), 'HFA')

    #create output_file LayerInfo
    li = models.LayerData('result', output_file, False)
    li.mapfile = mapfile
```

```
    # add output_file to mapfile
    models.add(map, li, saver.min, saver.max)

    # redirect
    return HttpResponseRedirect('/map/')
  except ValueError:
    return show_map(request, "Value error")
```

8.3.5.1.4 The models.py Module

The models.py module defines the functions that the views module uses to carry out operations, including the WLC analysis. It also contains several functions that are used to work with MapServer files.

```
import mapscript
from os import path, mkdir
from subprocess import call
from shapes import RasterLayer
from prep import prepare_data
import gdal
import numpy
from shutil import copy

def read_mapfile(map):
  """Reads a mapfile and maps layers to files."""
  return [LayerData(map.getLayer(i).name,
    map.getLayer(i).data,
    map.getLayer(i).type != mapscript.MS_LAYER_RASTER)
  for i in range(map.numlayers)]

class LayerData(object):
  def __init__(self, name, file, trans):
    self.name = name
    self.file = file
    self.trans = trans
    self.mapfile = None
  def __str__(self):
    return self.name

def add_mapfile_to_layers(layers, mapfile):
  for layer in layers:
    layer.mapfile = mapfile

class WlcFactory(object):
  """A factory that enables the data normalizer, wlc
  calculator, and data exporter to be swapped in and out.
  Create a new calculator and new factory to do different
  SDSS operations."""
```

```
  def create_layer_preparer(self, inputs):
    return Preparer(inputs, self)
  def create_calculator(self, inputs):
    return Calculator(inputs, self)
  def create_saver(self, width, height, geotransform, data):
    return Saver(width, height, geotransform, data)

class Calculator(object):
  """Reads each file, multiplies them by a supplied weight
  and adds them together."""
  def __init__(self, rasters, factory):
    self.__rasters = rasters
    self.__factory = factory
  def perform_calculations(self, weights):
    f = gdal.Open(self.__rasters[0])
    geotransform = f.GetGeoTransform()
    width = f.RasterXSize
    height = f.RasterYSize
    ret = f.ReadAsArray() * weights[0]
    for f, w in zip(self.__rasters[1:], weights[1:]):
      ret += gdal.Open(f).ReadAsArray() * w
    result_data = numpy.round(ret)
    return self.__factory.create_saver(width, height,
geotransform, result_data)

class Saver(object):
  """A simple class that uses GDAL to export raster data."""
  def __init__(self, width, height, geotransform, data):
    self.width = width
    self.height = height
    self.geotransform = geotransform
    self.data = data
  def save(self, output_file, gdal_driver):
    driver = gdal.GetDriverByName(gdal_driver)
    dst_ds = driver.Create(output_file, self.width, self.
height, 1, gdal.GDT_Byte)
    dst_ds.SetGeoTransform(self.geotransform)
    dst_ds.GetRasterBand(1).WriteArray(self.data)
  def getMin(self):
    return numpy.min(self.data)
  def getMax(self):
    return numpy.max(self.data)
  min = property(getMin)
  max = property(getMax)

class Preparer(object):
  """This essentially figures out how to get all the raster
files to have the same dimensions."""
```

```
  def __init__(self, inputs, factory):
    self.__inputs = inputs
    self.__factory = factory

  def prepare_rasters(self, basepath):
    layers = [RasterLayer(path.join(basepath, d.file)) for d
in self.__inputs]
    fixed = prepare_data(layers)
    return self.__factory.create_calculator(fixed)

  def prepare_temp_data(map, destination):
    """Set up the temporary data location for clipped
rasters and other output."""
    mkdir(destination)
    copy(map, destination)
    m = mapscript.mapObj(map)
    mkdir(path.join(destination, m.shapepath))
    return path.join(destination, path.split(map)[1])

  def clip(data, inpath, bottom, top, left, right, outpath):
    'Clips a set of raster files.'
    for d in data:
      call(['gdal_translate',
        '-projwin',
        left, top, right, bottom,
        path.join(inpath, d.file),
        path.join(outpath, d.file)])

  def read(map):
    "Read a mapfile and split it into raster and shape
files."
    # read mapfile
    data = read_mapfile(map)
    # split into rasters and shapes
    return [d for d in data if not d.trans], [d for d in
    data if d.trans]

  def add(map, output_file, minimum, maximum):
    "Add a new raster file to a mapfile."
    i = 0
    while i < map.numlayers:
      if map.getLayer(i).name == output_file.name:
        map.removeLayer(i)
      else:
        i += 1
    layer = mapscript.layerObj()
    layer.name = output_file.name
    layer.data = output_file.file
```

```
    layer.type = mapscript.MS_LAYER_RASTER
    layer.status = mapscript.MS_ON
    layer.dump = mapscript.MS_TRUE
    layer.classitem = '[pixel]'
    layer.setProjection('+init=epsg:26915')
    pixdiff = 255./(maximum - minimum)
    for x in range(minimum, maximum+1):
      grey = int(round((x-minimum) * pixdiff))
      style = mapscript.styleObj()
      style.color = mapscript.colorObj(grey, grey, grey)
      c = mapscript.classObj()
      c.setExpression('([pixel] = %i)' % x)
      c.insertStyle(style)
      layer.insertClass(c)
    map.insertLayer(layer)
    map.save(output_file.mapfile)
```

8.3.5.1.5 The prep.py Module

The prep.py module is used by the models module to normalize raster files so that each file has the same extent and cell size.

```
import os
from os import path
import subprocess

INTERSECT = 0
UNION = 1

SMALLEST = 2
LARGEST = 3

NEARESTNEIGHBOR = 4
BILINEAR = 5
CUBIC = 6
CUBICSPLINE = 7

def union(r1, r2):
  r1.unionRect(r2)
  return r1

def intersect(r1, r2):
  sofar, r1 = r1
  if not sofar or not r1.intersects(r2):
    ret = (False, r1)
  else:
    ret = (True, r1.intersect(r2))
  return ret
```

```
def prepare_data(layers, merge_strat = INTERSECT, pixel_size
= SMALLEST, resampling_method = NEARESTNEIGHBOR):
  """layers is list of layers (from QQIS).
  merge_strat is either INTERSECT or UNION, defined in this
  module.
  If INTERSECT is specified, the intersection of the specified
  layers is used.
  Otherwise the union is used.
  pixel_size is LARGEST or SMALLEST, defined in this module.
  If LARGEST is used, the pixel size is the largest from the
  specified files.
  The return value is a list of paths to files in the same
order as was given.
  These files will have the exact same extent and pixel size
  (hopefully!)
  If INTERSECT is used and the intersection of the layers is
  nothing, an
  exception will be thrown.
  resampling_method is NEARESTNEIGHBOR, BILINEAR, CUBIC,
  or CUBICSPLINE.
  It defaults to NEARESTNEIGHBOR. These are arranged from
  fastest to slowest.
  Currently this function does not deal with images from
  other coordinate
  systems."""
  rects = [l.extent() for l in layers]

  wsizes = [(e.xMax() - e.xMin())/l.getRasterXDim() for (e,
l) in zip(rects, layers)]
  hsizes = [(e.yMax() - e.yMin())/l.getRasterYDim() for (e,
l) in zip(rects, layers)]
  if pixel_size == LARGEST:
    wsize = max(wsizes)
    hsize = max(hsizes)
  elif pixel_size == SMALLEST:
    wsize = min(wsizes)
    hsize = min(hsizes)
  else:
    raise Exception("3rd argument (cell size) was
    nonsense!")
  if resampling_method == NEARESTNEIGHBOR:
    resamp = "-rn"
  elif resampling_method == BILINEAR:
    resamp = "-rb"
  elif resampling_method == CUBIC:
    resamp = "-rc"
   elif resampling_method == CUBICSPLINE:
    resamp = "-rcs"
```

```
  else:
    raise Exception("4th argument (resampling_method) was
    nonsense!")
  output = rects[0]
  if merge_strat == UNION:
    output = reduce(union, rects, output)
  elif merge_strat == INTERSECT:
    worked, output = reduce(intersect, rects, (True,
    output))
    if not worked:
      raise Exception("Intersection equals nothing!")
  else:
    raise Exception("2nd argument was nonsense!")
  xdim = int(round((output.xMax() - output.xMin())/wsize))
  ydim = int(round((output.yMax() - output.yMin())/hsize))

  ret = []
  for l in layers:
    # outpath = unique_name(temp_dir, path.split(str(l.
    source()))[-1])
    outpath = os.tmpnam()
    success = subprocess.call([ "gdalwarp",
      resamp,
      "-te",
      str(output.xMin()),
      str(output.yMin()),
      str(output.xMax()),
      str(output.yMax()),
      "-tr",
      str(wsize),
      str(hsize),
      str(l.source()),
      outpath])
    if success != 0:
      return False
      ret.append(outpath)
    return ret
```

8.3.5.1.6 The shapes.py Module

The shapes.py module determines the extent of a given raster file.

```
class RasterLayer(object):
  def __init__(self, path):
    from osgeo import gdal
    print path
    layer = gdal.OpenShared(path)
    gt = layer.GetGeoTransform()
    RasterXSize = layer.RasterXSize
```

```
    RasterYSize = layer.RasterYSize
    left = gt[0]
    top = gt[3]
    right = gt[0] + RasterXSize * gt[1] + RasterYSize *
    gt[2]
    bottom = gt[3] + RasterXSize * gt[4] + RasterYSize *
    gt[5]
    xmin = min(left, right)
    xmax = max(left, right)
    ymin = min(top, bottom)
    ymax = max(top, bottom)
    self.xdim = RasterXSize
    self.ydim = RasterYSize
    self.src = path
    self.ext = Rect(xmin, ymin, xmax - xmin, ymax - ymin)

  def extent(self):
    return self.ext

  def source(self):
    return self.src

  def getRasterXDim(self):
    return self.xdim

  def getRasterYDim(self):
    return self.ydim

class Rect(object):
  def __init__(self, l, t, w, h):
    self.top = t
    self.left = l
    self.width = w
    self.height = h

  def UnionRect(self, r2):
    left = min(self.left, r2.left)
    top = min(self.top, r2.top)
    width = max(self.left + self.width, r2.left + r2.width)
    - left
    height = max(self.top + self.height, r2.top + r2.height)
    - top
    return Rect(left, top, width, height)

  def intersects(self, r2):
    if self.left < r2.left:
      if self.left + self.width < r2.left:
        return False
```

```
    elif r2.left + r2.width < self.left:
      return False
    if self.top < r2.top:
      if self.top + self.height < r2.top:
        return False
    elif r2.top + r2.height < self.top:
      return False
    return True

  def intersect(self, r2):
    left = max(self.left, r2.left)
    top = max(self.top, r2.top)
    width = min(self.left + self.width, r2.left + r2.width)
    - left
    height = min(self.top + self.height, r2.top + r2.height)
    - top
    return Rect(left, top, width, height)

  def xMin(self):
    return self.left
  def xMax(self):
    return self.left + self.width
  def yMin(self):
    return self.top
  def yMax(self):
    return self.top + self.height
```

8.3.5.2 View Templates

8.3.5.2.1 The begin.html Template

```
<html xmlns="http://www.w3.org/1999/xhtml">
  <head>
    <link href="style.css" rel="stylesheet" type="text/css"
    />
    <script src="http://www.openlayers.org/api/OpenLayers.
    js"></script>
    <script src="layers.js"></script>
    <script src="util.js"></script>
    <script type="text/javascript">
    <!--
      function init(){
        initialize_map();
        var control = new OpenLayers.Control();
        OpenLayers.Util.extend(control, {
          draw: function() {
            this.box = new OpenLayers.Handler.Box(control,
{"done": this.notice});
            this.box.activate();
```

```
          },
          notice: function(bounds) {
            var ll = map.getLonLatFromPixel(new OpenLayers.
Pixel(bounds.left, bounds.bottom));
            var ur = map.getLonLatFromPixel(new OpenLayers.
Pixel(bounds.right, bounds.top));
          window.location = "clip/" + ll.lat + "/" + ur.lat +
"/" + ll.lon + "/" + ur.lon + "/";
        }
      });
      map.addControl(control);
    }
    -->
  </script>
  <title>Iowa browser</title>
  </head>
  <body onload="init()">
    <div id="map"></div>
    <div id="main">Please drag a box around your area of
    interest.</div>
  </body>
</html>
```

Begin.html is the starting point for this Web application. This template requests layers.js and util.js. When those two JavaScript files are loaded, it initializes the map. It has a custom OpenLayers map control that allows the user to select an area of the map and directs the browser to access the server's clip function. The clip function will redirect the browser to an address that uses the map.html template.

8.3.5.2.2 The map.html Template

```
<html xmlns="http://www.w3.org/1999/xhtml">
  <head>
    <script src="http://www.openlayers.org/api/OpenLayers.
    js"></script>
    <link href="/style.css" rel="stylesheet" type="text/css"
    />
    <script src="/layers.js"></script>
    <script src="/util.js"></script>
    <script type="text/javascript">
      <!--
        function init(){
          initialize_map();
          map.addControl(new OpenLayers.Control.
          MouseToolbar());
        }
        -->
```

```
      </script>
      <title>Iowa browser</title>
    </head>
    <body onload="init()">
      <a href="/">Start over</a>
      <br/>
      <div id="map"></div>
      {% if error %}
        <div>Please type numerical values in each field.</
        div>
      {% endif %}
      <form action="/wlc/" method="post">
        <table>
        {% for l in layers %}
          {% if not l.trans %}
          <tr>
            <td>{{ l.name }}:</td>
            <td><input type="text" name="{{ l.file }}" /></td>
        </tr>
        {% endif %}
      {% endfor %}
      </table>
      <input type="submit" value="Submit"/>
    </form>
  </body>
</html>
```

Map.html is very similar to begin.html, but it contains a dynamically generated form as well. When the user clicks the submit button, the browser will access the WLC function which will do the calculations and redirect the browser back to this page.

8.3.5.2.3 The layers.js Template

```
var jlayers = [
{% for l in layers %}
  {"name": "{{ l.name }}", "file": "{{ l.file }}",
"mapfile": "{{ l.mapfile }}", "trans": "{{ l.trans }}" },
{% endfor %}
];
```

This is a dynamically generated JSON file. It contains the data the browser needs to construct a map control that interfaces with MapServer.

8.3.5.2.4 The util.js Template

```
var map;
function initialize_map(){
  var tileSize = new OpenLayers.Size(512,512);
```

```
  map = new OpenLayers.Map('map', {maxExtent: new
OpenLayers.Bounds(190889.053, 4469040.985, 742389.050,
4823040.985 ), maxResolution: 512, minResolution: 1,
projection:"EPSG:26915", controls:[], tileSize: tileSize,
units:"m", numZoomLevels: 10 } );
  var olayers = [];
  var i;
  for (i = 0; i < jlayers.length; i++)
  {
    olayers = olayers.concat([new OpenLayers.Layer.
WMS(jlayers[i]["name"], "http://geotree2.geog.uni.edu:8080/
cgi-bin/mapserv?map=" + jlayers[i]["mapfile"] +
"&SERVICE=WMS&VERSION=1.1.1", {layers: jlayers[i]["name"],
format: "image/png", transparent: jlayers[i]["trans"]})]);
  }
  map.addLayers(olayers);
  map.addControl(new OpenLayers.Control.Scale());
  map.addControl(new OpenLayers.Control.LayerSwitcher({activ
eColor:"green", nonActiveColor:"#386cb0"}));
  map.addControl(new OpenLayers.Control.ZoomToMaxExtent());
  map.addControl(new OpenLayers.Control.PanZoom());
  center = new OpenLayers.LonLat(486096.5, 4843919.0);
  zoom = map.getZoomForExtent(map.maxExtent);
  map.setCenter(center, zoom);
}
```

The util.js template contains some JavaScript that is used by both begin. html and map.html. It creates an OpenLayers map control and initializes with data from the layers.js file.

8.3.5.2.5 The style.css Template

```
#map {
  width: 100%;
  height: 600px;
  border: 2px solid black;
  margin:10px;
}

a:hover {
text-decoration: underline;
}
```

The style.css template should be used to customize the appearance of the Web application.

8.4 Summary

This chapter provided a brief introduction to some of the important considerations and issues in developing Web-based SDSS. The chapter then provided two examples of the development of Web-based SDSS applications. These examples were meant to provide an introduction to possible technological configurations of Web-based SDSS using a commercial GIS server and an open source software configuration. The ArcGIS Server example is one that might be more likely utilized by those used to working with commercial ESRI desktop GIS software. The ArcGIS Server example is potentially easier to develop as there are significant online resources for potential developers. Because the open source example required experience with a variety of technologies, it is more likely that the developer would require a higher level of knowledge to be able to create such an example and even to know where to look for help and tutorial materials. The open source software example provided here would require expertise in a number of technologies including Linux, Python, Django, MapServer, and others. Both approaches would require some HTML and JavaScript knowledge. The open source solution was developed with entirely free software packages, while the ArcGIS Server example required the developer to have several pieces of commercial software (ArcGIS, Spatial Analyst, ArcGIS Server). Although they present different technological challenges, Web-based SDSS have advantages as compared to a desktop SDSS, including the fact that they allow a user to access the SDSS from anywhere in the world with an Internet connection. Secondly, the user does not need to invest in traditional GIS software to run spatial analyses. However, without sufficient knowledge in Web-based programming, building a Web-based SDSS can be fairly difficult.

References

Peng, Z-R., and M-H. Tsou. 2003. *Internet GIS: Distributed geographic information services for the Internet and wireless networks*. New York: John Wiley & Sons, Inc.

Rinner, C., and P. Jankowski. 2002. Web-based Spatial Decision Support—Technical Foundations and Applications." In *The encyclopedia of life support systems (EOLSS), Theme 1.9—Advanced geographic information systems*, edited by Claudia Bauzer Medeiros. Oxford, U.K.: UNESCO/EOLSS Publishers.

Rinner, C. 2003. Web-based spatial decision support: Status and research directions. *Journal of Geographic Information and Decision Analysis* 7(1):14–31.

Sugumaran, V., and R. Sugumaran. 2007. Web-based spatial decision support systems (WebSDSS): Evolution, architecture, and challenges. *Communications of the Association for Information Systems* 19:844–875.

9

SDSS Applications

Learning Objectives

- Gain an understanding of the width and breadth of SDSS applications by investigating a range of examples from various disciplines.
- Be exposed to a Web portal that allows the searching of an SDSS database. The Web portal and database allow researchers, managers, and developers to focus on specific research and management application examples of SDSS in their geographic, technical, or scientific domain.
- Learn how SDSS software are developed and applied by examining in detail numerous case studies from various application domains.

9.1 Introduction

Chapter 2 examined the history and evolution of spatial decision support system (SDSS) technology and applications. It highlighted the large number and wide range of SDSS applications that have been created over the last several decades. These SDSS applications have used various technologies and approaches to address spatial decision-making situations from a variety of disciplines or domains. The objective of this chapter is to both provide an overview of the types of SDSS applications that have been published in the literature highlighted and to provide detailed descriptions of SDSS applications from a range of application domains. This chapter will highlight the range of disciplines to which SDSS have been applied. This chapter offers planners, managers, developers, and other decision makers who might utilize geographical information science (GIS), decision support systems (DSS), and SDSS a rich source of information regarding the application areas where SDSS have been applied. The review in this chapter and the accompanying searchable database will allow researchers, managers, and implementers to focus on specific research and management application examples of SDSS in their geographic, technical, or scientific domain. Potential SDSS application developers will be able to investigate existing systems that might be similar to the system that they would like to develop. Specifically, they can see what kinds of technologies (e.g., GIS based, Web based), integration levels (i.e., tightly or loosely coupled), and software development environments have been used for developing SDSS.

The present review is purposefully broad historically, geographically, technologically, and as it pertains to discipline. This chapter will first detail the compilation of references from journals, conference proceedings, books and book chapters, the creation of a database to store information about these references, and the development of a Web portal for accessing this database. Then an overview of publications from specific disciplines or application domains will be provided. For each major domain area, at least one in-depth presentation of a case study will be given. These case study examples are meant to cover a variety of aspects including SDSS evolution, application domains, software integration techniques, models and platforms used, type of end users, and other relevant topics.

9.2 Reference Collection, Database Creation, and Web-Portal Development

9.2.1 Literature Compilation

As mentioned in Chapter 2, a thorough examination of the literature, particularly from journals, conference proceedings, books, and book chapters from 1976 to 2008, was compiled using multiple Web-based search engines, electronic libraries, and databases, which are given in Table 9.1.

The search focused on obtaining a wide variety of SDSS publications covering the diversity of application domains, publication types, study areas, and so on. Publications were included only if published in peer-reviewed journals, book chapters, or conference proceedings. Search terms used included *spatial decision support system, spatial and DSS, GIS-based decision support system, GIS and DSS, multi-criteria spatial decision, spatial decision making*, and *spatial decision support*. These search terms occasionally resulted in numerous publications that could more accurately be described as GIS applications and did not meet the definition of an SDSS. Only those articles that applied, reviewed, or developed systems that met the definition of an SDSS were included. The functional definition of

TABLE 9.1
Sources Used for Literature Compilations

Source	URL
Google Scholar	http://scholar.google.com
ESRI	http://training.esri.com/campus/library
Ingenta	http://www.ingentaconnect.com
Inspec®	http://www.engineeringvillage2.org
ISI Web of Knowledge™	http://www.isiwebofknowledge.com
Pion Publications Ltd.	http://www.pion.co.uk
Project MUSE®	http://muse.jhu.edu
ProQuest®	http://www.proquest.com/en-US/
ResearchIndex	http://www.researchindex.com
ScienceDirect®	http://www.sciencedirect.com
Scirus	http://www.scirus.com
Scopus™	http://www.scopus.com/scopus/home.url
SpringerLink	http://www.springerlink.com
WorldCat®	http://firstsearch.oclc.org

SDSS, as considered here, is an integrated model-based spatial software system that is capable of supporting a decision-making process. In addition to the Internet searches, relevant articles cited in located articles were also acquired. Especially useful in this regard were some of the review articles including Gould and Densham (1991) and Malczewski (2006). The final number of publications included in this review and in the database is 451.

9.2.2 SDSS Database Development

In order to systematically store the results of the literature search, a Microsoft Access database was developed. This database contains unique records for each of the articles reviewed. Relational tables containing the following information were entered in the database : lead author, year of publication, article title, publication type (e.g., journal, proceedings, book chapter), publication name, study area where SDSS was applied, the spatial data type(s) used (vector, raster), main application domain, secondary application domain, primary software used in SDSS, other software used, the coupling mechanism (loose, tight, integrated), primary users, software integration language(s), platform (desktop, Web-based, mobile), operating system, keywords, a URL, and the abstract as taken from the articles. Sometimes it was not possible to discern all of this information from a given article, and thus no value was added or "unspecified" was entered in the database.

9.2.3 Web Portal Development

In order to provide access to potential readers of this book, a Web-based portal was developed for searching the SDSS database and displaying results. The reason for the development of this portal was to provide existing and potential SDSS users and researchers (planners, managers, developers, educators, and others) with a searchable database covering SDSS-related publications. The Web-based SDSS database portal (Figure 9.1) developed in this project can be accessed at www.geotree.uni.edu/SDSSbook. The Web portal operates with a MySQL version of the publication database. The two main functions of this Web portal are to search the SDSS database and insert new records to the database for new SDSS articles (Figure 9.2). The portal allows users to search in a variety of ways, including by keywords, author names, publication source type, publication name, study area, generalized application domain, software employed, models used, and year published. Search results return a list of relevant articles. Web links to each of the articles online are also returned. A Web-based mapping interface developed using the MapServer software highlights countries of the world that match given search criteria by the user. So, for example, if a user is interested in urban SDSS applications using ArcGIS software, they can enter those criteria, and then countries for which SDSS of this type have been applied will be highlighted. A list of relevant publications will also be provided. A user can register with the Web portal manager and, after assignment of a password, can enter information concerning a new publication. He or she will be able to enter all of the relevant information into a Web form (Figure 9.2), and this will be inserted into the database after review by the Web site managers.

9.3 Publication Sources

As discussed previously, many decision problems covering a wide range of societal issues have a geographic component. With the continued evolution of more powerful and affordable computer systems, there has been, over the last few decades, a consistent increase in the application of spatial technologies to a wide range of disciplines and complex spatial decision situations. This wide-ranging activity is reflected in the number of publications registered in the database and discussed here. Even given the large number of publications, there are many further uses of SDSS that are not captured in this type of examination of generally academic literature. There are many applications by government agencies and commercial organizations that are not published in a manner in which they are easy

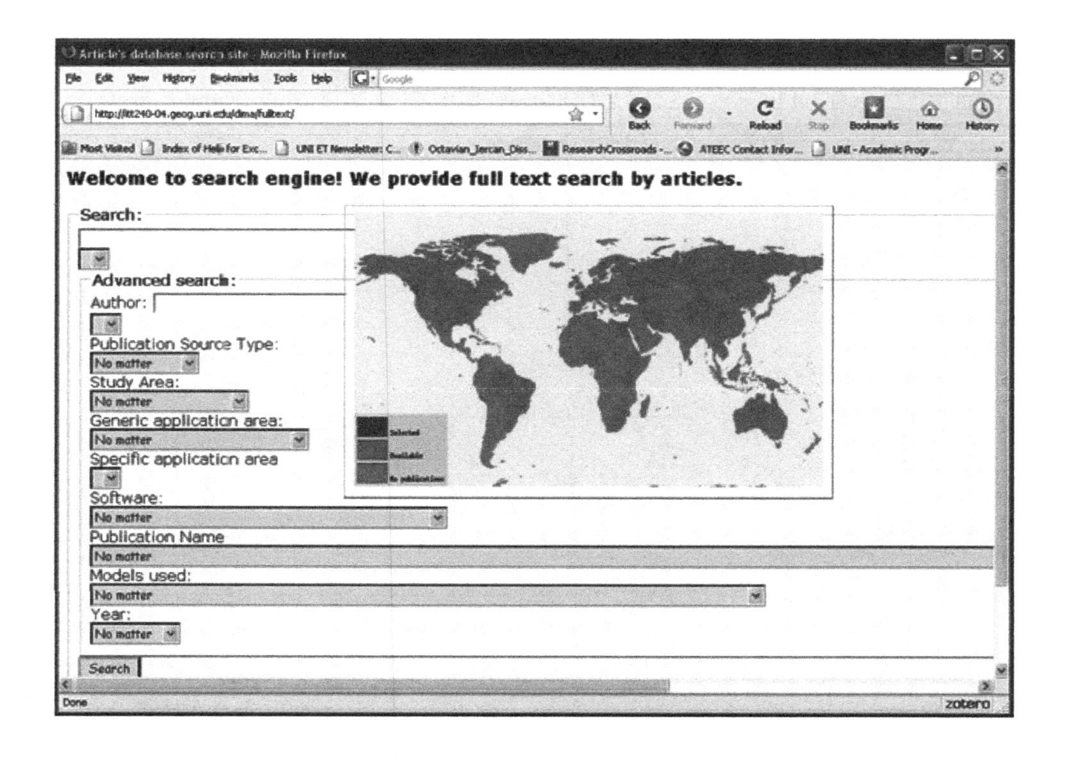

FIGURE 9.1
(See color insert following page 74.) SDSS Web portal interface.

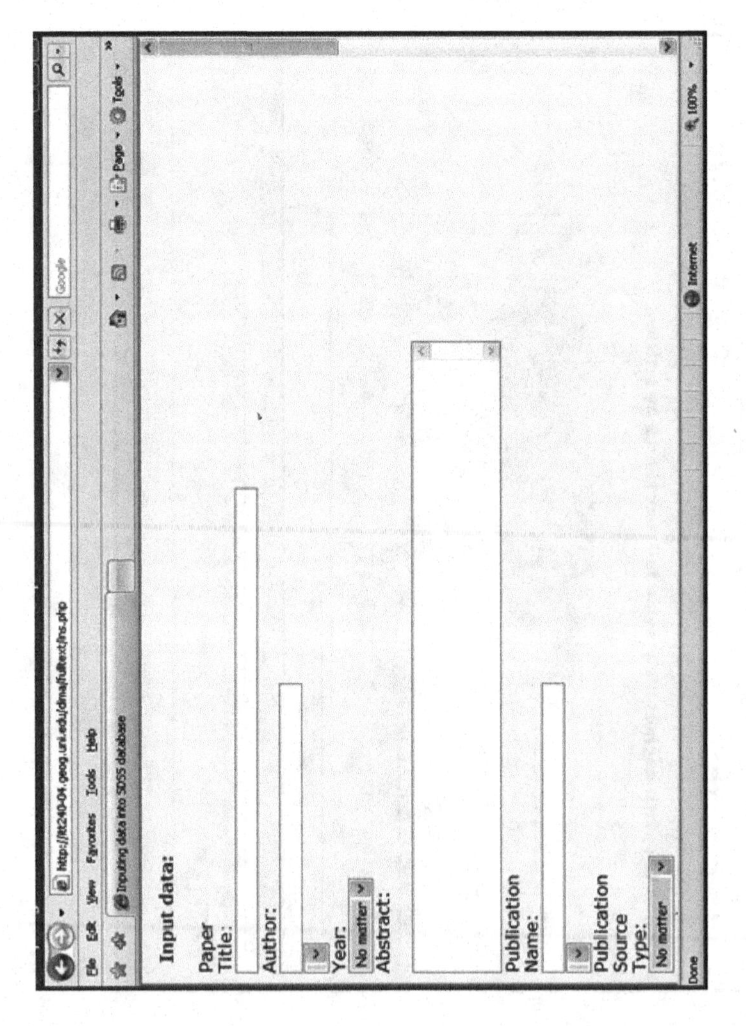

FIGURE 9.2
Web portal new publication data entry form.

to find using the search methods adopted here. Further Internet searches might uncover other applications (e.g., informal published reports), but these types of applications are not included here as the lack of standards in reporting would make it difficult to discern the quality of the publication. Table 9.2 provides a breakdown of the publication source, number of articles published in specific sources, and the percent of articles found in these publications.

Journals: The majority of publications were in peer-reviewed journals, followed by conference proceedings and book chapters. Articles were found in a large number of different journals. The articles appeared in publications dealing with a wide range of subject areas, including environmental, urban, agriculture, GIS, planning, ecology, water resources, and various others. Only six journals published at least 10 SDSS-related publications, while more than 40 other journals had more than one SDSS-related publication. Several GIS-related journals (*International Journal of*

TABLE 9.2

Summary of Publications Sources (All are Journals except the URISA 92 Annual Conference Proceedings)

Publication Source	Number of Articles	%
Environmental Modelling & Software	24	5.3
Computers, Environment and Urban Systems	19	4.2
Computers and Electronics in Agriculture	16	3.5
Decision Support Systems	12	2.7
International Journal of Geographical Information Science	12	2.7
Journal of Environmental Management	10	2.2
Environment and Planning B: Planning and Design	7	1.6
Ecological Modelling	6	1.3
Ecological Economics	5	1.1
Environmental Management	5	1.1
Landscape and Urban Planning	6	1.3
Transactions in GIS	5	1.1
URISA 92 Annual Conference Proceedings	5	1.1
Automation in Construction	4	0.9
Forest Ecology and Management	4	0.9
Journal of Geographical Systems	4	0.9
Land Use Policy	4	0.9
The International Archives of the Photogrammetry, Remote Sensing and Spatial Information Sciences	4	0.9
URISA Journal	4	0.9
Other	295	65.4
Total	**451**	**100.0**

Geographical Information Science, Transactions in GIS, Journal of Geographical Systems, Journal of Geographic Information and Decision Analysis) contained numerous articles. The top four journals explicitly state the terms *software, computers,* or *decision support systems* in their title. Given these journals' focus on technological approaches to specific application domains, such as the environment, urban studies, and agriculture, it is not a surprise that many authors submitted their articles to them. Also, given the very common use of GIS technologies in SDSS, it is not surprising that GIS-related journals and conferences are well represented. A possible reason that the non-GIS-specific journals lead the list is that authors come from various disciplines utilizing spatial technologies as a tool or means to an end within their own discipline. The leading journal, *Environmental Modeling and Software,* contained a wide range of SDSS applications focusing on water quality modeling, air quality emissions, ecological and land use management, aquaculture, and urban planning. This journal has been published since 1997, but interestingly, all articles included in this review were from 2002 and after. The dispersion of SDSS papers across various subject areas in the literature demonstrates both the broad nature of SDSS applicability and the fact that SDSS technologies are fairly young with no specific journals devoted to the subject area. In the case of more traditional DSS, there is a journal called *Decision Support Systems,* which has been published since 1985.

Conferences: Conference proceedings were a common source of publications in the early years of SDSS development. A significant number of publications appeared in proceedings from early Geographic Information Systems/Land Information Systems (GIS/LIS) conferences (1989–1995) as well as from ESRI International User Conferences. Otherwise, publications came from a very wide variety of conference proceedings focusing on geospatial technologies, natural resources, environmental modeling, decision science, agriculture, transportation, engineering, and others. There have been no single conferences dedicated to SDSS. Most of the conference proceedings relating to this topic come from GIS-based conferences (e.g., GIS/LIS conferences, ESRI User conferences, and International Society for Photogrammetry and Remote Sensing conferences). These conferences focus on GIS-based and remote sensing–based research and modeling, but commonly included SDSS applications or techniques.

Books and Book Chapters: The topic of SDSS has been addressed in chapters or sections of a number of books. However, there are no books that specifically cover the subject of SDSS as their main focus. Book chapters came from books on a variety of subjects, including the environment, planning, natural resources, business, and decision support systems. books are very topic specific, such as forest management (Kangas et al. 2008), land use change (Koomen et al. 2007), architecture and urban planning (van Leeuwen and Timmermans 2006), environmental scenarios (Alcamo

2008), and site suitability (Reitsma 1990). A number of technique-specific SDSS books focused on such topics as multiple criteria decision analysis (Malczewski 1999), Web-based SDSS (Peng and Tsou 2003), intelligent systems (Turban and Aronson 1998), and agent-based modeling techniques (Gimblett 2002). The following sections provide an analysis of different application domains with a few detailed case study examples.

9.4 SDSS Application Domains

A large number of publications were found and compiled in the SDSS database with a wide range of disciplines covered. Many of these publications described the development or application of an SDSS for a specific purpose, study area, and with specific stakeholders. In order to provide a useful summary, the publications were categorized into eleven generalized application domains (Figure 9.3) based on the major theme that appeared in the publication. Many publications encompassed aspects of several of the generalized application areas. For example, an urban application might have contained some transportation elements, while an SDSS meant to address pollution from various land uses might have analyzed agricultural practices. Efforts were made to identify the most fundamental goal or important aspect of the publications and then to assign the article to the corresponding generalized application area. A significant number of articles were general SDSS reviews or reviews of specific application domains or technologies.

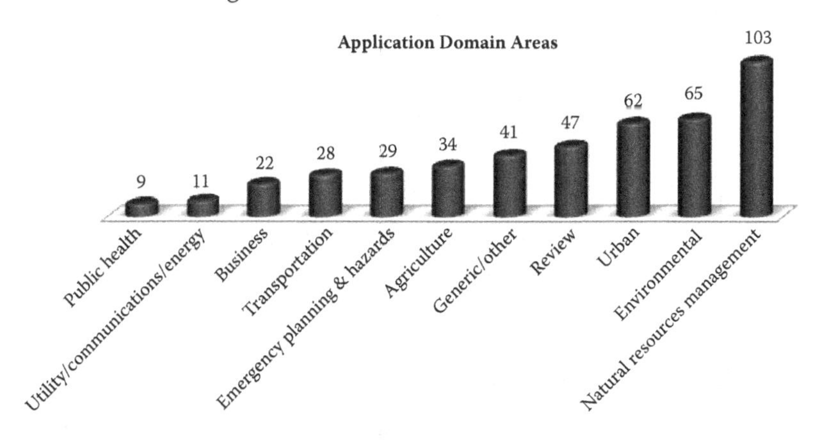

FIGURE 9.3
Number of articles by application domain for which the SDSS was applied.

Many articles could not be clearly classified into an application category or the SDSS discussed was generic in nature. These types of articles were categorized as generic/other. For example, Yeh and Qiao (2004a and 2004b) discussed the development of a knowledge-based planning support system (PSS) that can be used to build new models or utilize existing models for planning problems. Chakhar and Mousseau (2007) discussed a new algebra for carrying out multi-criteria spatial modeling. Rivest et al. (2005) examined techniques for merging business intelligence with geospatial technology. Vyas et al. (2007) investigated the use of spatial association rule mining (spatial ARM). Some articles did not fit clearly into any of the generalized application areas and were characterized as Others. These included SDSS dealing with electoral systems, mineral extraction, and poverty management. In the following sections, the authors attempt to highlight publications from some of the most common application areas. The examples provided are not exhaustive but demonstrate classical examples of SDSS for different applications that can be referenced by managers, developers, planners, and any other decision makers. These examples offer a rich source of information about different domains where SDSS is effectively implemented and used.

The rest of this chapter will focus on providing an overview of SDSS applications from the domains shown in Figure 9.3 and also a few selected in-depth examinations of SDSS examples from several of those domains. These in-depth case studies were at least partially chosen due to unique techniques or technologies that were utilized. The goal of this chapter is to show the diversity in application of SDSS to different domains along with detailed case studies of unique SDSS developed within these application areas. The more detailed discussion of several case study examples will specifically cover the application domain, the platform used, the models used, software integration techniques, and a description of the users or stakeholders involved. The purpose of these detailed descriptions is to give a broad and diverse sampling of practical and technical aspects of SDSS applications to a variety of domains. Table 9.3 highlights some of the case study examples selected for detailed discussion. The examples are listed in chronological order. The Implementation column is meant to indicate the spatial management/analysis software that was used. These examinations provide enough detail for a reader to get a feel for the entire process of conceiving, developing, and applying an SDSS in a particular application domain.

9.4.1 Natural Resources Management

The natural resources management category contained a wide range of publications describing applications of SDSS. The authors qualified a publication as falling in the natural resources management category if the

TABLE 9.3

Case Study Examples of Various SDSS from Different Application Domains

Application Domain	Purpose	Platform	Implementation	Author(s)
Natural Resources Management	Land Use Planning	Desktop Intelligent	ArcInfo	Zhu et al. 1996
Natural Resources Management	Habitat Site Development and Restoration	Desktop Collaborative	ArcView	Jankowski et al. 1997
Urban	Future City Development and Woodland Regeneration	Web-based Collaborative	Geotools	Carver et al. 2001
Urban	Land Use Change Impacts	Web-based Collaborative	MapServer	Sikder and Gangopadhyay 2002
Environmental	Environmental Sensitivity for Watershed Management	Web-based	ArcIMS	Sugumaran et al. 2004
Emergency Planning and Hazards	Construct Drought Indices	Desktop	GRASS	Wu et al. 2004
Business	Hotel Finder/ Bar Finder	Mobile SDSS	ArcPad	Rinner et al. 2005
Business	Housing Accessibility Analysis	Web-based	Custom	Neis et al. 2007
Emergency Planning and Hazards	Snow Removal Planning	Web-based Intelligent	ArcIMS	Sugumaran et al. 2007
Public Health	Health Care Allocation	Web-based	ArcGIS, ArcSDE, ArcIMS	Schuurman et al. 2008
Agriculture	Real time Crop Yield SDSS	Web-based	MapServer and ERDAS	Kaparthi and Sugumaran 2009

application of SDSS related to the purpose of management of land and water resources for the benefit of the public good with an emphasis on conservation or restoration of natural or seminatural resources. The most common applications of SDSS within the natural resources management category dealt with forestry or water management. Examples of forestry-related SDSS included those used for forest restoration planning (Hampton et al. 2003; Reynolds and Hessburg 2005; Stoms et al. 2004), afforestation planning (Gilliams et al. 2005), and management of forest pests such as spruce budworm (MacLean et al. 2000) and the eastern hemlock looper

(Power and Saarenmaa 1995). Examples of SDSS applications dealing with water resources included those for irrigation demand estimation (Leenhardt et al. 2004), watershed management (Choi et al. 2005; Dymond et al. 2004; Loi 2008), water pollution abatement management (Volk et al. 2008), and socioeconomic analysis for integrated watershed management (Hirschfeld et al. 2005). Applications dealing with biodiversity (Bolte et al. 2006; Romero-Calcerrada and Luque 2006; Wong et al. 2007), coastal and marine management and regulation (West 1999), habitat conservation (Larson and Sengupta 2004), ecological restoration (Trepel 2007), fisheries or marine conservation (Carrick and Ostendorf 2007; Wood and Dragicevic 2007), ecologically related land use planning (Sugumaran 2002; Geneletti 2007), and assessment of national park carrying capacity (Prato 2001) all were included in the natural resources management category. A sampling of SDSS applications in the natural resources management category is provided in Table 9.4.

Detailed Case Studies: The following three summaries of unique SDSS include one on land use planning, one regarding habitat site development and restoration, and one for woodland regeneration. These three SDSS are useful for illuminating the application of spatial decision support for various natural resource management problems, including an SDSS for use by an individual, a collaborative or group SDSS, and a public participatory SDSS.

The first example is the Islay Land Use Decision Support System (ILUDSS), which *combined a knowledge-base and modeling system with GIS for land use planning* on the island of Islay off the west coast of Scotland. Zhu et al. (1996) developed the stand-alone desktop system. The major purpose of this knowledge-based SDSS was for strategic land use planning in rural areas for use by land managers and planners. Both vector (ArcInfo coverages) and raster (ArcInfo grids) data types were incorporated in the modeling process and for visualization. Potential areas considered for development were constrained by spatial characteristics such as the proximity to roads, nature conservation areas, slope, and aspect. ILUDSS utilized spatial data including topography, socioeconomic data, and environmental and conservation area locations. These data were stored in ArcInfo coverages and an Oracle database storing ecological information. ILUDSS was developed using a combination of C Language Interface Production System (CLIPS; an expert system development tool), HARDY (a diagramming tool), and ArcInfo (Figure 9.4). The CLIPS tool was used for building query processing subsystems, the modeling subsystem, and for building and making inferences on rule bases. The spatial operations were carried out by ArcInfo, the intelligent components were developed with CLIPS, and HARDY was used to display, create, and edit land use models and to build the user interface. The software components were tightly integrated through a single interface with communication between the software applications hidden from the users. ILUDSS

TABLE 9.4

Natural Resources Management Application Examples

Purpose	Platform	Implementation	Author(s)
Strategic Land Use Planning	Desktop Intelligent	ArcInfo	Zhu et al. 1996
Modeling Carrying Capacity for National Parks	Theoretical	Theoretical	Prato 2001
Future City Development and Woodland Regeneration	Web-based Collaborative	Geotools	Carver et al. 2001
Identification of Nature Conservation Priorities	Desktop	ILWIS	Geneletti 2004
Evaluating Economic Incentives for Rainforest Restoration	Desktop	ArcView	Stoms et al. 2004
Groundwater Management Using Numerical Groundwater Modeling	Desktop	GRAM++ GIS	Kumar 2005
Watershed Management	Web-based	MapServer	Choi et al. 2005
Public Participatory Water Use Planning	Desktop Collaborative	ArcGIS	Jankowski et al. 2006
At-Risk Species and Habitat Management	Web-based	MapServer	Wong et al. 2007
Integrated River Basin Management	Desktop	Custom	Volk et al. 2007
Predicting Marine Mammal Habitats	Desktop and Web-based	ArcGIS and MapServer	Best et al. 2007
Planning Terrestrial and Marine Conservation Reserves	Desktop	ArcGIS	Crossman et al. 2007
Forest Harvesting	Desktop	ArcGIS	Zhang et al. 2008

contained multiple spatial analysis tools or models, including buffering, overlay, and reclassify, which were developed as Arc Macro Language (AML) programs. The results of the land suitability models showed areas suitable for specific land uses but not the most profitable areas for specific land uses. There were only three types of land use considered, and evaluation of land use potential only included physical suitability, proximity to desirable and undesirable landscape features, and the minimum area of each parcel of land. The ILUDSS is an example of a tightly coupled GIS-centric SDSS that utilized a knowledge or expert component. The system was applied for a specific purpose but could have been applied to other areas.

A second natural resources management application was developed by Jankowski et al. (1997) for *collaborative spatial decision-making*. This SDSS

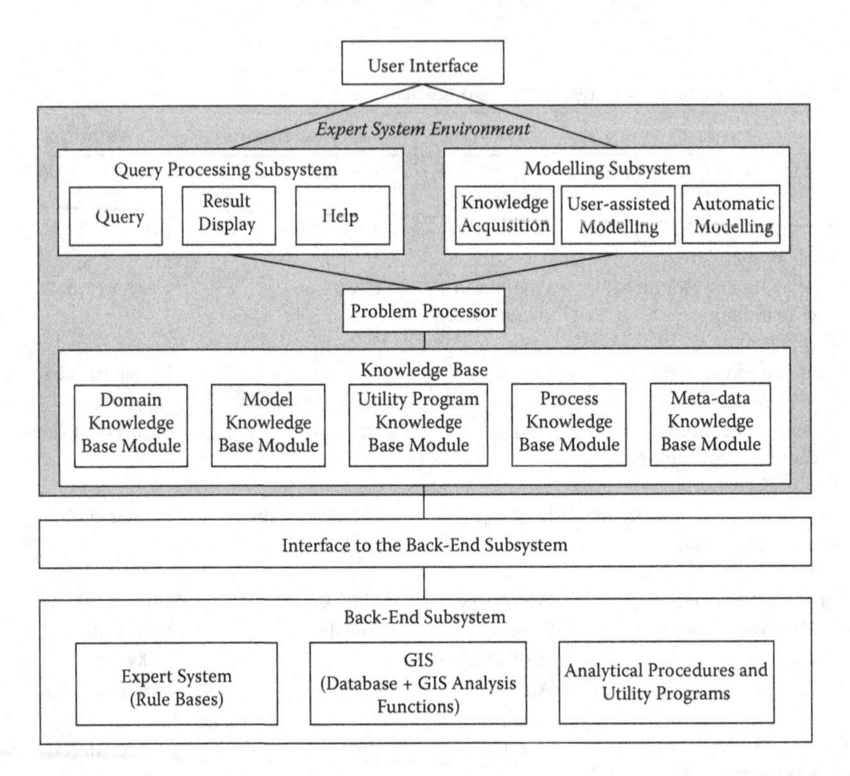

FIGURE 9.4
ILUDSS Architecture used. (Adapted from Zhu et al. 1996.)

was called Spatial Group Choice and was used in prioritizing habitat site development and restoration (Figure 9.5). The goals of the system included exploring and understanding the problem, articulating and sharing decision criteria and criteria preferences, evaluating solution alternatives, and negotiating a consensus solution. The area considered was the Duwamish Waterway area, a tributary to Elliott Bay in the Puget Sound Region of the state of Washington. Urban and industrial development over the last 150 years had resulted in degraded estuarine environments. Water and sediment pollution and habitat degradation for fish, aquatic life, birds, and mammals had occurred. Channeling/straightening, dredging and filling, construction of urban infrastructure, and shoreline and stream stabilization all affected the natural environment. Spatial Group Choice was used to facilitate the computer-supported interactions of small groups in meetings. Two modules exist in this SDSS—multi-criteria evaluation and interactive map visualization. The decision makers involved were from many organizations, including the National Oceanic and Atmospheric Administration (NOAA), City of Seattle, King County Department of Metropolitan Services, U.S. Fish and Wildlife Service, Washington State

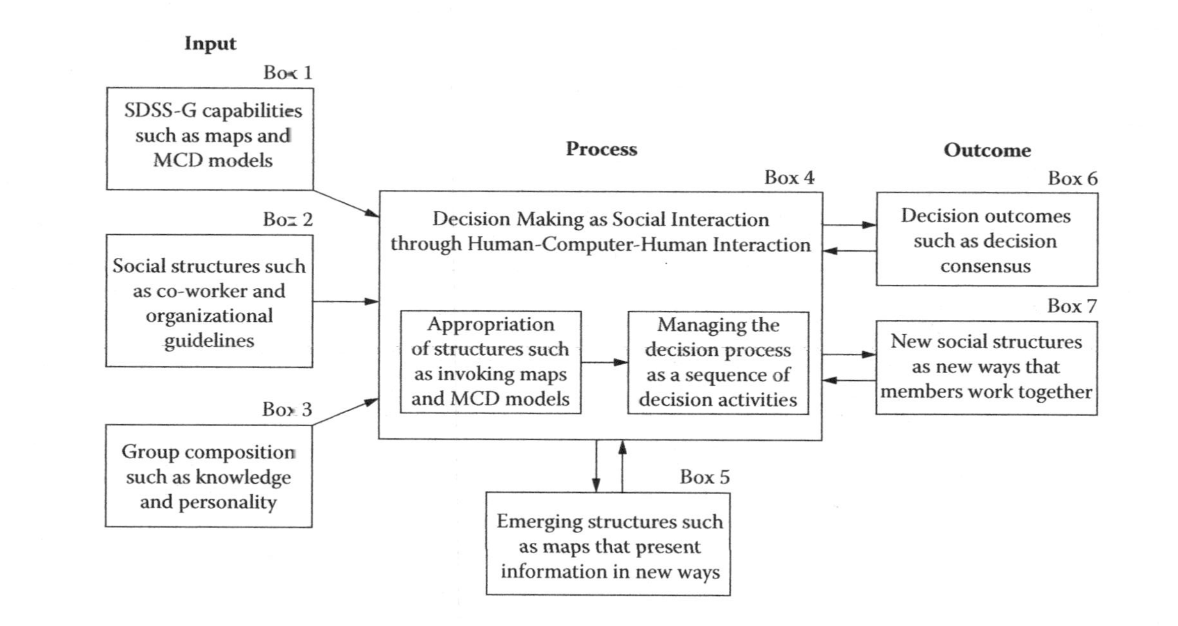

FIGURE 9.5
Overview of the decision structure used in the Spatial Group Choice application. (Jankowski, P., T. L. Nyerges, A. Smith, T. J. Moore, and E. Horvath. 1997. Spatial group choice: An SDSS tool for collaborative spatial decision-making. *International Journal of Geographical Information Science* 11 (6):577–602.)

Department of Ecology, the Muckleshoot Indian Tribe, the Suquamish Tribe, and the Duwamish Coalition, a public interest group with a mix of public and private members whose aims include sustainable economic growth. ArcView 2 was used in conjunction with Microsoft Windows for Work Groups in a loosely coupled configuration. The data consisted of orthophotos of the area along with coverages and thematic maps of the different proposed habitat restoration sites. The system worked as follows: five decision makers were given the task of selecting three sites for habitat development. Everyone was given a computer with all of the relevant data and seated around a public display. There was a facilitator to help encourage the group to reach a consensus. The ArcView module provided file management, presented background information about the region and attribute data for each site, displays for the ranking of sites, and a help menu. A database contained information on distance to nearest contaminated area, ecological suitability, estimated cost of development, existing and potential future land use, property ownership, proximity to nearest existing habitat, proximity to public access, proximity to nearest public facility, and the size of the site. Each user was given a list of attributes and had to decide (based on their stakeholder values) the importance of the various attributes. Then the different decision makers' criteria for site selection were compared and discussed by the entire group, and the criteria were then voted on (Figure 9.6). The results were shown to the group to help reach a consensus. This type of group SDSS was designed to be a flexible research tool used to discover the dynamics of collaborative spatial decision-making processes that make use of SDSS. The researchers saw room for improvement, including a tighter coupling with GIS, more techniques for expressing and visualizing preferences and prioritizing choice alternatives, and making the system flexible enough to work for group consensus when meetings are held at the same time but in different locations, and for when decision makers are at different locations and at different times. This SDSS was an influential early collaborative SDSS and public participatory GIS.

In the article titled "Public Participation, GIS, and Cyber Democracy: Evaluating On-line Spatial Decision Support Systems" (Carver et al. 2001) the use of a *Web-based public participatory GIS* (PPGIS) in decision making was discussed. Carver et al. presented a case study that used Web-based PPGIS for planning woodland regeneration in a national park in the United Kingdom. This application used spatial analysis and modeling results in a public participatory process that was developed for guiding woodland regeneration in the Yorkshire Dales National Park. This park covers 680 square miles (1,762 square kilometers) and contains many Sites of Special Scientific Interest, farmland, woodland, moorland, heathland, grassland, and heritage sites with a population of approximately 20,000 people. The purpose of the PPGIS was to collect data on the public's feelings on

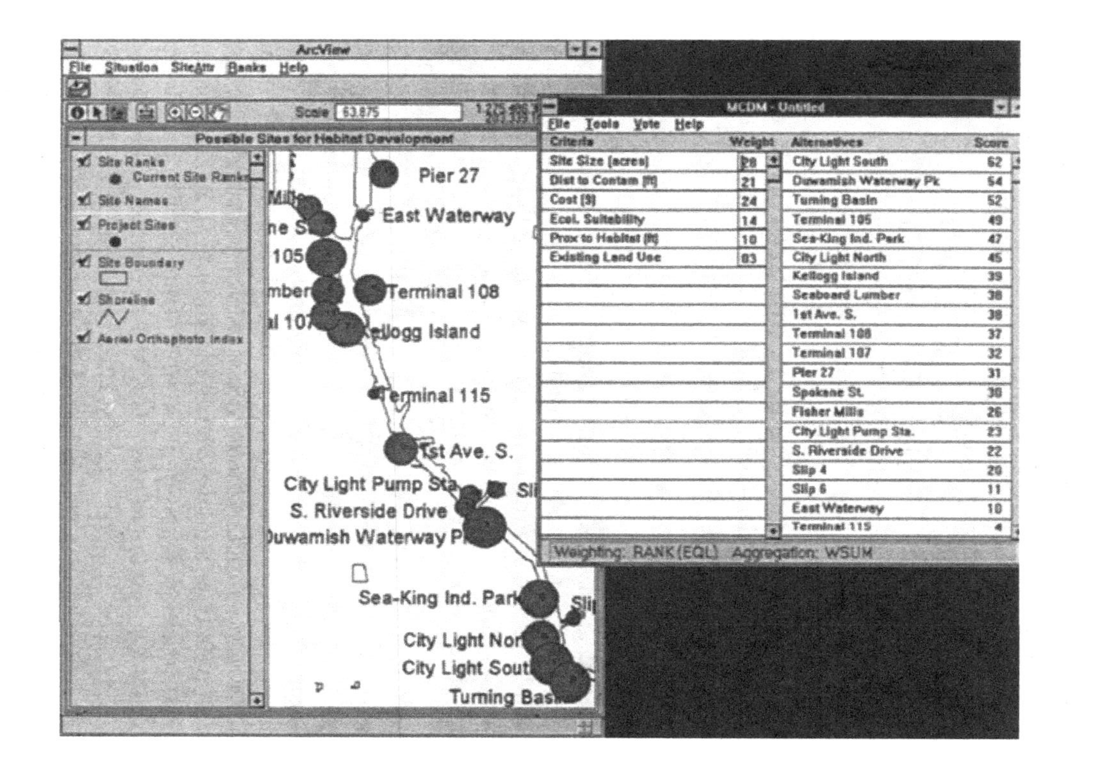

FIGURE 9.6
Display of multi-criteria analysis results in ArcView interface of Spatial Group Choice. (Jankowski, P., T. L. Nyerges, A. Smith, T. J. Moore, and E. Horvath. 1997. Spatial group choice: An SDSS tool for collaborative spatial decision-making. *International Journal of Geographical Information Science* 11 (6):577–602.)

locations of woodland regeneration and to involve them in the planning process. The suspected benefit of the PPGIS, as indicated by the authors, was that marginalized populations would have the ability to voice their opinions with the same respect as other, perhaps more resourceful populations. The Woodland Online Decision System (WOODS) was a Web-based GIS set up at four different locations in the Yorkshire Dales National Park Authority (Figure 9.7). WOODS allowed the users to model possible planning scenarios by identifying relevant factors and setting the importance of these factors in woodland regeneration. They were able to view resultant maps showing suitability of regeneration and allowed to construct new scenarios. The system was implemented using Java and JavaScript. Complications identified by the authors in this type of approach included lack of computer skills among some individuals, Internet access, and copyright issues with Ordnance Survey spatial data. Carver et al. (2001) demonstrated some of the problems and benefits of a Web-based PPGIS. The use of Web-based and public participatory techniques has continued to grow since this publication and will likely continue to do so in the future.

9.4.2 Environmental

The environmental application domain overlaps, in some cases, with agriculture and natural resource management. However, if the main purpose of the SDSS application concerned air, soil, or water pollution, then the application was considered environmental. A significant number of the environmental articles focused on water quality (Assaf and Saadeh 2008; Bennett and Vitale 2001; Vairavamoorthy et al. 2004). Many publications focused on pollution in surface water bodies with the scale of the SDSS application varying from local areas or single watersheds (Bunch and Dudycha 2004; Halls 2003) to large regions and entire countries (Ropke et al. 2004; Fassio et al. 2005). Although far fewer in number, there were multiple articles dealing with soil or land contamination (Chiueh et al. 1997; Salt and Dunsmore 2000). Finally, a few articles described SDSS use in relation to air pollution (Hunova 2001; Symeonidis et al. 2004). Table 9.5 lists some of the articles written about environmental SDSS applications.

Detailed Case Study: Sugumaran et al. (2004) developed a Web-based environmental decision support system known as WEDSS. This SDSS was created to help planners and the local government staff in the city of Columbia in Boone County, Missouri, to prioritize local watersheds in terms of environmental sensitivity using multiple criteria. The criteria used included slope, erosion potential, abundance of endangered species, buffers around the streams containing wooded areas, and types of land cover. Some of the data layers used included a digital elevation model (DEM), hydrography data, a parks dataset, a natural heritage dataset, soil survey, land cover, and remotely sensed imagery. A steering committee met and

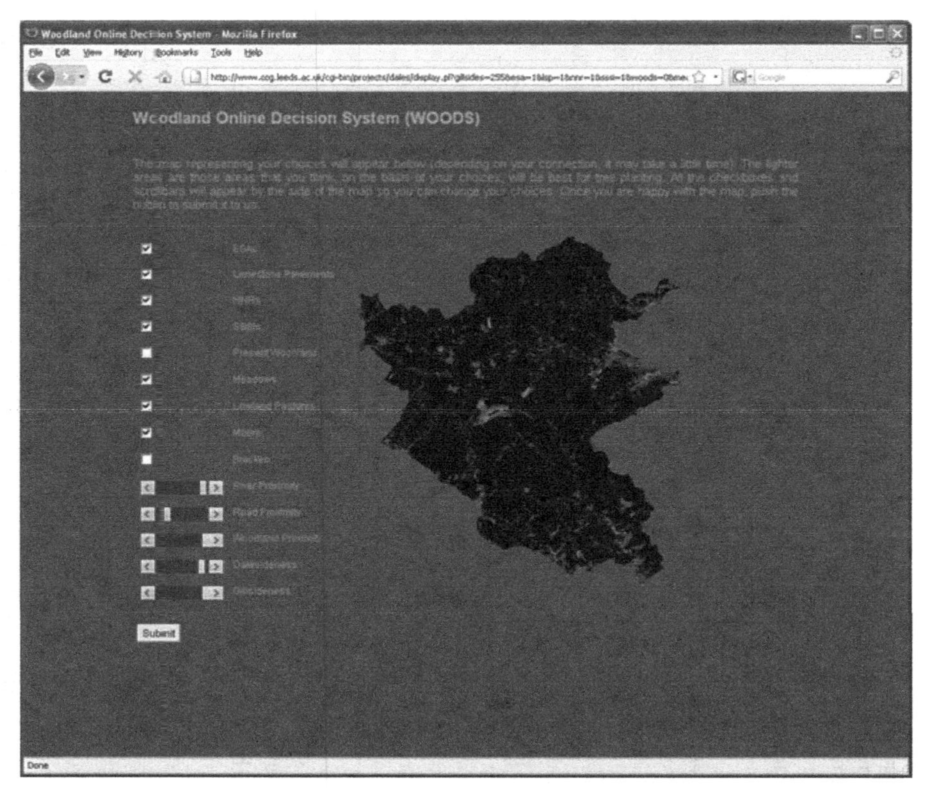

FIGURE 9.7
The Web-based interface of the Woodlanc Online Decision System. (Carver, S., A. Evans, R. Kingston, and I. Turton. 2001. Public participation, GIS, and cyberdemocracy: Evaluating on-line spatial decision support systems. *Environment and Planning B: Planning and Design* 28(6):907–921.)

TABLE 9.5

Environmental Application Examples.

Purpose	Platform	Implementation	Author(s)
Assessment of Agricultural Nonpoint Source Pollution	Desktop	GRASS	Srinivasan and Engel 1994
Assessing Soil Contamination	Desktop	ArcInfo	Chiueh et al. 1997
Postemergency Management of Radioactively Contaminated Land	Desktop	ArcView	Salt and Dunsmore 2000
Collaborative SDSS for Environment Design	Web-based Collaborative SDSS	ArcInfo and MapCafe	Medeiros et al. 2001
Environmental Impact Analysis of Forest Roads	Desktop	ArcView	Girvetz and Schilling 2003
Environmental Planning	Web-based	ArcView and ArcView IMS	Sugumaran et al. 2004
Watershed Management (Soil Erosion)	Desktop	ERDAS IMAGINE	Rao and Kumar 2004
Transport Emission Inventory System	Desktop	Custom (using MapX 4.5 OCX)	Symeonidis et al. 2004
Nuclear Waste Storage Site Selection	Web-based	Various Spatial Web Services	Huang and Sheng 2006
Remediation of Contaminated Land	Desktop	ArcGIS	Carlon et al. 2008

reached a consensus on the weighting of the different environmental criteria. However, using the SDSS from the Internet, any user could change the weights or even exclude certain environmental layers from the analysis. The results of using WEDSS informed planners and the local government of the areas that are the most environmentally sensitive in regard to watershed management. A Web browser, Microsoft Internet Information Server (IIS), and ArcView Internet Map Server (AvIMS) were the basis of the client/server model developed for WEDSS and functioned in the following manner: the user initiated a request, the Web server sent the request to the AvIMS, the analysis was performed, AvIMS created map images and data tables and sent them to the IIS Web server; then IIS formatted the output into HTML pages and served the content on the user's Web browser (Figure 9.8). Figure 9.8 includes a screen shot of the user interface with application demonstration. A potential limitation to this *Web-based SDSS implementing multiple criterion evaluation* was that there were only thirteen environmental criteria. The authors planned on collecting improved datasets and possibly adding additional environmental criteria.

9.4.3 Urban

There were a variety of specific applications within the urban category, including ones that addressed crime (Contino and Virgilio 2002; Kun 2006), housing (Johnson 2005; Barton et al. 2005; Baker 2008), land use and location planning (Pettit and Pullar 2004; Stevens et al. 2007; Taleai et al. 2007; Compas and Sugumaran 2004; Deal and Schunk 2004), policy analysis (Ballas et al. 2007), school redistricting (Armstrong et al. 1993; Casas et al. 2008), water distribution and assessment (Rao 2005), public participatory systems for urban planning (Sidlar and Rinner 2007), and locating industry (Eldrandaly et al. 2003), green spaces (Pelizaro 2005), landfills (Leao et al. 2004; Chang et al. 2008), and public parks (Zucca et al. 2008). In some of the publications categorized as urban, significant overlap occurred with other categories such as transportation or business. For example, Zagorskas and Turskis (2006) examined the influence of retail centers on city structure. A list of examples of urban-related SDSS applications as well as details on their focus, platform, and software implementation characteristics is presented in Table 9.6.

Detailed Case Study: Sikder and Gangopadhyay (2002) published an article titled "Design and Implementation of a Web-Based Collaborative Spatial Decision Support System: Organizational and Managerial Implications." This paper discussed their work on constructing a *collaborative Web-based SDSS* known as GEO-ELCA (Exploratory Land Use Change Assessment), which is used for land use change assessment and analysis of hydrological impacts of potential pollutants. GEO-ELCA was hosted on the Web in order that users could work collaboratively from different locations and at different times. The complex behavior of individuals and institutions in the planning of land use changes is unpredictable and could result in an overall land use scenario that is undesirable for everyone involved unless they are able to collaborate effectively. In GEO-ELCA, each user has transparent access to others' planning scenarios. Each user can change the land use of a given zone under the constraints of the zoning laws. This results in different visualizations of the pollutant distribution and in tabular summaries of annual pollutant load based on land use type. This allows each user to understand the global scenario through visualization of different land use suggestions. Additionally, this allows for each user to specify their individual preferences and also to see the differences and similarities between different proposed land use changes. The study area for which GEO-ELCA was applied was an open, undeveloped 34-square-kilometer area known as the Hackensack Meadowlands District located in northeastern New Jersey, previously the site of major landfills and decades of environmental destruction. Various what-if scenarios were investigated in order to assess the impact of potential land use changes. A series of rules determined where certain land use types could exist.

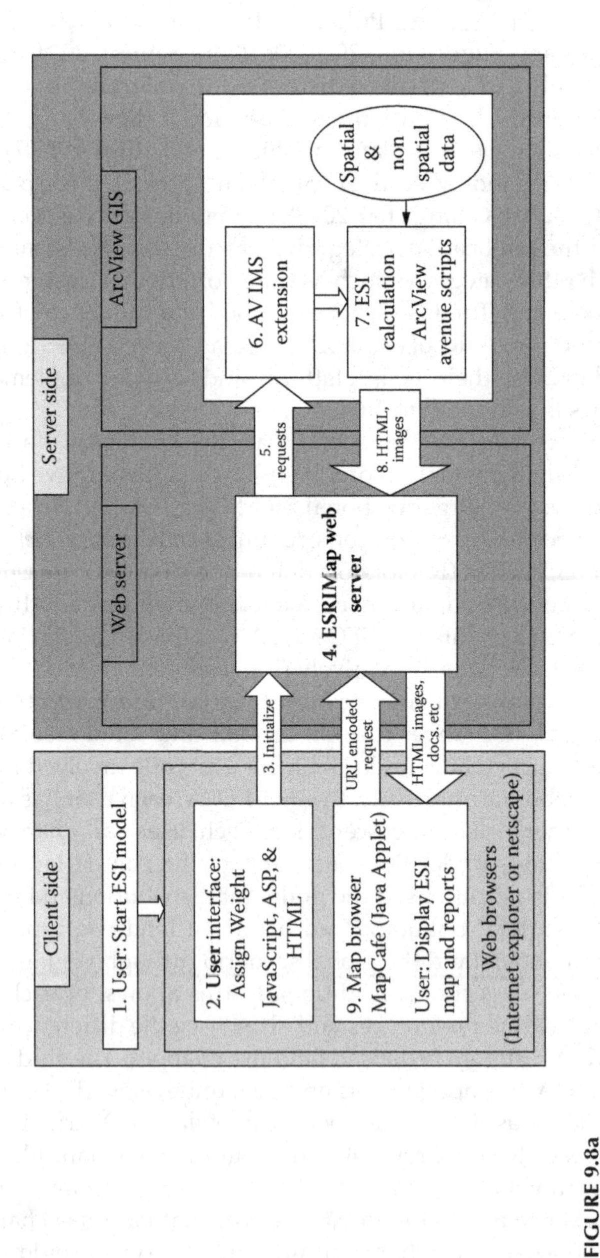

FIGURE 9.8a
The WEDSS architecture (above) and the WEDSS user interface (next page). (Sugumaran, R., J. C. Meyer, and J. Davis. 2004. A Web-based environmental decision support system (WEDSS) for environmental planning and watershed management. *Journal of Geographical Systems* 6(3):307–322.)

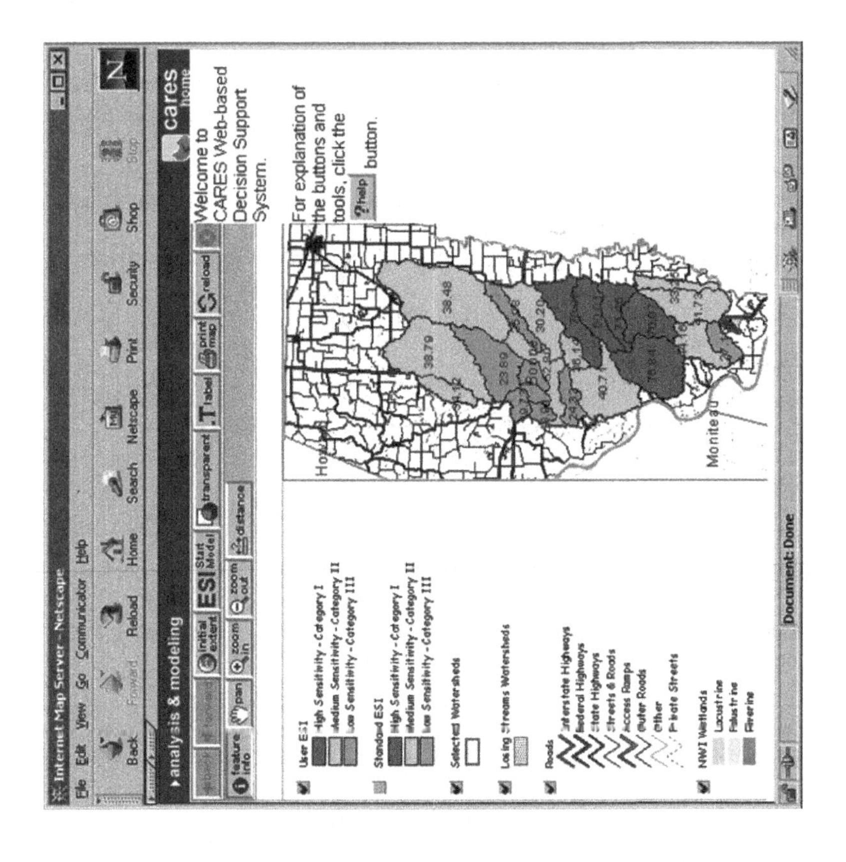

FIGURE 9.8b
(See color insert following page 74.) WEDSS user interface.

TABLE 9.6
Urban Application Examples

Purpose	Platform	Implementation	Author(s)
School Redistricting	Desktop	TransCAD	Armstrong et al. 1993
Land Use Change Assessment for Urban Planning Agencies	Web-based Collaborative	MapServer	Sikder and Gangopadhyay 2002
Industrial Site Selection	Desktop Intelligent	ArcGIS	Eldrandaly et al. 2003
Urban Growth Modeling	Desktop and Web-based	ArcView and ArcView IMS	Compas and Sugumaran 2004
Spatio-temporal Model for Demand and Allocation of Waste Landfills	Desktop	ArcView	Leao et al. 2004
Urban Land-use Planning Scenario Analysis	Desktop	ArcView	Pettit and Pullar 2004
Planning Public Housing Locations	Web-based Collaborative	Customized 3D Visualization System	Barton et al. 2005
Forecasting Urban Water Requirements	Desktop	ArcView	Rao 2005
Urban Green Space Planning	Desktop	MapObjects	Pelizaro 2005
Assisted Housing Mobility Counseling	Desktop and Web-based	ArcView and ArcIMS	Johnson 2005
Crime Analysis	Desktop and Web-based	ArcGIS and ArcIMS	Kun 2006
Multi-criteria Evaluation of Urban Quality of Life	Desktop	CommonGIS	Rinner, 2007
Site Selection for a Local Park Using Multi-Criteria Analysis	Desktop	ILWIS	Zucca et al. 2008

For example, the distance between two different land use elements was limited by a given rule. Land use changes were constrained by a variety of legal, environmental, and regulatory situations and by the desires of different organizations and individuals. The scale at which these land use changes were implemented (zonal and neighborhood) is not necessarily ecologically relevant, but the overall objective of land use change in the New Jersey application was to decrease the impact of pollutants on the environment. One limitation of the GEO-ELCA system is that it does not work as a mediator between individuals and groups, but instead acts as a method to help negotiate the proposed land use changes. There was little discussion of the technical development aspects of the GEO-ELCA system in their article.

9.4.4 Agriculture

There have been a wide range of SDSS applications for agricultural-related purposes. Among the reviewed publications, some described broad-based SDSS for individual farm management (Hey 1998; Jones and Taylor 2004). Others used SDSS to investigate agriculture patterns over geographic areas larger than a single farm (Matthews et al. 1999; Lagacherie et al. 2000). Land use planning or land use decision-making SDSS, with an emphasis on agricultural and rural lands, was applied by Matthews et al. (1999) in Scotland, Mwasi (2001) in Kenya, Roetter et al. (2005) in the Philippines, and Sengupta et al. (2005) in the United States. Agricultural land assessment with SDSS has been carried out with the Land Evaluation and Site Assessment (LESA) model at a local scale (Dung and Sugumaran 2005), with an expert system using fuzzy analysis techniques for agricultural land evaluation based on ecological and economic criteria at a small watershed scale (Nehme and Simoes, 1999), and using crop growth and erosion impact models at a global scale (Tan et al. 2004). Other examples of agricultural SDSS applications include those for evaluation of cropping patterns and water use strategies in Egypt (Abu-Zeid 1998), informing small farmers in their choice of forage crops in Nicaragua and Honduras (O'Brien et al. 2004), aiding in the management of environmentally sensitive livestock production in Iowa (Jain et al. 1995a and 1995b), management of agricultural areas to reduce nitrogen loads in Italy (Ianni et al. 2008), management of poultry litter to agricultural land for the purpose of reducing pollution levels in Alabama (Kang et al. 2008), modeling of crop yields based on imprecise soils data in France (Lagacherie 2000), insect management (Cohen et al. 2008; Deveson 2001), and drought risk assessment (Goddard et al. 2002). The most common journal for publication of agriculturally related SDSS applications has been *Computer and Electronics in Agriculture*. In general, publications dealing with agriculturally related SDSS have frequently been found in environmental management, modeling, planning, and software journals along with agricultural journals focusing on computers. Table 9.7 further highlights agricultural applications.

Detailed Case Study: A paper entitled "A Web-Based Agricultural Crop Condition and Yield Prediction Modeling System Using Real-Time Data" from Kaparthi and Sugumaran (2009) is used as an example for the agriculture application domain. This is a good example of a *near-real-time SDSS*. The goal of this project was to develop a decision support system that provides dynamic Web-based crop condition and yield estimations using near-real-time data throughout the growing season. This system uses 250-m resolution Normalized Difference Vegetation Index (NDVI) products generated from imagery collected from NASA's MODIS instrument. The system also uses soil moisture, surface

TABLE 9.7

Agriculture Applications Examples

Purpose	Platform	Implementation	Author
Rural Land Use Planning	Desktop	Smallworld	Matthews et al. 1999
Modeling Crop Yields Using Imprecise Soil Data	Desktop	ArcInfo and MapObjects	Lagacherie et al. 2000
Land Use Conflict Resolution in Fragile Ecosystems	Desktop	IDRISI	Mwasi 2001
Locust Management	Desktop and Mobile	ArcInfo and ArcView	Deveson 2001
Data Integration in a Farm SDSS	Desktop	ArcView	Jones and Taylor 2004
Farm-level Agronomic Decision Making	Web-based	ArcIMS	Sha and Bian 2004
Selection of Forage Species for Farmers	Desktop	MapObjects	O'Brien et al. 2004
Assessing Future Agriculture Land Use Changes at Global Scale	Desktop	ArcView	Tan et al. 2004
Agricultural Land Evaluation and Site Assessment	Desktop	ArcGIS	Dung and Sugumaran 2005
Confined Animal Feeding Operations Site Suitability	Desktop	ArcGIS	Sugumaran and Bakker 2007
Agricultural Pest Control Planning	Desktop	ArcGIS	Cohen et al. 2008
Precision Agriculture	Desktop	ArcGIS	Thorp et al. 2008
Real-time Crop Yield SDSS	Web-based	MapServer and ERDAS IMAGINE	Kaparthi and Sugumaran 2009

temperature, and precipitation data. The front end was implemented using PHP for Web-development and MapServer to enhance user interactivity and facilitate ease in the understandability of the results. The back end is supported by ERDAS Imagine software (Figure 9.9a). This system allows the user to define different datasets and predicts potential crop yields based on user inputs. Figure 9.9b is a screenshot of the user interface. Some of the limitations of this system are that it (a) uses a very simple regression model for prediction,(b) only uses four inputs, and (c) requires ERDAS Imagine software on the server side. It is, however, a good example of the potential for using near-real-time data for dynamic SDSS use.

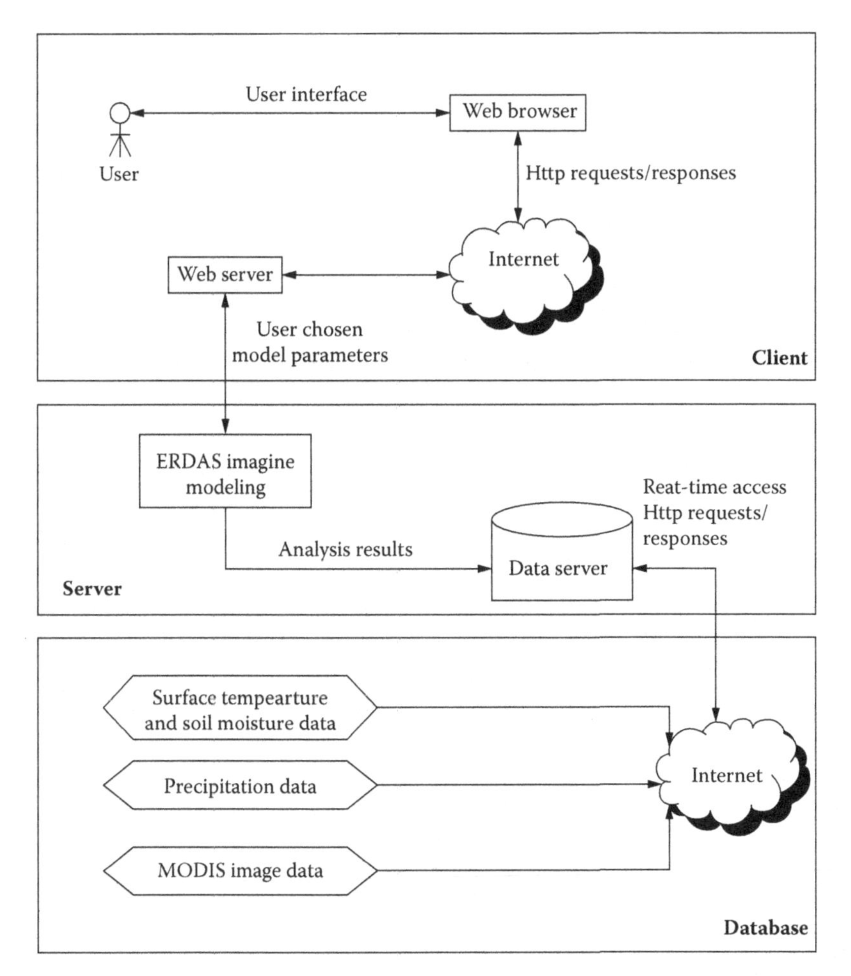

FIGURE 9.9a
Architecture used in real-time crop yield SDSS. (Kaparthi, P., and R. Sugumaran. 2009. A Web-based agricultural crop condition and yield prediction modeling system using real-time data. Paper presented at the Iowa Geographical Information Council Conference, Waterloo, Iowa.)

9.4.5 Utility/Communication/Energy and Transportation

A variety of SDSS applications have been carried out in the utility, communication, energy, and transportation sectors. The articles dealing with utility, communications, and energy included ones on bioenergy (Voivontas et al. 2001), water utility management (Sinske and Zietsman 2004), wireless broadband communications (Scheibe et al. 2006), and wind generation facilities (Lejeune and Feltz 2008; Monteiro et al. 2001; Ramirez-Rosado et al. 2008). Transportation applications included SDSS for modeling dynamic network congestion and route planning in Salt Lake City,

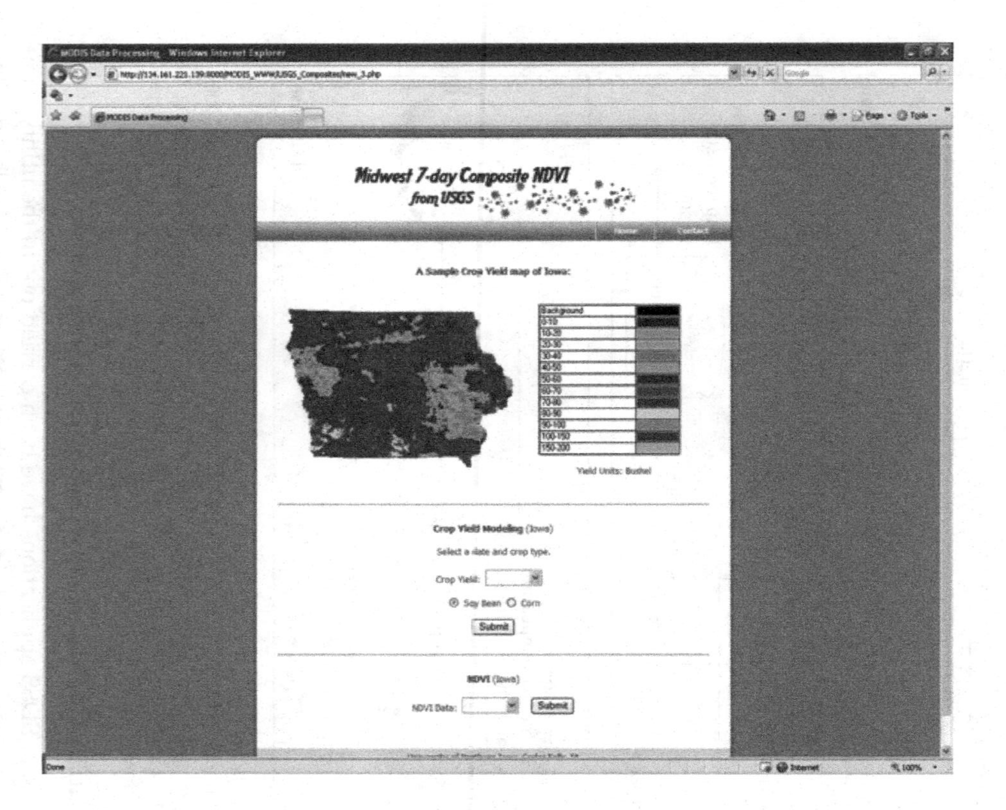

FIGURE 9.9b
(See color insert following page 74.) Screenshot of the user interface developed for the crop yield SDSS. (Kaparthi, P., and R. Sugumaran. 2009. A Web-based agricultural crop condition and yield prediction modeling system using real-time data. Paper presented at the Iowa Geographical Information Council Conference, Waterloo, Iowa.)

Utah (Wu et al. 2001), a transportation policy evaluation SDSS in Greece (Tsamboulas and Mikroudis 2006), railway network design in China (Kuby et al. 2001), a multi-vehicle multi-route SDSS for efficient trash collection in Portugal (Santos et al. 2008), and overweight vehicle permitting in Delaware (Ray 2007). Details of SDSS dealing with transportation, communications, and energy are shown in Table 9.8.

Detailed Case Study: The development and implementation of an SDSS for assisting in managing and organizing a city's snow removal operations is described. A *Web-based Intelligent SDSS* (WebISDSS) was created to help snow removal operations in Black Hawk County, Iowa (Sugumaran et al. 2007). This system is unique in that it incorporates weather data from the

TABLE 9.8

Utility/Communications/Energy and Transportation Application Examples

Purpose	Platform	Implementation	Author
Route Choice in Congested Urban Road Networks	Desktop	ArcInfo	Wu et al. 2001
Assesment of Biomass Potential for Power Production	Desktop	MapInfo	Voivontas et al. 2001
Solving Vehicle Routing Problems	Desktop	ArcView	Tarantilis and Kiranoudis 2002
Petroleum Investment Analysis	Desktop	ArcGIS	Yu et al. 2003
Analysis of Pipe-break Susceptibility of Municipal Water Distribution Systems	Desktop	ArcView	Sinske and Zietsman 2004
Locating Wireless Broadband Infrastructure	Desktop	ArcView	Scheibe et al. 2006
Transportation Policy Scenario Analysis	Desktop	MapObjects	Tsamboulas and Mikroudis 2006
Snow Removal Asset Management	Web-based	ArcIMS	Sugumaran et al. 2007
Site Selection of Distributed Generation Facilities for Renewable Energy	Desktop	ArcGIS	Tegou et al. 2007
Permitting and Routing for Oversize/ Overweight Vehicles	Web-based	Custom Software	Ray 2007
Vehicle Routing for Trash Collection	Desktop	ArcView	Santos et al. 2008
Evaluation of Adaptive Traffic Control Strategies	Desktop	GeoMedia Professional	Mudigonda et al. 2008
Analysis of Landscape Constraints for Wind Farm Development	Desktop	ArcGIS	Lejeune and Feltz 2008

Web and an intelligent expert system. This SDSS was developed for the planners and decision makers of Black Hawk County, Iowa, (and potentially other locations) to manage the removal of snow by determining shortest paths and prioritized routes and for allocating resources optimally. WebISDSS works by integrating road data with weather data and an intelligent component. This Web-based SDSS combines knowledge from snow removal experts with real-time weather conditions (precipitation, dew point, visibility) streamed from the Web. The expert system produces a set of suggestions. The suggestions inform the user whether or not to initiate snowplowing and de-icing, which materials to use, and how much of each material should be used. The interface is mainly menu driven with non-technical nomenclature so that anyone can use and understand the system. WebISDSS is organized into various areas or menus (Figure 9.10b). The Weather menu allows for the viewing of live or forecasted weather information. Also, the user can view live weather radar from NOAA NEXRAD. The Route menu has options for creating, deleting, and loading snowplow routes. Driving directions based on the shortest or quickest routes can be selected by the user. The Vehicle & Drivers menu allows for the assignment of drivers and vehicles. The Materials menu allows for inventory analysis, material assignment, and the editing of materials like salt and sand. A Help menu is also included. The system uses ArcSDE for data management and routing through its RouteServer extension, ArcIMS for Web mapping, and a rule-based expert system for the intelligent component developed with Visual Rule Studio software (Figure 9.10a). The authors point out that the system could be improved by utilizing ArcObjects with ArcGIS Server. Commercial weather data could be incorporated as well to generate more accurate snowplow routes. A visual display of road weather conditions could also be added to WebISDSS. This SDSS could eventually find its way to portable devices such as palmtops and cell phones.

9.4.6 Business

Publications detailing business SDSS applications have included site selection, retail location planning, real estate, consumer behavior, tourism planning, and investment analysis. Specifically, there were SDSS used to identify favorable locations for builders/developers (Ahmad et al. 2004), retail outlets (Clarke and Rowley 1995), warehouses using a multi-criteria evaluation method (Vlachopoulou et al. 2001), and a new ski run using an optimization model (Aerts et al. 2003). One real estate SDSS application used multi-criteria evaluation techniques with location, proximity, and direction as important spatial characteristics used for supporting consumers in home buying. An agent-based SDSS simulated consumer behavior for grocery shopping at a regional level in Sweden (Schenk et al. 2007). A participatory system for regional tourism planning

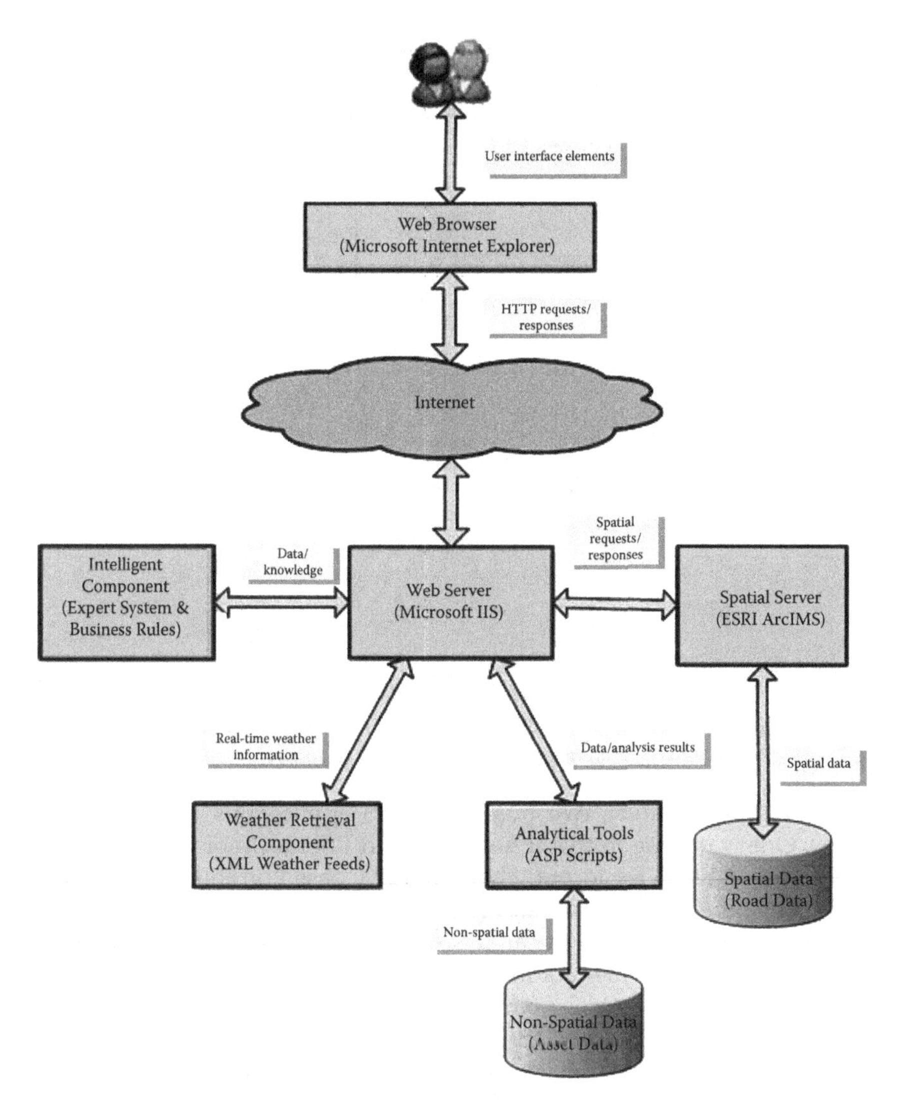

FIGURE 9.10a
Architecture of WebISDSS.

in Queensland, Australia, was developed using multi-criteria evaluation (Taranto 2007). An SDSS for guiding investment strategies that incorporated spatial information on physical, social, human, knowledge, and productive capital in London was described by Weber et al. (2006). As there were not a large number of examples, the authors have consolidated business examples with emergency planning and public health SDSS applications in Table 9.9.

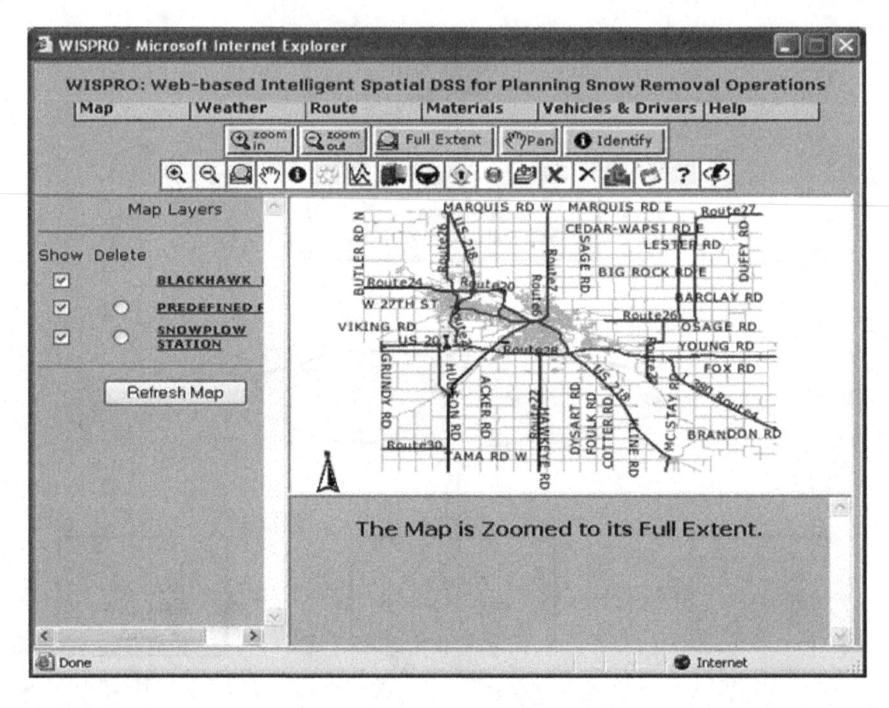

FIGURE 9.10b
(See color insert following page 74.) Web interface of the WebISDSS.

Detailed Case Studies: Two business SDSS applications are discussed below. One is unique in that it has embraced PDA technology to create a mobile-based SDSS for locating bars in a city based on the user's preferences. The other business application is unique because it is a Web-based SDSS developed for real estate analysis. The first SDSS has the potential to influence the development of mobile SDSS, while the second SDSS could potentially affect the price of real estate in a region of Germany.

Rinner et al. (2005) created two *mobile SDSS* for locating hotels and bars using ArcPad for mobile devices. The two applications are known as Hotel Finder and Bar Finder (Figure 9.11). This paper focused on the improvements made in the usability of Bar Finder as compared to Hotel Finder. Bar Finder works by having the user input his or her location and then select the criteria preferred in a bar, such as the price of beer, drinks, and soda, presence/absence of live music, cover charge, and handicap accessibility. These preferences may be changed in order to locate a bar acceptable to the patron. The novelty of this SDSS is the fact that it operates on a mobile device using multi-criteria evaluation (MCE) for location-based services. Another unique aspect is its potential commercial applications. Bar Finder uses a simple additive weighting method to determine which bar is best suited to the user's interests. The most important spatial data used is in

TABLE 9.9

Business, Emergency Planning and Hazards, and Public Health Application Examples

Purpose	Platform	Implementation	Author
Wildland Fire Prevention and Fighting	Desktop	GRASS	Guarnieri and Wybo 1995
Selecting Locations for Rural Health Practices	Desktop	ArcView	Jankowski and Ewart 1996
Emergency Evacuation	Desktop	ArcInfo	Pidd et al. 1996
Landslide Hazard Monitoring	Desktop	MapInfo	Lazzari and Salvaneschi 1999
Home-Delivered Services Planning	Desktop	ArcView	Gorr et al. 2001
Warehouse Site Selection	Desktop	MapObjects and ArcView	Vlachopoulou et al. 2001
Analysis of Fire Vulnerability in Wildland–Urban Interface	Desktop	ArcView	Tucek et al. 2003
Site Selection for Builders/Developers	Desktop	ArcView	Ahmad et al. 2004
Hotel and Bar Finder	Mobile SDSS	ArcPad	Rinner et al. 2005
Real-time 3D GIS for Emergency Response in Urban Environments	Desktop, Mobile, and Web-based Components	Custom Software (prototype)	Kwan and Lee 2005
Community Health Assessment	Desktop	Custom	Scotch and Parmanto 2006
River Levee Management	Desktop	ArcGIS	Serre et al. 2006
Urban Search and Rescue	Mobile SDSS	Custom Software	Heth et al. 2006
Epidemic Disease Prevention	Desktop and Web-based	ArcGIS and ArcIMS	Yang et al. 2007
Regional Tourism Planning	Desktop	ArcView	Taranto 2007
Health Care Allocations in Rural Areas	Web-based	ArcGIS, ArcSDE and ArcIMS	Schuurman et al. 2008

vector format and represents road networks and points of service. They used Visual Basic to create the customized applications in ArcPad software. The mobile SDSS incorporated the MCE technique of ordered weighted averaging. One limitation of Bar Finder is that the software is not built to automatically locate the user with GPS. Rather, the user has to mark his or her location on the map interface. One of the most significant limitations to the system, as considered at the time of publication, is the resolution of screens on mobile devices. The resolution and screen size is sufficient for most applications, but can become burdensome or confusing when displaying details of a large area on a background map.

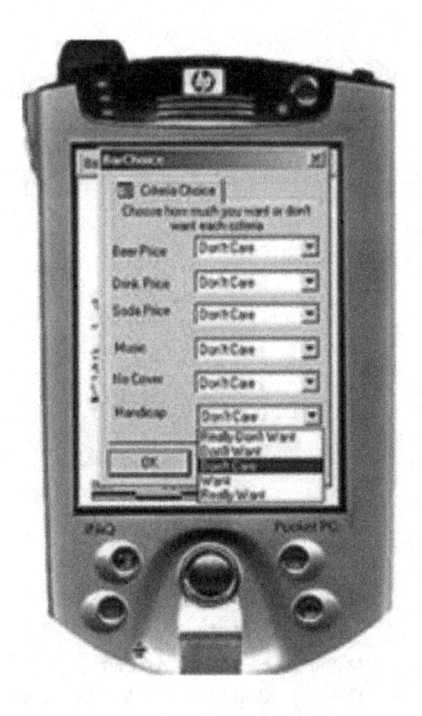

FIGURE 9.11
(See color insert following page 74.) User interface from the Bar Finder application. (Rinner, C., M. Raubal, and B. Spigel. 2005. User interface design for location-based decision services. Paper presented at 13th International Conference on GeoInformatics, Toronto.).

Neis et al. (2007) constructed an *SDSS for a Web accessibility analysis service based on the OpenLS route service.* The study area was the German state of Rhineland Palatinate (RLP). The major purpose was to develop a Web-SDSS that automates multi-criteria model building for user-specified, regionalized housing market analyses in RLP. The goals included analysis and characterization of the regional housing market situation in RLP, delineation of spatial boundaries of housing market segments, prediction of the future development of the RLP housing market, and continuous updates of the regionalized housing market. One of the main objectives in this development was to calculate the region (as a polygon) that can be reached from a given location within a specified time or distance (Figure 9.12). This system was developed for supporting political and economic decision makers. Detailed street networks (vector data) containing street type and traveling times were used. Menu-driven components are available in this Web-SDSS. This type of accessibility analysis tells what areas can be reached within a given time or distance and works very similarly to Find Service Area from ESRI's ArcGIS Network Analyst. Multiple attributes can be returned to the user, including a summary (number of

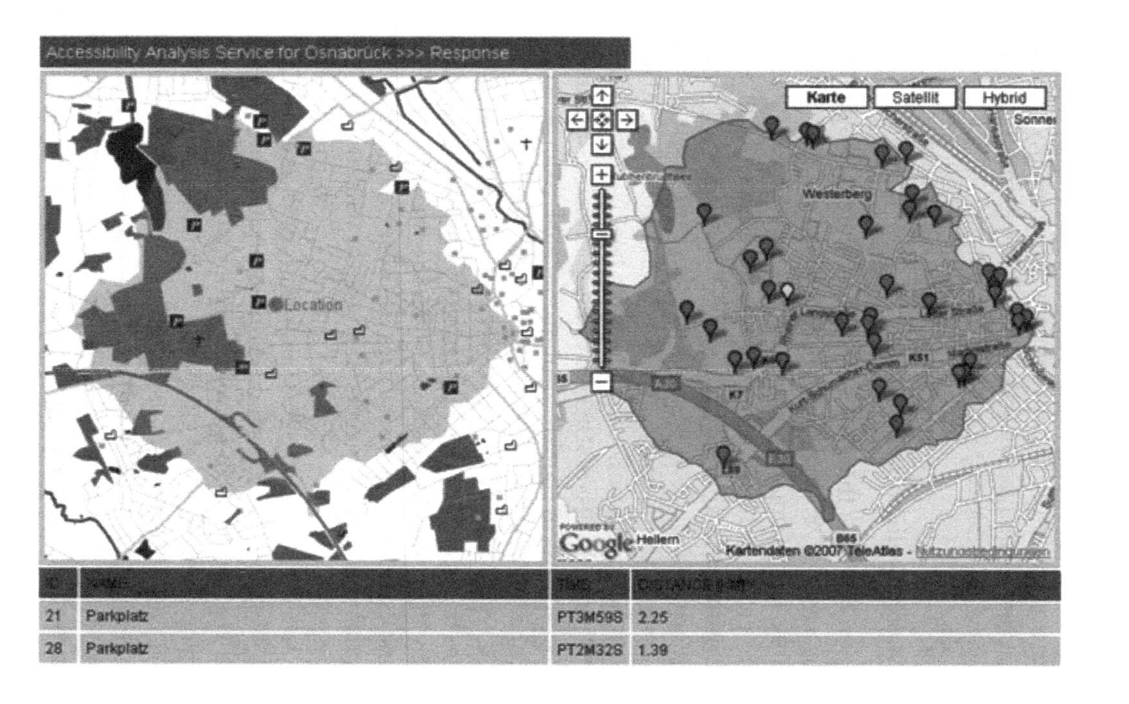

FIGURE 9.12
(See color insert following page 74.) Example of the analysis results from the accessibility analyses displayed in their Web page and Google Maps. (Neis, P., L. Dietze, and A. Zipf. 2007. A Web accessibility analysis service based on the OpenLS Route Service. Paper presented at the 10th AGILE International Conference on Geographic Information Science, Aalborg University, Denmark.)

locations in the accessibility area and bounding box), an output list (name, distance, and time), information about the locations within the accessibility area, the geometry of the accessibility area (polygons or relevant streets), and maps with the plotted accessibility area or the streets involved in this area (Figure 9.12). Limitations in the analysis possibilities include the inability to look at more than one point and one distance at a time. The authors discuss the possibility of improving the functionality of this SDSS to include the ability to analyze accessibility services for multiple points at once and to review multiple distances at the same time.

9.4.7 Other Major Application Domains

Other application domains that had a significant number of publications devoted to the use of SDSS included public health and emergency preparedness. Public health applications included an SDSS for tuberculosis management in cattle and deer in New Zealand (McKenzie et al. 1997), community health assessment using Spatial On-Line Analytical Processing in Indonesia (Parmanto et al. 2008), an application of a software called Dynamic Exploratory Cartography for Decision Support (DECADE) used for examining funding allocation for primary health services in Idaho (Jankowski 2001), a Web-based SDSS for making decisions on healthcare allocations in rural areas in Canada (Schuurman et al. 2008), and an SDSS for community health assessment (Scotch and Parmanto 2006).

The use of SDSS for a variety of emergency planning and hazard mitigation purposes has been described in various publications. Applications focusing on fire response included an SDSS for the assessment of the propagation and combating of forest fires in Greece (Bonazountas et al. 2007) and a system for land managers in charge of wildland fire prevention and fighting that used fire behavior modeling (Guarniéri and Wybo 1995). Several articles described SDSS methods for evacuation planning, including de Silva (2001), who discussed linking traffic simulation modeling and GIS. The use of SDSS for route planning of hazardous materials and emergency response was discussed by Boulmakoul et al. (1999), who relied on a Web-based system using real-time GPS information from trucks and route identification as well as evacuation modeling (Zografos and Androutsopoulos 2008). Other emergency- and hazards-related SDSS publications included an SDSS for managing possible landslides through the integration of real-time monitoring systems in the SDSS (Lazzari and Salvaneschi 1999), an SDSS for dealing with environmental contamination from nuclear power accidents (Gheorghe and Vamanu 1995), and an SDSS with 3D modeling capabilities for urban emergency response (Kwan and Lee 2005; Lee and Zlatanova 2008). Table 9.9 shows some of the example applications for business, emergency planning, and public health.

Detailed Case Studies: There are two unique examples provided in this category. The first publication discusses the development of drought indices for Nebraska using a *component-based SDSS* based on Geographic Resource Analysis Support Systems (GRASS), a free but traditional GIS software program. The second example details a Web-based SDSS for healthcare allocation decisions. The development of a component-based GIS using GRASS was the focus of research conducted by Wu et al. (2004). They introduced the idea of turning a traditional GIS, in this case GRASS, into a component-based GIS using the Common Object Request Broker Architecture (CORBA) environment. They discussed the development of an SDSS, the National Agricultural Decision Support System (NADSS) using this type of framework. This SDSS utilizes climate data (i.e., temperature and precipitation) for agricultural models, drought indices, and knowledge in the form of exposure analysis (impact of a natural hazard). NADSS transforms climatic data into drought maps using the standard precipitation index (SPI) and the Palmer Drought Severity Index (PDSI). Their example application included the creation of a PDSI map for Nebraska. CORBA serves as a means of communication between disconnected software components written in multiple computer languages and/or running on multiple computers. Also utilized in this study were the Distributed Component Object Model (DCOM) and JAVA remote method invocation (RMI). Figure 9.13 demonstrates the architecture used for NADSS. The uniqueness of this study was in creating a component-based GIS based on existing code rather than developing one from scratch.

The final case study example is from the public health area. Schuurman et al. (2008) created a Web-based graphical user interface (wGUI) to assist health policy makers and administrators in locating and allocating resources of time-sensitive services in rural regions of British Columbia. The wGUI provides information about the location of different medical services (emergency services, basic inpatient services, and core specialty services). The standards imposed by the British Columbia Ministry of Health dictate that 98 percent of the population should be able to access medical services in a specific amount of time. The wGUI is useful for planners to check if these standards are being met and to make decisions about hospital closures, hospital openings, and service reallocations in rural and remote regions of British Columbia. The wGUI system allows for the examination of populations affected by the removal or addition of hospital services. This interface was developed for health policy makers and administrators, specifically British Columbia Ministry of Health Medical Officers and Information Officers in charge of decision making about service provision. The five regional Health Authorities in the province would likely be users. British Columbia has an area of 364,764 square miles (944,735 sq km) with a population estimated to be over 4.4

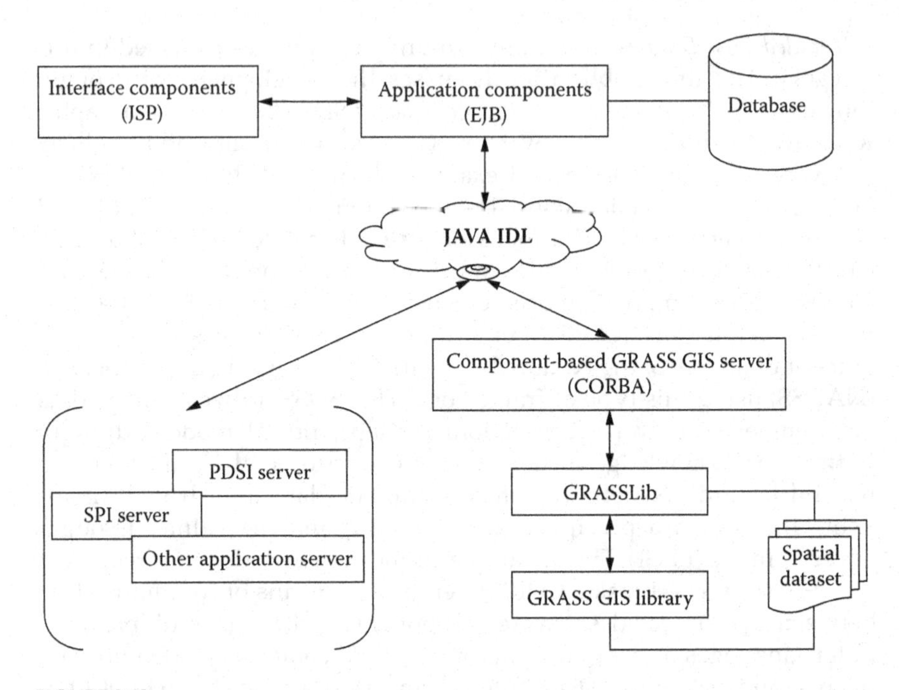

FIGURE 9.13
The general architecture of the NADSS built with the component-based GRASS GIS. (Wu, X., S. Zhang, and S. Goddard. 2004. Development of a component-based GIS using GRASS. Paper presented at the FOSS/GRASS Users Conference, Bangkok, Thailand.)

million people. Vector data incorporated included hospital locations (plus attributes on the availability of various health services), road networks, census block population information, Health Authority polygons, and Census Metropolitan Area data. These spatial data allowed the developers to derive an origin–destination cost table, a Health Authority population table, and three hospital catchment feature classes for one-, two-, and four-hour travel times. Initially stored in an Access geodatabase, the data was converted to an Oracle database for use in the wGUI. Based on the number of hours it takes for a person to travel to the nearest health facility, multiple catchments were derived showing how much of the population has access to health facilities for discrete intervals of one, two, and four hours of traveling time. At the time of writing, the service data only included trauma and maternity services, but others were being added. An ArcSDE Server communicated data between the GIS software and the Oracle database and also allowed for concurrent multiple users. The wGUI was built on a foundational ArcMap GUI that would allow the use of all of the included ArcMap tools in further analyses (Figure 9.14). ArcIMS software was used to publish the interactive map

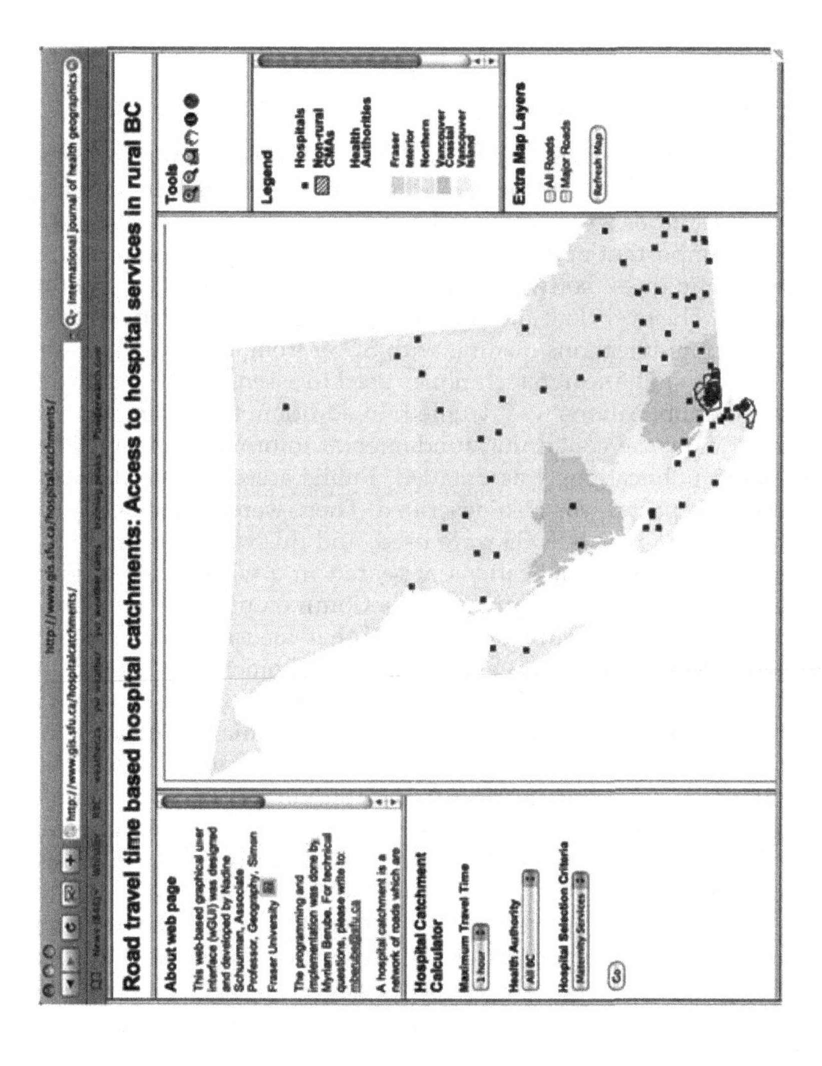

FIGURE 9.14
The wGUI interface.

on the Internet. Limitations dealing with this SDSS include the accuracy of population data, which is only updated every five years. The authors contend that the combination of a *spatially enabled Web-based graphical user interface to explore evidence-based resource allocations* is unique.

9.5 Summary

This chapter has provided an overview of application domains or disciplines in which SDSS have been most commonly applied. In addition, detailed case studies were provided from a variety of disciplines, which highlighted important application domains, technological approaches, modeling techniques, software integration methods, and the role of stakeholders. The overview and detailed case studies were derived from a collection of publications dealing with SDSS from approximately the past thirty years. The search techniques used to assemble this catalog of SDSS-related publications was detailed. In addition, the compilation of a relational database containing fundamental information about these SDSS-related publications was detailed. Public access to the database through a Web portal was also described. There were a range of application domains in which SDSS were used, and this was reflected in the fact that SDSS-related publications appeared in a wide range of journals, books, and conference proceedings. Common publication sources were journals or conference proceedings that focused on the use of computers in specific domains such as the environment, agriculture, or urban studies. Also, many publications appeared in journals, conference proceedings, or books with a focus on GIS or geospatial technology. We categorized the publications as falling into several main disciplines or domains. From greatest to least (based on number of publications), these domains were natural resource management, environmental, urban, agriculture, emergency planning and hazard response, transportation, business, utility/communications/energy, and public health. Practitioners from natural resources management and environmental disciplines were early adopters of GIS technology, which led to their involvement in SDSS. Other disciplines such as business have been a bit slower at adopting GIS and SDSS technology, but this is changing and it is likely there will be a continued increase in the use of SDSS in all disciplines discussed.

References

Abu-Zeid, K. M. 1998. A multicriteria decision support system for evaluating cropping pattern strategies in Egypt. In *Multiple objective decision making for land, water, and environmental management*, ed. S. A. El-Swaify and D. S. Yakowitz, 105–120. Boca Raton: Lewis Publishers.

Aerts, J., M. F. Goodchild, and G. B. M. Heuvelink. 2003. Accounting for spatial uncertainty in optimization with spatial decision support systems. *Transactions in GIS* 7(2):211–230.

Ahmad, I., S. Azhar, and P. Lukauskis. 2004. Development of a decision support system using data warehousing to assist builders/developers in site selection. *Automation in Construction* 13(4):525–542.

Alcamo, J. 2008. *Environmental futures: The practice of environmental scenario analysis.* Amsterdam: Elsevier.

Armstrong, M. P., P. Lolonis, and R. Honey, R. 1993. A spatial decision support system for school redistricting. *URISA Journal* 5(1):40–51.

Assaf, H., and M. Saadeh. 2008. Assessing water quality management options in the Upper Litani Basin, Lebanon, using an integrated GIS-based decision support system. *Environmental Modelling and Software* 23(10–11):1327–1337.

Baker, E. 2008. Improving outcomes of forced residential relocation: the development of an Australian tenants' spatial decision support system. *Urban Studies* 45(8):1712–1728.

Ballas, D., R. Kingston, J. Stillwell, and J. Jin. 2007. Building a spatial microsimulation decision support system. *Environment and Planning A* 39(10):2482–2499.

Barton, J., J. Plume, and B. Parolin. 2005. Public participation in a spatial decision support system for public housing. *Computers, Environment and Urban Systems* 29(6):630–652.

Bennett, D., and A. J. Vitale. 2001. Evaluating nonpoint pollution policy using a tightly coupled spatial decision support system. *Environmental Management* 27(6):825–836.

Best, B. D., P. N. Halpin, E. Fujioka, A. J. Read, S. S. Qian, L. J. Hazen, and R. S. Schick. 2007. Geospatial web services within a scientific workflow: Predicting marine mammal habitats in a dynamic environment. *Ecological Informatics* 2(3):210–223.

Bolte, J. P., D. W. Hulse, S. V. Gregory, and C. Smith. 2006. Modeling biocomplexity—actors, landscapes and alternative futures. *Environmental Modelling & Software* 22(5):570–579.

Bonazountas, M., D. Kallidromitou, P. Kassomenos, and N. Passas. 2007. A decision support system for managing forest fire casualties. *Journal of Environmental Management* 84(4):412–418.

Boulmakoul, A., R. Laurini, S. Servigne, and M. A. J. Idrissi. 1999. First specifications of a telegeomonitoring system for the transportation of hazardous materials. *Computers, Environment and Urban Systems* 23(4):259–270.

Bunch, M. J., and D. J. Dudycha. 2004. Linking conceptual and simulation models of the Cooum River: Collaborative development of a GIS-based DSS for environmental management. *Computers, Environment and Urban Systems* 28(3):247–264.

Carlon, C., L. Pizzol, A. Critto, and A. Marcomini. 2008. A spatial risk assessment methodology to support the remediation of contaminated land. *Environment International* 34(3):397–411.

Carrick, N. A., and B. Ostendorf. 2007. Development of a spatial decision support system (DSS) for the Spencer Gulf penaeid prawn fishery, South Australia. *Environmental Modelling & Software* 22(2):137–148.

Carver, S., A. Evans, R. Kingston, and I. Turton. 2001. Public participation, GIS, and cyberdemocracy: Evaluating on-line spatial decision support systems. *Environment and Planning B: Planning and Design* 28(6):907–921.

Casas, I., R. Garrity, D. Mandloi, M. Sunm, J. Weaver, R. Nagi, and R. Batta. 2008. A spatial decision support system combining GIS and OR tools to optimize boundaries and bus routes for a suburban school district. *OR Insight* 7(2):3–16.

Chakhar, S., and V. Mousseau. 2007. An algebra for multicriteria spatial modeling. *Computers, Environment and Urban Systems* 31(5):572–596.

Chang, N-B., G. Parvathinathan, and J. B. Breeden. 2008. Combining GIS with fuzzy multicriteria decision-making for landfill siting in a fast-growing urban region. *Journal of Environmental Management* 87(1):139–153.

Chiueh, P-T., S-L. Lo, and C-D. Lee. 1997. Prototype SDSS for using probability analysis in soil contamination. *Journal of Environmental Engineering* 123(5):514–519.

Choi, J-Y., B. A. Engel, and R. L. Farnsworth. 2005. Web-based GIS and spatial decision support system for watershed management. *Journal of Hydroinformatics* 7:165–174.

Clarke, I., and J. Rowley. 1995. A case for spatial decision-support systems in retail location planning. *International Journal of Retail & Distribution Management* 23(3):4–10.

Cohen, Y., A. Cohen, A., Hetzroni, V. Alchanatis, D. Broday, Y. Gazit, and D. Timar. 2008. Spatial decision support system for Medfly control in citrus. *Computers and Electronics in Agriculture* 62(2):107–117.

Compas, E., and R. Sugumaran. 2004. Urban growth modeling on the web: A decision support tool for community planners. *Papers and Proceedings of Applied Geography Conference* 27:255–269.

Contino, G., and G. Virgilio. 2002. A spatial decision support system for situational crime prevention in urban environment. Paper presented at the 5th AGILE Conference on Geographic Information Science, Palma, Spain.

Crossman, N. D., L. M. Perry, B. A. Bray, and B. Ostendorf. 2007. CREDOS: A conservation reserve evaluation and design optimisation system. *Environmental Modelling and Software* 22:449–463.

de Silva, F.N. 2001. Providing spatial decision support for evacuation planning: A challenge in integrating technologies. *Disaster Prevention and Management* 10(1):11–20.

Deal, B., and D. Schunk. 2004. Spatial dynamic modeling and urban land use transformation: A simulation approach to assessing the costs of urban sprawl. *Ecological Economics* 51(1–2):79–95.

Deveson, T. 2001. Decision support for locust management using GIS to integrate multiple information sources. Paper presented at Proceedings of the Geospatial Information & Agriculture Conference, Sydney, Australia.

Dung, E. J., and R. Sugumaran. 2005. Development of an agricultural land evaluation and site assessment (LESA) decision support tool using remote sensing and geographic information system. *Journal of Soil and Water Conservation* 60(5):228–234.

Dymond, R. L., B. Regmi, V. K. Lohani, and R. Dietz. 2004. Interdisciplinary web-enabled spatial decision support system for watershed management. *Journal of Water Resources Planning and Management* 130(4):290–300.

Eldrandaly, K., N. Eldin, and D. Sui. 2003. A COM-based spatial decision support system for industrial site selection. *Journal of Geographic Information and Decision Analysis* 7(2):72–92.

Fassio, A., C. Giupponi, R. Hiederer, and C. Simota. 2005. A decision support tool for simulating the effects of alternative policies affecting water resources: An application at the European scale. *Journal of Hydrology* 304(1–4):462–476.

Geneletti, D. 2004. A GIS-based decision support system to identify nature conservation priorities in an alpine valley. Land Use Policy 21:149-160.

Geneletti, D. 2007. An approach based on spatial multicriteria analysis to map the nature conservation value of agricultural land. *Journal of Environmental Management* 83(2):228–235.

Gheorghe, A. V., and D. Vamanu 1995. A pilot decision support system for nuclear power emergency management. *Safety Science* 20(1):13–26.

Gilliams, S., J. Van Orshoven, B. Muys, H. Kros, G.W. Heil, and W. Van Deursen. 2005. AFFOREST sDSS: A metamodel based spatial decision support system for afforestation of agricultural land. *New Forests* 30(1):33–53.

Gimblett, H. Randy. 2002. *Integrating geographic information systems and agent-based modeling techniques for simulating social and ecological processes*: New York: Oxford University Press.

Girvetz, E., and F. Shilling. 2003. Decision support for road system analysis and modification on the Tahoe National Forest. *Environmental Management* 32(2):218–233.

Goddard, S., S. Zhang, W. J. Waltman, D. Lytle, and S. Anthony. 2002. A software architecture for distributed geospatial decision support systems. Paper presented at Proceedings of the 2002 National Conference on Digital Government Research, Los Angeles, California.

Gorr, W., M. Johnson, and S. Roehrig. 2001. Spatial decision support system for home-delivered services. *Journal of Geographical Systems* 3:181–197.

Gould, M. D., and P. J. Densham. 1991. *Spatial decision support systems: A bibliography*. Paper presented at the National Center for Geographic Information & Analysis, Buffalo, New York.

Guarnieri, F., and J. L. Wybo 1995. Spatial decision support and information management application to wildland fire prevention: The WILFRIED System. *Safety Science* 20(1):3–12.

Halls, J. N. 2003. River Run: An interactive GIS and dynamic graphing website for decision support and exploratory data analysis of water quality parameters of the lower Cape Fear river. *Environmental Modelling & Software* 18(6):513–520.

Hampton, H., E. Aumack, J. Prather, Y. Xu, B. Dickson, and T. Sisk. 2003. Demonstration and test of a spatial decision support system for forest restoration planning. Paper presented at the 7th Biennial Conference of Research on the Colorado Plateau, Flagstaff, Arizona.

Heth, C. D., G. Dostatni, and E. H. Cornell. 2006. Spatial decision support for urban search and rescue and disaster management. Paper presented at the GIScience Conference, Münster, Germany.

Hey, P. 1998. SSToolbox: an agricultural spatial decision support system. Paper presented at Proceedings of the Eighteenth Annual ESRI User Conference, San Diego, California.

Hirschfeld, J., A. Dehnhardt, and J. Dietrich. 2005. Socioeconomic analysis within an interdisciplinary spatial decision support system for an integrated management of the Werra River Basin. *Limnologica—Ecology and Management of Inland Waters* 35(3):234–244.

Huang, L. X., and G. Sheng 2006. Web-services-based spatial decision support system to facilitate nuclear waste siting. Paper presented at Geoinformatics 2006: Geospatial Information Technology, Wuhan, China.

Hunova, I. 2001. Spatial interpretation of ambient air quality for the territory of the Czech Republic. *Environmental Pollution* 112(2):107–119.

Ianni, E., I. Ortolan, M.Scimone, and E. Feoli. 2008. Assessment of management options to reduce nitrogen load from agricultural source in the Grado. *Management of Environmental Quality: An International Journal* 19(3):318–334.

Jain, D. K., U. S. Tim, and R. W. Jolly. 1995a. A spatial decision support system for livestock production planning and environmental management. *Applied Engineering in Agriculture* 11(5):711–719.

Jain, D. K., U. S. Tim, and R. W. Jolly. 1995b. Spatial decision support system for planning sustainable livestock production. *Computers, Environment and Urban Systems* 19(1):57–75.

Jankowski, P., N. Andrienko, and G. Andrienko. 2001. Map-centered exploratory approach to multiple criteria spatial decision making. *International Journal of Geographical Information Science* 15(2):101–127.

Jankowski, P., and G. Ewart. 1996. Spatial decision support system for health practitioners: Selecting a location for rural health practice. *Geographical Systems* 3:279–299.

Jankowski, P., T. L. Nyerges, A. Smith, T. J. Moore, and E. Horvath. 1997. Spatial group choice: A SDSS tool for collaborative spatial decision-making. *International Journal of Geographical Information Science* 11(6):577–602.

Jankowski, P., T. Nyerges, S. Robischon, K. Ramsey, and D. Tuthill. 2006. Design considerations and evaluation of a collaborative, spatio-temporal decision support system. *Transactions in GIS* 10(3):335–354.

Johnson, M. P. 2005. Spatial decision support for assisted housing mobility counseling. *Decision Support Systems* 41(1):296–312.

Jones, M., and G. Taylor. 2004. Data integration issues for a farm decision support system. *Transactions in GIS* 8(4):459–477.

Kang, M. S., P. Srivastava, T. Tyson, J. P. Fulton, W. F. Owsley, and K. H. Yoo. 2008. A comprehensive GIS-based poultry litter management system for nutrient management planning and litter transportation. *Computers and Electronics in Agriculture* 64:212–224.

Kangas, Annika, Jyrki Kangas, and Mikko Kurttila. 2008. *Decison support for forest management*. New York: Springer.

Kaparthi, P., and R. Sugumaran. 2009. A Web-based agricultural crop condition and yield prediction modeling system using real-time data. Paper presented at the Iowa Geographical Information Council Conference, Waterloo, Iowa.

Koomen, E., J. Stillwell, A. Bakema, and H. J. Scholten. 2007. *Modelling land-use change*. New York: Springer.

Kuby, M., Z. Xu, and X. Xie. 2001. Railway network design with multiple project stages and time sequencing. *Journal of Geographical Systems* 3(1):25–47.

Kumar, A. 2005. Spatial decision support system foe groundwater management using numerical groundwater modelling approach in margajo watershed in Hard Rock Area of Jharkhand, India. Paper presented at the Map India 2005 Conference, New Delhi, India.

Kun, Y. 2006. The design and implementation of urban police spatial decision support information systems based on COM GIS technology. Paper presented at Geoinformatics 2006: Geospatial Information Technology, Wuhan, China.

Kwan, M-P., and J. Lee 2005. Emergency response after 9/11: The potential of real-time 3D GIS for quick emergency response in micro-spatial environments. *Computers, Environment and Urban Systems* 29:93–113.

Lagacherie, P., D. R. Cazemier, R. Martin-Clouaire, and T. Wassenaar. 2000. A spatial approach using imprecise soil data for modelling crop yields over vast areas. *Agriculture, Ecosystems & Environment* 81(1):5–16.

Larson, B. D., and R. R. Sengupta. 2004. A spatial decision support system to identify species-specific critical habitats based on size and accessibility using U.S. GAP data. *Environmental Modelling & Software* 19(1):7–18.

Lazzari, M., and P. Salvaneschi. 1999. Embedding a geographic information system in a decision support system for landslide hazard monitoring. *Natural Hazards* 20(2):185–195.

Leao, S., I. Bishop, and D. Evans. 2004. Spatial-temporal model for demand and allocation of waste landfills in growing urban regions. *Computers, Environment and Urban Systems* 28(4):353–385.

Lee, J., and S. Zlatanova. 2008. A 3D data model and topological analyses for emergency response in urban areas. In *Geospatial information technology for emergency response*, ed. S. Zlatanova and J. Li. London: Taylor & Francis Group.

Leenhardt, D., J. L. Trouvat, G. Gonzales, V. Perarnaud, S. Prats, and J. E. Bergez. 2004. Estimating irrigation demand for water management on a regional scale: I. ADEAUMIS, a simulation platform based on bio-decisional modelling and spatial information. *Agricultural Water Management* 68(3):207–232.

Lejeune, P., and C. Feltz. 2008. Development of a decision support system for setting up a wind energy policy across the Walloon Region (southern Belgium). *Renewable Energy* 33(11):2416–2422.

Loi, N. K. 2008. Decision support system (DSS) for sustainable watershed management in Dong Nai watershed, Vietnam: Conceptual framework and proposed research techniques. Paper presented at the Forest and Water in Warm Humid Asiz, IUFRO Workshop, Kota Kinabalu, Malaysia.

MacLean, D. A., W. E. MacKinnon, K. B. Porter, K. P. Beaton, G. Cormier, and S. Morehouse. 2000. Use of forest inventory and monitoring data in the spruce budworm decision support system. *Computers and Electronics in Agriculture* 28(2):101–118.

Malczewski, J. 1999. *GIS and multicriteria decision analysis*. New York: John Wiley & Sons, Inc.

Malczewski, J. 2006. GIS-based multicriteria decision analysis: A survey of the literature. *International Journal of Geographical Information Science* 20(7):703–726.

Matthews, K. B., A. R. Sibbald, and S. Craw. 1999. Implementation of a spatial decision support system for rural land use planning: Integrating geographic information system and environmental models with search and optimisation algorithms. *Computers and Electronics in Agriculture* 23(1):9–26.

McKenzie, J. S., R. S. Morris, C. J. Tutty, and D. U. Pfeiffer. 1997. EpiMAN-TB, a decision support system using spatial information for the management of tuberculosis in cattle and deer in New Zealand. Paper presented at the second annual conference of Geocomputation'97, Dunedin, New Zealand.

Medeiros, S. P. J., J. C. M. Strauch, J. Moreira de Souza, and G-R. B. Pinto. 2001. SPeCS-a spatial decision support collaborative system for environment design. *International Journal of Computer Applications in Technology* 14(4/5/6):158–165.

Monteiro, C., V. Miranda, I. J. Ramirez-Rosado, C. Morais, E. Garcia-Garrido, M. Mendoza-Villena, L. A. Fernandez-Jimenez, and A. Martinez-Fernandez. 2001. Spatial decision support system for site permitting of distributed generation facilities. Paper presented at 2001 IEEE Porto Power Tech Proceedings, Porto, Portugal.

Mudigonda, S., K. Ozbay, and H. Doshi. 2008. GIS-based decision support tool for the evaluation and selection of adaptive traffic control strategies on transportation networks. Paper presented at the Transportation Research Board's 87th Annual Meeting, Washington, D.C.

Mwasi, B. 2001. Land use conflicts resolution in a fragile ecosystem using multi-criteria evaluation (MCE) and a GIS-based decision support system (DSS). Paper presented at Proceedings of an International Conference on Spatial Information for Sustainable Development, Nairobi, Kenya.

Nehme, C. C., and M. Simoes. 1999. Spatial decision support system for land assessment. Paper presented at Proceedings of the 7th ACM international symposium on Advances in Geographic Information Systems, Kansas City, Missouri.

Neis, P., L. Dietze, and A. Zipf. 2007. A Web accessibility analysis service based on the OpenLS Route Service. Paper presented at the 10th AGILE International Conference on Geographic Information Science, Aalborg University, Denmark.

O'Brien, R., M. Peters, A.Schmidt, S. Cook, and R. Corner. 2004. Helping farmers select forage species in Central America: The case for a decision support system. Paper presented at CIAT (Centro Internacional de Agricultura Tropical), Cali, Colombia.

Parmanto, B., M. V. Paramita, W. Sugiantara, G. Pramana, M. Scotch, and D. S. Burke. 2008. Spatial and multidimensional visualization of Indonesia's village health statistics. *International Journal of Health Geographics* 7:30.

Pelizaro, C. 2005. *A spatial decision support system for the provision and monitoring of urban greenspace*. Amsterdam: Eindhoven University Press.
Peng, Z-R., and M-H. Tsou. 2003. *Internet GIS: Distributed geographic information services for the Internet and wireless networks*. New York: John Wiley and Sons.
Pettit, C., and D. Pullar. 2004. A way forward for land-use planning to achieve policy goals by using spatial modelling scenarios. *Environment and Planning B: Planning and Design* 31(2):213–233.
Pidd, M., F. N. de Silva, and R. W. Eglese. 1996. A simulation model for emergency evacuation. *European Journal of Operational Research* 90(3):413–419.
Power, J. M., and H. Saarenmaa. 1995. Object-oriented modeling and GIS integration in a decision support system for the management of eastern hemlock looper in Newfoundland. *Computers and Electronics in Agriculture* 12(1):1–18.
Prato, T. 2001. Modeling carrying capacity for national parks. *Ecological Economics* 39(3):321–331.
Ramirez-Rosado, I. J., E. Garcia-Garrido, L. A. Fernãndez-Jimãnez, P. J. Zorzano-Santamaria, C. Monteiro, and V. Miranda. 2008. Promotion of new wind farms based on a decision support system. *Renewable Energy* 33(4):558–566.
Rao, K. H. V. D. 2005. Multi-criteria spatial decision analysis for forecasting urban water requirements: A case study of Dehradun city, India. *Landscape and Urban Planning* 71(2–4):163–174.
Rao, K. H. V. D., and D. S. Kumar 2004. Spatial decision support system for watershed management. *Water Resources Management* 18(5):407–423.
Ray, J. J. 2007. A web-based spatial decision support system optimizes routes for oversize/overweight vehicles in Delaware. *Decision Support Systems* 43(4):1171–1185.
Reitsma, R. F. 1990. Functional classification of space: aspects of site suitability assessment in a decision support environment. PhD thesis, International Institute for Applied Systems Analysis, Laxenburg.
Reynolds, K. M., and P. F. Hessburg. 2005. Decision support for integrated landscape evaluation and restoration planning. *Forest Ecology and Management* 207(1–2):263–278.
Rinner, C. 2007. A geographic visualization approach to multi-criteria evaluation of urban quality of life. *International Journal of Geographical Information Science* 21(8):907–919.
Rinner, C., M. Raubal, and B. Spigel. 2005. User interface design for location-based decision services. Paper presented at 13th International Conference on GeoInformatics, Toronto, Canada.
Rivest, S., Y. Bedard, M-J. Proulx, M. Nadeau, F. Hubert, and J. Pastor. 2005. SOLAP technology: Merging business intelligence with geospatial technology for interactive spatio-temporal exploration and analysis of data. *ISPRS Journal of Photogrammetry and Remote Sensing* 60(1):17–33.
Roetter, R. P., C. T. Hoanh, A. G. Laborte, H. Van Keulen, M. K. Van Ittersum, C. Dreiser, C. A. Van Diepen, N. De Ridder, and H. H. Van Laar. 2005. Integration of systems network (SysNet) tools for regional land use scenario analysis in Asia. *Environmental Modelling & Software* 20(3):291–307.

Romero-Calcerrada, R., and S. Luque. 2006. Habitat quality assessment using weights-of-evidence based GIS modelling: The case of Picoides tridactylus as species indicator of the biodiversity value of the Finnish forest. *Ecological Modelling* 196(1–2):62–76.

Ropke, B., M. Bach, and H-G. Frede. 2004. DRIPS—a DSS for estimating the input quantity of pesticides for German river basins. *Environmental Modelling & Software* 19(11):1021–1028.

Salt, C. A., and M. C. Dunsmore. 2000. Development of a spatial decision support system for post-emergency management of radioactively contaminated land. *Journal of Environmental Management* 58(3):169–178.

Santos, L., J. Coutinho-Rodrigues, and J. R. Current. 2008. Implementing a multi-vehicle multi-route spatial decision support system for efficient trash collection in Portugal. *Transportation Research Part A: Policy and Practice* 42(6):922–934.

Scheibe, K. P., L. W. Carstensen Jr., T. R. Rakes, and L. P. Rees. 2006. Going the last mile: A spatial decision support system for wireless broadband communications. *Decision Support Systems* 42(2):557–570.

Schenk, T. A., G. Löffler, and J. Rauh 2007. Agent-based simulation of consumer behavior in grocery shopping on a regional level. *Journal of Business Research* 60(8):894–903.

Schuurman, N., M. Leight, and M. Berube. 2008. A Web-based graphical user interface for evidence-based decision making for health care allocations in rural areas. *International Journal of Health Geographics* 7:49.

Scotch, M., and B. Parmanto. 2006. Development of SOVAT: A numerical-spatial decision support system for community health assessment research. *International Journal of Medical Informatics* 75(10–11):771–784.

Sengupta, R., C. Lant, S. Kraft, J. Beaulieu, W. Peterson, and T. Loftus. 2005. Modeling enrollment in the Conservation Reserve Program by using agents within spatial decision support systems: an example from southern Illinois. *Environment and Planning B: Planning and Design* 32:821–834.

Serre, D., P. Maurel, L. Peyras, and Y. Diab. 2006. A spatial decision support system to optimize inspection, maintenance and reparation operations of river levees. Paper presented at the Joint International Conference on Computing and Decision Making in Civil and Building Engineering, Montreal, Canada.

Sha, Z., and F. Bian. 2004. An integrated GIS and knowledge-based decision support system in assisting farm-level agronomic decision-making. Paper presented at ISPRS Conference Proceedings of Commission VI, Istanbul, Turkey.

Sidlar, C. L., and C. Rinner. 2007. Analyzing the usability of an argumentation map as a participatory spatial decision support tool. *URISA Journal* 19(1):47–55.

Sikder, I. U., and A. Gangopadhyay. 2002. Design and implementation of a web-based collaborative spatial decision support system: organizational and managerial implications. *Information Resources Management Journal* 15(4):33–47.

Sinske, S. A., and H. L. Zietsman. 2004. A spatial decision support system for pipe-break susceptibility analysis of municipal water distribution systems. *Water SA* 30(1):71–80.

Srinivasan, R., and B. A. Engel. 1994. A spatial decision support system for assessing agricultural nonpoint source pollution. *Journal of the American Water Resources Association* 30(3):441–452.
Stevens, D., S. Dragicevic, and K. Rothley. 2007. iCity: A GIS-CA modelling tool for urban planning and decision making. *Environmental Modelling & Software* 22(6):761–773.
Stoms, D. M., K. M. Chomitz, and F. W. Davis. 2004. TAMARIN: A landscape framework for evaluating economic incentives for rainforest restoration. *Landscape and Urban Planning* 68(1):95–108.
Sugumaran, R. 2002. Development of an integrated range management decision support system (WebSDSS) for environmental planning and watershed management. *Computers and Electronics in Agriculture* 37(1–3):199–205.
Sugumaran, R., and Bakker, B. 2007. GIS-based site suitability decision support system for planning confined animal feeding operations in Iowa. In *Emerging spatial information systems and applications*, ed. B. N. Hilton, 219–239. Hershey, PA: Idea Group.
Sugumaran, R., S. Ilavajhala, and V. Sugumaran. 2007. Development of a web-based intelligent spatial decision support system WEBSDSS: A case study with snow removal operations. In *Emerging spatial information systems and applications*, ed. B. N. Hilton, 184–202. Hershey, PA: Idea Group.
Sugumaran, R., J. C. Meyer, and J. Davis. 2004. A Web-based environmental decision support system (WEDSS) for environmental planning and watershed management. *Journal of Geographical Systems* 6(3):307–322.
Sugumaran, V., and R. Sugumaran. 2007. Web-based spatial decision support systems (WebSDSS): Evolution, architecture, and challenges. *Communications of the Association for Information Systems* 19:844–875.
Symeonidis, P., I. Ziomas, and A. Proyou. 2004. Development of an emission inventory system from transport in Greece. *Environmental Modelling & Software* 19(4):413–421.
Taleai, M., A. Sharifi, R. Sliuzas, and M. Mesgari. 2007. Evaluating the compatibility of multi-functional and intensive urban land uses. *International Journal of Applied Earth Observation and Geoinformation* 9(4):375–391.
Tan, G., R. Shibasaki, and K. Matsumura. 2004. Development of a GIS-based decision support system for assessing land use status. *Geo-Spatial Information Science* 7(1):72–78.
Tarantilis, C. D., and C. T. Kiranoudis. 2002. Using a spatial decision support system for solving the vehicle routing problem. *Information & Management* 39(5):359–375.
Taranto, T. J. 2007. Using spatial information to aid decision-making: Case study of developing a participatory geographic information system for regional tourism planning. *Journal of Spatial Science* 52(2):23–34.
Tegou, L. I., H. Polatidis, and D. A. Haralambopoulos. 2007. Distributed generation with renewable energy systems: The spatial dimension for an autonomous grid. Paper presented at the 47th conference of the European Regional Science Association, Paris, France.
Thorp, K. R., K. C. DeJonge, A. L. Kaleita, W. D. Batchelor, and J. O. Paz. 2008. Methodology for the use of DSSAT models for precision agriculture decision support. *Computers and Electronics in Agriculture* 64:276–285.

Trepel, M. 2007. Evaluation of the implementation of a goal-oriented peatland rehabilitation plan. *Ecological Engineering* 30(2):167–175.

Tsamboulas, D. A., and G. K. Mikroudis. 2006. TRANS-POL: A mediator between transportation models and decision makers' policies. *Decision Support Systems* 42(2):879–897.

Tucek, J., J. Skvarenina, J. Mindas, and J. Holecy. 2003. *Catalogue describing the fire vulnerability of landscape structures in the Slovak Paradise National Park*. Paper presented at the International Scientific Workshop on Forest Fires in the Wildland-Urban Interface and Rural Areas in Europe: an Integral Planning and Management Challenge, Athens, Greece.

Turban, Efraim, and Jay E. Aronson. 1998. *Decision support systems and intelligent systems*. Englewood Cliffs, NJ: Prentice-Hall.

Vairavamoorthy, K., J. M. Yan, H. Galgale, S. Mohan, and S. D. Gorantiwar. 2004. A GIS-based spatial decision support system for modelling contaminant intrusion into water distribution systems. Paper presented at People-Centred Approaches to Water and Environment Sanitation, WEDC International Conference, Vientiane, Laos.

van Leeuwen, J. P., and Harry J. P. Timmermans. 2006. *Innovations in design and decision support systems in architecture and urban planning*. New York: Springer.

Vlachopoulou, M., G. Silleos, and V. Manthou. 2001. Geographic information systems in warehouse site selection decisions. *International Journal of Production Economics* 71(1–3):205–212.

Voivontas, D., D. Assimacopoulos, and E. G. Koukios. 2001. Assessment of biomass potential for power production: A GIS based method. *Biomass and Bioenergy* 20(2):101–112.

Volk, M., J. Hirschfeld, A. Dehnhardt, G. Schmidt, C. Bohn, S. Liersch, and P. W. Gassman. 2008. Integrated ecological-economic modelling of water pollution abatement management options in the Upper Ems River Basin. *Ecological Economics* 66(1):66–76.

Vyas, R., L. K. Sharmaand, and U. S. Tiwary. 2007. Exploring spatial ARM (spatial association rule mining) for geo-decision support system. *Journal of Computer Science* 3(12):882–886.

Weber, P., D. Chapman, and M. Hardwick. 2006. "London Calling"—a spatial decision support system for inward investors. Paper presented at the 9th AGILE international conference on geographic information science, Visegrád, Hungary.

West, L. A. 1999. Florida's marine resource information system: A geographic decision support system. *Government Information Quarterly* 16(1):47–62.

Wood, L. J., and S. Dragicevic. 2007. GIS-based multicriteria evaluation and fuzzy sets to identify priority sites for marine protection. *Biodiversity and Conservation* 16(9):2539–2558.

Wong, I. W., R. Bloom, D. K. McNicol, P. Fong, R. Russell, and X. Chen. 2007. Species at risk: Data and knowledge management within the WILDSPACE(TM) decision support system. *Environmental Modelling & Software* 22(4):423–430.

Wu, X., S. Zhang, and S. Goddard. 2004. Development of a component-based GIS using GRASS. Paper presented at the FOSS/GRASS Users Conference, Bangkok, Thailand.

Wu, Y. H., H. J. Miller, and M. C. Hung. 2001. A GIS-based decision support system for analysis of route choice in congested urban road networks. *Journal of Geographical Systems* 3(1):3–24.

Yang, K., S. Peng, Q. Xu, and Y. Cao. 2007. A study on spatial decision support systems for epidemic disease prevention based on ArcGIS. In *GIS for health and the environment*, ed. W. Cartwright, G. Gartner, L. Meng, and M. P. Peterson, 30–43. New York: Springer.

Yeh, A. G. O., and J. J. Qiao. 2004a. Component-based approach in the development of a knowledge-based planning support system (KBPSS). Part 1: The architecture of KBPSS. *Environment and Planning B: Planning and Design* 31(4):517–537.

Yeh, A. G. O., and J. J. Qiao. 2004b. Component-based approach in the development of a knowledge-based planning support system. Part 2: The model and knowledge management systems of KBPSS. *Environment and Planning B: Planning and Design* 31:647–671.

Yu, Y., X. Wu, F. Wang, X. Luo, F. Li, and G. Zhang. 2003. The application of GIS in petroleum upstream investment decision making. Working paper, Chinese University of Geosciences, Wuhan, China.

Zagorskas, J., and Z. Turskis. 2006. Multi-attribute model for estimation of retail centres influence on the city structure. Kaunas City case study. *Technological and Economic Development of Economy* 12(4):347–352.

Zhang, Y., P. K. Barten, R. Sugumaran, and J. DeGroote. 2008. Evaluating forest harvesting to reduce its hydrologic impact with a spatial decision support system. *Applied GIS* 4(1):1–16.

Zhou, J., and R. Golledge 2007. Real-time tracking of activity scheduling/schedule execution within a unified data collection framework. *Transportation Research Part A: Policy and Practice* 41(5):444–463.

Zhu, X., R. J. Aspinall, and R. G. Healey. 1996. ILUDSS: A knowledge-based spatial decision support system for strategic land-use planning. *Computers and Electronics in Agriculture* 15(4):279–301.

Zografos, K. G., and K. N. Androutsopoulos. 2008. A decision support system for integrated hazardous materials routing and emergency response decisions. *Transportation Research Part C* 16:684–703.

Zucca, A., A. M. Sharifi, and A. G. Fabbri. 2008. Application of spatial multi-criteria analysis to site selection for a local park: a case study in the Bergamo Province, Italy. *Journal of Environmental Management* 88(4):752–769.

10

SDSS Challenges and Future Directions

Learning Objectives

- Understand key technological, technical, social, and organizational challenges in SDSS development.
- Understand the likely future directions of SDSS development.

10.1 Introduction

In order to address complex multidisciplinary issues with spatial dimensions, spatial decision support systems have been designed, developed, and implemented for a variety of different application domains over the last several decades. The large number of scientific publications (451 publications in the database compiled by the authors and described in Chapters 2 and 9) has demonstrated the importance of SDSS in spatial decision making. The tremendous growth in spatial decision support system (SDSS) applications has at least partially been facilitated by computer hardware and networking advancements, the increased access and availability of spatial and nonspatial data, Web software, and progression in software such as geographic information systems (GIS), modeling, expert systems, and application programming interfaces. In compiling literature for this book, more than 450 publications explicitly dealing with SDSS were reviewed from the literature from a variety of disciplines including geographic, computer, decision, physical, biological, and social science as well as in business and government publications over the past 30 years. The number of publications has grown greatly since the mid-1990s with the publication rate still growing.

Despite the fact that SDSS are increasingly accepted tools in spatial decision-making processes, the successful design, development, delivery, and use of SDSS still presents many challenges. Many of the SDSS developed over the past 30 years have been either prototypes, conceptual frameworks, or utilized only in academic exercises. Based on the review of publications that was carried out for this book, there is sometimes little evidence that SDSS have been utilized to aid real spatial decision-making situations or that any given SDSS has been used repeatedly. There are numerous reasons for the limited success of spatial decision support systems in real-world situations. The ability to successfully execute cross-discipline collaborations to solve complex spatial problems is often a very difficult endeavor. There are many important technical and technological considerations that influence the successful implementation of SDSS. Questions that need to be effectively addressed include:

- What is the best software platform to employ?
- What spatial models would help address the problem?
- How are user-friendly interfaces developed?
- How are outputs evaluated or validated?
- What technologies are appropriate?
- Are there sufficient resources available?

These questions are often not properly considered early enough in the process, leading to unsuccessful outcomes.

Numerous authors have addressed some of the overall challenges faced in trying to successfully apply SDSS (Uran and Janssen 2003; Vonk et al. 2005; Geertman 2006; Rutledge et al. 2008; Van Delden et al. 2009). Several authors discussed SDSS challenges within a particular application domain. For example, Thakuriah et al. (2008) described the challenges restricting the uptake of SDSS in transportation planning, and Uran and Janssen (2003) analyzed the reasons for success or failure of SDSS by comparing five coastal zone and water management SDSS examples. Uran and Jannsen (2003) considered factors including how alternatives were specified, how users executed the process of using the SDSS, how the output was presented, how the evaluation of the results was supported, and whether the SDSS did what it was meant to do. De Silva (2001) discussed the difficulties of integrating simulation modeling and GIS within spatial decision support systems for evacuation planning. Newman et al. (2000), Lynch et al. (2000), and Cox (1996) all examined reasons for the low adoption of decision support systems in the agricultural domain. Some of the reasons suggested by these studies included the complexity of software systems, lack of field testing, limited computer ownership among producers, no end user input preceding and during development of the decision support systems (DSS), and the users' lack of understanding the modeling components.

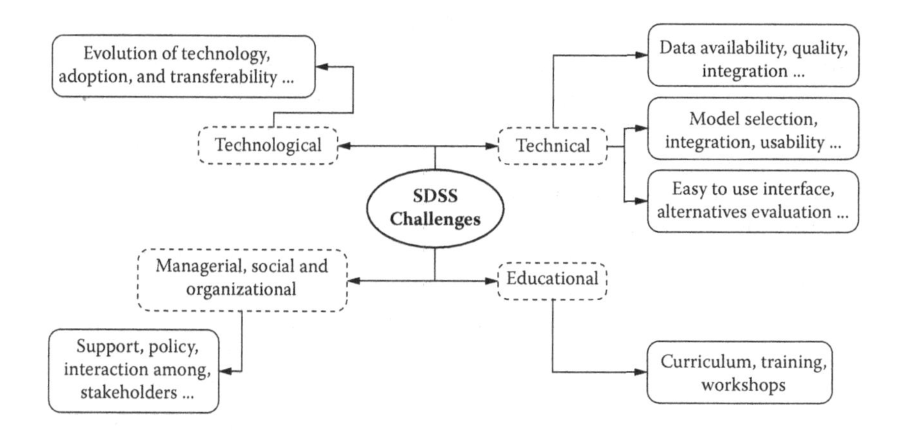

FIGURE 10.1
Overall SDSS challenges.

This chapter summarizes the key challenges faced in SDSS design, development, implementation, use, and adoption. These challenges will help end users and developers in understanding the issues related to SDSS development and effective use. We have classified SDSS challenges into four major categories: (1) technical, (2) technological, (3) social, managerial, and organizational, and (4) educational (Figure 10.1).

10.2 Technical Challenges

There are many technical challenges that limit the effectiveness of SDSS. There have been a wide variety of different SDSS developed but a lack of generic, flexible, and easy-to-use SDSS (Gao et al. 2004). Most SDSS developments have been carried out by piecing together components into a specific system, which is not necessarily useful in other situations. Each of these developments has to address specific issues related to the individual components of SDSS. In the following sections, we will discuss technical challenges specific to spatial database management, model management, and dialog management components of SDSS. Many of these challenges are similar for any GIS project, but are also crucial for SDSS.

10.2.1 Spatial Data Management Component Challenges

Some of the important challenges of the spatial data management component include spatial data availability, quality and quantity of data, and integration of diverse datasets. Spatial data availability is one of the critical issues in successful implementation and use of SDSS. Too much or too little data availability poses a problem in decision making (Sahota and Jeffrey 2005). As too much information becomes available, the task of integrating everything related to a particular issue becomes a challenge. On the other hand, a lack of available data, particularly in developing countries, is a major barrier in the use of SDSS (Hall 1996). An iterative process must take place in which data requirements are defined and gaps in availability are identified. During this process, decisions must be made regarding the willingness to invest time and resources to acquire existing data or to collect new necessary data. These types of decisions must be made in the SDSS development process in conjunction with decisions on model, knowledge, and interface components. In addition to availability, spatial data quality is important. Any successful SDSS applications must consider important factors in relation to spatial data such as scale, accuracy, and spatial and temporal resolution. Improperly using spatial data can introduce error and uncertainty to potential SDSS users if they are presented with results in the form of maps without the proper context. For example, if coarse-resolution data on soils is used with high-resolution data on land cover and topography in an SDSS using environmental modeling, some misleading results might occur. Similarly, an SDSS for locating new businesses at a local level would require detailed spatial data representing demographic characteristics of the population. The geospatial experts involved in SDSS development and application are responsible for properly managing and utilizing spatial data. Every SDSS should have some sort of system for creating metadata that will document the data inputs and outputs as well as processes that take place when utilizing the SDSS.

Another major challenge is the compatibility of spatial data for effective access, transfer, integration, and reuse by many users and programs (Williamson 2004; Sugumaran and Sugumaran 2007). There is a wide variety of spatial data models (e.g., vector, raster, TIN) and software, vendor, or government agency–specific data formats (e.g., coverage, shapefile, AutoCad Drawing Interchange Format [DXF], MapInfo TAB, National Transfer Format, Topologically Integrated Geographic Encoding and Reference System [TIGER], Personal Geodatabase, File Geodatabase, Geography Markup Language [GML], GRID, IMG, GeoTIFF, MrSID, BIL) in existence. The variety of spatial data models and formats in use has always created issues in combining data from a variety of sources in an SDSS. This becomes a more significant concern if an SDSS is being developed so it can be used by a large number of potential users. When developing an SDSS for use by a large number or wide variety of potential users, the developers must consider developing utilities for file conversion to allow the use of a variety of spatial data as inputs. Although in many ways spatial analysis software such as GIS and image processing programs have improved in their ability to read various formats, there is continued growth and development of new data formats. For example, the Environmental Systems Research Institute (ESRI) shapefile became one of the most common vector data formats in the late 1990s and into the 2000s, and many applications included functionality to import and export them. However, in recent years, ESRI has moved on to the more advanced but versatile personal and file geodatabases. So although many GIS and related software will be able to handle shapefiles, fewer will be able to read the geodatabase structures. The evolution and changing of data formats is always a concern in any custom GIS or SDSS development as the software may need to be updated to take into account changes in spatial data formats over time. The issue of data compatibility has been important in geographic information science for a long time. The Open Geospatial Consortium (OGC) is an international nonprofit consensus standards organization that attempts to set standards that "allow geospatial content and services to be seamlessly integrated into business and civic process, the spatial web and enterprise computing" and "to facilitate the adoption of open, spatially enabled reference architectures" (http://www.opengeospatial.org/ogc/vision). The Open Geospatial Consortium created the Geography Markup Language (GML) Encoding Standard, which is an XML grammar for geographic features. This is an open format for the exchange of geographic data. Although there is momentum for greater data sharing and open data sources, there are still a wide variety of spatial data formats, and any SDSS development must consider the potential variety of spatial data that might need to be handled by the SDSS.

Ever greater volumes of spatial data are being produced and utilized in GIS and SDSS, and this introduces unique challenges. The actual volume or size of a spatial dataset depends mainly on the size of the area for which data was collected and the amount of detail captured. As time passes, more and more spatial data become available, often with much more detail being captured. Greater amounts of data open up application possibilities but also introduce issues that are of great importance for SDSS development. For example, remotely sensed imagery (i.e., from satellite or airborne sensors) can be extremely useful in the visualization and analysis of ground conditions in an SDSS, and an ever larger amount of imagery is becoming available in a more timely fashion. However, using these datasets requires an understanding of the amount of computing power necessary. In the past, 2- to 5-meter resolution orthophotography might have been common, but now in some places 50-cm resolution or better is the norm. While this increased resolution is valuable, it also means that computing resources have to be much more powerful in order to handle the tremendous data volume. Another good example is Light Detection and Ranging (LiDAR)-derived elevation data. In some places, high-resolution (~1 m) elevation data is becoming available, whereas before, the best available data was low-resolution (~30 m). The amount of detail available in the LiDAR data opens up new opportunities but also presents spatial processing and visualization challenges as the data volume can be hundreds of times greater. During the SDSS development process, a thorough and careful evaluation of potential spatial data sources must be performed in order to discover the data that best serves the purposes of the SDSS application.

10.2.2 Model Management Component

The modeling component is crucial to any SDSS and is what often separates the use of GIS from that of SDSS. There are many challenges to successfully implementing the modeling component within an SDSS. The developers must decide between selecting existing models and developing new spatially explicit modeling techniques. They must ensure that, in either case, the modeling techniques will meet the requirements defined by the range of issues being addressed and can be properly understood by the range of stakeholders in the process.

10.2.2.1 Model Selection or Development

The economic, social, physical, chemical, biological, and other processes that occur in the real world are very complicated and difficult to reproduce in concise representation or models. However, in order to develop effective spatial decision support tools, it is often necessary that at least some of these types of processes are represented through models or modeling techniques. In addition to representing processes, there are models that can utilize the processing power of computers in order to relieve some of the cognitive load on the human decision maker. There is a wide variety of modeling techniques that have been used in SDSS for these purposes. Many of these methods (Analytic Hierarchy Process [AHP], ordered weighted averaging [OWA], weighted linear combination [WLC], genetic algorithms, etc.) were discussed in Chapter 4. Especially common in SDSS applications have been multi-criteria decision models that have been incorporated to provide decision aids to SDSS users. The development of models is often an expensive and tedious process (Wang and Cheng 2006). Thus, it is often preferable from a time and resources perspective to be able to use or adapt an existing model (Rutledge et al. 2007). The relevant participants in the SDSS process, such as the scientists or modelers, need to carefully evaluate requirements of the SDSS and investigate how different models might help to meet those requirements. They need to be in communication with end users during the selection of the model to guarantee that the proper models for the spatial decision-making situation are being considered and chosen. This communication includes descriptions of the assumptions a given model makes, the strengths and weaknesses of a given model, and the costs and benefits of all possible models (Van Delden 2009). Indeed, Van Delden (2009) listed the faith that users have in the models used in an SDSS as crucial to success. This faith is improved when peer-reviewed, calibrated, and validated models are used. Uran and Janssen (2003) pointed out that uncertainty in model output and doubt in the appropriateness of the model to the decision situation as important reasons for underutilized SDSS. The modeling expert must also treat the nonexpert potential users as crucial to the process of defining a final configuration (Van Delden 2009). It is important for a team environment to be built in which individuals, such as the modeling expert, do not feel a need to isolate themselves and protect their specific turf by keeping others uninformed. Consideration of their applicability to the given study area is necessary, as some models might have been developed for a specific geographic area that has certain types of databases available for model parameterization. Only if commensurate data are available for the given study area should the model be considered.

After choosing a model and prior to implementation, the expert (scientist or modeler) needs to work with end users to explain and provide documentation for the reason behind the selection of a particular model. In addition, they should provide detailed documentation for carrying out the necessary steps for operating the model so that users can understand and effectively use the model in the SDSS. This documentation must also include guidance for how outputs from the modeling within the SDSS can be analyzed and interpreted. Uran and Janssen (2003) indicated that SDSS often fail because there are various tabular and graphical outputs but no guidance on how to interpret these results. Although integrating existing model algorithms can save time and resources by limiting development time, there is less flexibility than when models and algorithms are developed from scratch. The developers and experts must weigh the potential gain in developing their own modeling components versus the cost in time and resources to do so.

10.2.2.2 Model Integration

A key challenge is to integrate the modeling component with other components within the SDSS. According to Choi (2004), one of the technical challenges in developing successful SDSS is linking models to other components within an SDSS. Often, there are significant differences in components that make it necessary to develop compromise solutions in linking different software packages. For example, modeling algorithms might utilize temporal data that is not readily supported in either GIS databases or programs, forcing the developers to develop bridges between the two different programs. In general, GIS software is becoming more effective at handling temporal data. This should lead to more elegant solutions for spatiotemporal modeling in SDSS in the future. In Chapter 5, methods for linking multiple pieces of software by loose, tight, and full coupling were described. The selection of the most suitable technique depends at least partially on the amount of time and resources the SDSS developers want to devote to building the system. Greater investment in development time is more likely to produce tightly coupled or fully embedded systems. These systems usually provide a more user-friendly software experience as compared to loosely coupled systems. Rutledge et al. (2007) found that potential end users considered a single integrated software package to be most attractive. A single software system has the advantage of greater portability, as end users do not have to install multiple software packages. Loosely coupled systems likely will require more expertise by the potential user in handling the various software components of the SDSS. The choice of integration or coupling levels will likely depend on the developer's knowledge about the model, his or her level of programming experience, and the availability of software. A final important aspect is defining how the modeling component will be maintained and supported in the future. If an existing third-party model is incorporated, then it is important to plan if and how updates to that model or other software that work with the modeling software (e.g., GIS) will be handled in the future.

10.2.2.3 Model Usability and Interpretation of Results

As the modeling components of an SDSS are often technically and scientifically challenging aspects in SDSS development and use, it is important to develop mechanisms to facilitate their effective usage. The application of the modeling components in an SDSS is not always going to be carried out by a modeling expert. Thus, in these cases, the modeler or scientist who is involved in the SDSS development needs to communicate with the end users from the beginning of the project in order to develop the SDSS with modeling components that will be accessible to the end users (Van Delden 2009). By involving the end users from early stages, they can be made aware of underlying assumptions, advantages, and limitations of any models. This knowledge and involvement can help to build trust among end users as to the validity of the modeling component. In addition, the end users can also provide insight on the modeling component's relation to the policy or organizational context in which the SDSS is supposed to operate (Van Delden 2009). Some of these lessons were learned in the application of a Web-based SDSS by one of the authors of this book. In this experience, end users were not involved at early stages but had the model component explained subsequent to SDSS development. The end users reported being interested in the model results, but not understanding the modeling context itself. This led to a misunderstanding of model results as being *the solution* as opposed to one of a range of possible solutions depending on how the model weights were parameterized. The end users also had trouble seeing how to translate policy drivers into model weights (Compas and Sugumaran 2004). Effective design of software interfaces will aid in the efficient use of the modeling components within the SDSS. This includes effective design of output mechanisms (maps, reports, tables, etc.) that communicate results of modeling operations efficiently. Also, extensive training should take place for the use of the SDSS with sufficient time presented for the modeling component. Thorough documentation in the form of user manuals and online help should also be provided. The training should begin before the final product is presented. Early training and input sessions can determine if the modelers and end users are on the same page as far as the functionality of the model and the way it will be operated from within the SDSS.

10.2.3 Dialog Management Component Challenges

10.2.3.1 User Interfaces

The software environment or user interfaces that users interact with can have a significant impact on their reactions and impressions in regard to the SDSS. Consequently, significant implications as to the success or failure of SDSS adoption derive from the characteristics of the dialog management component. If end users find the system unintuitive and difficult to use, then they will become frustrated. Van Delden (2009) listed SDSS ease of use as one of eight elements that likely dictate the success of an SDSS implementation. He indicated that ease of use is dictated by whether the user interface is quick and simple to use and provides easy access to all functions. Evers (2007) noted that unintuitive or unfriendly user interfaces are one of the four main reasons why DSS are not successful. Compas and Sugumaran (2004) pointed out that in their Web-based SDSS, users noted that the Analytical Hierarchical Process (AHP) model user interfaces were not intuitive and did not help them understand the modeling component. Uran and Jannsen (2003) specifically indicated that one of the problems in SDSS is that many alternatives need to presented, both during input parameterization and output analysis, which can cause problems in presenting these alternatives effectively. The goal of software design should be user interfaces that are intuitive, easy to use, and understandable. Malczewski (1999) highlighted some main considerations for user interface design, including that use is intuitive, there are mechanisms for users to recover from errors, there is efficient flow of information between the user and the application, and that users should be aware of the processes being carried out during interaction with the software. The key to developing effective user interfaces is an iterative development process, as discussed in Chapter 7, in which prototypes are presented to end users in order for them to provide useful critiques. Through iterative testing and development, more user-friendly interfaces will result in the final product.

In the early years of SDSS development, many of the software utilized, such as ArcInfo GIS, used command-line user interfaces. These applications required expertise in the proper syntax and thus generally limited their effective use by those more intimately involved in the policy and operational aspects of the decision-making process but not in the GIS software. As GIS and other software evolved to have more graphical user interfaces (GUI), a larger number of organizations from various disciplines adopted these technologies, and in time, more development of SDSS occurred. An SDSS developed with easier-to-use interfaces allows a greater number of end users to be involved in the development and application of the SDSS. However, graphical user interfaces in component programs do not guarantee a user-friendly SDSS. In the development process, the potential level of users must be accounted for. For example, Van Delden (2009) mentioned that they had two sets of user interfaces in their SDSS for future regional planning in New Zealand. The first was meant for policy analysts and was used for defining important scenarios, running scenarios, visualizing indicators, and comparing results. The second was meant for modelers and could be used for updating data, fine-tuning parameters, and accessing and changing model parameters. The idea of dual interface is a good technique because it provides mechanisms for greater flexibility for the modelers while protecting against unintended actions of those without necessary expertise in the modeling or database aspects.

10.2.3.2 Output Presentation and Evaluation

An important aspect of any SDSS is the presentation of output and scenario alternatives to users. The presentation of alternatives in the form of various outputs including maps, graphs, tables, 3D visualizations, and reports is important in guiding the decision-making process. It is important for the user to be provided with a manageable amount of information. A lack of output presentation can potentially limit the end user's ability to make informed decisions. However, too much information in the form of various outputs can be detrimental to a user's ability to meaningfully interpret results for decision making. During the development process, the effective design of output templates (e.g., map layouts) is important. Uran and Jannsen (2003) stressed the need for cartographically sound maps to be presented. Cartographic elements that should be included on any map include a legend, scale bar, title, metadata about the map (producer, date of production, data sources, etc.), and a caption explaining the critical message of the map. The SDSS should include functionality for producing maps based on templates that contain this level of information. It is important in maps, graphs, and tables that an excess of information is not presented. Uran and Jannsen (2003) gave an example in which an SDSS presented a map of raster data with over 15 classes represented in the legend. They suggested something on the order of six classes is more reasonable for users to understand. In GIS-based SDSS, it might be useful to provide output in the form of map compositions and also to provide the resulting spatial datasets so users can carry out further spatial analysis or investigation using these outputs. Effective presentation of outputs is especially difficult when there are multiple time steps. In these situations, it is important to be prudent in the choice of outputs presented in order to not overwhelm potential end users. New and improved temporal capabilities in GIS should help facilitate effective spatiotemporal presentation of output in GIS-based SDSS. An example of this type of tool is the Tracking Analyst extension in ArcGIS software.

One of the general weaknesses in many SDSS is that sufficient deliberation and decision output structures are not provided. It could be that an SDSS might provide attractive outputs that do not actually move the decision process forward. The lack of support for analyzing and evaluating the output generated by the system was the major shortcoming in five SDSS reviewed by Uran and Janssen (2003). The lack of proper spatial evaluation mechanisms limited the effective utilization of the SDSS that they reviewed. They cited an SDSS that presented an index of species for different geographic areas. The calculation of this index was flawed and not explained clearly for users to properly evaluate the measure. Evers (2007) listed the lack of capabilities for evaluating alternatives as a major reason for the failure of SDSS. Uran and Janssen (2003) also indicated that an SDSS must allow the user the ability to create and test alternatives but that the complexity must not overwhelm the user. They gave an example of a river management SDSS in which the user has to enter an impractical number of spatiotemporal inputs. In this example, it was impossible for the user to comprehend the translation by the SDSS of input parameterization to resulting outputs because the number of inputs was overwhelming. There is a need for at least some capabilities for testing the sensitivity of the models in an SDSS to input parameters.

10.3 Technological Challenges

Technological advances will continue to influence the development of SDSS as we have seen in Chapter 2 (SDSS evolution). The growth in computing power has helped fuel the development of SDSS. Modern desktop computers are more powerful than mainframe computers decades ago and generally are strong enough to run desktop SDSS effectively (contingent on data volumes as mentioned above). The constraints of desktop SDSS are often more related to data compatibility and availability (discussed previously) as well as software issues. Indeed, Vonk et al. (2005) noted that hardware concerns were minimal in relation to the adoption of planning support systems (PSS; a subset of SDSS). Very important to SDSS is the issue of integration and compatibility between programs. Proper planning is necessary to plan for the effective integration of various programs in a single SDSS. The appearance of more generic SDSS software (e.g., IDRISI capabilities, ArcGIS Model Builder, Geonamica, OpenSDSS) provide more opportunity for the development of SDSS using only one software package. If not using one of these generic SDSS, an evaluation process needs to be carried out in order to choose the proper software configuration for an SDSS. Equally important, if the SDSS is not meant for only a single or one-time application, is for a plan to be developed for maintenance and updating of the system for the future. Many SDSS are composed of several separate programs. The use of an SDSS on an individual machine is thus jeopardized when any of the integrated programs is updated to a new version. If there is no one to account for these updates, then it is likely the SDSS will no longer be used due to incompatibility problems between programs. Two successful planning (and commercial) SDSS—INDEX and CommunityViz—have gone through numerous updates to accommodate and take advantage of changes in the versions of underlying GIS software. This type of maintenance and update regime is more amenable to a fully integrated approach with commercial software in which there is financial incentive to continually update the software. Van Delden (2009) stressed the importance of maintenance and software support for model and data updates. In some of the cases for the SDSS he discussed, the maintenance aspect was part of a contract in which the Research Institute for Knowledge Systems was tasked with developing, maintaining, and updating the SDSS based on their Geonamica SDSS development framework. It is fairly uncommon to see repeated publications detailing applications of the same SDSS. This is in part due to the fact that these SDSS quickly become dated due to the lack of maintenance and updating.

As has been highlighted throughout the book, there has been an evolution beyond only desktop SDSS to Web-based, mobile, and distributed SDSS platforms. While the latter two are still fairly uncommon, Web-based SDSS have become quite common. The suitability of the Web as a medium for implementation of spatial decision making has increased mainly because of advantages such as platform independency, reductions in distribution costs and maintenance problems, ease of use, greater access, and its mechanisms for sharing of information by the worldwide user community (Sugumaran and Sugumaran 2007; Peng and Tsou 2003). However, the design and implementation of SDSS over the Web are subject to a unique set of technological issues and constraints. These include performance, technology integration, security, and interoperability concerns (Sugumaran and Sugumaran 2007).

When developing a Web-based SDSS, consideration of existing IT infrastructure (e.g., database management system) is important. It only makes sense to understand the limiting factors of the existing infrastructure and also to take advantage of those aspects of the existing infrastructure that would be beneficial. Effective integration with the existing IT infrastructure can lower the cost of maintenance and speed up the implementation process (Peterson 1998). This is particularly important for larger enterprises or organizations. This has not always been easy as spatial databases have not always fit nicely into enterprise databases. However, with the greater incorporation of spatial data in many database management systems, this has become less of an issue. Ray (2007) provided an example that highlights these points. He detailed a Web-based SDSS for overweight vehicle permitting in the state of Delaware. He pointed out that an earlier version of the SDSS failed because it could not integrate with their Web platform or share data with other applications in their infrastructure. The subsequent SDSS design was incorporated as part of the greater IT infrastructure and has been very successfully utilized by the Delaware Department of Transportation.

Hardware capabilities have improved greatly over the years, making performance issues less prominent for desktop SDSS. Desktop personal computers today are many orders of magnitude more powerful than when GIS and SDSS software were first being developed. With greater data volumes being included in applications, greater computing power is necessary. However, with Web-based applications, there are still certainly performance issues. The performance of Web-based SDSS can be constrained at several levels including the server (processing power, multi-thread functionality, etc.), the client machine, and the network infrastructure (speed of Internet connection or bandwidth). As SDSS applications utilize large datasets in the form of raster and vector spatial data, the consideration of these constraints is important. In addition, there are often complex modeling routines that can be processor intensive. Functionality for communicating the progress of the SDSS application is necessary and must be accounted for in the SDSS development and programming process. Great improvements in networking infrastructure, including bandwidth improvements, have facilitated and will continue to facilitate Web-based SDSS. The field of Web-based SDSS is still fairly early in development but can be expected to continue to grow and make up a greater percentage of total SDSS development.

There have been advancements in standards that facilitate software interaction and data compatibility, and these advancements can facilitate SDSS development. These types of standards are especially important in relation to Web-based SDSS and Web mapping applications. Standards from organizations such as the OGC have moved forward data compatibility as well as standards for mapping and spatial analysis services. The OGC, a consortium with members from government, business, and academia, has a mandate to develop and foster interoperability specifications in regard to geographic information and services. A list of standards developed by the OGC can be found at http://www.opengeospatial.org/standards/is. Three important spatial interoperability standards in regards to Web applications include the Web Map Service (WMS), Web Feature Service (WFS), and Web Coverage Service (WCS). The WMS is a specification that provides a simple HTTP interface for requesting georegistered map images (JPEG, PNG, etc.) from one or more distributed geospatial databases. The WFS is a specification providing an interface for retrieving geographic features across the Web. The WCS provides standard interfaces and operations that enable interoperable access to geospatial information. The MapServer Web-based SDSS from Chapter 8 used the WMS standard with OpenLayers. ArcGIS Server can publish using these services and can consume these types of services. Many existing Web-based SDSS are not based on OGC standards and do not easily interface with other products. The greater availability of standards-based geospatial data and services will provide guidelines and building blocks for Web-based and component-based SDSS development in the future.

Other challenges in the development of Web-based and distributed SDSS include security and privacy issues as well as quality of service issues. Depending on data distribution models, certain datasets might be proprietary or nonpublic, and therefore access to them must be restricted. In addition, the frequent transfer of files between server and client introduces issues of viruses, hackers, and bottlenecks on the network. In a distributed SDSS environment in which data and services are accessed from various locations on the network, there are multiple nodes that could become bottlenecks. Hence, the architecture must include technical solutions to combat potential disruptions.

Finally, a major issue is that of the transferability and adaptability of a given SDSS. There are issues for both desktop and Web-based SDSS in this regard. There have been many SDSS developed that were never adapted for use in other geographic areas due to a lack of flexibility built into the system. Often these systems were built for specific geographic areas that had unique spatial datasets, and the developers of the applications did not program the applications for incorporating dissimilar datasets (e.g., different formats, different attributes). If the flexibility for inclusion of datasets that are similar, but not exactly alike, is not built into a system, it means that the user has extra responsibility for transforming their specific datasets for use in the SDSS. This limited flexibility is going to be somewhat of an inherent problem in Web-based SDSS in that it is difficult for users to add their own data to an application. For example, in our ArcGIS Server Web-based example, the user could not add datasets. Functionality could be built for this, but then one would have to account for upload and security issues as well as for the handling of spatial data issues such as projections.

10.4 Social, Policy, and Organizational Challenges

SDSS development can result in a perfectly sound technological integration of components, but the SDSS could still be unutilized because important organizational or social issues were not considered. The social, policy, and organizational contexts surrounding SDSS development and adoption are crucial to the success or failure of any SDSS. Vonk et al. (2005) formally examined the bottlenecks that have limited the use of planning support systems in spatial planning through surveys of planning professionals. They identified human, organizational and institutional, and technical issues as being the major bottlenecks. There are no exact recipes or guidelines to follow when undertaking an SDSS development and implementation process, but there are certain pitfalls to be avoided and goals to be strived for. There are many different possible configurations of individuals and organizations that could be involved in an SDSS development and application process. They could range from quite a small SDSS example that only has a few stakeholders, such as the example we provided in Chapter 7 for the ArcGIS desktop SDSS, to very complicated ones such as that described by Rutledge et al. (2008) in which a long-term planning SDSS, which included many stakeholders from different organizations and the public, was described. Each configuration will have some unique aspects, but there are general lessons that have been learned and objectives that will lend a greater chance of success.

There is a wide range of stakeholders that could be involved in a spatial decision-making process. These individual stakeholders will fall into some general categories. In Chapter 4, we listed four general categories of stakeholders including the decision maker/end user, developer, domain expert, and analyst. These stakeholders could be present in a single organization or might be spread over several organizations. In addition, one individual could fulfill more than one stakeholder role. For example, a GIS analyst might serve as the programmer/developer and also as the analyst who carries out simulations in the SDSS, analyzes outputs, and helps summarize information from the SDSS for the decision makers. Regardless of the circumstance, and as we emphasized in Chapter 7, it is important to carry out the SDSS development process in an iterative fashion in which all of the stakeholder groups are included during the iterations. This point about inclusion of stakeholders from the beginning of the process and through the iterative process has been made by numerous authors. For example, Van Delden (2009) stressed that end user interaction from the early stages of the development process is crucial to connect the system to the policy context and also to build a feeling of ownership in the SDSS among users. Uran and Janssen (2003) identified limited involvement of end users in the development phase as a major reason for the failure of SDSS. Evers (2007) stressed that an iterative and interdisciplinary development process with the future user group is very important for the success of an SDSS project. Rutledge et al. (2007) explained that one of the key organizational considerations for the successful development of a new SDSS is to develop a partnership between researchers and end users so ideas are shared and developed prior to the start of the project.

The early involvement of end users can help to alleviate some of the potential issues that might subsequently arise. One goal Van Delden (2009) stressed is the need for an SDSS to become embedded within an organization or institution. Thus, early in the process it is necessary to begin to define where the system will be based within an organization. By clearly defining responsibilities within an organization in relation to the SDSS, some potential problems can be avoided. For example, involving the potential analysts or users in early stages is likely to alleviate their reluctance to take on new work as they will feel ownership in the process. Also, clear definition of roles can help reduce potential resentment regarding responsibilities within an organization. By involving all stakeholders in early stages, the goal is to foster proponents among all stakeholder groups. Van Delden (2009) believed it was of key importance to have proponents of the SDSS at all levels of the organization. Without support from individuals at decision-making or management levels, it will be unlikely that a decision support system will be incorporated into important decisions (Sahota and Jeffrey 2005). People from these levels have to dedicate the necessary resources for the SDSS development and implementation process to have a chance of success. Having proponents at other levels helps to ensure that proper communication between stakeholders takes place and that proper attention is paid to all aspects of the development process. Sometimes it is difficult to establish and maintain the level of interest and involvement depending on the development model being used. For example, when software development is outsourced to an outside group or company, extra effort must be made to establish and maintain interaction among the stakeholders. Evers (2007) stressed that a communication strategy must be established with multiple channels for communication.

10.5 Educational Challenges

In order for SDSS to become an integral part of organizations involved in spatial decision-making situations, there is a need for education and training activities at a variety of levels. On the most basic level, a general awareness of the nature and potential of SDSS must be established. Many individuals from disciplines that might benefit from the use of SDSS, or at least geospatial technologies, are unaware of the potential benefits. These individuals come from business, government, and academic institutions. Vonk et al. (2005) found that a lack of awareness and experience with PSS blocks widespread usage and adoption in spatial planning practices. This is at a broader conceptual level. At a more concrete level, development and use of an SDSS can require a variety of knowledge and skills. There is a need for skills in geospatial data management and analysis, software design and computer programming, and modeling, as well as knowledge in decision science concepts and specific domains or disciplines for which an SDSS would potentially be applied. Although individuals are unlikely to possess this breadth of knowledge and skills, having a key stakeholder familiar with all aspects is useful as he or she will be able to understand the whole picture and communicate potential benefits with all other stakeholders. This key stakeholder can help develop training plans to ensure that the necessary skills are then represented.

There is a growing understanding of the benefits of using geospatial technologies in a variety of domains. The accessibility of freely available mapping platforms over the Web (Google Earth/Maps, Bing Maps, OpenStreetMap, etc.) is making the general public aware of the availability and usefulness of spatial data. In addition, an explosion in the use of recreational and vehicle-based Global Positioning Systems (GPS) units and mapping systems have increased interest in the use of spatial information. These technologies can foster a more widespread interest in geography and spatial information, and this will help in the process of having spatial data become a common component in institutional or organizational data structures. This transition is already well established with many relational database management systems, including support for spatial data (e.g., Oracle, SQL Server). However, the greater awareness of the value of spatial data does not mean an automatic translation into the effective use of technologies such as SDSS. Formal training in relevant areas is necessary to ensure proper usage.

A lack of widespread understanding and capabilities of SDSS has been a contributing factor to their somewhat ineffective use. In 1996, Hall said that decision-making processes, especially in developing countries but also in developed countries, did not make use of possible tools such as SDSS for planning purposes because there was a lack of trained and knowledgeable personnel. Two out of the three top bottlenecks to the usage of PSS identified by Vonk et al. (2005) were experience within the planning organization and user awareness of the potential of PSS. They believed that the activity in the PSS development community stays in that community with ineffectual advertising and distribution of the systems. They also stated that other bottlenecks come from the attitude concerning the systems as being "black boxes" and difficult to operate. They believed it is necessary to carry out more marketing of these types of systems to get them embedded in more planning processes. The adoption of new techniques or technologies can be very slow. Thakuriah et al. (2008) mentioned that adoption of new planning technologies in the transportation sector is slow, with an example of how some agencies continue to use a process instituted in 1962. They argue for more exposure to the use of SDSS through publications in practitioner publications and conference presentations.

Spatial decision-making processes and tools, such as SDSS, need to be included in university and college curricula. Thakuriah (2008) stressed a need for curriculum changes to incorporate SDSS in order to take advantage of the impending turnover in the workplace (i.e., older workers retiring). Content regarding SDSS and courses dealing with spatial decision making are becoming more common. An Internet search revealed a number of courses being taught at universities throughout the world. For example, the University of Twente in the Netherlands offers a distance learning course called Spatial Decision Support Systems (Figure 10.2). The Department of Geography at San Diego State University offers a course called Spatial (GIS) Decision Support Methods. The Institute for Advanced Education in Geospatial Sciences at the University of Mississippi has a course called Decision Support Systems that covers DSS in general and SDSS specifically. The Geography Department at Adam Mickiewicz University in Poland offers a course called Introduction to Spatial Decision Support Systems. Although these types of courses usually fall in geography departments, there should be efforts made to make students from other disciplines aware of the applicability to their own disciplines.

In addition to developing a greater level of knowledge in general for spatial decision-making techniques and technology through academic curriculum, there is a need for specific training opportunities for those stakeholders who might play a role in such processes. Training in this sense could take different forms. Workshops or training sessions on the general applicability of decision support tools in disciplines such as business or transportation planning would be useful for making potential decision makers aware of the techniques and tools available. These types of workshops or symposia could take place at academic conferences or trade shows. An example of such a workshop was Visualization, Analytics, and

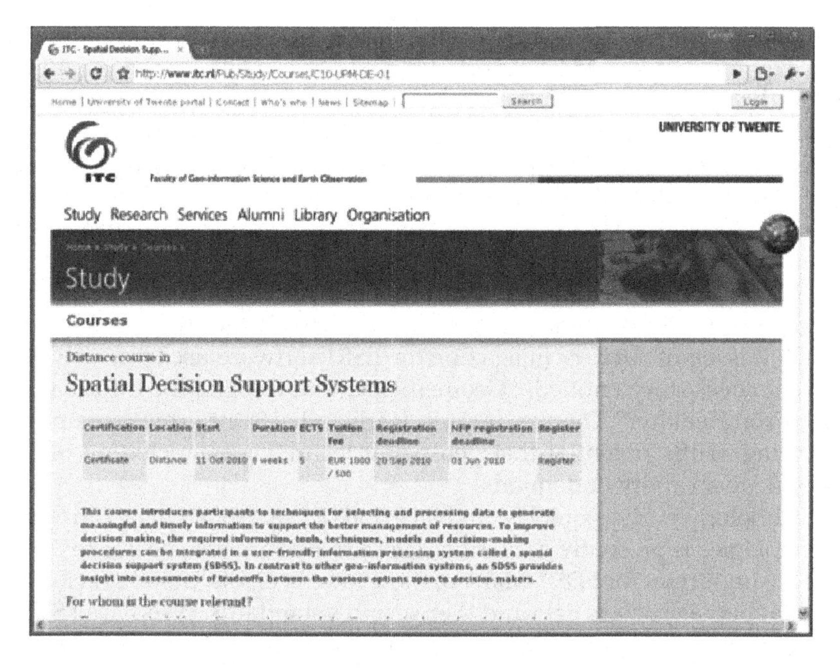

FIGURE 10.2
The Website describing the Spatial Decision Support Systems course offered by the University of Twente (http://www.itc.nl/Pub/Study/Courses/C10-UPM-DE-01).

Spatial Decision Support, which took place at the GIScience conference in Münster, Germany, in 2006. This workshop led to a special issue of the *International Journal of Geographic Information Science* called Geovisual Analytics for Spatial Decision Support. More specifically, sufficient training sessions must be built into any SDSS development and implementation process. Hall (1996) proposed that training programs were necessary for five major groups of users: policy makers to be made aware of the potentials and limitations of DSS, decision makers in the field who should have a general understanding of the DSS components, programmers to gain a higher level of technological competence, technicians for data collection and entry, and educators for developing knowledge in students.

10.6 Future Trends and Directions

SDSS and related technologies, like many other areas of information systems, continue to evolve at a rapid pace. The number of publications detailing SDSS developments has grown rapidly since the 1990s, and this growth will likely continue. The range of disciplines for which SDSS are utilized has expanded. It should be expected that SDSS use will expand greatly in those disciplines in which the value of SDSS has only been recently realized. For example, in the database of publications that were detailed in Chapters 2 and 9, 17 of 22 publications that qualified as having business as the primary discipline were published in the 2000s. This trend will continue as location-based services become more commonplace and established in business models. Similarly, the majority of public health-related SDSS publications occurred within the last four years. Highlighting the recent interest in geospatial technologies in public health was the winter 2010 issue of *ArcUser* (magazine for ESRI software users), which had a front cover story entitled, "Geomedicine: Can Geographic Information Keep You Healthy?" These examples clearly indicate the growth in understanding of the potential that geospatial technologies, including SDSS, provide to a variety of domains.

Technology evolves quickly, and a period of important change in SDSS technologies is presently occurring. These trends are not specific to SDSS but are important for SDSS. Many of these have been discussed previously in this chapter, such as data and Web services standards. Web-based SDSS applications have become more common, and their frequency should continue to increase in the future. Potential developers of Web-based SDSS should gain an understanding of some of the standards from the OGC. Important data standards include the GML standard, which allows geographic features to be expressed in Extensible Markup Language (XML) grammar. The Keyhole Markup Language (KML) standard, which was accepted by the OGC after submission by Google, is also an important standard. Web mapping (WMS) and GIS service (WFS) standards are also important and will play a role in future SDSS developments. Given these standards, it also must be noted that the majority of spatial data held are still in vendor-specific data formats such as shapefiles, geodatabases, CAD files, and so on. Another important development in SDSS applications will be the use of real-time or near-real-time datasets. Wang and Cheng (2006) mentioned that the Canadian Geospatial Data Infrastructure (CGDI) provides a lot of spatial data and information that can be accessed in real time from distributed sources. Remote sensing data are becoming more widely available in a timely fashion and will be more frequently built into SDSS, such as the near-real-time crop yield prediction system discussed in Chapter 9. Improvements in spatiotemporal data handling will be reflected in more realistic modeling frameworks within SDSS. Distributed GIS or geospatial analytic services will become available in the future. It will be possible to consume these services from a network-connected desktop application, but they will more commonly be part of Web-based or cloud computing configurations. These services might be provided as a software service in which the user pays for every use. It must be noted that although the frequency of Web-based SDSS developments has increased, they are still less common than desktop applications. Desktop generic SDSS software such as IDRISI, Geonamica, and OpenSDSS will play a bigger role in the future of desktop developments as they provide tools for easily developing SDSS functionality.

10.7 Summary

In this chapter, we have attempted to provide an overview of some of the technical, technological, organizational, and educational challenges involved in SDSS development and implementation. We finished the chapter by mentioning some of the broad future trends and directions that can be expected in SDSS.

There are many issues to be considered in relation to spatial database construction, model development, and dialog design for an SDSS application. Spatial database concerns include data availability, quality, formats, resolution or scale, compatibility with modeling formats, and so on. As with other aspects of the SDSS process, iterative review of data concerns must take place until data requirements are met. In the future, spatial data standards such as GML and KML will alleviate some issues with data compatibility. The modeling component must be defined for an SDSS as part of the iterative process in which stakeholders come to a consensus regarding the modeling components. With the inclusion of modeling components in GIS software and generic SDSS, it will be easier to build fully embedded SDSS in the future. The design of user interfaces should be part of an iterative process in which end users are allowed to test, critique, and help improve the product.

More important to the success of an SDSS than technical issues are social and organizational concerns. As many authors have expressed, it is important for decision makers and end users to be involved in the SDSS development process from an early stage. SDSS development should take place in an iterative process in which decision makers, developers, domain experts, and analysts play a continuous role and in which communication is ongoing. Greater understanding of geospatial data and technologies, such as SDSS, fostered by curricula at universities, workshops at conferences, and training within organizations will lead to more effective use of SDSS within various application domains.

There are many technological changes occurring that are affecting and will continue to affect the SDSS field. Primary among these is the greater availability of mapping and geospatial processing services available through the Web. In recent years, the number of Web-based

SDSS has risen, and this should be expected to continue. In the future, there will be more systems that use data and software services from distant computers in SDSS configurations. These services might be available as free components or as pay-for-service software components. In addition, mobile applications will utilize these services in true mobile SDSS configurations. Although these types of changes are occurring, it should still be expected that desktop SDSS will continue to be developed and utilized.

References

Choi, Y. Y. 2004. A framework for web-based SDSS for watershed management. Working paper, US Environmental Protection Agency. http://www.epa.gov/waterspace/partners_aboutsdss.html (accessed November 2009).

Compas, E., and R. Sugumaran. 2004. Urban growth modeling on the web: A decision support tool for community planners. *Papers and Proceedings of Applied Geography Conference* 27:255–269.

Cox, P. G. 1996. Some issues in the design of agricultural decision support systems. *Agricultural Systems* 52:355–381.

De Silva, F. N. 2001. Providing spatial decision support for evacuation planning: A challenge in integrating technologies. *Disaster Prevention and Management* 10(1):11–20.

Evers, M. 2007. Requirements for decision support in integrated water resources management. Paper presented at the REAL CORP 007 Conference, Vienna, Austria.

Gao, S., D. Sundaram, and J. Paynter. 2004. Flexible support for spatial decision-making. Paper presented at the 37th Annual Hawaii International Conference on System Sciences, Honolulu, Hawaii.

Geertman, S. 2006. Potentials for planning support: A planning-conceptual approach. *Environment and Planning B: Planning and Design* 33(6):863–880.

Hall, P. A. 1996. Use of GIS based DSS for sustainable development: Experience and potential, http://www.qub.ac.uk/mgt/papers/devel/hall.html.

Lynch, T., S. Gregor, and D. Midmore. 2000. Intelligent support systems in agriculture: Can we do better? *Australian Journal of Experimental Agriculture* 40:609–620.

Malczewski, Jacek. 1999. *GIS and multicriteria decision analysis*. New York: John Wiley & Sons, Inc.

Newman, S., T. Lynch, and A. A. Plummer. 2000. Success and failure of decision support systems: Learning as we go. *Journal of Animal Science* 77:1–12.

Peng, Z-R., and M-H. Tsou. 2003. *Internet GIS: Distributed geographic information services for the Internet and wireless networks*. New York: John Wiley and Sons.

Peterson, K. 1998. Development of spatial decision support systems for residential real estate. *Journal of Housing Research* 9(1):135–156.

Ray, J. J. 2007. A web-based spatial decision support system optimizes routes for oversize/overweight vehicles in Delaware. *Decision Support Systems* 43(4):1171–1185.

Rutledge, D., M. Cameron, S. Elliott, T. Fenton, B. Huser, G. McBride, M. McDonald, M. O'Connor, D. Phyn, J. Poot, R. Price, F. Scrimgeour, B. Small, A. Tait, H. van Delden, M. E. Wedderburn, and R. A. Woods. 2008. Choosing regional futures: challenges and choices in building integrated models to support long-term regional planning in New Zealand. *Regional Science Policy & Practice* 1(1):85–108.

Rutledge, D., G. McDonald, M. Cameron, G. McBride, J. Poot, F. Scrimgeour, R. Price, D. Phyn, H. van Delden, B. Huser, B. Small, L. Wedderburn, and T. Fenton. 2007. Development of spatial decision support systems to support long-term, integrated planning. In *MODSIM 2007 International Congress on Modelling and Simulation. Modelling and Simulation Society of Australia and New Zealand*, ed. L. Oxley and D. Kulasiri, 308–314. Christchurch, New Zealand: Modelling and Simulation Society of Australia and New Zealand Inc.

Sahota, P. S., and P. Jeffrey. 2005. Decision-support tools: Moving beyond a technical orientation. *Engineering Sustainability* 158(3):127–134.

Sugumaran, V., and R. Sugumaran. 2007. Web-based spatial decision support systems (WebSDSS): Evolution, architecture, and challenges. *Communications of the Association for Information Systems* 19:844–875.

Thakuriah, P., P. S. Sriraj, P. Metaxatos, I. Minocha, and T. Swarup. 2008. Economic and social urban indicators: A spatial decision support system for Chicago area transportation planning. Paper presented at the Association of American Geographers Conference, Boston, Massachusetts.

Uran, O., and R. Janssen. 2003. Why are spatial decision support systems not used? Some experiences from the Netherlands. *Computers, Environment and Urban Systems* 27(5):511–526.

Van Delden, H. 2009. Lessons learnt in the development, implementation, and use of integrated spatial decision support systems. Paper presented at the 18th World IMACS/MODSIM Congress, Cairns, Australia.

Van Delden, H., P. Luja, and G. Engelen 2007. Integration of multi-scale dynamic spatial models of socioeconomic and physical processes for river basin management. *Environmental Modelling and Software* 22:223–238.

Vonk, G., S. Geertman, and P. Schot. 2005. Bottlenecks blocking widespread usage of planning support systems. *Environment and Planning A* 37:909–924.

Wang, L., and Q. Cheng. 2006. Web-based collaborative decision support services: Concept, challenges and application. Paper presented at ISPRS Technical Commission II Symposium, Vienna, Austria.

Williamson, I. P. 2004. Building SDIs—The challenges ahead. Paper presented at the 7th International Conference on Global Spatial Data Infrastructure, Bangalore, India.

Index

E

F

G

H

I

J

K

L

M

N

O

Printed in the United States
by Baker & Taylor Publisher Services